TITAN
Exploring an Earthlike World

Second Edition

SERIES ON ATMOSPHERIC, OCEANIC AND PLANETARY PHYSICS

Series Editor: F. W. Taylor *(Oxford Univ., UK)*

Vol. 1:   TITAN: The Earth-Like Moon
          Athena Coustenis & Fredric W. Taylor

Vol. 2:   Inverse Methods for Atmospheric Sounding: Theory and Practice
          Rodgers Clive D.

Vol. 3:   Non-LTE Radiative Transfer in the Atmosphere
          M. López-Puertas & F. W. Taylor

Vol. 4:   TITAN: Exploring an Earthlike World (2nd Edition)
          Athena Coustenis & Fredric W. Taylor

Series on Atmospheric, Oceanic and Planetary Physics — Vol. 4

TITAN

Exploring an Earthlike World

Second Edition

Athena Coustenis

Paris-Meudon Observatory

Fredric W Taylor

University of Oxford

With illustrations by **D J Taylor**

World Scientific

NEW JERSEY · LONDON · SINGAPORE · BEIJING · SHANGHAI · HONG KONG · TAIPEI · CHENNAI

Published by

World Scientific Publishing Co. Pte. Ltd.

5 Toh Tuck Link, Singapore 596224

USA office: 27 Warren Street, Suite 401-402, Hackensack, NJ 07601

UK office: 57 Shelton Street, Covent Garden, London WC2H 9HE

British Library Cataloguing-in-Publication Data
A catalogue record for this book is available from the British Library.

Series on Atmospheric, Oceanic and Planetary Physics — Vol. 4
TITAN (2nd Edition)
Exploring an Earthlike World

ISBN-13 978-981-270-501-3
ISBN-10 981-270-501-5

Typeset by Stallion Press
Email: enquiries@stallionpress.com

Contents

Prologue xiii

Foreword xvii

1. Introduction   1

    1.1   Early History   1
    1.2   Titan in Mythology   5
    1.3   Space Exploration of the Solar System   8
    1.4   The 20th Century, Before Voyager   13

2. The Voyager Missions to Titan   16

    2.1   Space Missions to the Saturnian System   16
    2.2   Voyager Observations of Titan   20
    2.3   Atmospheric Bulk Composition   22
    2.4   Vertical Temperature Structure   22
    2.5   Energy Balance and the Temperature Profile
          in the Thermosphere   26
    2.6   Atmospheric Composition   28
    2.7   Photochemistry   31
    2.8   Cloud and Haze Properties   34
    2.9   Speculations on the Surface and Landscape of Titan
          from Voyager   35
    2.10 The Aftermath of Voyager   36

3. Observations of Titan from the Earth   38

    3.1   Introduction   38
    3.2   Space Observatories   38
          3.2.1   Hubble Space Telescope   39
          3.2.2   The James Webb Space Telescope   40
          3.2.3   Infrared Space Observatory   41
    3.3   Ground-Based Observatories   44
          3.3.1   Mauna Kea Observatories   44
          3.3.2   The European Southern Observatories   46

	3.3.3	The University of Arizona and Steward Observatory Telescopes 48
	3.3.4	Radio Astronomy 49
3.4	Earth-Based Studies of Titan 52	
	3.4.1	Occultations of Titan 52
	3.4.2	The Radar Search for Oceans, Seas or Lakes 54
	3.4.3	Spectroscopic Measurements of Titan's Albedo 57
	3.4.4	Imaging Titan's Atmosphere in the Near-Infrared 64
	3.4.5	Imaging the Surface 66
3.5	Ground-Based Observations and Cassini–Huygens 71	

4. Cassini–Huygens: Orbiting Saturn and Landing on Titan   72

4.1	Introduction 72	
4.2	The Spacecraft and its Systems 73	
4.3	Scientific Objectives 77	
4.4	The Long History of the Cassini–Huygens Mission 79	
4.5	Departure for the Saturnian System 82	
4.6	Journey to Saturn and Orbit Insertion 84	
4.7	Huygens Descends onto Titan 85	
4.8	Experiments and Payloads 89	
	4.8.1	The Scientific Instruments on the Orbiter 89
	4.8.2	The Scientific Instruments on the Probe 99
4.9	Touring the Saturnian System 107	
	4.9.1	Observations of Saturn 107
	4.9.2	The Icy Satellites, and Saturn's Rings 109
	4.9.3	Saturn's Magnetosphere and Titan 112
4.10	Being Involved: Scientists and Instrument Providers 113	
4.11	Reaping the Benefits 115	

Colour Plates   117

5. Titan's Atmosphere and Climate   129

5.1	The Climate on Titan 129	
	5.1.1	Atmospheric Pressure Profile 129
	5.1.2	Atmospheric Thermal Structure 131
	5.1.3	Troposphere 133
	5.1.4	Stratosphere 135
	5.1.3	Mesosphere 136
	5.1.6	Thermosphere 136
	5.1.7	Exosphere 136
5.2	Radiation in Titan's Atmosphere 137	
	5.2.1	Solar and Thermal Radiation 137

		5.2.2	Energy Balance and Surface Temperature	137
		5.2.3	Model Temperature Profile	138
		5.2.4	Radiative Equilibrium Temperature Profile	139
	5.3	Remote Atmospheric Temperature Sounding		141
	5.4	Titan's Ionosphere and its Interaction with the Magnetosphere of Saturn		144
	5.5	Climate Change on Titan		147

6.   Chemistry and Composition   150

	6.1	Titan's Chemical Composition		150
	6.2	The Bulk Composition of the Atmosphere		154
	6.3	Ionospheric Chemistry		155
	6.4	Trace Constituents in the Neutral Atmosphere		157
		6.4.1	Stratospheric Composition Measurements with Cassini	160
		6.4.2	Vertical Distributions	175
		6.4.3	Spatial Variations	177
		6.4.4	Temporal Variations of the Trace Constituents	180
	6.5	Photochemistry		181
		6.5.1	Hydrocarbons	185
		6.5.2	Nitriles	188
		6.5.3	Oxygen Compounds	189
		6.5.4	Condensation Efficiencies	192
		6.5.5	Aerosol Production	193

7.   Clouds and Hazes   194

	7.1	Introduction and Overview		194
	7.2	Terrestrial Clouds and Precipitation		197
	7.3	Visible Aspects of Titan's Haze		198
	7.4	Size and Vertical Distribution of the Haze Particles		201
		7.4.1	Haze Vertical Profiles	203
		7.4.2	Haze Opacity Spatial Variations	205
	7.5	Tropospheric Condensate Clouds		206
	7.6	Thermal and Dynamical Interactions with the Haze		210
	7.7	Observational Evidence on the Aerosol Composition		213
	7.8	Laboratory Simulations of Haze Materials		217
		7.8.1	Chemical Composition of Tholins	218
		7.8.2	Optical Properties of Tholins	221
	7.9	Microphysical Models of Titan's Haze		223
		7.9.1	Organic Haze Production	223
		7.9.2	Fractal Models and Scattering Properties of the Haze	227
	7.10	Discussion and Conclusion		229

8.   Atmospheric Dynamics and Meteorology   231

      8.1   Introduction   231
      8.2   Dynamics of Planetary Atmospheres   232
      8.3   Titan's General Circulation   237
      8.4   Zonal Motions   238
      8.5   The Meridional Circulation   243
            8.5.1   The Hemispherical Asymmetry   244
            8.5.2   The Polar Vortex   245
      8.6   Vertical Motions   247
      8.7   Waves, Tides and Turbulence   250
      8.8   The Weather Near the Surface   251
      8.9   Does Lightning Occur on Titan?   255

9.   The Surface and Interior of Titan   258

      9.1   Introduction   258
      9.2   Remote Sensing of the Surface   259
      9.3   Huygens Takes a Plunge   261
      9.4   Naming Distant New Places   266
      9.5   Evidence for Geological Activity   269
            9.5.1   Albedo Variations   269
            9.5.2   Craters   270
            9.5.3   Mountains and Volcanoes   273
            9.5.4   Dunes   274
            9.5.5   Lakes   275
      9.6   The Nature and Composition of the Surface   277
      9.7   The Interior of Titan   281

10.   Titan's Origin and Evolution in the Solar System   286

      10.1 Introduction   286
      10.2 Relations Among Solar System Bodies   287
            10.2.1 The Formation of the Solar System   289
            10.2.2 The Terrestrial Planets and Titan   290
            10.2.3 Titan and the Outer Planets   297
            10.2.4 Titan and the Other Saturnian Satellites   304
            10.2.5 Titan and Europa   318
            10.2.6 Nitrogen Atmospheres in the Outer Solar System   318
      10.3 Titan's Origin and Evolution   321
            10.3.1 Evolutionary Models for Titan's Atmosphere   324
            10.3.2 Origin of the Atmospheric Components   325
      10.4 Titan and Life   329
      10.5 Open Questions   330

11. Beyond Cassini/Huygens: The Future Exploration of Titan 332

    11.1 Returning to Titan 332
    11.2 Titan as a Target of Astrobiological Interest 334
    11.3 Science Drivers and Measurements Needed 335
    11.4 Advanced Titan Mission Concepts 337
    11.3 Technology Requirements 339
    11.6 Mission Architecture and Design 339
    11.7 Getting to Titan: Launch and Propulsion 340
    11.8 The Voyage of the *Titania* 341
    11.9 Explorers on Titan 343

Glossary and Acronyms 350

References and Bibliography 358

Index 391

CHRISTIANI
HUGENII
ΚΟΣΜΟΘΕΩΡΟΣ,

SIVE

De Terris Cœlestibus, earumque ornatu,

CONJECTURÆ.

AD

CONSTANTINUM HUGENIUM,

Fratrem:

GULIELMO III. MAGNÆ BRITANNIÆ REGI,
A SECRETIS.

HAGÆ-COMITUM,
Apud ADRIANUM MOETJENS, Bibliopolam.
M. DC. XCVIII.

Humans have searched for another Earth practically since coming into existence, at least since we first began to realise that the points of light in the sky were large objects very far away, and the science of astronomy was born. In the modern era, every possible approach, from telescopes and space missions to philosophical arguments, is being used to probe the existence of Earthlike planets elsewhere in the universe. The ultimate question is whether there is another world sufficiently like ours, with the same temperature conditions and atmospheric composition, where we could live in comfort, or even where life could have developed in a form that we would readily recognise.

In the last thirty years or so, space missions and Earth-based telescopes have shown us that our own Solar System is mainly composed of uninhabited, and essentially uninhabitable, worlds. There is plenty of variety, ranging from giant balls of primordial gas with no solid surface (the outer planets) to objects that have no substantial atmosphere like Mercury and the minor planets; or like our Moon and almost all of the other satellites in the Solar System. Mars has enough atmospheric density to exhibit phenomena like polar caps, clouds, and dust storms, but the composition is mostly CO_2, and therefore unfit for human respiration. Venus and Mercury are too hot and dry, and so on. Since the prospects for life are not good in our immediate neighbourhood, it is not surprising then that scientists have turned to the rest of the Universe, outside the Solar System.

Recently we have been witnessing the regular discoveries, through modern techniques ranging from adaptive optics and radio astronomy to gravitational microlensing, of protostellar disks, and of a rapidly increasing number of new extrasolar planets and planetary systems around distant stars like Beta Pictoris, Fomalhaut, 51 Pegasus, HD209458, and Tau Bootis. As we get better at searching, we increase our chances of finding another Earth some day, but it will not be nearby or easy to reach, even with robot probes. In our Solar System, in particular, the chances for life outside Earth, long ago considered excellent, receded as exploration revealed the details of the non-Earthlike environments on the most Earthlike planets, Mars and Venus. It made the surprise all the greater when we found, only relatively recently, that there is another place in the Solar System that *does* have an Earthlike atmosphere, albeit a very cold one. Strangely, this is not on any of the planets, but on a satellite, the only one among the dozen or so very large moons in the Sun's family for which this is true.

Titan, Saturn's biggest moon, and (by a narrow margin) the second in size among the satellites in our Solar System, has been known for a long time to have a substantial atmosphere. The Catalan astronomer José Comas Solà claimed in 1908 to have observed limb darkening on Titan. This is the effect whereby the solar light reflected back to Earth by Titan's limb shows a stronger attenuation than that from its centre, which usually implies the presence of a substantial atmosphere. Confirmation came from spectroscopic observations by Gerard Kuiper in the 1940s, but it was not until the Voyager 1 spacecraft visited Titan in 1980 that the composition and the surface pressure were found to be so similar to those on Earth. Furthermore, complex organic chemistry is active there, producing multiple layers of orange-coloured haze, which render the atmosphere opaque in the optical range of wavelengths. It took a long time and a lot of effort to get a glimpse of the surface, and we still do not know today exactly what its composition is, except that the crust beneath must be mainly water ice.

To address the many questions asked about Titan over the centuries since its discovery, a series of space probes has been developed and dispatched towards this intriguing body. Pioneer 11 arrived first, in 1979, followed by Voyager 1 a year later. The scientific understanding of Titan as a planet-like object that emerged from the analysis of Voyager data was improved by ongoing ground-based observations, using increasingly more powerful optical and spectroscopic techniques, such as radar and adaptive optics, and advanced platforms on Earth-orbiting space observatories. At the same time as new data were acquired and studied, new theoretical models were being developed to account for these observations, new theories proposed and debated, and old or repeated measurements re-analysed. All of these have yielded formidable results, and posed new questions, over the past several years.

The latest envoy to Titan, a large and sophisticated international space mission called Cassini/Huygens, was launched in the year 1997. It arrived in July 2004, and started gathering new measurements from an orbit around Saturn that was designed to permit multiple Titan encounters. The Huygens probe descended in Titan's atmosphere on January 14, 2005 and recorded breathtaking data, revealing an astounding new world, and the most distant one to be landed on by a human-made machine.

The aim of this book is to bring together a general overview of our current understanding of all aspects of Titan, at a comprehensive and scientific level, but in terms basic enough to be mostly accessible to the non-specialist as well. We begin with a history of Titan studies, covering Earth-based facilities and programmes and leading into the early space missions. The Voyager mission, in particular, engulfed the scientific community with large amounts of data nearly twenty years ago, some of which is still being analyzed. The torrent of data from Cassini, with 40 times as many flybys, not to mention the Huygens *in situ* science, dwarfs even that from Voyager. We describe how, along with concurrent ground-based observations and theoretical modelling, Voyager, and now Cassini, have revolutionised our view of the outer Solar System. Titan's place in the Saturnian system, its structure and composition, its unique atmosphere, and the extent to which it really resembles the Earth, are the main themes. As a secondary objective, we hope to show the reader that the close look astronomers have been afforded since 2004 with their very expensive Cassini/Huygens space mission to the Saturnian system is paying dividends. An up-to-date synthesis of current Titan knowledge, and the remaining big questions, should give everyone the possibility to better appreciate the Cassini/Huygens discoveries.

We have tried to cover most of what is known about Titan today, and some informed speculation as well, while keeping the account as simple as possible. Our experience tells us that interest in Titan and its exploration, as evidenced by the huge popular reaction to the Cassini space mission, extends far beyond the small international group of professional planetary scientists. Thus, we have tried to make the book accessible to all, assuming only a basic familiarity with physics and astronomy. In the chapters that deal with more technical subjects, it has been necessary to use some more complicated concepts and words, in order not to leave out important knowledge or key questions. Where scientific terminology could not be avoided, the more specialised terms are defined in the text, in footnotes, and in the glossary, for the general reader. We hope that interested non-scientists will persevere as far as they can, and then move on to the next chapter where a more basic level is again resumed. For those who, on the other hand, want more detail, or wish to read about the original research of which this book is a summary, we include references and guides to further reading in a comprehensive bibliography at the end of the book.

Finally, we would like to acknowledge friends and colleagues who have helped with the text, either directly or by communicating their work on Titan. Particularly valuable inputs and comments on the draft manuscript came from David Luz, Patrick Irwin, Emmanuel Lellouch, Ralph Lorenz, Conor Nixon, Robert Samuelson, Tetsuya Tokano, Daniel Gautier and Thomas Widemann. For help with illustrations, proof-reading and other helpful comments, we further thank Iannis Dandouras, Mathieu Hirtzig, Tom Krimigis, Panagiotis Lavvas, Alberto Negrão, Hasso Niemann and Véronique Vuitton. We are grateful to the EUROPLANET Consortium for funding part of our meetings to work on this project, and to the Observatoire de Paris for access to historical material. We are also grateful to many Cassini–Huygens scientists who provided guidance and information on their work. In particular we thank Dennis Matson and Jean-Pierre Lebreton, Cassini–Huyens project managers who kindly read the book in advance and wrote a foreword.

The original figures in this volume were drawn and montages created by Dr. D. J. Taylor, to whom we extend our deepest gratitude.

Athena Coustenis
Fred W Taylor
Paris and Oxford, April 2008

Titan is Saturn's largest moon. It was discovered by the Dutch Astronomer Christian Huygens in 1655. For almost three centuries thereafter, Titan remained nothing more than a dot of light in the sky. Then, in 1994, Gerard Kuiper discovered that its atmosphere contained methane! It immediately became a world to explore and a destination in the space era.

Titan is shrouded by a thick atmosphere with a blanket of organic haze that hides the surface. The haze is a product of UV photolysis of methane in the upper atmosphere. In the early 80s, Voyager investigated Titan but was unable to see through to the surface. However, Voyager confirmed that the atmosphere served as a big chemical factory producing many complex organic compounds. This made Titan one of the most fascinating bodies in the solar system.

In 1999, while the international Cassini–Huygens mission was on its voyage to Saturn, the authors of this book published their first comprehensive review of the knowledge of Titan. They are two of the scientists most knowledgeable about this fascinating moon of Saturn. As a result of Cassini–Huygens' arrival around Saturn in 2004, and the Huygens probe landing on Titan's surface on January 14, 2005, a giant advance has been made in our understanding of Titan. Thus, it is timely to issue a new textbook. Athena Coustenis and Fred Taylor have come up with a superb revised version of their book. The new edition gives an excellent and up to date account of our knowledge about Titan. It provides a comprehensive review of this peculiar solar system object which bears many similarities to Earth albeit under very different conditions, where methane plays the role that water plays on Earth. The latest results are described and include Earth-based observations, laboratory work and modelling in addition to the Cassini–Huygens observations.

The book tells the history of the exploration of Titan. It includes the most recent ideas about the processes that govern Titan. It shows extremely well the synergy of *in situ* robotic observations and space-based and Earth-based telescopic observations. It also shows the importance of laboratory work and modelling. All of these approaches are needed to make progress in our understanding of Solar System objects. The book is an excellent reference for students with some general background in the field of planetary sciences and for new planetary scientists looking for a comprehensive book on Titan. After reading the book, you may decide to stick with Titan for the rest of your career! The authors of the book are currently active in the Cassini–Huygens mission and have been involved in Titan research for several decades. They have mentored young scientists who are now Titan researchers themselves.

Titan has become a well-known object in the Solar System. It is the most Earthlike world we know of. This book puts this extraordinary solar system object on our Solar System map for additional exploration. The closing chapter describes new concepts for the future exploration of Titan and gives us an irresistible invitation to return.

Jean-Pierre Lebreton
ESA Huygens Project Scientist

Dennis Matson
NASA Cassini Project Scientist

For more than a hundred years, those of us in the speculative fiction business have been speculating like mad about Titan, Saturn's largest moon. Back in 1894, in Journey in Other Worlds, John Jacob Astor wrote of a group of travelers whose spacecraft crossed the orbit of Titan on its way to Saturn. (On Saturn, the travelers wore their winter clothes, as it was rather cold.) Over the passing years, science fiction authors have written of Titan as a mining colony (Arthur C. Clarke's Imperial Earth, Alan E. Nourse's Trouble on Titan) and as a source of aliens (Philip K. Dick's The Game-Players of Titan). These days, authors seem more inclined to consider the possibilities of life on Saturn's largest moon (Stephen Baxter's Titan and Michael Swanwick's Hugo Award-winning novelette "Slow Life.")

Pat Murphy & Paul Doherty, *Fantasy & Science Fiction*

1.1 Early History

The discovery that the giant planets of the outer Solar System had large moons came quickly following the invention of the telescope in 1610. Galileo observed Jupiter and found the four biggest Jovian satellites, which he called the Medici stars, now known to us as the Galilean satellites: Io, Europa, Ganymede and Callisto. In the four decades from 1610 to 1651, Italian, French and English astronomers identified seventeen giant planet satellites, or more than half of the total known before the space age began. Then, from 1671 to 1684, Cassini found four more in the Saturnian system, the satellites we now know as Iapetus, Rhea, Tethys and Dione.

Next came Herschel, discovering in two years (1787–88) two Uranian satellites (Titania and Oberon) and two more Kronian (i.e. Saturnian) satellites, Mimas and Enceladus. Finally, four more moons were discovered thanks to the observational skills of Lassell from 1846 to 1851: Triton, the large satellite of Neptune, Hyperion of Saturn (a co-discovery with Bond from the USA), and two moons of Uranus, Ariel and Umbriel.

Thus, European astronomers developed a trend by which they discovered, every sixty to one hundred years or so, in short intense periods, four at a time, the satellites of the Solar System, until in the mid-nineteenth century they mostly lost interest and turned to new astronomical topics, leaving the rest of the satellite discoveries to their American colleagues. Near the beginning of that first period of satellite discoveries, one of them did not conform to the four-by-four rule and was for many years believed to be the biggest satellite of all. They called it Titan. It was not, in fact, until *Voyager* measured Ganymede's radius with precision that Titan lost its crown as the largest moon in the Solar System, and then only by a very small margin. It still *looks* bigger than Ganymede, however, because its disk is extended more than 100 km by the thick, hazy atmosphere.

The story of the first detection of Titan is a classic of its kind. On the night of March 25, 1655, a novice Dutch astronomer, Christiaan Huygens, pointed his telescope at Saturn. It was the first professional telescope he had ever had in his possession, and was by no means an exceptional one, with just a 4 m long refractor and an enlarging capacity factor of 50. He saw a small star 3 arc minutes away from the planet. This was not even the first time someone had noticed this object: Hevelius from Poland, and Wren in England had perceived it previously in the night sky, but not knowing what to make of it — and star catalogues being scarce at the time — they had believed it to be just another star.

Huygens did not make this mistake: he almost immediately guessed it was a satellite, and confirmed his guess a few days later when the 'star' had moved. Thus, a Dutchman discovered, right in the middle of the period of Italian, French and English supremacy, a Saturnian satellite. It is true that Huygens was not one of those who discovered four satellites, and in fact he never even noticed Tethys and Dione. But what he did discover was enough to make him famous, because Titan was soon realised to be an important object by virtue of its great size. Huygens believed Titan to be the biggest satellite of all, greater even than Ganymede, the

Figure 1.1   Giovanni Domenico Cassini, 1625–1712 (Observatoire de Paris).

Figure 1.2   Christiaan Huygens, 1629–1695 (Observatoire de Paris).

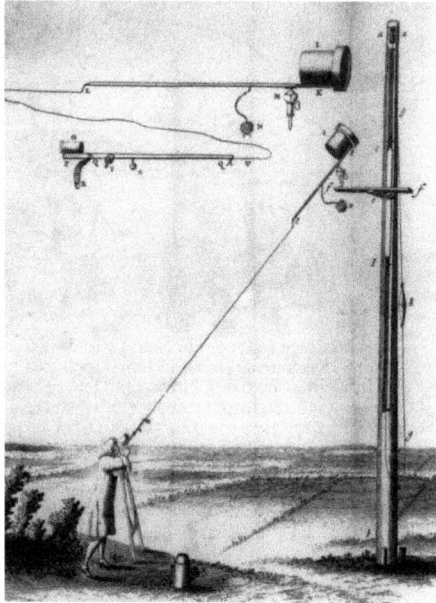

Figure 1.3   Huygens at his telescope (Observatoire de Paris).

largest satellite of Jupiter, the king of planets. The error was understandable: Titan's atmosphere increases the apparent size of the object, so the disc as seen from Earth is larger than any other satellite, the others being airless, or nearly so. This attribute is the reason for the name it was given, following a proposition by Herschel, who

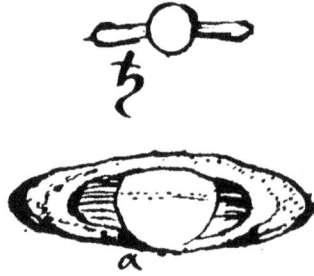

Figure 1.4   Huygens' first recorded observations of Saturn in March 1655 (above) and a sketch by Cassini made on January 18, 1691 (below) (Observatoire de Paris).

Figure 1.5   Saturn — the original (Observatoire de Paris).

had suggested using the names of gods associated with Saturn for his satellites, according to mythology:

As Saturn devoured his children, his family could not be assembled around him, so that the choice lay among his brothers and sisters, the Titans and Titanesses. The name of Iapetus seemed indicated by the obscurity and remoteness of the exterior satellite, Titan by the superior size of the Huyghenian, while the three female appellations [Rhea, Dione, and Tethys] class together the three intermediate

Cassinian satellites. The minute interior ones seemed appropriately characterised by a return to male appellations [Enceladus and Mimas] chosen from a younger and inferior (though still superhuman) brood.

This proposal was favourably received by the astronomical community. Titan has certainly turned out to deserve its important name, not only because of its size but also because of its substantial atmosphere similar to the Earth's. This latter, unique property remained undiscovered, however, until almost 300 years after Titan itself was first observed.

After Solà's doubtful claims of having observed an atmosphere around Titan in 1908 (doubtful because he had ascertained the same thing for the Galilean satellites), Sir James Jeans decided in 1925 to include Titan and the biggest satellites of Jupiter in his theoretical study of escape processes in the atmospheres around solar system objects. His results showed that Titan could have kept an atmosphere, in spite of its small size and weak gravity, if low temperature conditions that he evaluated as between 60 and 100 K (kelvins*) or -213 to -173 C (Centigrade) prevailed. In this case, a gas of molecular weight higher than or equal to 16 could not have escaped Titan's atmosphere since the satellite's formation. The constituents which could have been present in non-negligible quantities in the mix of gas and dust particles that condensed to form the Solar System and which, at the same time, satisfy Jean's criterion, are: ammonia, argon, neon, molecular nitrogen and methane. Ammonia (NH_3) is solid at the estimated Titan temperature and could therefore not substantially contribute to its atmosphere. The others, however, are gaseous within this same temperature range. Methane (CH_4), unlike argon, neon and molecular nitrogen, exhibits strong absorption bands in the infrared spectrum, which make it relatively easy to detect.

This analysis obviously invited a spectroscopic search for methane absorption bands in the light reflected from Titan. We shall see in due course that it took about 20 years to get around to doing that; but first let us dwell a moment more on the background to the name invited by Herschel's words.

1.2 Titan in Mythology

Hesiodos, a Greek epic poet and historian, Homer's contemporary, relates in his great poem "*Theogony*" ($\Theta\varepsilon o\gamma o\nu\iota\alpha\cdot$: Gods' birth) the hierarchy among the ancient Greek Gods and develops the cosmogonical and theogonical conceptions that prevailed in ancient times.

According to him, in the beginning all was Chaos ($\chi\acute{\alpha}o\varsigma$: nothing). Then came Gaia ($\gamma\alpha\iota\alpha\cdot$: Earth), universal mother and nurse of all beings, firm and offering unshakeable eternal support. Eros ($\acute{\varepsilon}\rho\omega\varsigma$: love) followed, the most handsome among immortals, he who in his sweetness brings happiness to Gods and humans. He is the moving force that brings together and unites all elements. Gaia gave birth first to Ouranos ($o\upsilon\rho\alpha\nu\acute{o}\varsigma$: the sky), who covered her with his celestial sphere and became

---

*For definitions of this and other technical terms, see the Glossary at the end of the book.

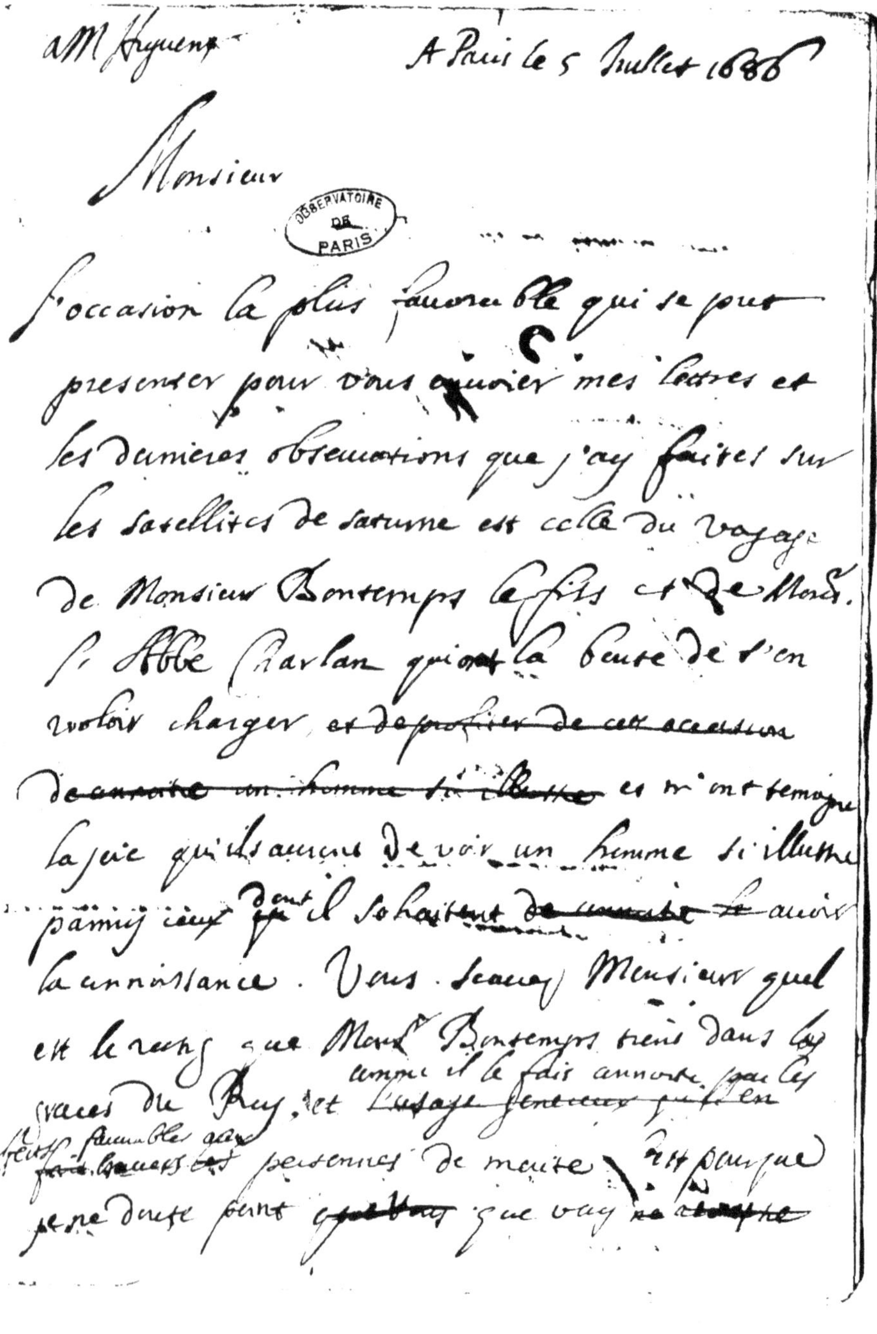

Figure 1.6   Fragments of letters from Cassini to Huygens, written in 1686 (left) in French and 1691 (right) in Italian (Observatoire de Paris).

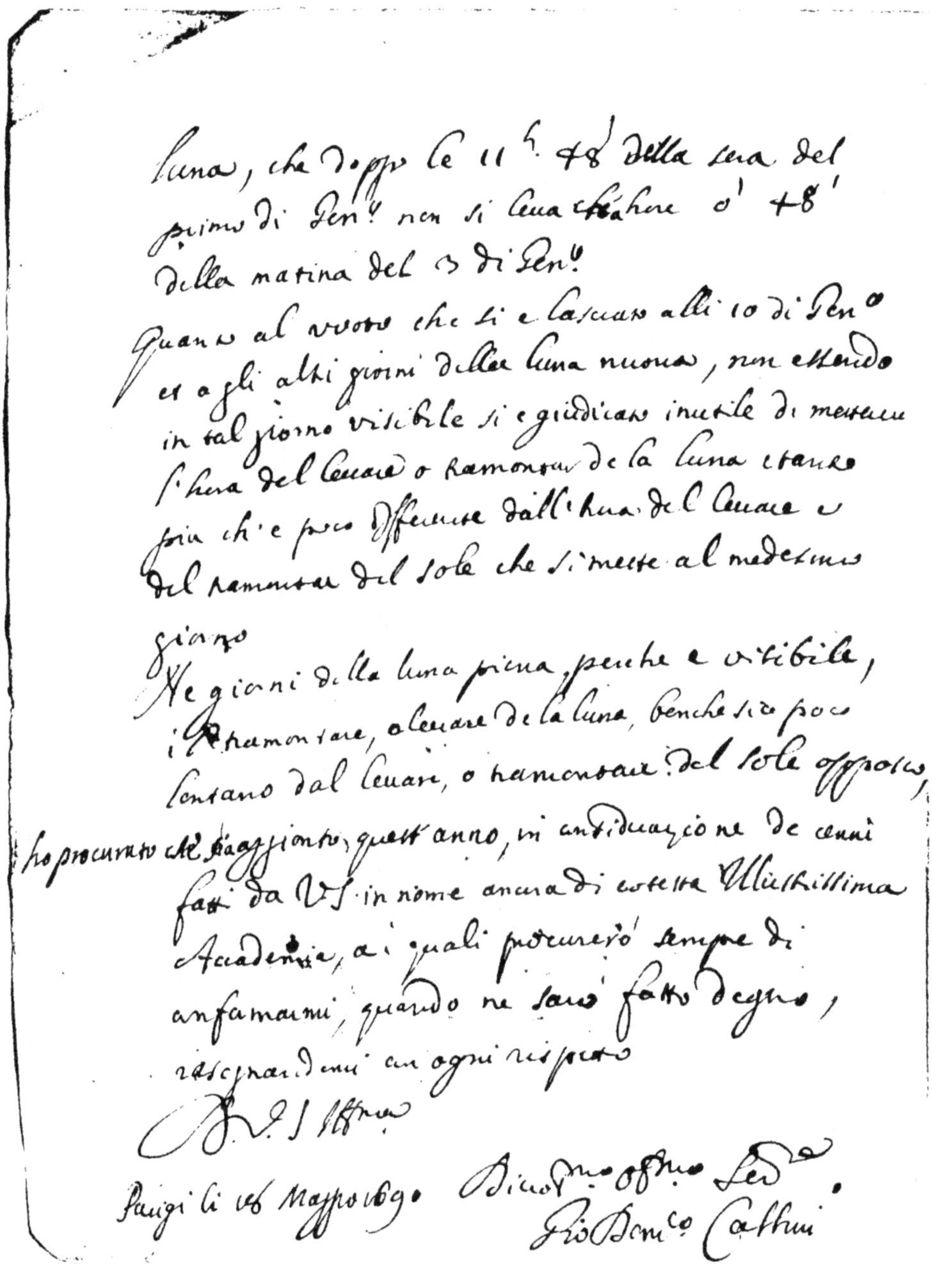

luna, che doppo le 11ͪ 48 della sera del
primo di Gen? non si leua chehore ò 48'
della matina del 3 di Gen?
Quanto al vuoto che si e lasciato alli 10 di Gen°
et agli altri giorni della luna nuoua, non essendo
in tal giorno visibile si e giudicato inutile di mettere
l'hora del leuare o ramontar de la luna estando
foia ch'e poco differente dall'hora del leuare e
del ramontar del sole che si mette al medesimo
giorno
He giorni della luna piena, penche e visibile,
il ramontare, oleuare de la luna, benche sia poco
lontano dal leuare, o ramontare del sole opposto,
ho procurato che siaggiunto, quest'anno, in consideracione de cenni
fatti da V.S. in nome ancora di cotesta Illustrissima
Academia, ai quali procurerò sempre di
anfamarmi, quando ne sarò fatto degno,
rassegnandomi an ogni rispetto

Parigi li 16 Maggio 1690.   Diuotissimo obligatissimo Ser.
                              Gio. Dom.co Cassini

Figure 1.6 (*Continued*)

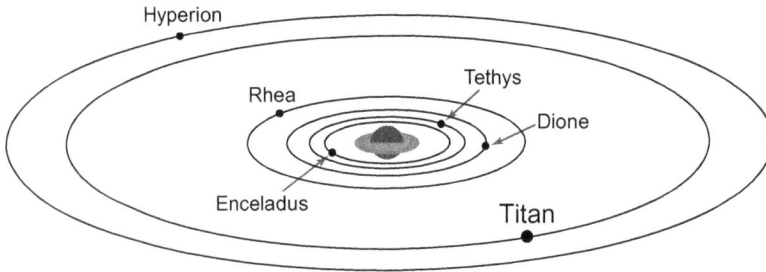

Figure 1.7   Saturn and the largest of its satellites, roughly to scale; Titan is about twenty Saturn radii from the planet, and has about one-twentieth the diameter of the gas giant.

her husband. From this very prolific union, the first dynasty of Gods was born, in spite of the fact that Ouranos sent his children down the abyss of the Earth for fear that they might steal his power.

The couple's first children were the twelve Titans: six males and six females. Their brothers and sisters were the three Cyclops with one eye, the Ecatochires, with a hundred hands and fifty heads, and the Giants. Ouranos' fears came true when the youngest male Titan, Kronos (Saturn to the Romans), assassinated his father, with Gaia's help, and took his place on the throne of Gods. But Kronos inherited his father's curse: for fear of losing his place, he devoured the children his wife Rhea, the youngest of the female Titans, brought into the world. Rhea was not pleased by this procedure. She asked Gaia's and Ouranos' help to save her last son, Zeus (known as Jupiter to Latins). She managed to hide him in a cave in Crete where a goat, by the name of Amalthea, fed him. When the goat died, Zeus put on the animal's impenetrable skin to protect his body, and decided to go against his father, after freeing the rest of the Titans from the Abyss to help. He was surprised to find that they preferred to fight on Kronos' side. With the Ecatochires and the Cyclopes, Zeus managed to beat the Titans and Kronos, after 12 years of violent battle. He finally overcame his father and took the first place in the "modern" Greek Pantheon.

Titan then was not one mythological figure, but a group of twelve. Their mighty power kept them in key positions in ancient religions and the literary and historic community never lost interest in them. Nowadays, the satellite Titan is a new object of awe, challenging the astronomical community to uncover its mysteries.

1.3 Space Exploration of the Solar System

The Sun is a small yellow star at a distance of 300,000 trillion kilometres from the centre of the Milky Way Galaxy. It appeared relatively recently, at a time when many of the first stellar generations were already extinct, and, ever since its formation, has followed an orbit around the Galactic centre, like hundreds of millions of others. It needs 225 million years to complete the circuit, so since its birth 4.5 billion years ago, the Sun has completed a full orbit around the centre of the Galaxy only 20 times, accompanied on its trek by 8 planets and thousands of smaller bodies. We now know that many other planetary systems exist in the cosmos, but among these, only the third planet in the Sun's family is known to have experienced the formation of life.

During billions of years of evolution, increasingly complex beings filled the oceans, invaded the air, and walked the continents of the Earth. A few million years ago, some of these beings stood upright, and used their hands to manufacture tools and machines that allowed them to invent a better, more comfortable way of living. These men and women developed efficient means of communication between themselves and eventually were established as the dominant beings on the planet. They soon became aware of the existence of thousands of millions of bright points of light in the night sky, turning around the centre of the universe, which was to them the Earth itself. The nature of these lights was a mystery to them, and excited their curiosity for a long time. Were they signs from the Gods, luminous stones, or holes in a black celestial sphere? Were they other worlds, and if so might they bear inhabitants, strange, extraterrestrial forms of life that turned their eyes to their own skies?

Little by little, people saw beyond the myth and the fear inspired by natural phenomena, eventually realising that neither their Earth, nor their Sun, was at the centre of the Universe. They also understood that the bright spots on the sky were distant luminous suns, much farther away than anything our ancestors could imagine. In the fullness of time, the inhabitants of Earth began to realise their dreams of travelling into space. It was less than 50 years ago that they managed to send the first man into space, and about 40 years ago that they succeeded in landing on our closest companion in space, the Moon. Today, the heirs to these pioneers, we can actively explore our Solar System in depth and are actively embarked upon the continuation of our cosmic adventure. Stimulated by the images returned by the early space missions, in the process of placing a permanent settlement of people on an orbital station around the Earth, and soon on the Moon and Mars, our generation is mostly not aware that the first logical argument for the support of space travel was given in the second century AD by a Syrian philosopher who wrote in ancient Greek. Lucian of Samosata was broadly travelled for his day, and wrote caustic essays that criticised the important figures of his time. For many centuries, his 'Dialogues' caused great tension between him and the powerful men of his country.

"Icaromenippos" (or "A journey through the clouds") was written in 167 AD, and consists of dialogue in which the hero, Menippos, tells his friend how he explored the sky looking for the truth and what he discovered. His explanation as to why he attempted this adventure is extremely modern. He begins by saying that one morning he opened up his eyes and realised that earthly goods were only temporary, and that important things in life had to do with the mysteries of the Universe. He therefore tried to learn something about the birth and the meaning of the Cosmos. But the more he tried to think it out, the more he ran into great difficulties, until he reached the conclusion that he had to ask for help and consult with the wise philosophers.

Menippos: *"Thus I chose the best men I could find, judging as well as possible by the sharpness in their eye, the paleness of their skin and the length and thickness of their beard. I have to admit that these men inspired confidence, had a grand air of wisdom about them and seemed to be knowledgeable of all the secrets in the sky. So,*

I put myself in their hands and paid a handsome sum in advance for their wonderful teaching. I sat there and waited for them to impart with me their science of subtle reasoning, hoping that in the end I would understand how the Universe works".

Soon enough though, Menippos is disappointed by the philosophers who can offer him no valuable information, as he admits to his friend: *"Well, my friend, you'll laugh when I tell you what sort of crooks these so-called 'philosophers' are, what kind of strange fantasies they have in their heads, and how cunning they are in setting up fictional stories. My teachers, who are nothing more than you and I after all, can see no farther than their own noses! Most of them are blind, either because they are so old, or because they do nothing all day long. This does not stop them from talking about what lies at the end of the Universe, or about the circumference of the Sun. They are capable — or so they say — of describing in detail the shapes and sizes of the Moon and the other stars farther away, and even say they can reach them! These same people who know not what the distance between Athens and Megara might be, discuss seriously about the distance of the Sun to the Moon.[...] One hears them swear that the Sun is a ball of fire, that the Moon is inhabited and that the celestial bodies must be covered with water, because the Sun absorbs the humidity from the sea lifting it up in the air, like water from a well, and redistributing it to all the world, thus providing us with drinking water!"*

"Such stupidity!" exclaims his friend, after these fantastic descriptions of the Solar System and the beautifully (and incidentally correctly) defined cycle of the water and clouds in the Earth's atmosphere. *"They really are crooks, all these astronomers".*

Menippos: *"And this is not all. Imagine, my friend, that these gentlemen manage to have arguments about whether the Universe has borders, therefore being closed, or whether it may have no limits. Some of them even insist that there is an infinite number of other worlds and fight with those who claim there is only the Earth".*

After these incredible announcements of the crazy ideas of the philosophers of the second century AD, Menippos reaches the inevitable conclusion:

Menippos: *"And so it was that, almost driven to madness and having lost all hope to find final answers to my questions here on Earth, I decided that the only solution was to get a pair of wings and to go and see for myself what happens in the sky".*

In these words, somewhat liberally translated by the authors, we find the best and most valid reason for space travelling: the scientific approach of searching and discovering the reality in nature.

Besides "Icaromenippos", Lucian made another trip to space in his "True Story" (which, despite the name, was a pure creation of his imagination), where he describes how he sailed through the Atlantic and — crossing a storm — found himself on the Moon. There he is mixed up in a battle between the Moon people and the inhabitants of the Sun. The Moon loses the war and Lucian is carried away to be a prisoner on the Sun, where he lives through some more wild and eccentric adventures, worthy precursors of modern sci-fi, before returning finally back to Earth.

Lucian was the pioneer among ancient astronauts in literature. Significantly, the idea of space travel has never seemed to present an insurmountable obstacle for ancient philosophers or their descendants, who include present-day science fiction writers, and even a good number of professional mathematicians and physicists. Still, tempting though the idea was of reaching strange worlds, a trip beyond the Earth was not an easy venture in practice. How could one attempt it? The ancient philosophers thought perhaps that with bird-like wings, by the force of formidable storms, through dreams or with the help of demons, man could hope to discover other worlds. More recently, the trip has been made, in the imagination, in balloons, with strange gravity-defeating machines, or by using great cannons.

Today we believe that the only practical way to launch a person or a machine into space is with a rocket. But this method has only fairly recently been adopted and never appeared in literature before the 20th century, when Constantin Tsiolkovski, a Russian theoretical physicist, published his famous book on "The exploration of space by action-reaction machines". Tsiolkovski once said: "Our planet is the birthplace of humanity, but one doesn't live one's whole life in the cradle". The practical realisation of rocket travel is however due to pioneers like Robert Goddard, who launched his first prototype in 1926 and continued with the construction of liquid-fuelled rockets, which, towards the end of 1930 reached a height of a few kilometres.

The large-scale construction of rockets devoted to space exploration might have remained the hobby of a few inspired inventors, working almost alone and with few resources, if the Second World War had not intervened. Despite the great misery and disaster it brought, this global madness produced the V2 rocket of Werner von Braun and his associates. This terrible weapon of revenge was later to become a powerful resource in the service of the space programme.

The first day of glory in real space flight belongs to the Russians: on October 4, 1957, the hundredth anniversary of the birth of Tsiolkovski, the legendary team of Sergei Korolev sent the first artificial satellite, Sputnik 1, into orbit. On April 12, 1961, Yuri Gagarin became the first man to leave the human cradle, spending 108 minutes, the time required by the Vostok spacecraft to perform a complete orbit around the Earth, at a height of a few hundred kilometres.

"Man", says American anthropologist Ben Finney, "is an exploring animal". Exploration is important to us. We are ready to confront dangers and disasters, disappointments and the lessons of humility, in order to learn something about the appearance and evolution of life on our planet. It is mainly for this reason that manned missions are pursued. The Americans made a great leap forward when Neil Armstrong became the first man to step on the Moon on July 21, 1969. This was the first of seven trips to the Moon, which made up the challenging but also costly Apollo programme. Apollo ended in 1973, epoch of the Skylab. The Russians, in the meantime, developed space stations like Salyut (1971) and Mir (1986), primarily to study the reaction of man to life in space. During the 1980s, the Space Shuttle made its debut with great success, a programme that continues into the era of the International Space Station, having transcended the tragic accident with Challenger in 1986. In spite of the vagaries of economics and war, plans are well advanced to send more spacecraft to the Moon and beyond, eventually following in the steps of the Voyager missions which have now left the Solar System altogether.

The exploration of the planets began, logically, with missions to our closest neighbours in the 1960s. Since then, unmanned spacecraft have visited all of the planets, some of them many times. Mercury, the closest to the Sun, was studied by the NASA spacecraft Mariner 10. Mercury is named after the Greek god of commerce and travel, the Roman messenger of the gods. It is slightly bigger than our Moon, and in many ways quite similar to it, because it has no atmosphere and the surface is heavily cratered by impacts by meteorites. The difference in temperature between the sunlit and dark sides is enormous — a blistering 450°C on the bright side, while in the shade of the other hemisphere the temperature plummets to −170°C. Future human explorers will need more than breathing gear if they are to live on Mercury without burning or freezing.

Venus was explored by NASA with Mariners 2, 5 and 10 and the orbiter and multiprobes of Pioneer Venus. The Soviet Union sent Veneras 7 to 16, as well as Vega 1 and 2. NASA's Magellan, launched in 1989, returned astounding radar images of the surface of Venus. ESA's Venus Express arrived near the planet in April 2006 and is returning new views of the planet, including the first close views of the huge double vortex at the south pole. Long admired for its beauty as the morning, and the evening, star, the high temperatures and densities (nearly 500°C and 100 bars) prevailing in the choking carbon dioxide atmosphere render the surface a hellish prospect for humans.

Not surprisingly, Earth's Moon has received special attention, with, among others, the Russian Luna sample return missions, the manned Apollo 11–17 landings, and prospect of a permanent base sometime in the next few decades. The Moon of course has no atmosphere, and offers a desert-like, hostile surface to travellers, so the base when it comes will resemble a land-bound version of the space station, replete with life support systems.

Mars was visited by the American Mariner 4, 6, 7, 9 and Viking 1 and 2 missions, and more recently Odyssey, Reconnaissance Orbiter and Mars Express, which are addressing basic questions about the geology, atmosphere, surface environment, history of water and potential for life on Mars. We know today that there are no artificial channels on the surface of Mars, and no Martians, although microbial life below the surface is still not ruled out. Missions such as Pathfinder have confirmed that conditions on the red planet were once very different, warmer and wetter, but it is also certain that today Mars is hostile to most forms of terrestrial life. Its atmosphere is very thin, cold and unbreathable, while its surface is constantly bombarded by cosmic and ultraviolet rays. To last long on the surface of Mars, a person would have to use a special heated suit with breathing gear, like the ones used by Apollo astronauts on the Moon.

The outer gas giant planets were visited by the Voyager missions in the 1980s. Jupiter, Saturn, Uranus and Neptune formed from the primitive gas nebula, which also formed the Sun. Because they are massive and cold, they have thick atmospheres, in which even the lightest molecules, hydrogen and helium, are retained. Thus, they are very different from the terrestrial planets, not only in having atmospheres of different composition in which the pressures are very high, but also in having no surfaces in the usual sense. The system formed by Pluto and its satellite Charon, both small, light and distant bodies, does not belong in either the terrestrial or giant planet categories, but instead resemble more the icy satellites of the outer Solar System or the planetesimals of the Kuiper belt. On August 24, 2006, Pluto was demoted from a planet to a "dwarf planet" according to IAU's resolution 5A.

The giant planets have many large satellites between them, several of them larger than the smaller planets. However, all but one are without significant atmospheres. They are mainly composed of ices, and have surfaces that are heavily marked by meteoritic impacts. The exception of course is Titan, in the Saturnian system, which not only is nearly Earth-sized but also has an atmosphere that has the same major constituent as ours and much the same surface pressure. In many ways, Titan is like a fifth terrestrial planet; cold, and lifeless almost certainly, but nevertheless one of the more relevant places in the known universe as a seductive place for humans to explore.

1.4 The 20th Century, Before Voyager

Before the Voyager missions to the outer Solar System, ground-based observations of the satellites of the outer planets had produced only fairly scanty information, so much so that the smaller of those known today had not been found at all, and virtually nothing definite could be said about those whose existence had been known for three hundred years.

With respect to Titan, the main scientific concern was focused on its atmosphere. The first formal proof of its existence came only after the Second World War, in 1944, when Gerard Kuiper, of the University of Chicago (and originally a Dutchman, like Huygens) discovered spectral signatures on Titan at wavelengths longer than $0.6\,\mu$m (microns), among which he identified two absorption bands of methane at $6190\,\text{Å}$ and $7250\,\text{Å}$. By comparing his observations with methane spectra taken at low pressures in the laboratory, Kuiper derived an estimate of the amount of methane on Titan: 200 metre-amagats.

Kuiper searched for similar behaviour in the spectra of other Saturnian satellites. But his data, obtained in 1952, showed differences between Titan and the other satellites in the intensity observed in the ultraviolet and visible continuum, as well as in the methane bands, which were absent except on Titan. Kuiper concluded Titan was a unique case in the Saturnian system due to the presence of an atmosphere, of such a composition that it gave the satellite an orange colour.

In the years that followed, in spite of much interest, it proved difficult to make significant further progress in exploring or comprehending Titan's atmosphere. By 1965, a consensus had still not been reached on a value for the ground temperature, in the presence of contradictory radio and infrared measurements that ranged from 165 to 200 K. From 1972 to 1979, a number of scientists concentrated their efforts on seeking a better estimate for the methane abundance and the surface pressure, using observations made in the 1-to-2 μm infrared spectral region. Limb darkening was finally unambiguously observed in 1975, consistent with an optically thick atmosphere.

At about this time, Laurence Trafton, from the University of Texas, conducted observations of the $3\nu_3$ spectral band of methane at 1.1 μm, in which he found unexpected strong absorption, indicating either a methane abundance at least 10 times higher than that inferred by Kuiper, or a broadening of the CH_4 bands induced by collisions with molecules of another as yet undetected, but quite abundant, gas in the atmosphere. In either case, the intensity of the absorption band is a function of the methane abundance and of the local atmospheric pressure. By comparing the weak absorption bands of methane in Titan's visible spectrum with spectra of Jupiter and Saturn, in which these bands have almost identical absorption strengths, Lutz and his colleagues derived from Trafton's measurements a 320 m-amagat abundance for methane, and an estimate for the effective pressure on Titan of about 200 mbar. The immediate consequence of this result was that methane suddenly became just a minor atmospheric component, since even 1.6 km-amagat (i.e. 1600 m-amagats, Trafton's highest estimate) could only correspond to a surface pressure of about 16 mbar.

By 1973, observations of the satellite's low albedo and of the positive polarisation of the reflected light, confirmed the presence of a thick, cloudy atmosphere, with the cloud particles present up to high altitudes. Theoretical considerations suggested that two sorts of aerosols were expected to co-exist in Titan's atmosphere: clouds of condensed CH_4, and a photochemical fog of more complex condensates. The latter would arise as a result of methane photolysis, that is dissociation by sunlight, mostly at ultraviolet wavelengths. The fragments of methane, CH_2, CH_3, etc., combine, leading to the production of a variety of polymers that condense to form oily droplets. Something similar happens on Earth in the photochemical smog engendered by terrestrial road traffic. In 1975 again, Gillett found evidence in Titan's thermal emission spectrum not only of methane (CH_4), but also of ethane (C_2H_6) at 12.2 μm, monodeuterated methane (CH_3D, at 9.39 μm), ethylene (C_2H_4, at 10.5 μm) and acetylene (C_2H_2, at 13.7 μm).

Trafton had also announced in 1975 a tentative identification of a spectral feature of molecular hydrogen, H_2, in the spectrum of Titan, for which he had evaluated an abundance of 5 km-amagat. In spite of much effort directed at the detection of NH_3, the observers of the time failed to produce more than upper limits, which got lower and lower with successive measurements, suggesting that if any of this gas existed on Titan, it must either have been photodissociated, with subsequent production of N_2 and H_2, or else trapped on the surface as ammonia ice.

Close examination of Titan's spectrum had already revealed at this time that the continuum absorption decreased with frequency, suggesting that the aerosol became more transparent at longer wavelengths. This led to the assumption that it might be possible, at certain frequencies in the near infrared, to probe all the way down to the satellite's surface. At short wavelengths, down to about 2200 Å, the brightness remains nearly constant, suggesting that the aerosol is uniformly mixed at high altitudes. The measurements say little about the nature of the aerosols, their composition for example, but the fact that they are present in the atmosphere makes all attempts to interpret spectroscopic observations extremely dependent on assumptions about the cloud properties.

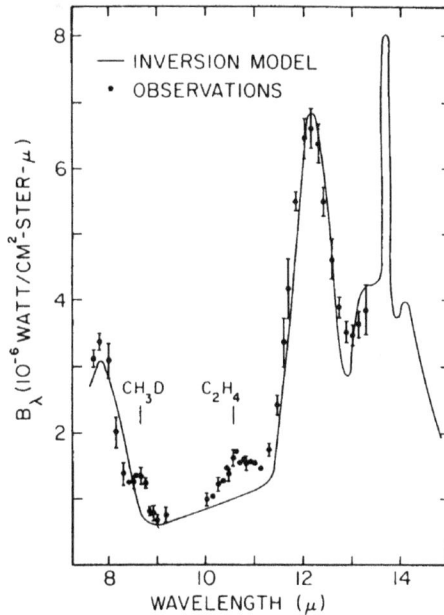

Figure 1.8 One of the first ground-based observations of Titan's infrared spectrum (Gillett, 1975).

A detailed set of assumptions, used to provide the context in which new data can be understood, is what scientists call a 'model'. Before the Voyager encounter, there were two principal models in contention for explaining the observations as they then stood. The first, suggested by Danielson in 1973 and completed by Caldwell in 1977, favoured methane as the main component (about 90%) of the atmosphere, and predicted surface conditions of $T = 86\,K$ at a pressure of 20 mbar, with a temperature inversion in the higher atmospheric levels demonstrated by the presence of emission features of hydrocarbon gases in the infrared spectrum of Titan. The second model, based on work by Lewis in 1971 and developed by Hunten in 1977, started with the assumption that dissociation of ammonia should produce molecular nitrogen, which is transparent in the visible and infrared spectrum, in large quantities. In this model the surface temperature and pressure would be quite high (200 K and 20 bars). These high temperatures on the ground could be explained by a pronounced greenhouse effect, resulting essentially from pressure-induced opacity in hydrogen at wavelengths longer than 15 μm. As on the Earth, and other planets (most notably Venus), this opacity blocks the thermal emission from the lower atmosphere and surface, creating a build-up of heat in the lower part of the atmosphere.

Just prior to the Voyager encounter, Owen and Jaffe made radio telescope observations with the newly-completed Very Large Array in New Mexico, and obtained the emission temperature of the surface finding a value of $87 \pm 9\,K$, a range that includes the modern value. They even suggested that conditions on Titan might support oceans of methane, an idea that was ahead of its time, but the paper failed to get the attention it deserved as it was published during the excitement of the Voyager encounter.

CHAPTER 2

The Voyager Missions to Titan

An amoeba of blackness leaked out from the zenith, obscuring Sun and blue sky....
The same sandy beach was beneath her feet; she dug her toes in. Overhead... was
the cosmos. They were, it seemed, high above the Milky Way Galaxy, looking down
on its spiral structure and falling toward it at some impossible speed...... A network
of straight lines appeared, representing the transportation system they had used.
It was like the illuminated maps in the Paris Metro.

Carl Sagan, *Contact*

2.1 Space Missions to the Saturnian System

The previous chapter gives an idea about where the situation stood before Voyager 1
flew by Titan in 1980. Voyager was not the first visitor from Earth to the Saturnian
system, as the ringed planet had been visited by small, unmanned Pioneer probes
in 1979. Nor was Voyager to be the last such visitor, but a quarter of a century
elapsed before Cassini/Huygens picked up the quest, with nothing but Earth-based
observations to fill the gap.

Pioneer 11 was a spin-stabilised spacecraft designed primarily to carry out a
first reconnaissance of Jupiter in the early 1970s. The 258 kg craft was launched on
April 5, 1973, and encountered the largest planet in the Solar System in December
1974. This encounter was such a success that Pioneer, still functioning well, was
re-targeted to use the gravity assist from Jupiter to send it on to a Saturn encounter
five years later, in September 1979.

One of the goals of the Pioneer Saturn encounter was to check the environment
in the vicinity of the planet's extensive system of rings. This was not just a scientific
objective, but aimed partly to blaze the trail that the larger and more sophisticated
Voyager spacecraft were to follow. It was not known, until Pioneer survived a passage
through the ring plane, whether there were any particles in the plane outside the
visible rings that might cause a hazard to spacecraft. Pioneer 11 was targeted at the
distance from Saturn — 2.9 planetary radii — that Voyager 2 would have to follow
if it were to go on to Uranus, which in the event it successfully did.

The Pioneer 11 trajectory carried it across the orbit of Titan one day after its
closest approach to Saturn, on September 2, 1979, at a distance from the satellite
of 363,000 km, much too far for Pioneer's relatively simple instruments to gather

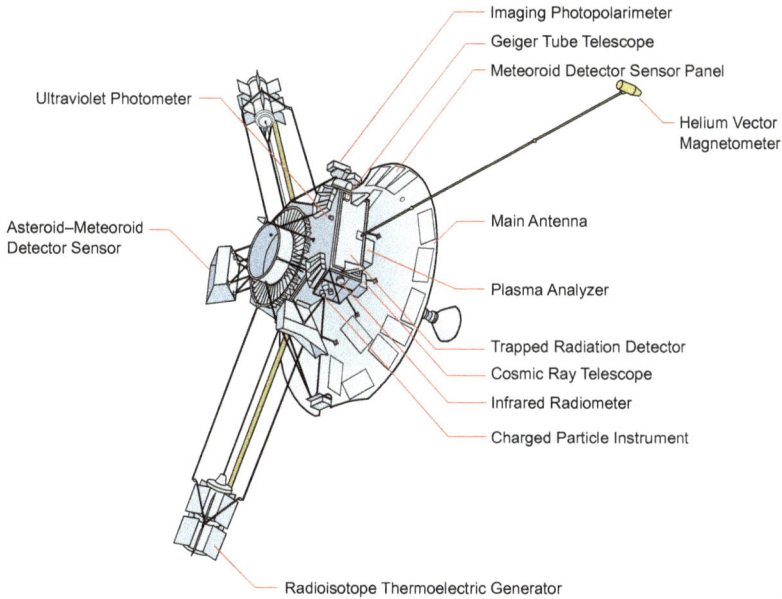

- Imaging Photopolarimeter
- Geiger Tube Telescope
- Meteoroid Detector Sensor Panel
- Ultraviolet Photometer
- Helium Vector Magnetometer
- Asteroid–Meteoroid Detector Sensor
- Main Antenna
- Plasma Analyzer
- Trapped Radiation Detector
- Cosmic Ray Telescope
- Infrared Radiometer
- Charged Particle Instrument
- Radioisotope Thermoelectric Generator

Figure 2.1   Pioneer 11 encountered Saturn on September 1, 1979. It flew within 13,000 miles of Saturn and took the first close-up pictures of the planet (NASA).

Table 2.1   Characteristics of the Pioneer and Voyager missions.

	Pioneer 11	Voyager 1	Voyager 2
Launch	April 5, 1973	Sept. 5, 1977	Aug. 20, 1977
Titan encounter	Sept. 2, 1979	Nov. 12, 1980	Aug. 27, 1981
Closest approach to Titan (km)	363,000	4,394	663,385

images or much useful data of any kind. However, it was the first man-made object to enter the realm of Saturn, and it showed the way was safe for Voyager. (Although, in fact, Voyager 1 made its closest encounter with Titan before passing through the ring plane.)

The Voyager 1 and 2 missions were the first large, stabilised spacecraft to travel to the outer Solar System. Both were launched in 1977, Voyager 2 actually a few days before Voyager 1, but the faster trajectory of the latter got it to Jupiter first. Each of the two spacecraft executed several thousand instructions without error, controlled from the Voyager operations centre at the Jet Propulsion Laboratory (JPL) in Pasadena, California, USA, which received tracking, engineering and science data via the stations of the Deep Space Network. The Voyager 1 encounter with Jupiter took place on March 5, 1979, while Voyager 2 swung past the giant planet on July 9 of that same year. In November 1980, Voyager 1 encountered Saturn and Titan. Voyager 2 arrived in the Saturnian system in August 1981, some nine months later.

Although not without interest, the data relative to Titan obtained by Voyager 2 are not as extensive as those taken by Voyager 1, because the closest approach distance of Voyager 2 was more than 100 times greater.

From an intellectual point of view, the encounter with Titan had been inevitable since the time of Titan's discovery by Huygens more than three centuries earlier. It became a step closer in practice in 1918, when the Soviet academic Yu. V. Kondratyuk devised the principle of using gravity assistance to space flight for the first time. The idea in itself is simple enough: by flying past a planet from behind, a spacecraft can acquire some extra velocity with respect to the Sun, which allows it to accelerate on to another destination. By a series of gravity assists, space missions can go farther and faster than would otherwise have been possible with our limited rocket technology. Half a century later, two JPL engineers discovered an opportunity for a launch in the year 1977, which coincided with a rare alignment of the outer planets which would not occur again before the middle of the twenty-second century. By grasping this opportunity to use gravity assists at Jupiter, NASA was able to send mission to the more distant giant planets, including Saturn, much sooner than it had otherwise planned.

The trajectory of Voyager 1 around Saturn allowed extensive coverage of the planet's atmospheric characteristics, of the rings and the icy satellites. The closest approach to Saturn at a distance of 126,000 km took place on November 12, 1980, the same day as the closest approach to Titan, 6969 km (4394 miles) from the satellite's centre. The orbital plane of Titan was crossed from north to south, the spacecraft trajectory inclined with respect to the orbital plane at about 8.7°, at a speed with respect to the satellite of $17.3 \, \mathrm{km \, s^{-1}}$.

The Voyager 2 spacecraft, destined originally to travel only to Jupiter and Saturn, was performing so well that its mission was dramatically extended, again using the gravity assist technique to make the journey feasible. Its trajectory was directed past Saturn in such a way that it was accelerated towards Uranus, where it arrived on January 24, 1986, and then again on to Neptune, where the encounter took place on August 24, 1989. The spacecraft then dived below the ecliptic at an angle of about 48°, leaving the Solar System behind at a speed of about 470 million kilometres a year. Voyager 1 ended its planetary mission with Titan, and has also left the Solar System, rising above the ecliptic plane at a rate of about 520 million kilometres a year, at an angle of about 35°. As of April 2006, Voyager 1 is at 12.32° declination and 17.114 h right ascension, in the constellation of Ophiuchus, while Voyager 2 is at −52.51° declination and 19.775 h right ascension, placing it in the constellation Telescopium. Voyager 1, the most distant human-made object in the cosmos, reached 100 astronomical units from the Sun on Tuesday, August 15, 2007 at 12.13 UT, meaning it is over a hundred times more distant from the Sun than Earth is.

The Voyager spacecraft itself is a three-axis stabilised platform with a total mass of 800 kg, including the science instruments at about 105 kg. A 3.7 m antenna is used for telecommunications and radio science. A very advanced machine for its era, Voyager was capable of operating with a high degree of autonomy at vast distances

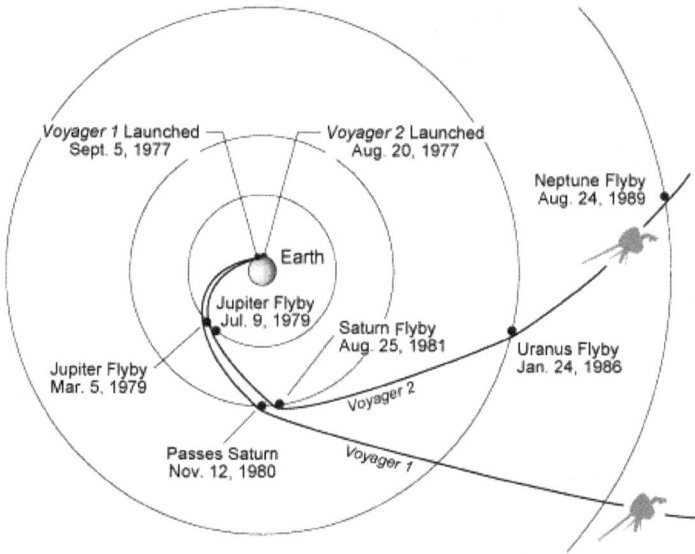

Figure 2.2   Voyagers 1 and 2 followed similar courses to Saturn and then separated. Because of its trajectory, designed to reach Uranus, Voyager 2 did not encounter Titan (NASA/JPL).

from the Earth. Its high-resolution television camera could read the headlines of a newspaper from a distance of one kilometre, with the gas jets controlling its position and alignment providing remarkable stability. Three radioisotope thermoelectric generators (RTGs) use nuclear power to produce 7000 W of heat, which is converted into 370 W of electricity and used to operate the spacecraft, its transmitter, and the scientific payload. Before leaving the Earth, the fuel tank was filled with more than 100 kilos of liquid hydrazine. This propellant was used for attitude control, course corrections, and to position Voyager so that the instrument pointing platform can observe its targets. Three onboard computers carry out instructions from Earth but can also operate the spacecraft autonomously for long periods of time, with what now seems like the very limited capability to store about 538 million bits of data on-board.

Each Voyager carries the same eleven scientific instruments, four of them mounted on the movable scan platform so they can be pointed at specific targets. The latter are the imaging experiment, consisting of boresighted narrow- and wide-angle cameras; the infrared interferometer spectrometer and radiometer (IRIS); the ultraviolet spectrometer (UVS); and a photopolarimeter-radiometer. Six other instruments are used to study fields, particles and waves in interplanetary space and near planets, including magnetometers, a plasma detector, a low-energy charged particle detector, a plasma wave detector, a planetary radio astronomy instrument, and a cosmic ray detector. In addition, the spacecraft's radio antenna doubles as a radio telescope for scientific investigations.

Figure 2.3   The Voyager spacecraft: the optical instruments are on the articulated scan platform shown at the top; the cylinder below and to the left of the long magnetometer boom is one of the radioisotope power generators (NASA/JPL).

The Voyagers may be the most prolific space probes ever launched, having visited all four of the giant planets and their many satellites and having returned a mass of data taken by the cameras and other scientific instrument onboard. This Planetary Grand Tour, as it was called, was possible thanks to a rare geometric arrangement of the outer planets that only occurs once every 176 years.

2.2 Voyager Observations of Titan

Titan's visible appearance at the time was unexciting — an orange ball, completely covered by thick haze which allowed no visibility of the surface. These "smog" particles form a layer that enshrouds the entire globe of Titan and stretches from the surface to an altitude of about 200 km, with the altitude of unit vertical optical depth in visible light at about 100 km. Views of the limb from Voyager showed the presence of a detached haze layer at 340–360 km altitude with large, irregular dark particles.

The most obvious global feature seen by Voyager in the haze cover was a difference in the brightness of the two hemispheres, about 25% at blue wavelengths, falling to a few percent at ultraviolet and at red wavelengths. This so-called north-south asymmetry is apparently related to the circulation of the atmosphere pushing haze

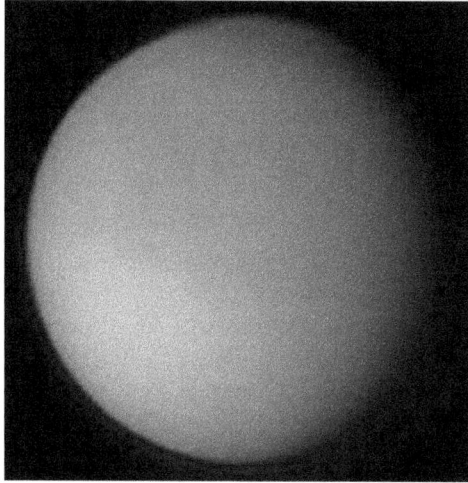

Figure 2.4   The first close-range image of Titan from the Voyager spacecraft showed little detail, just a hint of the north-south atmospheric asymmetry and a dark polar collar. The surface of Titan is hidden under a deep haze layer (NASA/JPL).

and gases from one hemisphere to the other. The asymmetry has been observed to reverse — when the Hubble Space Telescope (HST) first observed Titan in 1990, a little over a quarter of a Titan year after the Voyager encounters, the northern hemisphere was found to be brighter than the south. Whereas Voyager only observed up to red wavelengths, HST can image Titan in the near-infrared. At these wavelengths, the asymmetry is reversed, and indeed is somewhat stronger than in the visible. This is due to the wavelength dependence of the atmospheric brightness (bright at short wavelengths due to Rayleigh scattering, dark in the near-infrared due to methane absorption) and the haze (which seems to be dark in blue and bright at red and longer wavelengths, by analogy with synthetic haze material generated in the laboratory). The visible bright hemisphere has less haze than the darker one.

The limited photometric data we had from 1970 until the Cassini arrival suggests that the hemispheric contrast varies smoothly, and that limb-darkening is also strongly wavelength-dependent. The disk at UV and violet wavelengths is fairly flat, while it is near-lambertian (coefficient $\sim$1.0) at green and red wavelengths and shows limb brightening in the near infrared. Voyager also saw a dark ring around the north (winter) pole. This feature, part of a polar hood extending from 70° to 90° north latitude, is most prominent at blue and violet wavelengths, and it has since then been suggested that it may be associated with lack of illumination in the polar regions during the winter (since the subsolar latitude goes up to 26.4°), and/or subsidence in global circulation.

The data from the brief encounter covered only a few hours, as Voyager was travelling past Titan at a relative speed of over 17.3 kilometres per second, but it immediately clarified a lot of questions, and of course raised many others. In particular,

a combination of radio occultation, infrared spectroscopic and ultraviolet observations from the spacecraft came out in favour of a model like Hunten's, albeit with a
lower surface pressure and temperature of about 1.5 bars and 100 K, respectively.

During the rush that usually follows the arrival of a space mission at a planet,
especially when it is the first visit to a new world, scientists work on tight time
schedules to extract new findings as quickly as possible from masses of data, in
order to inform the public, their sponsors and the media who are eager for news.
Once the initial stir subsides, longer-term, more thorough analyses are undertaken,
generally based on improved laboratory measurements and computer models, and
designed to extract all of the new information. The following sections discuss the
results obtained from Voyager and serve as an introduction to the much more detailed
information about Titan's atmosphere, surface, and interior that is being provided
by the Cassini–Huygens mission, as described in the following chapters.

2.3 Atmospheric Bulk Composition

The first reasonably complete picture of the basic nature of Titan's atmosphere turned
out to be a combination of the two most popular pre-Voyager models. Molecular
nitrogen, N_2, was detected by the ultraviolet spectrometer as the major component
of the atmosphere, at about 95%, with methane as the next most abundant molecule,
with abundances determined by the Infrared Interferometer Spectrometer (IRIS) to
be 0.5–3.4% in the stratosphere and from 4–8% at the surface. Traces of hydrogen and of various moderately complex organic gases, consisting of several hydrocarbons, nitriles and CO_2 were found, with just one firm detection of a condensate
(C_4N_2) at first. Some simple oxygen compounds like CO_2, were also observed
by IRIS.

2.4 Vertical Temperature Structure

From the Voyager radio-occultation experiment, the surface temperature of Titan
was found to be 94 ± 1.5 K, or $-179°C$, at a pressure of about 1.5 bar, 50% higher
than Earth. A more precise value for Titan's radius was also derived: 2575 ± 2 km.
These, and the other main properties of Titan as established at the time, are collected
in Table 2.2.

In the radio-occultation experiment, the refraction of the radio beam from the
spacecraft was measured as it flies behind (or emerges from) the planetary disk, as
seen from Earth, provided density profiles as a function of altitude near the equator.
These can be converted into temperature versus mean molecular weight vertical
profiles in the atmosphere. From an analysis of the rate at which Voyager's radio
signal was attenuated by the atmosphere as the spacecraft passed behind the satellite,
and the opposite effect when it re-emerged, two vertical refractivity profiles were

Table 2.2   Physical characteristics of Titan obtained from Voyager 1 data, shown
in familiar, terrestrial units. The length of Titan's day and month are the same, as
they are for Earth's Moon.

Mass	1.346×10^{23} kg (0.0226 of Earth)
Equatorial radius	2,575 km (0.202 of Earth)
Mean density	1.88 gm cm^{-3} (1.88 of water)
Mean distance from Saturn	1,221,850 km (20.32 Saturn radii)
Mean distance from Sun	1.422×10^{9} km (9.546 times Earth's)
Sunfall	1.1% of Earth's
Orbital period	15.945 Earth days
Rotational period	Same as above
Titan day (period of rotation)	15.945 Earth days
Titan month (period around planet)	0.584 Earth months
Titan year (period around Sun)	29.46 Earth years
Mean orbital velocity	5.58 km s^{-1}
Orbital eccentricity	0.0292
Orbital inclination	0.33°
Escape velocity	2.65 km s^{-1}
Visual geometric albedo	0.21
Magnitude (V_o)	8.28
Mean surface temperature	94 K (−179°C)
Atmospheric pressure	1496 ± 20 mbar (1.5 of Earth's)

obtained, near Titan's equator. These were converted into vertical density distributions, which in turn were used to calculate vertical pressure profiles, and finally to obtain temperature profiles as a function of altitude over the range 0 to 200 km. The uncertainty in the temperature at 200 km was as much as 10–15 K, because profiles determined by radio occultation depend on the assumed atmospheric composition, in particular the proportions of nitrogen, methane and argon. The most likely value for the molecular weight of the air on Titan was found to be around 28 amu, the value that would correspond to nearly pure N_2, but could be as high as 29.4 amu (assuming the perfect gas law), and still be within the limits of experimental error. This left scope for the possible presence of a heavier component than nitrogen in Titan's atmosphere. Based on cosmological abundance data, the presence of several percent of argon was suggested. The formal result from Voyager IRIS data was that the argon abundance could be up to approximately 7%, but the gas escaped detection until the arrival of Cassini, and then was found only in trace amounts. Along with uncertainties in composition, calibration errors, deviations from the ideal gas law due to the low temperature, and so on, meant that the radio occultation data were reliable only up to about 100 km. At higher altitudes, the thermal structure became more and more sensitive to the initial conditions adopted in the integration of the refractivity profiles.

In the troposphere the temperature was found to fall from the surface value to a minimum of about 71 K, at the level known (by analogy with Earth) as the

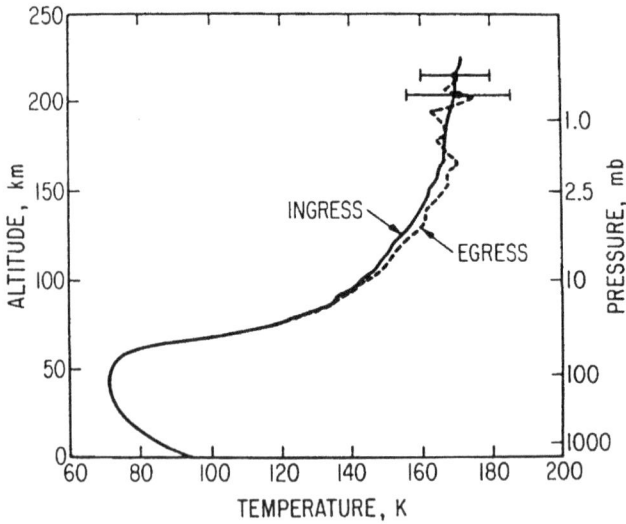

Figure 2.5 Titan's atmospheric temperature profile, as inferred from Voyager radio occultation measurements (Lindal *et al.*, 1983).

tropopause, located around 40 km in altitude. A profile in which temperature stops falling with increasing height and starts to rise again, is common in planetary atmospheres. It manifests itself in the infrared spectrum by the presence of emission, as opposed to absorption, bands of the more strongly absorbing minor components. The atmospheric temperature on Titan is everywhere higher than the condensation temperature of molecular nitrogen, making it improbable that nitrogen clouds could form. Condensation of methane, however, can occur if the methane stratospheric mixing ratio exceeds 1.6%, and the surface pressure and temperature conditions found by Voyager were consistent with the presence of methane in liquid form. It was tempting for the scientists examining the data to imagine that methane rain accumulates in a vast ocean on the surface. However, they noted that the vertical gradient of temperature, called the 'lapse rate', observed near the surface, was close to the value expected for a dry nitrogen atmosphere, indicating that methane saturation probably did not apply, at least at the particular place and time probed by the Voyager occultation. The precise CH_4 abundance was difficult to determine; the best estimate was a height-dependent value that varied from 2.5% up to 6%.

Another candidate for the main component of lakes or seas on Titan is ethane, which is one of the main photochemical products of methane and which is more stable as a liquid under the temperature and pressure conditions on the surface. A global ethane ocean, 1 km deep, in which methane is dissolved in appreciable quantities, as well as traces of nitrogen and other atmospheric condensates, including higher hydrocarbons like propane and acetylene, was suggested by Jonathan Lunine of the University of Arizona in Tucson and his colleagues in 1983. The Voyager-determined lapse rate changes near a height of 3.5 km, an observation which could be interpreted as marking the boundary between a convective region near the surface and a radiative equilibrium zone higher up, although it could also be the sign of the bottom of a methane cloud with a clear mixture of nitrogen and methane gas below.

Two additional data sets were available to constrain the temperature structure on Titan. Firstly, the Voyager ultraviolet spectrometer observed an occultation of the Sun by Titan, obtaining a value for the air density at 1265 km of $2.7 \pm 0.2 \times 10^8$ molecules cm^3. This translates to a temperature of 186 ± 20 K at that level, and an average temperature of 165 K in the 200 to 1265 km altitude range. The UVS experiment also allowed the detection of a methane mixing ratio of $8 \pm 3\%$ around 1125 km, and placed the homopause level at around 925 ± 70 km. Secondly, the occultation of the star 28 Sgr by Titan was observed from places as widely dispersed as Israel, the Vatican, and Paris on July 3, 1989. This rare event provided information in the 250–500 km altitude range, including a temperature value of 183 ± 11 K near 450 km.

In principle, a combined analysis of the radio-occultation atmospheric refraction data with the infrared spectral radiance data from IRIS allows the temperature profile to be retrieved from the ground up to about 200 km, along with estimates of the abundances of the major components (N_2, CH_4 and H_2) that affect the distribution of infrared opacity sources. In reality, the usual approach is to assume a given composition of the atmosphere (about 98% of nitrogen, about 2% of methane and 0.2% of hydrogen) and then to infer the temperature profile by finding the temperature profile which fits the measured spectrum assuming this composition. IRIS data in the methane band at $1304\,cm^{-1}$ probe the 0.01 to 10 mbar (about 150 to 450 km) atmospheric region, and the retrieval can be extended to include the thermal structure in the thermosphere, up to 800 km, if the RSS and IRIS data are augmented by the ultraviolet spectrometer (UVS) solar occultation measurements.

Outside the equatorial region, where no radio occultation data were available and only the ν_4 CH_4 band was available to infer temperature, the results were restricted to the upper stratosphere and lower mesosphere. Assuming that methane is uniformly mixed in the atmosphere, with a methane mole fraction of 1.8%, nominal temperature profiles were retrieved for seven different latitudes in the 0–450 km altitude range by extrapolating downwards so that the various retrieved stratospheric profiles all join smoothly to a single tropospheric profile. Obviously, this is not completely satisfactory, since the tropospheric and surface temperatures also vary latitudinally, but we had no information about that with Voyager. Also, the thermal structure derived through inversion of the radiative transfer equation, using methane as an opacity source, did not exactly join the radio-occultation temperature profile in the zone where the latter was thought reliable. With that caveat, the retrieved temperature profiles revealed a larger temperature decrease from the equator to the north pole than from the equator to the south pole, with the south pole 2–3 K, and the north pole up to 20 K, colder than the equator. The Voyager encounter took place at the time of the northern spring equinox on Titan and the equator-to-north-pole temperature gradient, is expected to reverse as the seasons change, with a lag due to the thermal

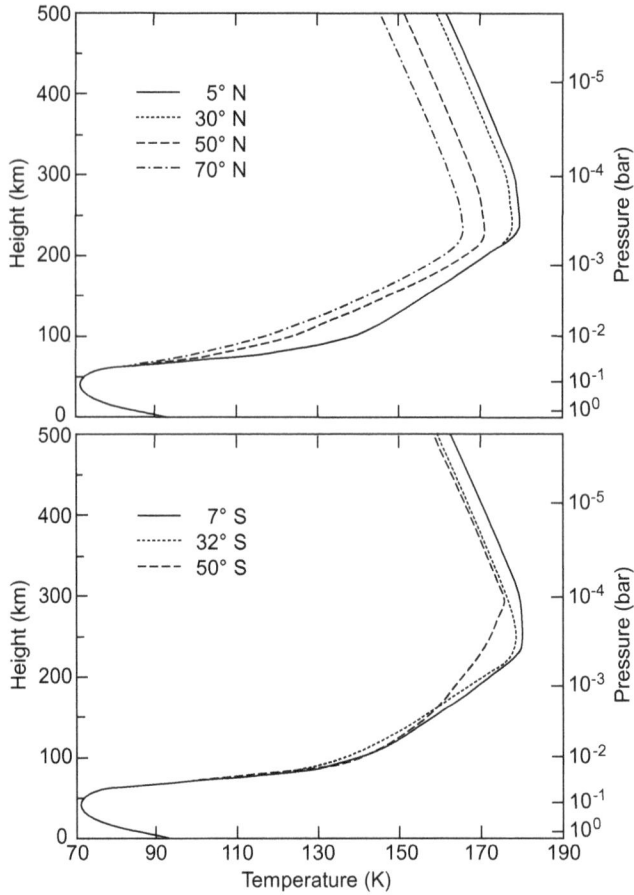

Figure 2.6   Temperature profiles retrieved from Voyager data, at different latitudes. The top set is for the northern hemisphere, at the latitudes shown; the bottom set for the southern hemisphere (Coustenis and Bézard, 1995).

and dynamical inertia of the atmosphere. The observed cooling in the north polar region could also be the result of an enhanced concentration of infrared emitters (gases and possibly aerosols), although this would require the hemispheres to have long-term compositional differences for some unknown reason.

2.5 Energy Balance and the Temperature Profile in the Thermosphere

With no significant internal energy source of its own, Titan is expected to be in overall energy balance with the Sun. In addition, the individual layers of the atmosphere need to be in equilibrium with the surrounding layers, the Sun, and the surface. Computer calculations of this balance are a well-established way of predicting the temperature

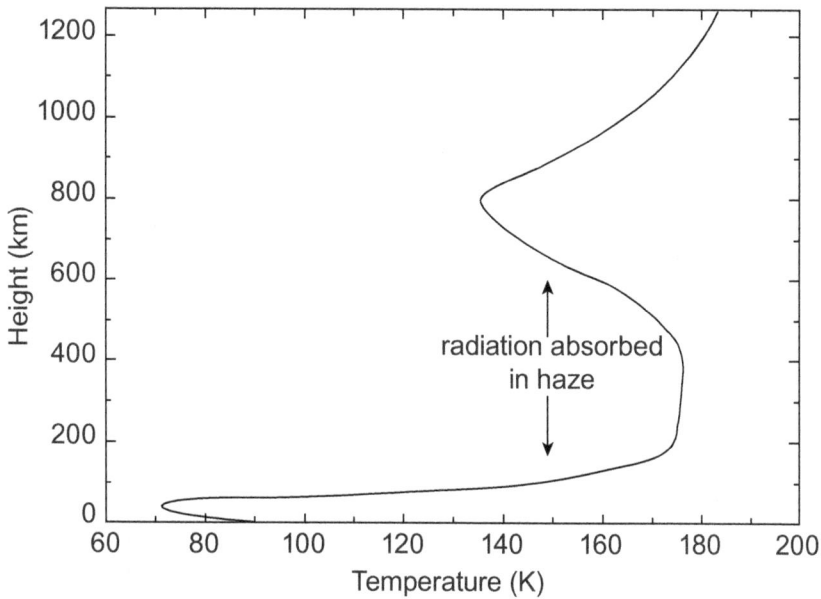

Figure 2.7 Model temperature profile for Titan, including the thermosphere, derived from energy balance considerations (Lellouch *et al.*, 1990).

structure on bodies with atmospheres. In 1983, Robert Samuelson developed an analytical radiative equilibrium model for Titan that was vertically homogeneous and that considered the radiation balance in three broad spectral intervals. His results confirmed pre-Voyager models, which gave a temperature inversion on Titan as a result of strong absorption of solar UV in the stratosphere, and penetration to near the surface of longer wavelength solar visible radiation. To obtain agreement with Voyager temperature data, Samuelson needed new sources of opacity in the height regions near 20 and 65 km, and he suggested these might be due to possible condensation clouds of methane and C_2H_2–C_2H_6–C_3H_8, respectively. He also identified the wavenumber range from 400 to 600 cm^{-1} as a thermal infrared "window", a spectral region of relatively high transparency throughout the atmosphere. The existence of such a window, at wavelengths where Titan's surface is emitting strongly, reduces the greenhouse effect on Titan and is responsible for cooling the surface by about 9 K. The stratospheric haze, which absorbs efficiently at short wavelengths, blocks much of the incoming solar radiation, but is more transparent in the thermal infrared and does not 'close' the window.

In 1984, Friedson and Yung undertook more detailed calculations of the thermal balance of individual layers in Titan's atmosphere to infer the temperature profile for the whole 0–1300 km altitude range. They solved simultaneously the equations of heat transfer and of hydrostatic equilibrium, using the UV occultation measurements of density and temperature as boundary conditions, taking into account the sources of energy (solar radiation and a contribution from electron precipitation into the magnetosphere) and the cooling through emission of the minor components (dominated by acetylene in the thermosphere). The heat is transported downwards by molecular conduction as far as the mesopause level at about 736 km and 110 K. Later, in 1990, the problem was revisited by Lellouch and colleagues, following the discovery of some errors in the solar heating profile and in the expressions of the collision-induced absorption rates in the earlier work. New thermal balance profiles were derived and tested for a match with the emission observed in the ν_4 methane band, and for consistency with the Voyager UVS temperature-density point at 1265 km. The temperature profile that satisfies these two conditions consists of a "warm" stratospheric region (mean temperature about 175 K) up to 500 km, to compensate for a cold mesopause (135 K) at 800 km. The associated methane mole fraction is 1.7%. The results of the retrievals for the lower atmosphere impose an additional constraint at 200 km of $T = 173 \pm 4$ K, and provide at the same time the continuity of the thermal profile down to the surface. The model thermal structure in Titan's thermosphere is extremely dependent on the solar heating efficiencies and on the relaxation rates of the cooling agents (mainly acetylene). Given the lack of information on these parameters, one may, in theory, obtain several solutions. The only firm conclusion of this exercise seems to be the existence of a temperature minimum at the mesopause, due to the very efficient cooling resulting from radiative emission in the ν_5 band of C_2H_2.

In 1991, Roger Yelle published realistic models of the thermal profiles for the upper atmosphere of Titan (0.1 to 10^{-2} nbar), including non-LTE effects (that is departed from local thermodynamical equilibrium conditions), heating/cooling in the rotation-vibration bands of the main hydrocarbons and HCN and aerosol heating. These remained the most precise and complete knowledge on Titan's thermal structure until 2004 when the Cassini/Huygens mission began to greatly add to our knowledge through in situ measurements at the equator and from data acquired by the experiments on board the orbiter (see Chapter 4).

2.6 Atmospheric Composition

The IRIS infrared spectrometer aboard Voyager 1 studied the emission from Titan's atmosphere with a field-of-view of 0.25°, which gave relatively high spatial resolution. The spectra taken by IRIS cover the spectral region from 200 to 1500 cm^{-1} with a resolution of 4.3 cm^{-1}. The data confirmed nitrogen as the main constituent, allowed the abundance of other minor components to be determined, and identified several other gases not previously detected. The presence of the simple hydrocarbons methane (CH_4), acetylene (C_2H_2), ethylene (C_2H_4) and ethane (C_2H_6) was confirmed, while the signatures of hydrogen cyanide (HCN), an important 'prebiotic' molecule (that is one of the building blocks of amino acids), and two other nitriles, cyanoacetylene and cyanogen (HC_3N and C_2N_2), were found. Abundances were also estimated by comparison with laboratory spectra for some more complex hydrocarbons: diacetylene (C_4H_2), methylacetylene (C_3H_4), propane (C_3H_8) and monodeuterated methane (CH_3D). Finally, carbon dioxide (CO_2) was found in the IRIS spectra at $667\,cm^{-1}$. The Voyager instrument did not cover the part of the spectrum where carbon monoxide (CO) might have been detected. The search for this common and widely-occurring gas was continued by other means, and it was eventually discovered in 1983 from ground-based observations in the near infrared. Its abundance was estimated to be about one part in ten thousand (a mixing ratio of 10^{-4}), in agreement with the photochemical model predictions at the time.

Some of the gases which were observed directly or inferred to be in the atmosphere of Titan, including molecular nitrogen, carbon monoxide and molecular hydrogen — also argon, although it was not detected by Voyager — are expected to be uniformly mixed throughout the lower atmosphere without undergoing phase changes. Since methane may condense in the troposphere, it may not be vertically uniform, and its abundance may increase with temperature down to the surface. After methane, the next most abundant hydrocarbons are ethane, acetylene and propane. Each of these organics and the less abundant ones could condense at some level in the lower stratosphere and precipitate out, which would restrict the amount present as a gas to smaller amounts below this level.

Latitudinal variations were observed for the least abundant molecules, while the most abundant gases appeared to be approximately constant from equator to pole. Cyanoacetylene and cyanogen (HC_3N and C_2N_2) showed an obvious enhancement at high northern latitudes, indeed they could not be detected equatorwards of about 60°N. Again, this is probably a manifestation of seasonal effects; Titan's northern hemisphere was in early spring, suggesting these species are most abundant in the darkness of the polar winter.

Voyager 2 flew by Titan nine months after Voyager 1 (on August 27, 1981), 170 times further from Titan's surface than Voyager 1. Because the Titan year is nearly 30 Earth years long, nine months is equivalent to only about a week in terms of the change of seasons on Titan, so the Voyager 2 encounter was also not long after the spring equinox, when the northern hemisphere was emerging from winter into spring. The Voyager 2 IRIS instrument, identical to that on Voyager 1, took a total of 115 infrared spectra, most of them between 15°S and 60°N latitude and at emission angles lower than 50°. The projected field of view on Titan's disk was of course much larger for Voyager 2's more distant encounter, and covered more than half Titan's diameter, allowing for only two locally independent locations to be used for atmospheric composition and temperature determinations. The results nevertheless confirmed the temperature variations in latitude found by Voyager 1, and all of the molecules previously found in the Voyager 1 data were detected by Voyager 2 IRIS, with the exception of C_2N_2, HC_3N, CH_3D, C_2H_4 and C_4N_2. The absence of HC_3N, C_4N_2 (which had been detected only in the solid phase) and C_2N_2 signatures in the later encounter was undoubtedly due to the fact that these molecules were detected by Voyager 1 only in horizontal viewing measurements near 70°N, that is, in the particularly favourable limb geometry conditions. The other species have weak bands in the spectrum so that they did not rise above the noise level.

Ethane, acetylene and propane, which are the most abundant carbon-containing compounds after methane, mono-deuterated methane and carbon monoxide in Titan's atmosphere, were found to be lacking any significant compositional variations in latitude, in accordance with Voyager 1 results. This tends to confirm that these molecules are homogeneously mixed in Titan's atmosphere from pole to pole. In the Voyager 2 analysis, methylacetylene (C_3H_4) and diacetylene (C_4H_2) tend to increase (by a factor of 2) near the northern region with respect to the equator. This tendency, although less marked, confirms the Voyager 1 results, in which the C_3H_4

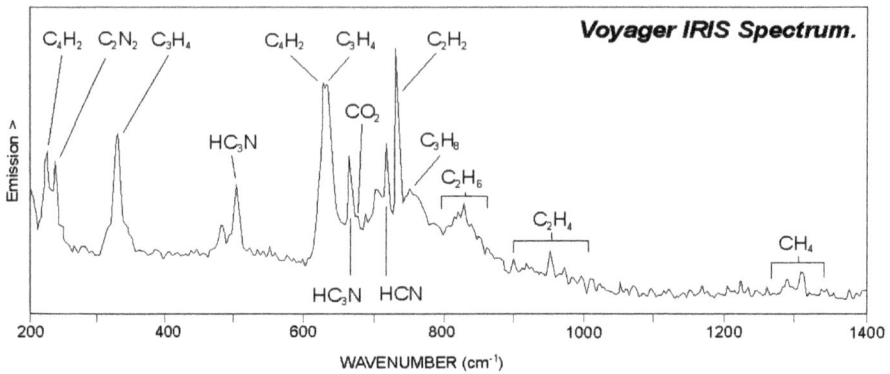

Figure 2.8   An infrared spectrum obtained by Voyager-IRIS viewing the north polar region of Titan, showing the emission features of several of the molecules discussed in the text.

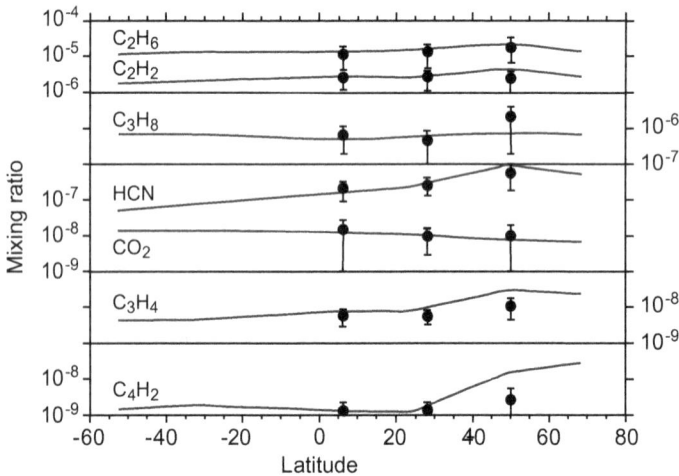

Figure 2.9   Mean molecular abundance of gases in Titan's stratosphere observed with Voyager 2 (Letourneur and Coustenis, 1993).

abundance factor of increase was between 3 and 4. The C_4H_2 mixing ratio in the Voyager 1 data is a factor of about 20 higher near the north pole.

HCN showed a steady increase in abundance in the Voyager 1 data from pole-to-pole by a factor of about 12 (of which a factor of about 4 from the equator to 50°N latitude). In the Voyager 2 data, the HCN mixing ratio at high latitudes is only about two times higher than at the equator. The much larger projected field of view of the Voyager 2 observations is probably responsible for the less marked enhancement.

In summary, Voyager 1 and Voyager 2 data show that the northern polar regions were associated with enhanced abundances for the nitriles and some hydrocarbons at the time of the Voyager encounters, probably because the north polar region was just coming out of the long winter darkness, having accumulated maximum nitrile abundances. Carbon dioxide (CO_2) was the only gas showing a possible decrease at the time, rather than increase, near the north pole.

Argon was not detected in Titan's atmosphere prior to Cassini–Huygens, because it shows no emission lines in the infrared spectrum and because, as we now know, the amount present is very small. A long search took place, driven by cosmogonic arguments that suggested a substantial presence was likely. The estimates of the mean molecular weight of Titan's atmosphere from the Voyager radio science experiment which allowed for an element heavier than nitrogen to be present in a substantial amount, originally thought to be as high as 27%. The expected argon resonance lines at 1048 and 1067 Å in Titan's EUV dayglow, due to solar and photoelectron excitation processes, did not appear in the Voyager UV spectra, and this led to the imposition of an upper limit of 14% by D. Strobel and colleagues in 1992. The solar occultation data from the same instrument, and a spectrum of the north polar region dayglow, combined to reduce the upper limit on the argon mixing ratio at the tropopause to less than 10%. Thermal considerations using Voyager radio-occultation measurements next led R. Courtin and colleagues to further lower the upper limit on the argon mole fraction, to 6%. These authors also used the IRIS intensity ratio derived in 1981 to check their analysis and found that only 5% of argon produced the best fit. Samuelson and his co-workers reanalysed the Voyager infrared and radio-occultation data and found an upper limit of 7% for argon. A search for the spectrum of H_2–Ar Van der Waals (loosely bound) molecules at 2.1 μm using ground-based observations was unsuccessful, possibly due to inadequate spectral resolution and atmospheric interference. Another approach tried unsuccessfully was to look for the argon fluorescent line at about 3 keV with an X-ray spectrometer on an Earth orbiting observatory. The Cassini–Huygens GCMS detection of Argon (see Chapter 6) was a happy ending to this long search.

2.7 Photochemistry

The discovery of a molecular nitrogen atmosphere on Titan by the combined infrared, ultraviolet and radio occultation experiments on Voyager brought new light to our understanding of photochemistry in Titan's atmosphere. The formation of ethane and acetylene through methane photolysis, as well as the further catalytic dissociation of methane by acetylene producing polyacetylenes, had been advocated even before the encounter, but the involvement of molecular nitrogen in hydrocarbon chemistry had not been widely considered. The first model of Titan's photochemistry based on the Voyager 1 observations and on photochemical reactions in a N_2-CH_4 atmosphere was proposed in 1984 by Y. Yung and colleagues of the California Institute of Technology. According to this, the atmosphere in the beginning contains only the parent molecules: molecular nitrogen, methane (probably contained in the volatiles trapped at the time of Titan's formation) and water (probably from a small but fairly regular meteoritic source, raining in from space).

Photochemistry occurs on Titan because sunlight in the ultraviolet part of the spectrum has enough energy to break up the methane molecules in the upper atmosphere; this is called UV photolysis. Energetic particles, mainly electrons, also rain down on Titan from Saturn's magnetosphere. The satellite has little magnetic field to deflect these particles, and they also dissociate molecules, including nitrogen, so these additional products become available to join in the photochemical 'soup'. The interaction of these molecules with ultraviolet radiation, energetic particles and cosmic rays in the thermosphere and mesosphere produces species such as CO, CO_2, C_2H_2, C_2H_4, and CH_3C_2H, as well as hydrogen cyanide (HCN, a prebiotic molecule) and other nitriles (HC_3N, C_2N_2, etc.). The major source of HCN according to this model is N_2 dissociation by magnetospheric electron impacts in the thermosphere, while the hydrocarbons mainly result from methane photolysis.

The photochemical fragments of methane nitrogen and the other simple molecules present can recombine as larger molecules, which continue to grow until they form quite large and complex molecules. Just how large and how complex is one of the big questions that remains outstanding to this day. These heavier molecules diffuse downwards to the regions where the lowest temperatures are found and further condensation, coalescence, transport and mixing can occur. The resulting suspended solid or liquid particles of condensed hydrocarbons and other complex molecules are responsible for Titan's characteristic orange colour, produced by a cycle somewhat like that which produces the man-made "smog" which occurs over many of the Earth's cities.

In parallel with these model predictions, B. Khare, C. Sagan and T. Reid at Cornell University performed laboratory spectroscopy to discover the optical properties (refractive index and absorption coefficient) of the organic compounds that are produced in the laboratory when mixtures of methane and other gases are subjected to electrical discharges. Of course, energetic sparking for a few days in a jar is not the same thing as irradiation by low levels of ultraviolet photons and bombardment by energetic particles over millennia, but all of these processes involve breaking up methane and nitrogen molecules and recombining their products, perhaps many times over.

The result is a mixture of materials of an oily or tarry nature, including liquids as well as solids, which the experimenters lumped together for convenience under the name "tholin", which, according to C. Sagan, comes from the Greek "θωλός", meaning "muddy". The exact composition of the solids made in the laboratory experiments is difficult to determine, and anyway varies in experiments that use different starting mixtures of gases under different conditions of pressure and discharge, as discussed further in Chapter 7. Nevertheless, the tholins manufactured in the laboratory were shown to provide a close match to the colours and light-scattering properties of Titan's orange upper-atmospheric haze. This haze, rather like Earth's ozone layer, absorbs sunlight at the blue end of the spectrum and results in heating of the middle regions of the atmosphere.

As a result of these parallel lines of enquiry, backed by the Voyager results, the idea of photochemical production of Titan's orange haze became well established. It has an interesting and puzzling corollary, however. The production of the haze, and the movement downwards of the particles as they become too large to stay suspended, must have been going on for all of Titan's history. Calculations suggest that the products that have condensed out and precipitated to the surface should have accumulated to a depth of about 1 kilometre over the whole globe in the age of the Solar System. At the same time, this process should have reduced the amount of methane in Titan's atmosphere, indeed calculations show that the amount of methane present now would be completely used up in about 10 million years. In that relatively short time, it would all be converted to heavier hydrocarbons, nitriles and other organics and to spare hydrogen. The last of these diffuses upward and escapes to space, the others end up on the surface as frozen or liquid deposits. A replenishing source for methane is therefore required in order to account for what we see today. The possibilities include the direct delivery of methane into the atmosphere from external sources such as comets. More likely, there could be a supply of methane inside Titan, which escapes regularly into the atmosphere by a kind of volcanic activity or outgassing from Titan's surface (see Chapter 9). Most intriguingly, since the amount of methane in the lowermost part of the atmosphere is close to saturation, the need for a reservoir supports the idea of liquid methane forming part of lakes or seas on the surface, in direct contact with the atmosphere.

According to this scenario, much favoured in the 1980s when Titan's surface still remained virtually unseen, deep seas of liquid ethane, methane, propane and possibly other liquid materials probably existed there. The vapour pressures of ethane and propane are too low to permit recycling back into the atmosphere, and the larger 'tholin' molecules are probably solid, or at least very thick and viscous, at Titan's surface temperature. However, the most volatile part of the ocean, methane, could be released into the atmosphere. As it rises into the cooler upper troposphere, some may condense and be cycled back to the surface as liquid methane rain, while some passes through the cold trap at the tropopause and diffuse upward to replenish the upper atmosphere. There it would be decomposed and eventually precipitated back to the surface as higher hydrocarbons and nitriles, falling as a mixture of outlandish tar, oil and snow.

2.8 Cloud and Haze Properties

The images taken by the cameras aboard Voyager 1 showed Titan covered by deep haze layers that enshroud the entire globe. The droplets appeared to be small and well-spaced, a haze rather than a cloud, so it was deduced that visibility inside the atmosphere would be quite good over small and medium distances. It is because the main layer is so deep — extending from close to the surface to an altitude of about 200 km — that it obscures our view of the surface from outside. Voyager scientists saw the subtle difference in the brightness of the two hemispheres, and hypothesised that it was probably caused by seasonal effects in the production of the haze or in the dynamical processes which keep the droplets aloft. The local peak in the density of aerosol particles about 100 km above the main layer, at an altitude of around 340–360 km, shows clearly as a separate layer in the Voyager pictures.

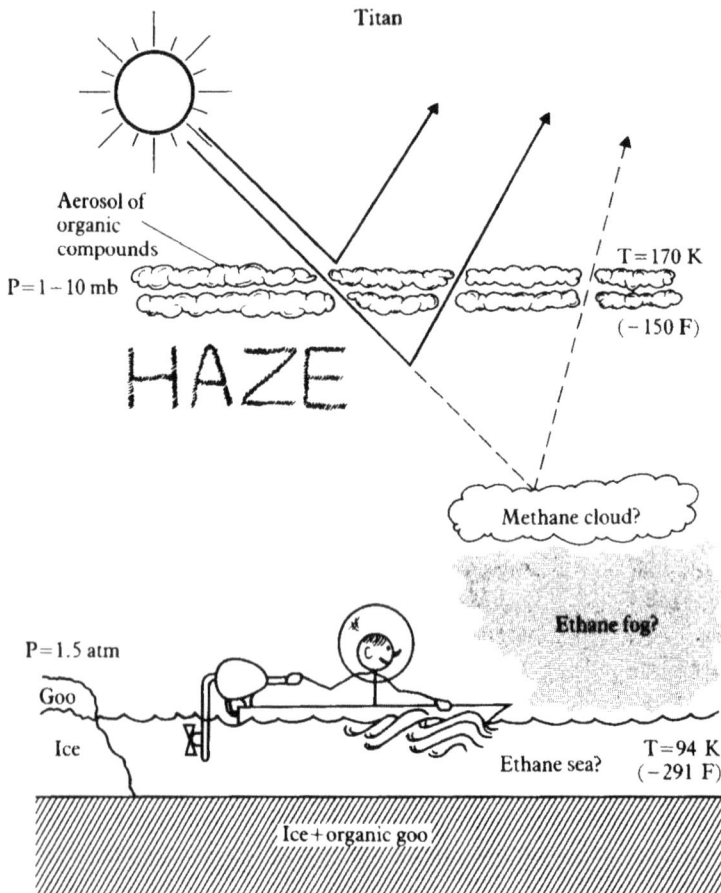

Figure 2.10   A sketch from 1988 of the possible scenario on Titan, with haze, clouds, rain, and a variegated surface with an intrepid explorer! This imaginative view is by David Morrison and Toby Owen.

In 1983 K. Rages and J. Pollack at NASA AMES Research Center investigated the properties of the aerosols from high-phase-angle Voyager images and found the particle radii to be between 0.2 and 0.5 μm. However, the degree of polarisation of the light scattered from the haze was found to be incompatible with spherical, liquid drops; they suggested they might be irregular conglomerates of smaller solid particles stuck together. The existence of solid material would imply that the photochemical processing of methane and its products goes on to the point where quite complex organic materials are produced, since these are most likely to be solid rather than liquid or gaseous. The first Titan haze models to include fractal aggregate particles composed of several tens of small (0.06 μm in radius) monomers were shown to produce strong linear polarization. 45 monomers would compose aggregates with an effective radius of about 0.35 μm, matching the Voyager observations.

2.9 Speculations on the Surface and Landscape of Titan from Voyager

Voyager's cameras could see no signs of the surface through the continuous blanket of aerosol in the upper atmosphere. In fact, the early evidence suggested that Titan's atmosphere was too opaque to permit a view of the surface until very long (radio or radar) wavelengths were reached. This misapprehension came about because Voyager had no near-infrared capability, and was understandable in terms of the knowledge available at the time the mission was planned, but turned out to be crucial, as Titan's haze is in fact optically thin at wavelengths just slightly longer than visible light. Earth-bound telescopes and the Hubble Space Telescope used this post-Voyager discovery to provide spectacular near-infrared views of the surface.

Based on the pre-Cassini knowledge, a number of features were hypothesized for Titan's landscape. However, it was really very difficult to speculate on what surface features on Titan might be like before the arrival of Huygens. The usual approach was to work on the basis of landforms found throughout the Solar System, but because Titan is such a unique object it would be easy to be misled by such thinking. Some parallels could be drawn, primarily with other cold, icy satellites, but also perhaps with Venus, when we consider the effect of the dense, optically thick atmosphere.

Following Voyager, there was quite a strong case for liquids in some locations on the surface of Titan. It seemed equally probable that there would be some craters. Other bodies in the Solar System, especially in its outer reaches, have been heavily bombarded and Titan should be no exception. However, the craters could have been modified, and in some cases virtually obliterated, by a combination of processes (see Section 9.5.2).

Voyager also had little to say with certainty about Titan's interior, although it did provide an improved density estimate, and stimulated the production of more

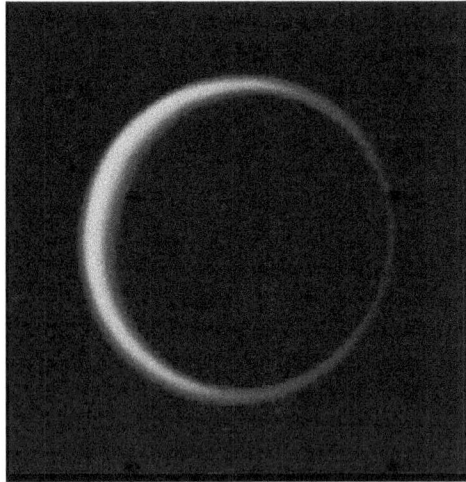

Figure 2.11   Crescent Titan viewed by Voyager 2 in August 1981 from a distance of around 200,000 kilometres, from a position almost directly looking back toward the Sun. The effect of the thick atmosphere can be seen in the refraction of sunlight around the edge of the disk to produce a continuous ring of light (NASA/JPL).

elaborate theoretical models, the most recent of which are described in Chapter 9. Titan's density falls midway between the two largest satellites of Jupiter, Ganymede and Callisto. From the abundances of the elements in the solar neighbourhood and the composition of primordial material in the present-day Solar System, as represented by comets, all three are believed to consist principally of iron, silicates and water ice, with smaller amounts of other elements and compounds. The simplest model which could be imagined for the solid body of Titan, which explains its low density and recognises cosmogonic considerations — that is, what we know about the universe and the formation of the Solar System — is one which has a rocky core, covered by a thick mantle of water ice. To give the observed mean density of $1.88 \, \mathrm{gm \, cm^{-3}}$, the depth of the ice would have be about half the radius of Titan.

2.10 The Aftermath of Voyager

One consequence of the success of the Voyager encounters was a sharp increase in the level of interest in Titan. Enough was revealed, especially about the atmosphere, to stimulate a considerable number of theoretical and modelling studies, most of which raised further questions which went on to define the goals for Cassini–Huygens. Some of the most intriguing ones were summarised in the first edition of this book, after Voyager but before Cassini, as follows:

• Where did the atmosphere come from, and why is it unique in the outer Solar System?

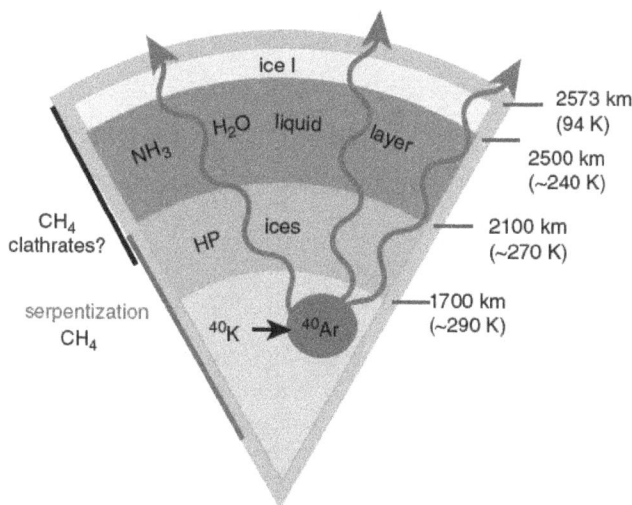

Figure 2.12 A model of the interior of Titan, showing the layering which may be present, and the escape of argon from the deep interior. The detection of ^{40}Ar by the Huygens probe is evidence for this kind of outgassing, and by implication methane must be escaping also (Atreya *et al.*, 2006).

- Does the composition vary from place to place? Are there condensable species which give rise to clouds and rain?
- What is the degree of complexity achieved by the chemistry on Titan? Are species like amino acids forming?
- What is the nature of the surface, its composition and topography? Are there seas, lakes or underground reservoirs?
- How much does the temperature vary on the surface of Titan, and in the atmosphere, across the globe?
- What is the circulation of the atmosphere? How strong are the winds and storms, do thunder and lightning occur? What is the effect of the seasons on weather?

We will be dealing with these questions in detail in the chapters that follow, and the progress made with the Cassini–Huygens mission, a massive investment by the US and European space agencies and the world scientific community. Since Voyager, there has also been an invigorated and successful ground-based observing programme, and new observations by Earth-orbiting telescopes, in particular the Hubble Space Telescope and the Infrared Space Observatory. Among other achievements, these produced the first images of Titan's surface, showing intriguing details on the surface, and identified a number of additional atmospheric species, providing more data to stimulate the production of improved atmospheric models.

"...Galileo's intellectual heir, Christiaan Huygens, the landed son of a Dutch diplomat who made science his life..., had divined that the "moons" Galileo observed at Saturn were really a ring, impossible as that seemed at the time. Huygens also discovered Saturn's largest moon, which he named Titan.... But Huygens couldn't be tied to the telescope all the time. He had too many other things on his mind. It is even said that he chided Cassini, his boss at the Paris Observatory for the director's slavish devotion to the daily observing."

Dava Sobel, *Longitude*

3.1 Introduction

In the two decades that followed the Voyager discoveries on Titan, up to and during the development, launch and flight of the Cassini–Huygens mission, astronomers turned to ground-based observatories and to Earth–orbiting artificial satellites in order to continue to probe what was now seen as a fascinating world. Breakthroughs were still to come despite the large distances involved, and these would help to optimize the return from the new space mission. In particular, new telescopes and pioneering observing techniques allowed a better understanding of the nature of Titan's surface well before the powerful Huygens instruments landed there. We review here the observational tools used to acquire data between Voyager and Cassini, and discuss how all of this was applied to our enhanced understanding of Titan's atmosphere and surface at the end of the 20th and in the first years of the 21st centuries.

3.2 Space Observatories

In the years since Voyager, technological progress has meant that observations from space through telescopes orbiting the Earth are now an affordable means of studying astronomical objects, including the planets and satellites of the Solar System. The Hubble Space Telescope is only the first of a long series of such observatories in space, which allow astronomers to avoid the atmospheric perturbations that affect even the highest terrestrial observatory, and to achieve higher quality data. From November 1995 until early 1998, another space telescope, the Infrared Space

Observatory (ISO), operated in geocentric orbit and yielded information on a great variety of astronomical objects. Both telescopes have observed Titan and the data have provided a source of valuable information on the satellite in the near-infrared and visible range.

3.2.1 *Hubble Space Telescope*

The Hubble Space Telescope is a 2.4 m reflecting telescope complemented by science instruments including three cameras and two spectrographs and fine guidance sensors primarily used for astrometric observations. These achieve a spatial resolution of about 0.1 arcsec, rarely attainable from the Earth.

The Space Telescope Imaging Spectrograph (STIS) operates across the spectral range from the ultraviolet (0.115 μm) through the visible red and the near-IR (1 μm). As a spectrograph, it spreads out the light recovered by the telescope so that it can be analysed to study the chemical composition, the temperature structure, radial and rotational velocity and magnetic fields of astronomical objects. STIS uses three detector arrays, all with a 1024 × 1024 pixel format: a caesium iodide photocathode for wavelengths from 115 to 170 nm, caesium telluride for 165 to 310 nm, and a Charge Coupled Device (CCD) for 305 to 1000 nm. The field of view is 25 × 25 arcsec for the shorter wavelengths and 50 × 50 arcsec for the CCD. The Near Infrared Camera and Multi-Object Spectrometer (NICMOS) provides the capability for infrared imaging and spectroscopic observations of various targets. NICMOS operates in the 0.8 to 2.5 μm range with a resolving power up to 100,000. The Faint Object Camera (FOC), built by the European Space Agency, uses two complete detector systems and is very sensitive: it can observe objects fainter than magnitude 21.

After the telescope was transported into space in 1990 by the crew of the space shuttle Discovery (STS-31) and set onto a 600 km orbit around the Earth, it was found to suffer from spherical aberration problems. Its primary mirror is 2 μm — only 0.0002 cm — too flat at the edge. Rescue missions in December 1993 and February 1997 successfully fitted parts which restored the expected performance of the telescope. The original Wide Field/ Planetary Camera was replaced by WF/PC2 (pronounced Wifpic 2), which is actually 4 cameras in one. The kernel consists of an L-shaped trio of wide-field sensors and a smaller, high resolution planetary camera tucked in the square's remaining corner. The Corrective Optics Package was installed, replacing the High Speed Photometer, its purpose being to correct the effects of the primary mirror aberration on the Faint Object Camera. Later instruments were designed with their own corrective optics.

Each HST orbit lasts about 95 minutes, not all of which is spent observing, since some is used for housekeeping functions. The observing plans are designed by the Space Telescope Science Institute at Johns Hopkins University in Baltimore, Maryland. Then the Goddard Space Telescope Operations Control Center takes

Figure 3.1   The Hubble Space Telescope during deployment (NASA).

over and produces the operations schedule to be sent to the onboard computers and executed. Data are broadcast from HST to the ground stations immediately or stored on tape and downlinked later. If an observer wishes, they can examine images and raw data within minutes for a quick-look analysis. The Science Institute is responsible for the data reduction and first processing (calibration, editing, etc.) and distribution to the scientific community. Competition is very active to obtain time on HST, with an acceptance score of about one proposal out of ten.

J. Caldwell and colleagues first used the Hubble Space Telescope to observe Titan in 1990. Subsequently, in 1994 and 1995 HST was used with the Wide Field/Planetary Camera and broad filters in the visible and near-IR range. The Titan images were part of an effort to probe Titan's lower atmosphere and surface using 'windows' in the spectrum where methane is weakly absorbing. The 1994 HST data in the window at $0.94\,\mu$m were the first to show surface features. They included a bright equatorial 'continent' on Titan's leading hemisphere, which explained the brightness fluctuations previously observed by spectroscopists as the satellite rotated. The HST results on Titan are more extensively discussed in the following sections.

3.2.2 The James Webb Space Telescope

NASA's planned follow-on to the successful Hubble Space Telescope was known as The Next Generation Space Telescope before it was named after a former administrator of the agency. The new telescope has a stated goal of "finding clues that

Figure 3.2 One of the conceptual designs for the James Webb Space Telescope, successor to the Hubble. It is likely to be placed in orbit either at the Lagrangian point L2, or in a 1×3 astronomical unit ellipse (NASA).

remain hidden by time and discovering how our universe evolved", to be addressed by observing the universe at much earlier times than currently achieved, to find the chemical makeup of early galaxies and the shape of the very early universe. The search for extra-solar planets will form a further step toward answering the question as to whether we are alone in the universe.

The JWST is scheduled for launch in 2013. The instruments it will carry are designed to work primarily in the infrared spectrum, using a primary mirror 6.5 m in diameter. Maintained at a temperature of 40 K or lower, NASA says it will be up to 1,000 times more sensitive than any existed or planned facility in the near-infrared (0.8–5 μm) region. To reduce stray light, it has a large sunshield and orbits a million miles from Earth, probably in the second Lagrangian point L2 where the gravitational attractions of Earth and Moon are equal, or in an elliptical orbit which passes close to the Earth.

3.2.3 *Infrared Space Observatory*

ISO was launched on November 17, 1995, from Kourou in French Guyana. It operated for more than two years in Earth orbit, with a 24 hr period, a perigee of 1000 km, and an apogee of 70,000 km. It consists of a helium-cooled 60 cm telescope and four scientific instruments: a camera, an imaging photopolarimeter and two spectrometers, SWS and LWS, dedicated respectively to short- and long-wavelength observations that cover the entire thermal spectrum between them from 2 to 200 microns, with resolving power ranging from 1,000 to 20,000. Two different observing modes are available on each spectrometer: Grating mode (mean resolution of about 0.5 cm^{-1}) and Fabry-Pérot mode (resolution between 0.01 and 0.05 cm^{-1}).

Figure 3.3   The Infrared Space Observatory (ESA).

The grating operated in the 2 to 45 μm wavelength range with a resolving power $\lambda/\Delta\lambda$ of 2000 and in the 45 to 200 μm range (50–220 cm^{-1}) with $\lambda/\Delta\lambda = 200$. The Fabry-Pérot acquired data from 15 to 35 μm (285–665 cm^{-1}) with a resolving power of 20,000 and observes the 45 to 200 μm range with $\lambda/\Delta\lambda = 10,000$. The noise equivalent power of the detectors was very low, so ISO benefited from extremely high sensitivity in a part of the spectrum that had not been explored before. However, due to the faint signal recovered and to the interference of stray light from Saturn, it was essentially the spectral region from 200 to 1500 cm^{-1} (7 to 50 μm), as with Voyager, that proved exploitable in the SWS, and, in addition, the 2.5–4.5 micron region.

The observing programme for ISO included about 20 hours devoted to observations of Titan. The first SWS observations were obtained on January 10, 1997 in the 220–340 and 600–1500 cm^{-1} ranges. Titan does not fill the ISO field-of-view (which is 100 arcsec for LWS and 30 arcsec for SWS, while Titan is only 0.8 arcsec in diameter); therefore the data are all averages over the whole disk. However, ISO spectroscopy — thanks to the higher resolution — made it possible to obtain more precise abundance and temperature profiles, to detect new molecules and to probe the atmosphere at lower levels than Voyager. The Titan SWS and PHT-S ISO data were processed and rendered exploitable by an expert team at the ISO Data Center of the ESA Satellite Tracking Station at Villafranca del Castillo, led by A. Salama and B. Schultz.

It is not simple to compare ISO and Voyager or Cassini results, since the flyby missions made measurements at close approach, therefore acquiring information on local parts of the satellite. A space observatory in Earth orbit, such as ISO, sees Titan as a small disk in a wider field-of-view, recovering information only as a disk average. In spite of that, ISO provided important insights into the satellite's atmosphere,

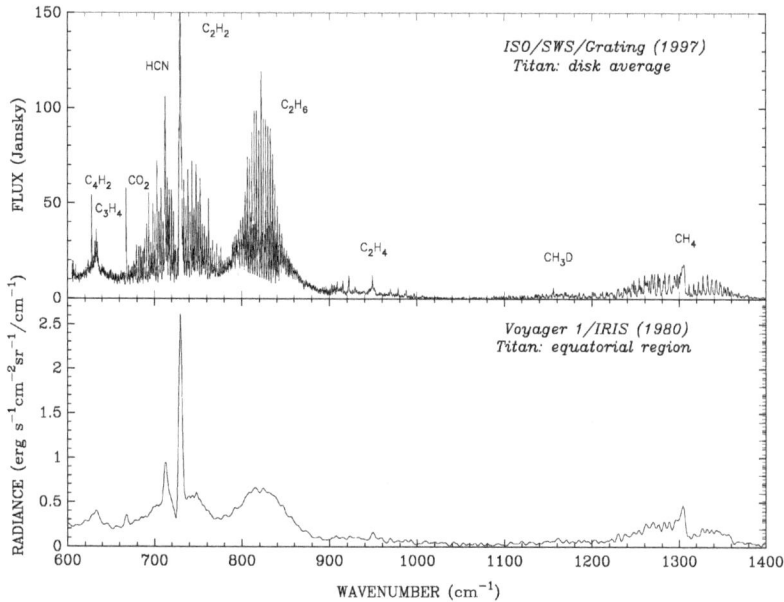

Figure 3.4   Pre-Cassini space infrared observations of Titan: ISO data from 1997 compared with Voyager, from 1980. Where the Voyager /IRIS spectrometer brought local information on Titan's disk, the ISO/SWS spectrometer provided much higher spectral resolution to compensate for the lack of spatial resolution.

mainly because it possesses a much higher spectral resolution than IRIS. At the spectral resolution of the ISO SWS/Grating mode, most of the molecular bands are resolved, allowing information on the mixing ratio of the constituent as a function of altitude to be retrieved. Also, the contributions of different species with bands in close locations can be separated because of the high resolving power. This gives more precise information on the mean molecular abundances of the atmospheric gases and also on the disk-average temperature profile. The ISO measurements mostly pertain to low latitudes because the limb contribution is not very important in the ISO field-of-view.

Despite the improved spectral resolution, ISO detected essentially the same hydrocarbon and nitrile molecules as seen by Voyager. However, it also detected for the first time the presence of water vapour in the higher parts of the atmosphere on Titan. This was a very exciting discovery because it addresses the mystery of the origin of the oxygen seen as CO and CO_2 in Titan's atmosphere and also, at the same time, gives an estimate of the water influx value and so some insight into the complicated issues involved with the transport of icy grains in the Saturn system (see Chapter 6). The observations of H_2O concentrated on narrow spectral regions at around $40\,\mu$m, in the rotational band of water vapour where the strongest lines were expected to occur. Another first tentative detection by SWS was that of benzene (C_6H_6) at $674\,cm^{-1}$. Thus, ISO observations nicely complemented the Voyager data and offer important additional information to be combined with the recent Cassini results, as discussed extensively in the following chapters.

3.3 Ground-Based Observatories

Titan has been observed regularly from the Earth since Kuiper made his famous detection of methane in 1944, with a surge of new findings in the mid-seventies by Trafton, Gillett and others. In more recent years, since 1990 or so, long-term observing campaigns in the United States and Europe have concentrated on those regions of the spectrum where methane absorption is weak and observations probe down to the lower troposphere and surface. Increasingly sophisticated instruments operating mainly in the near-infrared, but also in the thermal and millimetre spectral ranges, have allowed more precise determinations of temperature and the detection of new molecules in the atmosphere. The technique of adaptive optics (AO), which corrects for the distorting effect of the atmosphere, allows ground-based near-IR observations to resolve Titan's disc and obtain some spatial resolution on the surface by using spectroscopy and imaging. Attempts have been made to follow variations in the atmosphere and on the surface as a function of Titan's position in its orbit, as we discuss below.

The astronomical observatories that have been used to study Titan are located all over the world. In the Northern Hemisphere, the most important (and most impressive) site for ground-based astronomical observations is located on top of the mountain called Mauna Kea on the big island of Hawaii. In the southern hemisphere, the ESO (European Space Observatory) facilities at La Silla, Chile, offer a serious counterpart for Hawaii in the north. Besides its observatories in the Canary Islands, Europe maintains a large radio-antenna in the Sierra Nevada, and Russia and other Eastern European countries have major sites devoted to astronomy. Much of our current knowledge of the Solar System has been provided by these facilities, a testimony to the dedication and determination of the teams of engineers and telescope operators who maintain and improve the instrument capabilities and ensure the correct performance of the telescopes.

3.3.1 Mauna Kea Observatories

Mauna Kea ("White Mountain") is a dormant volcano on the Big Island of Hawaii, located about 300 km from the Hawaiian capital city, Honolulu. The highest point in the Pacific Basin and the highest island-mountain in the world, Mauna Kea rises 9750 m (32,000 ft) from the ocean floor to an altitude of 4205 m (13,796 ft) above sea level, which places its summit above 40 percent of the Earth's atmosphere. The broad volcanic landscape of the summit area is made up of cinder cones on a lava plateau.

The atmosphere above Mauna Kea is extremely dry — an important asset when detecting infrared and sub-millimetre radiation through the atmosphere — and

Figure 3.5   The Mauna Kea Observatories in Hawaii.

cloud-free, so that the proportion of clear nights is among the highest in the world. Since it is surrounded by thousands of miles of flat and relatively thermally-stable ocean, there is less turbulence in the upper atmosphere and less light-reflecting dust in the air than at most locations on Earth. Its distance from city lights ensures an extremely dark sky, which is also dry and free from atmospheric pollutants.

Many international telescopes have been built at the highest point of this (hopefully) extinct volcano, resulting from world-spanning collaborations between nations. There are currently twelve telescopes — more than on any other single mountain peak — plus the Hawaii Antenna of the Very Long Baseline Array. For the purposes of infrared astronomy, the Canadian French Hawaiian Telescope (CFHT), the United Kingdom Infrared Telescope (UKIRT), the Subaru (Japanese National Large Telescope), the University of Hawaii (UH), the Infrared Telescope Facility (IRTF) and the Keck Telescopes have the instruments best adapted for spectroscopy and/or imaging with adaptive optics. Other telescopes on Mauna Kea include the Submillimetre Array (SMA), The James Clerk Maxwell Telescope (JCMT), the California Institute of Technology 10.4 m Submillimetre Observatory (CSO), and the Gemini Northern 8 m Telescope.

Several of the instruments on Mauna Kea have been used to study Titan, by means of spectroscopy and imaging in the search for information on the atmosphere and surface of the satellite. They include the Hokupa'a AO system, which consists of a 36-element curvature wave front sensor and bimorph mirror, TEXES, the 8–25 μm high-resolution grating spectrograph at the IRTF, and the 3.6-metre optical/infrared CFHT. The CFHT has been used since 1990, first for spectroscopic measurements, and then for adaptive optics imaging of Titan. The twin telescopes at the W. M. Keck Observatory are among the world's largest, at eight stories tall and 300 tons in weight, with a primary mirror ten metres in diameter composed of 36 hexagonal segments that work in concert as a single reflector. The Keck I telescope began science observations in May 1993; Keck II began in October 1996. The adaptive optics system has a wave front sensor with 289 actively controlled subapertures, requiring a brighter star for operation than Hokupa'a but providing better image quality in poor seeing, ideal for Titan observations.

3.3.2 *The European Southern Observatories*

The telescopes based at La Silla and Paranal in Chile compete with Hawaii in offering a large number of clear nights per year and a location away from artificial light and dust sources. La Silla is a 2400 m mountain on the southern edge of the Atacama desert, about 600 km north of Santiago, in the province of Elqui. Its summit houses one part of the European Southern Observatory (ESO), a set of more than 15 astronomical instruments devoted mainly to exploring the southern celestial hemisphere. The mountain's name (La Silla) means "The Saddle", after its shape, and by coincidence the sinuous road leading up to the Mauna Kea Observatories is also called "The Saddle Road". Apart from a 15 m diameter parabolic antenna devoted to radio observations, the largest telescopes on La Silla include a couple of 3.6 m optical reflectors, the rest ranging in aperture from 2.2 m to half a metre. Among others, the NTT 3.6 m telescope and the adaptive optics system (ADONIS) it houses were used for observing Titan in the imaging mode in the first studies related to its surface.

Figure 3.6   The European Southern Observatory telescopes in the Chilean desert.

Figure 3.7   Close-up view of the ESO Paranal Observatory, with the control building in the fore-ground. Also visible are the railway tracks on which the auxilliary telescopes move, and the individual observing stations.

Several years ago, the ESO Council decided to build the Very Large Telescope at Cerro Paranal, in the Atacama desert. The VLT consists of four 8.2 m telescopes and several moving 1.8 m Auxiliary Telescopes, working independently or in combined mode as an interferometer. In this latter case, the total light collecting power of the telescope equals that of a 16 m diameter single telescope, making the VLT the world's largest and most advanced optical telescope, with unprecedented optical resolution and collecting surface area. Currently, eleven instruments including two interferometric instruments are in operation at the VLT and scientific observations are being carried out including observations of binary stars, circumstellar disks, protostars, brown dwarfs and extrasolar planets. Among the VLT instruments used to study Titan is ISAAC, an IR (1–5 μm) imager and spectrograph with two arms, one equipped with a 1024 $\times$ 1024 Hawaii Rockwell array, and the other with a 1024 $\times$ 1024 InSb Aladdin array from Santa Barbara Research Center. Another device, NAOS-CONICA, provides adaptive optics assisted imaging, imaging polarimetry and spectroscopy in the 1–5 μm range. It also includes a Fabry-Perot unit in the 2–2.5 μm range.

The adaptive optics system, NAOS, is equipped with both visible and infrared wavefront sensors. It contains five dichroics, which split the light from the telescope between CONICA and one of the wavefront sensors. CONICA is the infrared camera and spectrometer attached to NAOS and is equipped with an Aladdin 1024 $\times$ 1024 pixel InSb array detector. It contains several wheels carrying masks, slits, filters, polarizing elements, grisms (devices that combine a diffraction grating and a wedge prism to disperse the light without deviating the central wavelength), and several cameras, allowing diffraction limited operation across the full wavelength range. The UVES spectrometer was also used to characterise the zonal winds on Titan.

3.3.3 *The University of Arizona and Steward Observatory Telescopes*

The Department of Astronomy at the University of Arizona and its associated research division, Steward Observatory, form one of the finest centres for astronomical studies in the world with some world-class telescopes. New light detectors and giant telescope mirrors are a catalyst for breakthroughs in optical and infrared astronomy.

The current telescope list includes the 10 m Sub-Millimetre Telescope and the 1.8 m Lennon Reflector on Mt. Graham; a 4.5 m equivalent aperture multi-mirror telescope (MMT) Reflector on Mt. Hopkins; the 2.3 m Bok Reflector and a 0.9 m Reflector on Kitt Peak; the 1.6 Bigelow Reflector and a 0.4 m Schmidt on Catalina; the 1.5 m NASA reflector and a 1 m Reflector on Mt. Lemmon; and a 0.5 m reflector on the university campus, collectively equipped with a full range of instrumentation for photometry and spectroscopy in the visible and the infrared. The Bok and MMT reflectors have been used to study Titan's atmospheric transmittance in the near-infrared range.

Figure 3.8   The multi-mirror telescope MMT on the summit of Mount Hopkins, as photographed by H. Lester.

3.3.4 *Radio Astronomy*

Radio astronomy covers the electromagnetic spectrum from the far infrared (500 μm) to kilometre wavelengths, a range that includes several terrestrial atmospheric windows which are relatively free of atmospheric absorption so that radio emission from Titan can reach ground-based instruments. In the millimetre and submillimetre windows, the emission from cool objects such as planets, and molecules and dust in space, can be detected. The absorption of radio waves by the ionosphere becomes more important as wavelength increases until, at wavelengths longer than about 10 m, the ionosphere becomes opaque to incoming signals.

Before that limit is reached, at wavelengths longer than about 20 cm (1.5 GHz), irregularities in the ionosphere distort the incoming signals, causing scintillation. As at optical and infrared wavelengths, sophisticated signal processing can be used to correct for these effects, so that the effective angular resolution and image quality is limited only by the size of the instrument. To derive the physical properties of cold objects such as Titan, or to identify new molecules, the instruments at the various telescopes described below measure the spectral energy distribution of molecular emission using heterodyne receivers that amplify signals in such a way that high resolution spectroscopy becomes feasible.

3.3.4.1 *IRAM*

Institut de RadioAstronomie Millimétrique has the world's largest telescope operating at wavelengths between 0.8 and 3.5 mm (frequencies between 350 and 80 GHz).

Figure 3.9   The IRAM Radio Telescopes with a view of the 15 m antennas located at the Plateau de Bures in the French Alps.

It received its first millimetric light in May 1984, and was opened to the astronomical community the following year. Mainly the responsibility of France, Germany and Spain, it operates two major facilities: a 30 m diameter telescope on Pico Veleta in the Sierra Nevada of southern Spain, and an array of five 15 m diameter telescopes on the Plateau de Bure in the French Alps. The 30 m telescope at Pico Veleta has been used to make heterodyne observations of Titan that have allowed A. Marten and colleagues the precise determination of the vertical profiles of CO and HCN. In 1992 the first detection of a more complex nitrile than any of those detected by Voyager, CH_3CN (acetonitrile) was reported in the millimetre range around 220.7 GHz at 0.1–MHz resolution. More recently, several observational campaigns produced maps of Titan in the lines of CO, HCN, HC_3N and CH_3CN with a spatial resolution of 0.6", about half the apparent size of Titan's disk including the extended atmosphere. Vertical profiles for these constituents were derived at altitudes above 400 km. The same data contains information about winds in Titan's upper atmosphere, from the Doppler shifting of the spectral lines, thus complementing the wind field observations made by Cassini–Huygens.

3.3.4.2 The Very Large Array (VLA)

The National Radio Astronomy Observatory's (NRAO) Very Large Array is located in the plains of San Agustin, west of Socorro, New Mexico, at an elevation of 2,124 m. It has operated since January 1981 at various bands between 300 and 50,000 MHz, corresponding to wavelengths of 90 to 0.7 cm. The VLA consists of 27 antennas each 25 metres in diameter, arranged in a Y pattern to form the equivalent of a dish

Figure 3.10   The National Radio Astronomy Observatory (NRAO) Very Large Array (VLA) antennas in New Mexico.

130 metres in diameter, steerable at a rate of 40° per minute in azimuth and 20 in elevation. The maximum achievable resolution at the highest frequency of 43 GHz is 0.04 arc seconds, the angle subtended by a golf ball 150 km away.

The VLA normally operates as an interferometer, multiplying the data from each pair of telescopes together to form interference patterns that are subsequently analysed by taking the Fourier transform of the signal to make maps. It was used as the receiver for a signal transmitted to Titan by the NASA Goldstone radio telescope in California, in order to recover radar signal returns and study the satellite's surface (see Chapter 9).

3.3.4.3 The Very Long Baseline Array (VLBA)

The VLBA is similar to VLA in that it is made up of a system of radio-telescope antennas, each with a dish 25 m in diameter and weighing 240 tons. However, its ten dishes are deployed over a baseline of more than 5000 miles, from Mauna Kea in Hawaii to St. Croix in the U.S. Virgin Islands. This provides astonishing spatial resolution, the equivalent of reading a newspaper in Los Angeles from New York.

3.3.4.4 The Atacama Large Millimetre Array (ALMA)

ALMA is a major new facility under construction since 2003 for a mm and sub-mm wavelength telescope, the largest and most sensitive in the world, with sixty-four 12 m antennas located at an elevation of 5,000 m (16,400 feet) in Llano de Chajnantor, in the Chilean Andes, east of the Atacama Desert. It operates in all atmospheric windows between 10 mm and 350 microns and achieves a spatial resolution of 10 milliarcseconds, 10 times better than the VLA and the Hubble Space Telescope. The plan is for ALMA to be completed in 2010. On March 10, 2007, an official ceremony took place at 2,900 m altitude on the site of ALMA to mark the completion of the structural works. This operations support facility site will become the operational centre of one of the most important ground-based astronomical facilities on Earth.

3.3.4.5 The Square Kilometre Array (SKA)

SKA is an international project to develop a telescope to provide two orders of magnitude increase in sensitivity over existing facilities at metre to centimetre wavelengths, a goal that requires a telescope with 1 million square metres of collecting area — one hundred times more than the VLA. The SKA will be an interferometric array of 30 stations each with the collecting area equivalent to a 200 m diameter telescope, and 150 stations each with the collecting area of a 90 m telescope, distributed over distances out to 3000 km or more. This high angular resolution capability will allow imaging of faint emission from the interstellar medium of distant galaxies, as well as the surface of stars, planets, and the active nuclei of galaxies.

3.4 Earth-Based Studies of Titan

Over the last 50 years a wide range of ground-based observatories, techniques, and powerful tools have been applied to the study of Titan. Here we review some of the most prominent advances in our understanding of the satellite that have resulted from these programmes.

3.4.1 Occultations of Titan

Stellar occultations are another indirect means to probe Titan's atmosphere and recover information on the thermal structure, winds and other parameters. For instance, the atmospheric oblateness due to the zonal winds can be constrained from the analysis of the central flash, the increase of the signal at the centre of the shadow that forms when the star is behind Titan, due to the focusing of the atmospheric rays at the limb.

On July 3, 1989, Saturn and Titan passed in front of the 5th magnitude star 28 Sagitarii. The occultation was observed, among other places, from Paris Observatory at Meudon, Pic du Midi Observatory, Israel, the Vatican, and Catania Observatory. This rare event provided information on Titan's stratosphere in the 250–500 km altitude range. A mean scale height of 48 km at 450 km altitude was inferred, allowing the mean temperature between 149 and 178 K to be constrained at that level (about 3 μbar). Titan's shadow centre passed within about 20 km south of Meudon. The central flash observed there provided a unique opportunity to constrain the apparent oblateness of Titan at the 0.25 mbar level (about 250 km of altitude), and gave a value which may be as high as 0.014. Mean temperatures around 180 K were found, in agreement with models of Titan's mesosphere.

The occultation allowed an estimate of the optical depth of the haze at heights between 300 and 450 km above the surface. Two haze layers were detected, a lower layer, present globally, and a higher haze layer present only northward of about 20°S. This is the opposite of the asymmetry noted from Voyager and Pioneer data, where the high detached haze layer is present in the southern hemisphere and at low northern latitudes. The occultation data were also used to estimate the opacity of the haze at an altitude of 318 km above the ground: tangential optical depths of around 0.1 for the low haze and 2 for the upper haze were derived for a reference wavelength of 0.5 μm. The occultation measurements found the haze to be situated one pressure scale height higher than what Voyager found, with the scale heights of the haze layers themselves both smaller than those deduced from the spacecraft data. No discrepancy is necessarily implied — the 28 Sgr occultation occurred almost at northern midsummer, while the Voyager encounter occurred during Titan's northern spring, so seasonal effects could account for the difference.

The stellar occultation data showed that Titan's atmosphere is not spherically symmetric, but has an oblateness of about 0.017 at an altitude of about 250 km (0.25 mbar), which may be due to the rapid wind speeds around the equatorial zone

Occultation of 28Sgr by Titan

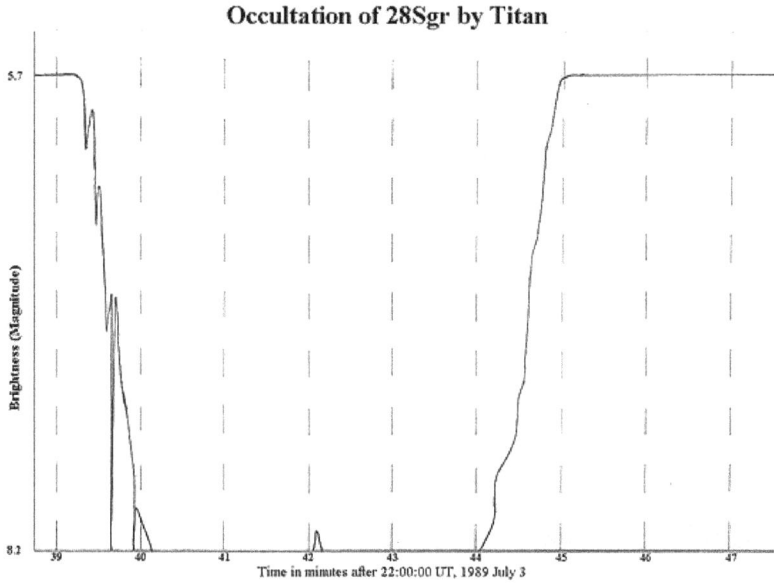

Figure 3.11   Occultation of 28 Sagittarius by Titan, 3 July 1989 (Sicardy *et al.*, 1990).

(see Chapter 5). It also suggested a slight systematic enhancement of the haze opacity on the lit side of Titan. Regional variations in haze opacity at small scales may be required to satisfy all the observations, which are no doubt associated in some way yet to be understood with equator-to-pole temperature gradients, dynamical effects, more efficient radiative cooling at higher latitudes, seasonal and spatial variations of the solar flux, and the latitudinal variations observed in the composition of some of the minor constituents (see Chapters 6 and 7).

Furthermore, the 28 Sgr occultation revealed fast zonal winds, up to $180 \, \mathrm{m \, s^{-1}}$ in the South, and close to $100 \, \mathrm{m \, s^{-1}}$ at mid latitudes. Other occultations occurred on December 20, 2001 and November 14, 2003. They seem to suggest a seasonal variation with respect to 1989. In 2001 a strong $220 \, \mathrm{m \, s^{-1}}$ jet was located at 60°N, with lower winds extending between 20°S and 60°S, and much slower motion at mid-latitudes. Last but not least, temperature inversions have been detected in the stellar occultation data, suggesting a stratified upper atmospheric structure, as confirmed later by the Huygens/HASI experiment. Inversion layers were present close to 510 km altitude in the HASI and the 2003 occultation data, and at 425 and 455 km in 1989 occultation light curves, with vertical wavelengths on the order of 100 km.

With the Cassini spacecraft in orbit in the Saturn system, spatial observations of Titan occultations are made possible. Solar and stellar occultations are observed by the VIMS (Visible and Infrared Mapping Spectrometer) instrument. While Earth-based observations are mainly refractive occultations, Cassini observes absorbing occultations. Absorption is due to gaseous constituents of the atmosphere and to the haze layers. With VIMS, absorption spectrum are available at each altitude sounded.

The main features observed in those data are methane and CO bands, while haze
absorption fixed the continuum level.

3.4.2 *The Radar Search for Oceans, Seas or Lakes*

As we saw in Chapter 2, after Voyager the prevailing view from the sum of observations and theoretical considerations was that the processes that give rise to the haze
and cloud materials, although not understood in detail, are likely to lead to precipitation onto the surface. Since at least some of these were expected to be liquids, and
since the process of accumulation had gone on for a long time, the idea of an extensive ocean was an attractive one. Despite this expectation, well before Cassini and
Huygens arrived, ground-based data was indicating that Titan has a predominantly
solid surface. We need to look more closely at the arguments, and the recent history
of how they have evolved, for clues to resolve this paradox.

The fundamental driving force in the long-term evolution of Titan's atmosphere
is the photolysis of methane in the stratosphere to form higher hydrocarbons and
aerosols. The photochemical products falling out of the atmosphere act to decrease
the methane vapour pressure. The predominant photochemical product, ethane, is

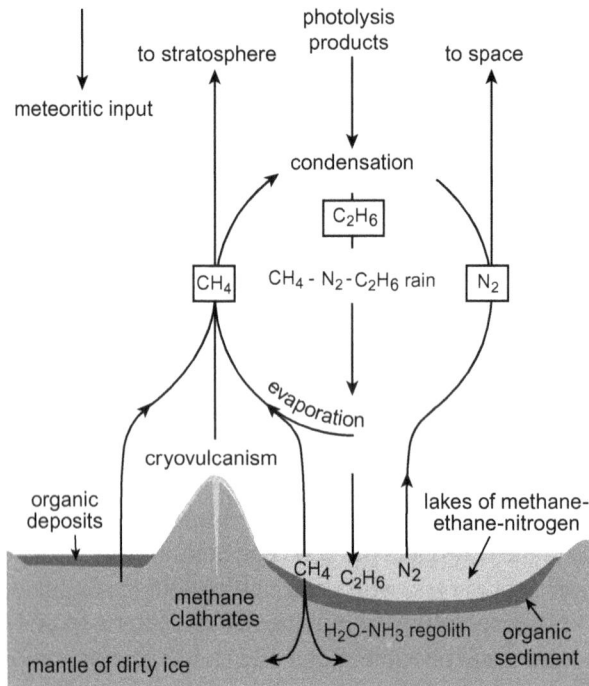

Figure 3.12   A schematic of the relationship between the atmosphere and bodies of liquid on the
surface of Titan. The main processes involved are methane (CH_4) evaporation, photolysis and condensation, and the production, condensation and fallout of ethane (C_2H_6).

Table 3.1   Pre-Cassini results for Titan's surface oceans, for two extreme assumptions about the surface temperature and atmospheric composition, compiled from several models.

Surface temperature	92.5 K	101 K
Nitrogen	1.8%	6%
Methane	7.3%	83.4%
Argon	0	5.6%
CH_4 abundance (lower troposphere)	1.55%	21.1%
Ocean composition: ethane + propane	90%	5%
Current depth	695 m	9.4 km

liquid at 94 K, and has a vapour pressure much lower than methane, allowing the two to mix in solution. Nitrogen, carbon monoxide and liquid products of methane photolysis other than ethane, such as propane, would also dissolve and accumulate on Titan's surface. Products that are solid at the temperatures on Titan, such as acetylene, would sink to the bottom.

The resulting "ocean" could serve as both the source and the sink of for the photolysis cycle. In numerical models, its depth ranges from 500 m to 10 km, and it contains a mass of nitrogen comparable to the atmospheric abundance. In the absence of outgassing from the crust or an external supply of volatiles, the ocean composition evolves to become more ethane-rich, as methane is photolysed over geologic time. The atmosphere responds to the change in ocean composition with a corresponding change in gaseous composition and spatially averaged cloud composition. The detailed solutions found in modelling studies lie within the extremes in Table 3.1.

The first experimental result which led to reservations about these theoretical studies, and the first remote sensing technique to be used to study Titan's surface, came from the detection of radar reflections from Titan by D. Muhleman and his colleagues at the California Institute of Technology. The first experiment was conducted in 1991, with another successful one in 1995. Titan is the most distant object detected by radar to this date. Muhleman and colleagues used NASA's radio telescope at Goldstone in California to send a signal from Earth, and the National Radio Astronomy Observatory's Very Large Array in Socorro, New Mexico, as a receiver at 3.5 cm. These wavelengths are unaffected by the presence of haze or clouds in the atmosphere of Titan, but the returning signal is sensitive to the type of materials present on the surface from which it bounces. The 1991 high reflectivity values reported (which led theoreticians originally to consider Titan as a very bright object, analogous to the Jovian satellites) were corrected in an update published in 1995, in which Muhleman *et al.* gave a revised value of 0.15 for the reflectivity of the Titan bright terrain. This value was confirmed by 1992 Goldstone radar observations by another group. The darker terrain of Titan showed generally lower radar reflectivity, albeit with marginal statistical significance.

The echoes of the signal sent from Earth showed that Titan is not covered with a deep global ocean of ethane-methane, because such an ocean, if devoid of suspended particulates and deeper than a few hundred metres, is a very poor reflector of radar signals. It should return only a small percentage (about 0.02) of the energy originally transmitted, whereas the radar cross sections obtained vary between 5% and 25%. The higher end of the range is then considerably smaller than the reflectivity of the largest Jovian satellites, Europa, Ganymede, and Callisto, which range from 30% to 90%. Even if large particles of some kind could remain suspended in the postulated ocean on Titan, perhaps stirred by the wind, the value of the reflectivity could not approach the Callisto value. The lower end of this range indicates that some of Titan's dark terrain could be covered with considerable amounts of solid or liquid hydrocarbons. The bright terrain, on the other hand, as far as the radar data can tell, has considerably smaller reflectivity than even Callisto, compatible with a rocky surface like those found in the inner solar system. Other radar parameters, such as the polarisation of the reflected signal, also indicate that Titan's surface is different from those of the Galilean satellites. Microwave emissivity measurements for Titan's surface resemble the Moon's rocky, silicate-type more than the icy Jovian satellites. This seems incompatible with the high surface albedo values found in the near-infrared, described in the next section, which are indicative of an icy component, except perhaps for the low surface albedo reported near $0.94\,\mu$m by some investigators, which could be compatible with the presence of a silicate absorption band. Titan appeared then as a unique object, more radar-reflective than if it were covered by a global hydrocarbon ocean or by tholin material, but still rather dark outside the brighter terrains at 60–160 degrees longitude.

A mixture of ice and rock on the surface could perhaps reconcile the various observations. In 2003, D. Campbell and colleagues collected radar measurements with the Arecibo Observatory in Puerto Rico, which showed a specular component at 12 of 16 of the regions observed, which was globally distributed in longitude at about 26°S. This was interpreted as indicative of dark, liquid hydrocarbon extends on Titan's surface. However, this was challenged in 2005 by R. West and colleagues whose observations failed to find any such signatures and proposed instead that flat solid surfaces could be causing the radar evidence.

As the idea of a global or deep ocean began to appeared to be incompatible with the latest observations, new theoretical studies showed that tidal forces would have dissipated Titan's orbital eccentricity of 0.03 long ago if there was such an ocean. The most compelling evidence of all against a deep global ocean on Titan with the first images of its surface, which showed extensive non-homogeneous and recurrent features clearly incompatible with a surface covered by liquid. On the other hand, a surface that is completely dry — either exposed ice or ice buried under layers of solid photochemical products, soil or rock — would not provide the source of methane nor the sink for the liquid products of the photochemistry (notably ethane and propane). Some early models attempted to reconcile these two requirements. D. Stevenson of Caltech suggested that the hydrocarbon ocean may be stored in

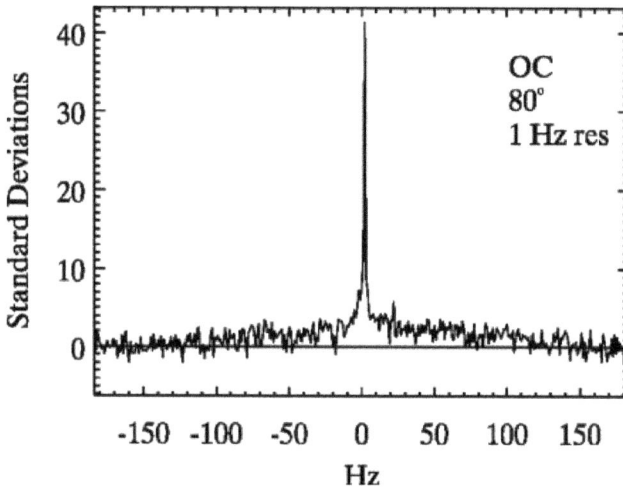

Figure 3.13   A radar echo spectrum of Titan taken in 2002 at Arecibo, showing the expected (OC) sense of received circular polarization, with a specular component at 0 Hz, interpreted by Campbell and colleagues in 2003 as proof for liquid surfaces on Titan.

a porous, uppermost few kilometres of methane clathrate or water ice "bed rock". The surface in this model is porous enough (about 20% to a depth of about one-half kilometre, not unrealistic) to allow all the required ethane and methane to pass into a subsurface "aquifer" in which the liquid is stored. The model satisfies the tidal constraints because the confined liquid cannot attain large velocities. One argument against such a porous surface is that hardened regolith upper layers enriched in organics (similar to "caliches" seen in Earth's deserts) may prevent ethane dribbling down into the regolith.

A qualitatively different model has the hydrocarbons which originate in the haze deposited as solids and thus permanently removed from the atmosphere. The atmospheric methane supply is replenished by outgassing from the interior, gaining access to the atmosphere through occasional release during cryovolcanic activity or through upward diffusion. The biggest problem with this sort of concept comes in explaining how the photochemical products can be buried in the mantle while leaving water ice exposed on the surface, which the spectroscopic evidence (see next section) seems to require.

3.4.3 Spectroscopic Measurements of Titan's Albedo

Observations in the near-infrared offer an extremely useful tool for attempting to solve the ambiguities in the models by obtaining observations that could be interpreted in terms of quantitative models of the structure and composition of Titan's lower atmosphere and surface. The spectral albedo (that is, the reflectivity at different wavelengths, as a function of orbital position, measured as Titan rotates) was

Table 3.2　Microwave properties of Titan and other solar system bodies.

Property	Titan terrain		Europa	Ganymede	Callisto	Moon
	Bright	Dark				
3.5 cm same sense	0.08 ± 0.08[a]	0.03 ± 0.0[a]	1.40 ± 0.23[b]	0.9 ± 0.10[b]	0.40 ± 0.04[b]	0.006[b,g]
3.5 cm opp. sense	0.16 ± 0.10[a] 0.15 ± 0.05[c] 0.15 ± 0.04[d]	0.13 ± 0.03[a] 0.10 ± 0.04[c]	0.91 ± 0.13[b]	0.6 ± 0.10[b]	0.32 ± 0.02[b]	0.07[g]
Polarisation ratio	0.5[a]	0.3[a]	1.43 ± 0.23[b]	1.40 ± 0.1[b]	1.22 ± 0.08[b]	0.1[b]
Microwave emissivity	0.85[c]	0.85[e]	0.4[f]	0.5[f]	0.8[f]	0.95[c]

[a]Averages of $60° < $ LCM $ < 160°$ (bright region) and LCM $ > 160°$ (dark region) data from Muhleman et al. (1995), by Lorenz and Lunine (1997). [b]Ostro et al. (1992). [c]Muhleman et al. (1995). [d]Goldstein and Jurgens (1992). [e]Grossman and Muhleman (1992), [f]Muhleman et al. (1991). [g]Pettengill (1965).

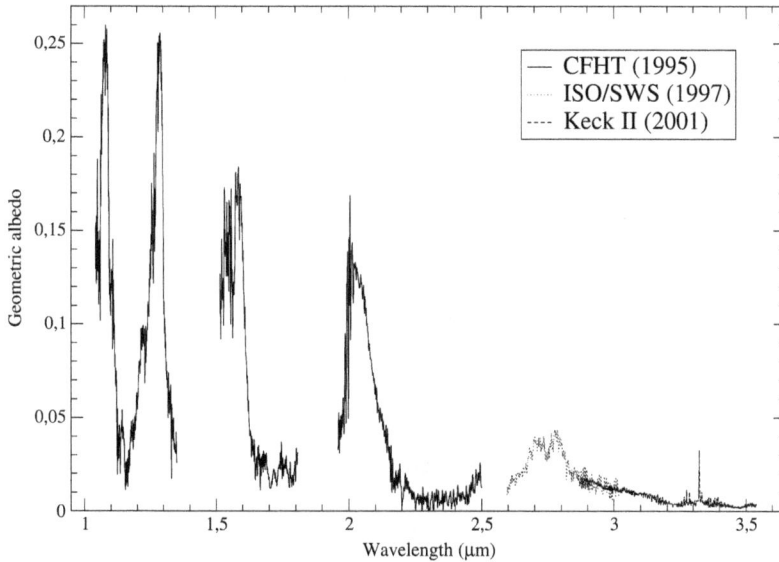

Figure 3.14   Titan's albedo exhibits several strong absorption bands, but also "windows" where the methane absorption is weak enough to allow the lower atmosphere and surface to be probed. The data shown here come from two observatories in Hawaii and (in the 2.75 micron region, where Earth-based observations are difficult) the Infrared Space Observatory, ISO.

the first quantity observed and is one of the most useful for probing the haze and surface properties.

The near infrared spectrum of Titan (0.8 to 5 μm), like those of the giant planets, is dominated by the absorption bands of methane. Where the methane absorption is weak, clear regions or "windows" permit the detection of radiation from the deep atmosphere and, in some cases, from the surface. Also, the haze scattering effect is small in the infrared, compared to the visible. Thus, the principal atmospheric windows, through which Titan's lower atmosphere and surface can be observed, are near 4.8, 2.8, 2.0, 1.6, 1.28, 1.07, 0.94 and 0.83 μm in wavelength, or 12,050, 10,640, 9300, 7810, 6250, 5000, 3500, and 2050 cm^{-1} in wavenumber. In between the windows, unlike the case of the giant planets, the "dark" regions are of interest since the solar flux is not totally absorbed but instead is scattered back through the atmosphere by the stratospheric aerosols, especially at the shorter wavelengths. Modelling of the albedo derived from ground-based spectra of Titan can therefore indicate the properties of the haze present in the atmosphere.

The haze is optically thick at visible wavelengths, and gas opacities are negligible, so the haze dominates the albedo and completely hides the surface. At wavelengths longer than 0.6 μm, the scattering extinction efficiency of the particles decreases due to the increasing ratio of wavelength to particle size. Also, if the data derived from tholins produced in laboratory simulations applies, the haze material becomes virtually non-absorbing. Thus, the haze becomes progressively more transparent at longer wavelengths and the surface properties increasingly affect the observed albedo, except in the strong CH_4 absorption features which become prominent longwards of 1 μm. Therefore, in the near-infrared spectral region, the albedo is determined mainly by the methane amount present and the surface properties, although the contribution of the haze opacity is not completely negligible even at far infrared wavelengths.

Several different groups have independently provided practical demonstrations that the surface of Titan can be probed by measuring the geometric albedo in the 1–2.5 μm region. First came the observations by U. Fink and H. P. Larson in 1979, using the Kitt Peak National Observatory 4 m telescope. D. Cruikshank and J. Morgan observed Titan in 1980 and looked for a 32-day variation in the albedo. Griffith and Owen used Titan's near-infrared spectrum to investigate Titan's surface in 1991, using the first detailed radiative transfer models of the near-infrared spectrum produced to deduce that the surface albedo is inconsistent with a global ocean, invoking instead the presence of 'dirty' water ice on the surface, and leaving open the possibility that cloud cover could partly account for the observed reflectivity in the atmospheric windows. C. Griffith and her co-workers of the State University of New York in Stony Brook using the CGAS array spectrometer at the IRTF Telescope in Hawaii, M. Lemmon and his colleagues at the University of Arizona using the GeSpec instrument at the Steward Observatory 2.3 m telescope and at the Multiple Mirror Telescope, and A. Coustenis and colleagues from Paris Observatory using the Fourier Transform Spectrometer at the 3.6 m Canadian French Hawaiian Telescope. An additional window near 2.75 μm, undetectable from the Earth, was first observed by the Infrared Space Observatory in 1997.

The observations all agreed in showing that the geometric albedo of Titan, measured over its 16-day orbit, shows significant variations indicative of a brighter leading hemisphere (facing the direction of Titan's orbital motion) and a darker trailing one. The leading side corresponds to Titan's Greatest Eastern Elongation at about 90° LCM (Longitude of Central Meridian — a geographical longitude of about 210°), when Titan rotates synchronously with Saturn; the trailing side is near 270° LCM. At conjunctions, that is on the hemispheres facing Saturn and its opposite, the albedo was similar, of intermediate values between the maximum appearing near 120° ($\pm$20°) LCM and the minimum near 230° ($\pm$20°). These albedo variations could have been due to (a) tropospheric cloud variation, if they were uncorrelated with the rotation period, or (b) surface properties variations, if correlated with Titan's orbital period of 15.945 days. Because they were measured independently over five years and they were recurrent, observers agreed that the features must be correlated with Titan's surface morphology. Titan's surface was then demonstrated to be heterogeneous, which was already an important result in that it ruled out a global ocean or indeed any uniform coverage by a single material, such as water ice. The variable brightness behaviour was subsequently identified in images as being mainly due to a single large bright area near the equator.

Titan's geometric albedo lightcurve

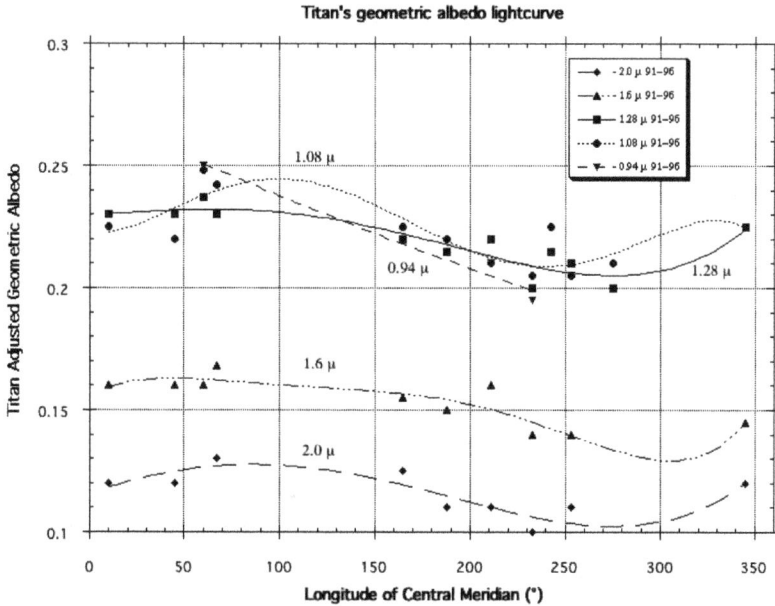

Figure 3.15   Light curve of Titan's geometric albedo with data taken between 1991 and 1996 with the FTS of the CFHT telescope, showing the hemispheric asymmetry, with a brighter leading (at 90° LCM) than trailing (at 250° LCM) side (Negrão *et al.*, 2006).

The next step — to deduce information on the real nature of the surface from the infrared spectrum of sunlight reflected from the surface of Titan — is very model-dependent and in particular requires precise knowledge of the non-surface contribution. This is obtained using a computer code, which includes such atmospheric properties as the production rate, and microphysics of the haze particles, the amount and distribution of methane in the atmosphere, and the optical properties of both. The methane absorption coefficients used as one of the model's input parameters have a major impact on the calculated geometric albedo and consequently inferred surface albedo. These come from theoretical calculations and laboratory measurements by a number of different researchers, and are assembled in widely accessible compilations known as databases. The STDS database, for instance, is built from a theoretical model of the methane ($^{12}CH_4$ and $^{13}CH_4$) rotation-vibration interactions and transition moments, with parameters fitted against laboratory high-resolution absorption spectra. It includes all the methane absorption lines between 1.62 and 7.69 μm, although modelling of the methane absorption spectrum at wavenumbers lower than 2.2 μm is still not fully satisfactory because of the large complexity of the rotation-vibration interactions in these regions, rendering the analysis very difficult.

Most recent work has used band models based on laboratory measurements, which include all of the multitude of weak lines that cannot be observed individually but which make a significant contribution in total. The laboratory measurements cannot reproduce the low temperatures and the large path lengths on Titan, but

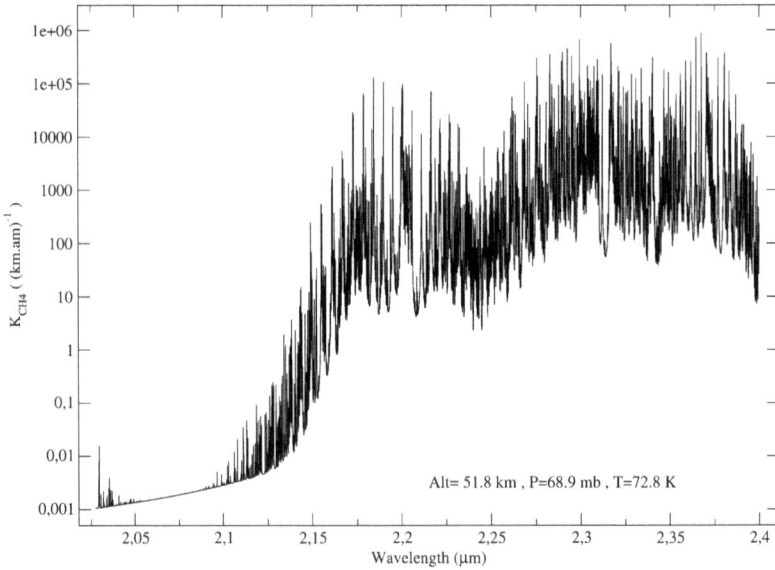

Figure 3.16   Methane absorption in the 2-micron range from calculations by V. Boudon and colleagues (Univ. de Bourgogne, France).

approximate these with some data using large methane concentrations, to give total path lengths typical of those found on Titan, and others at low temperatures, in one case down to 100 K. Combining the two provides an acceptable simulation of Titan's atmosphere.

The surface spectrum for Titan between 0.9 and 2.5 μm typically shows three distinct spectral regions: at 1–1.6 μm the surface albedo is about two times higher than at 2.0 μm or at 0.94 μm. Recent measurements at 2.9 and 5 μm place the associated surface albedos below 0.1, at around 0.03 and 0.07 respectively. If the exact values are not completely agreed upon, the various teams who have applied models to Titan data in order to model the surface spectrum from a large set of spectra, both from the ground or with Cassini–Huygens instruments, agree that there seem to exist at least two lower-albedo regions in the spectrum near 1.6 and 2 μm with respect to the continuum near 1 μm. The most probable reason that the surface albedo of Titan is significantly higher at 1.075–1.28 μm than at 1.6–2.0 μm is the presence of water ice on Titan's surface. Since, so far as we can tell at present, moving from one hemisphere to the other changes the total brightness at all wavelengths rather than altering the shape of the spectrum, the extra species responsible for the orbital variations may be spectrally neutral.

Laboratory-generated organic tholins like those thought to be present in the haze material show a neutral and fairly bright spectrum in the near IR and a tendency to get darker towards the visible end of the spectrum, and so represent a plausible cause of the high absolute albedos and the 0.94 μm depression. The water ice spectrum and the tholin reflectance, if combined, can in fact produce a reasonable simulation of Titan's surface albedo. However, we would expect the tholins to be distributed more or less uniformly with longitude if they are solids which have precipitated onto the surface from the global haze, and then they would not produced the observed orbital variations. Even so, it is interesting that the spectrum of Titan matches the decrease in the tholin spectrum from 1.3 to 0.9 μm most strongly for the darker, trailing side. The fact that this decrease is more evident on the dark side may be diagnostic of the presence of more tholin/organic material on the surface on that side of Titan than on the brighter, leading side. This might mean that there is a second source of tholins, not necessarily the same material but with broadly similar spectral properties, that originates outside of Titan. One is reminded of the dark, reddish coating on the leading side of Iapetus and other airless satellites, that looks like dark material swept up from space. Quite how such behaviour would work on Titan, with its thick atmosphere, is hard to see, and other possibilities are not rued out. For instance, the surface patterns, and hence the orbital variations, may be due to longitudinal differences in the ice morphology: high and low regions, covered with fresh or old, soft or hard, big or small particles, or even to the physical properties of the ice itself, like strength, viscosity, etc.

Observations of other satellites in the near-infrared range support the presence of water ice on Titan. There are depressions observed in the spectra of Hyperion and Callisto near 1.6 and 2 μm, due to water ice bands, and the Hyperion spectrum is in general consistent with that of Titan over the whole 1–2 μm region. This is

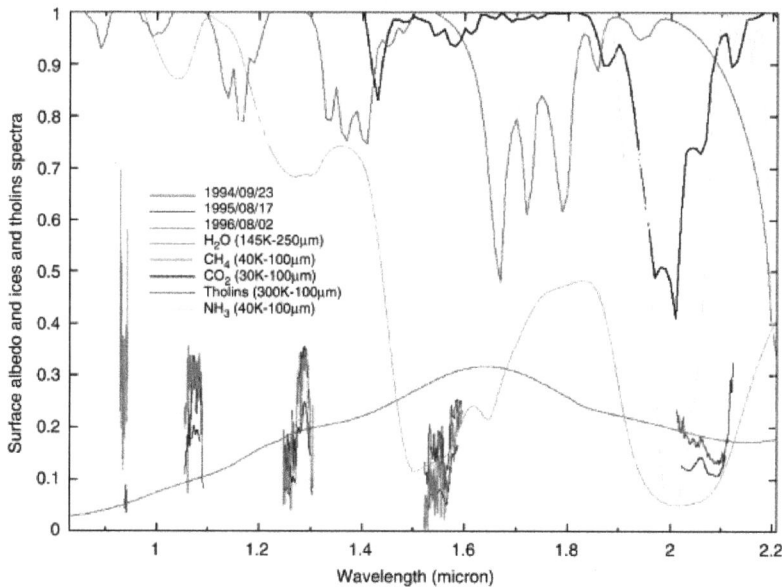

Figure 3.17 Titan's surface spectrum. The main features — bright around 1.2 μm and darker around 0.94 and 2.0 μm — are compared here with several possible surface constituents, and found to be compatible with water ice covered with a layer of condensed hydrocarbons (Negrão et al., 2006).

reproducible if water ice were the dominant constituent on Titan's surface, as has also been deduced from radar reflectivity and radio emissivity measurements. The water ice bands are present in Titan's surface spectrum at all longitudes.

However, although water ice must be present in large amounts, a single surface component cannot explain the low albedo at $0.94\,\mu$m, or the variations in brightness which Titan exhibits during its orbit around Saturn. From the other relevant ices that could be found on Titan's surface, CO_2 and CH_4 ice spectra show a behaviour compatible with what is observed in the general shape of the 1–$2.5\,\mu$m region. The NH_3 spectrum does not help the fit of the surface data either. Hydrocarbon ice (such as CH_4 or C_2H_2) has absorption bands near 1.65 and 2.3 μm and its presence cannot be excluded. So far, however, only water ice can be definitely described as a reasonable candidate from all the observations, and its presence was confirmed in the recent Huygens observations by DISR.

3.4.4 Imaging Titan's Atmosphere in the Near-Infrared

In 1994, two sets of data taken independently and with different methods were the first to clearly show in images of Titan's surface the heterogeneity that had been inferred from the near-IR and radar light curves. Extensive quasi-permanent features, which were too bright to be hydrocarbon liquid, were graphically revealed by the adaptive optics technique using the ADONIS camera at the 3.6 m ESO Telescope in Chile, and by the Hubble Space Telescope. The first HST images produced maps at 0.94 and 1.08 μm, with contrasts due to surface features of about 10%, prompting a search to identify spectrally-distinct surface units, which may indicate regions of different composition, and to detect and monitor atmospheric features. Prominent among the latter is the north-south asymmetry observed by Voyager, believed to be due to a planet-wide variability in the haze properties, such as spatial distributions of particle size, number density, or optical properties.

In the season when Voyager observed Titan (northern spring) the southern hemisphere had an albedo about 25% brighter than the north at blue wavelengths, with the contrast between the hemispheres smaller at green and violet wavelengths, and the UV and orange albedo asymmetry lower still. More recent Hubble Telescope images, using the WFPC-2 camera, clearly showed the north-south asymmetry, especially at blue and green wavelengths, to be similar to that observed by Voyager, two seasons earlier. In the near-infrared (between 0.94 and 1.07μm) these images showed a strong north-south asymmetry and brightening towards the lower limb. Spatially resolved images of Titan from ground-based telescopes, using adaptive optics at longer wavelengths, showed the same north-south asymmetry when Titan was observed in the methane bands, for instance at 2.2 μm. At 0.889 μm, in a methane band, HST images showed an asymmetry having a structure opposite to that observed at visible wavelengths, due to the optical properties of the aerosols, accumulated in the south at the current epoch. Indeed, the north-south asymmetry appears to have reversed in the methane band, as would be expected from seasonal effects as the sub-solar point moves north and south over a Titan year, due to the 26.4° obliquity of the Saturnian system.

The HST and Voyager images at visible, UV and near-infrared wavelengths allow the determination of limb-darkening coefficients, which measure the fall-off in brightness from the centre of Titan's disk to the edge, as a function of wavelength. These coefficients provide important information on the spectral characteristics of the haze. Limb darkening is observed to increase from violet to red wavelengths, giving the impression that the disk of Titan shrinks at longer wavelengths. In contrast, limb brightening is observed at $0.889\,\mu m$, in the strong methane absorption band, because the lower atmosphere is opaque here, whereas the optically thin haze in the upper atmosphere is bright. The data tell us that a change in the concentration of aerosol particles between 70 and 120 km altitude could explain the observed asymmetry on Titan. However, changes in the photochemical production of aerosols cannot be responsible, because the time constants associated with the aerosol formation and accumulation into optically thick layers are much longer than the seasonal period. More likely, the aerosol is being moved by the circulation of the atmosphere (meridional and vertical winds), as discussed further below. In this case, the boundary where the albedo contrast occurs, currently situated between 10 and 20°N, should also move with season. However, there is no conclusive evidence that these movements do take place in the expected direction.

The hemispheric asymmetry can also be measured in the thermal infrared. For wavenumbers longer than $600\,cm^{-1}$, the relatively warm stratospheric haze is the main source of the continuum. This has been used in an analysis of Voyager 1 IRIS data taken at different latitudes to determine that there is an apparent increase in the haze optical depth of about 2.5 near the north pole with respect to the equator. At visible wavelengths, the increased opacity in the infrared in the north corresponds to darkening in that hemisphere. This stratospheric north-to-south enhancement in the haze opacity, associated with an equivalent increase in gaseous abundances, can explain the colder temperatures found at high northern latitudes, as this may be caused by more efficient radiative cooling, although dynamical effects may also be involved.

Another phenomenon which was first reported in 2001 from AO data taken in 1998, was an East-West asymmetry, with a brighter morning limb found on Titan on several occasions. This dawn haze enhancement could be due to a deposition of condensates during the Titan night (8 Earth days, though the super-rotation of Titan's atmosphere would lead to shorter nights for stratospheric clouds), manifesting itself in a morning haze enhancement phenomenon at stratospheric altitudes. Indeed, most of the images showing this phenomenon are mainly probing the stratosphere (between 40 and 311 km). All teams observing Titan with adaptive optics at the time of the Huygens descent agree there is evidence of this diurnal effect.

In addition to the global asymmetries, images of Titan's atmosphere showed additional discrete bright areas, mainly near Titan's south pole, sometimes above a fine bright southern polar limb. Clouds have been invoked to interpret these features, including a very bright one, located very close to the south pole, that has attracted growing attention in the past decade. This feature was first discernible in speckle images of Titan published in 1999 and has since been extensively observed in different filters and by different teams. It is particularly visible in $2.12\,\mu$m images, where the upper troposphere of the southern limb is probed. Small clouds have been reported near mid-latitudes in observations from the Keck and Gemini Observatories. The clouds cluster near 350°W longitude, 40°S latitude and cannot be explained by a seasonal effects but may be linked to surface features, the most exciting possibility being that they are initiated by the plumes escaping from cryovolcanoes. Bright features are relatively rare in the north, apart from the bright northern limb due to the north-south anomaly.

The large polar vortex, a meteorological system over Titan's south pole, seemed to vary in shape and brightness with time. It was later demonstrated by observations from the Cassini orbiter that its shape changed, like a large ring of thin clouds or haze, deforming, with little bright clouds trapped within it, appearing and disappearing randomly, sometimes disintegrating, something which could not be resolved with Earth-based systems. The latter nevertheless allowed some monitoring of this phenomenon, which has apparently diminished in intensity and brightness, becoming quite marginal in the Cassini and ground-based images taken since 2005.

3.4.5 *Imaging the Surface*

Earlier HST observations had hinted at the ability to sound the surface of Titan, since a number of its filters sample the 940 nm window, but had been thwarted by the spherical aberration in the primary mirror. When the new Wide-Field Planetary Camera (WFPC-2) was installed, which corrected this optical problem, Titan could be resolved as about 20 pixels across, with the point-spread function about 3 pixels across. In the filters used, a large proportion of light still comes from scattering by haze, but the near-full longitudinal coverage of the HST dataset allowed this to be determined and removed (since the haze is longitudinally-invariant).

The team led by P. Smith of the University of Arizona that produced the first HST images in which features were discernible on Titan's surface worked in the $0.94\,\mu$m window, observing the bright leading and dark trailing sides, with a large bright region, about the size of Australia, at 110°E and 10°S, as well as a number of less bright regions. A coarse map was also produced at red wavelengths ($0.673\,\mu$m) — albeit somewhat blurred, since the haze optical depth is of the order of 3 at this wavelength, so each photon is reflected a number of times as it fights its way through the atmosphere. Subsequent HST data with the NICMOS camera have confirmed the initial findings with more extensive mapping at 1.6 and $2.0\,\mu$m, and identified spectrally-distinct surface units, which may indicate regions of different composition. The contrast in the HST images was about 10–20%.

At about the same time, the images taken using the adaptive optics ADONIS camera at the 3.6 m ESO Telescope at Chile showed the same bright region at the equator and near 120 degrees orbital longitude, but also revealed a north-south hemispheric asymmetry apparent on Titan's darker side. Diffraction-limited images were thus obtained regularly since 1994 at 1.3, 1.6 and 2.0 μm with narrow band filters centred on the methane windows and in the wings of the absorption bands, which allowed for 50–300 independent spatial elements to be distinguished on Titan's disk (for instance at 2.0 μm a typical adaptive optics resolution would be 0.1 arcsec).

Adaptive optics is now a generally-adopted method and such systems exist in almost all the large Earth-based telescopes. Prior to the Cassini encounter, the adaptive optics system in Hawaii and its equivalent at the VLT in Chile were used to scrutinize Titan and produce surface maps at different wavelengths. The Keck, Subaru and Gemini AO systems have also been applied to Titan and returned some of the most interesting images of the satellite. The contrast in the adaptive optics images can achieve 50% under good observing conditions.

The NACO data on Titan were taken with narrow-band filters in the terrestrial atmospheric windows near 1.3, 1.6, and 2.0 μm, which probe the surface of Titan.

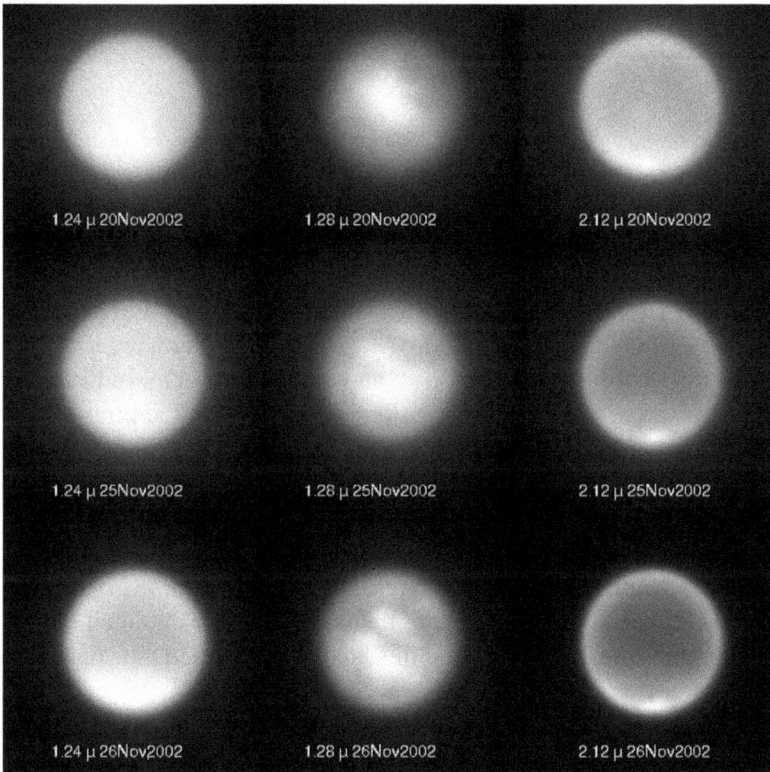

Figure 3.18   Images of Titan taken with the adaptive optics system NACO at the Very large Telescope of ESO in Chile in 2002. The different wavelength filters allow probing of various altitude levels from the atmosphere to the surface, hence the variable appearance. At 1.08 and 1.28 μm, Titan's surface is observed; elsewhere, the north-south asymmetry in the atmosphere evolves with a brighter northern pole at higher altitudes. At 2.12 μm, the bright spot on the south pole is due to a large meteorological system present from 2000 until 2005 (Gendron *et al.*, 2004).

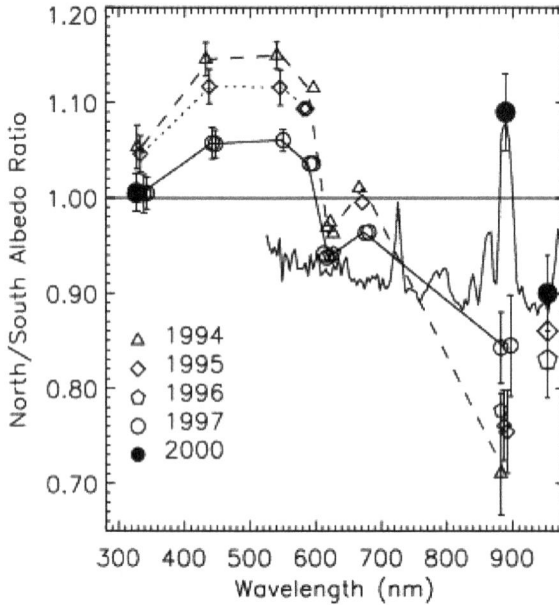

Figure 3.19   The wavelength dependence of the north/south albedo ratio on Titan from HST measurements over different periods in time, showing how the North-South asymmetry reverses at green and yellow wavelengths (Lorenz *et al.*, 1999).

Maps of Titan's surface at different wavelengths show very little variation, and always feature a bright equatorial region. At 2.0 μm, the images are less affected by scattering (about one-third) than at 1 μm, and the contrast achieved is then higher. The spatial resolution, before deconvolution is applied, is at the limit of diffraction (0.13 arcsec), and after deconvolution is very similar to that of the HST. Spectroscopically resolved images, recorded with a circular variable filter in adaptive optics, at 2.10 μm, were found to be quite similar to the 2.0 μm images, not showing the strong absorption by liquid hydrocarbons such as ethylene and ethane (C_2H_4, C_2H_6) that would have been expected in the presence of large hydrocarbon lakes in the dark regions. Of course, as supporters of the lake hypothesis pointed out, the lakes could contain impurities that confound this analysis, provided they have the proper spectral behaviour.

Even from the first set of data, it became apparent that the brighter leading hemisphere of Titan found in the spectra was associated with a large feature located near the equator centred near 114° LCM and extending over 30° in latitude and 60° in longitude. This bright spot we now know as Xanadu could be due to differences in relief, but it has a spectral behaviour that suggests otherwise. Like other bright spots visible in the S-W region (near 25°S) and near 30°N, it appears bright at all investigated wavelengths (0.9, 1.1, 1.3, 1.6 and 2.0 μm). It was also demonstrated that Titan's surface was much more complex than initially thought and that the "dark" hemisphere, was — fortunately because it was soon found out that the Huygens probe was not going to land where initially scheduled, close to the bright region, but rather on the trailing side — not all that dark, but also showed some fine structure with bright areas.

Early observers were tempted to attribute the bright 'continents' to the presence of mountainous areas on the surface, or alternatively depressions in the form of cratering produced by impacts. Higher regions certainly would look brighter, just by virtue of their height above some of the obscuring atmosphere. However, calculations show that even the effect of an unrealistically high plateau covering all of Titan's bright side and reaching up to 30 km in altitude, cannot account for the total observed albedo increase from one hemisphere to the other. Such a large asymmetry on the surface would alter Titan's centre of mass and the satellite would slowly turn to point the feature toward Saturn, which is not where the bright feature is found. This could mean that the feature formed relatively recently, in geological terms, but it seems much more likely that the brightness difference may be due to a compositional variation. Part of the brightness could be caused by moderately high mountains

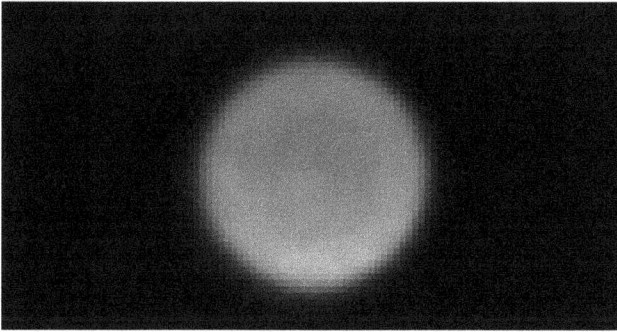

Figure 3.20   Titan's disk in the near-infrared, imaged with the HST, showing north-south asymmetry with a bright southern hemisphere, indicative of aerosol concentration in that region (Smith, Lemmon *et al.*, 1996).

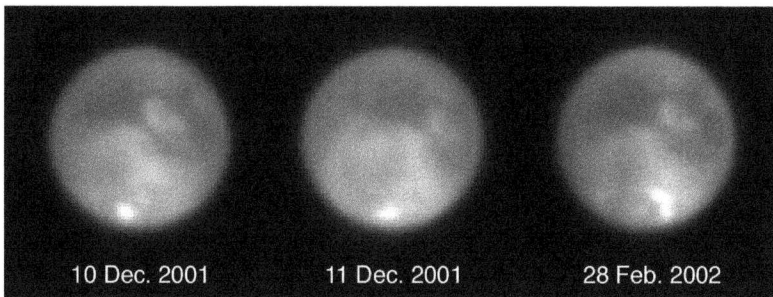

10 Dec. 2001   11 Dec. 2001   28 Feb. 2002

Figure 3.21   Keck images of Titan showing the cloud-like system in the south polar region (Roe *et al.*, 2002a).

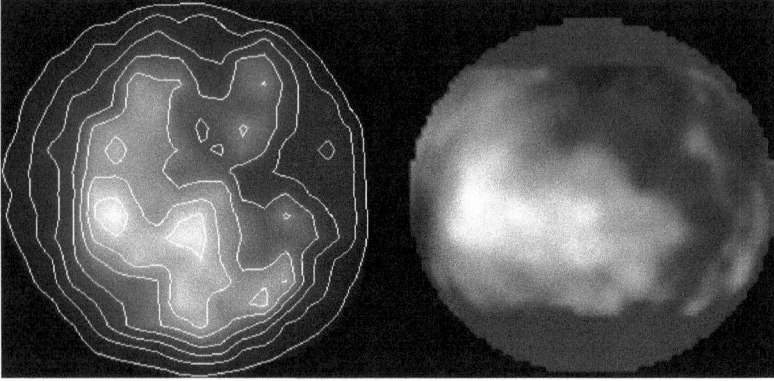

Figure 3.22   Images of Titan's surface at $2\,\mu$m (left) and $0.94\,\mu$m (right), obtained with the HST and ADONIS, respectively, in 1994. Note the large bright equatorial region located on the leading hemisphere (Coustenis and Lorenz,1998).

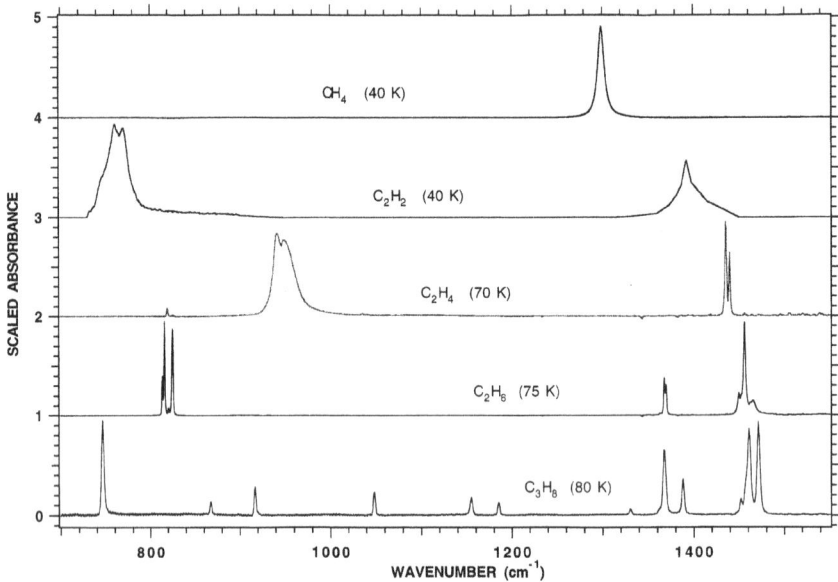

Figure 3.23   Laboratory spectra of some ices which may occur on Titan, measured by Schmitt and colleagues (Coustenis *et al.*, 1999).

(up to about 4 or 5 km would not be impossible, according to the above analysis) and the rest to a difference in their surface covering from the surrounding plains: mountains covered with fresh snow of water or methane, perhaps? Methane turns to ice at temperatures lower than 90 K, which is possible on top of a mountain, even near the equator, where the lapse rate is about 1 degree K per km. Other possible candidates in their icy form can be found on Titan and might be responsible for the observed images and spectra of the surface. Alternatively, methane rain might wash the high ground free of the dark hydrocarbon mud that probably coats much of Titan's surface.

3.5 Ground-Based Observations and Cassini–Huygens

With the arrival of the Cassini–Huygens mission in 2004, one might ask if ground-based measurements are still profitable. Even a powerful mission like Cassini can use complementary information from the ground (or from space for that matter: post-Cassini missions are already being discussed) and of course the Huygens probe explored the surface at one single area. Ground-based measurements acquired simultaneously with the orbiter's observations mean that the Cassini–Huygens data may be extrapolated to the whole disk surface. In particular, ground-based observations can provide measurements at solar phase angles not attained by Cassini (e.g., small phase angles); complementary observations for regions that are unlit (all flybys include global images and spectra before and after the close encounters; regions that are unlit can be observed within a few days on the ground); data during times the spacecraft is observing other objects to look for time-variable phenomena (cloud formation and decay); and measurements at wavelengths that are not included in spacecraft instrumentation. Campaigns performed during the descent of the Huygens probe in Titan's atmosphere and its landing on the surface on January 14, 2005, connected the *in situ* observations by the probe with the large coverage provided by high-technology ground-based runs. The observations covered radio telescope tracking of the Huygens signal at 2040 MHz, studies of the atmosphere and surface of Titan with spectroscopy and imaging, and some attempts to observe radiation emitted during the Huygens probe entry into Titan's atmosphere, which failed. The probe's radio signal was, on the other hand, successfully captured by a network of terrestrial radio telescopes, allowing scientists like D. Luz and colleagues to recover a vertical profile of wind speed in Titan's atmosphere from 140 km altitude down to the surface. O. Witasse supervised these campaigns for ESA and in his own words: "Ground-based observations brought new information on atmosphere and surface properties of the largest Saturn's moon.... The ground-based observations, both radio and optical, are of fundamental importance for the interpretation of results from the Huygens mission".

CHAPTER 4

Cassini–Huygens: Orbiting Saturn and Landing on Titan

"You thought I should find nothing but ooze" he said. "You laughed at my explorations, and I've discovered a new world!"

H.G. Wells, *In the Abyss*

4.1 Introduction

We are still in the era of the preliminary exploration of the Solar System, seeking to understand the basic nature of the planets and their satellites. Before Cassini/Huygens, Saturn had been visited only by fast 'flyby' spacecraft that spent only a few days in the vicinity of Titan and did not approach very closely, let alone land. Cassini is the first artificial satellite of Saturn, following a successful orbit insertion manoeuvre after a nearly seven-year journey. It has already spent several years carrying out the first detailed survey of the planet, its rings and satellites, and made multiple close approaches to Titan. An extended mission, recently approved, will assure operations until 2010, provided the spacecraft survives.

Huygens, a large probe equipped with parachutes, was carried by Cassini to Saturn and then ejected on a trajectory that resulted in a near-perfect landing on Titan. On the way down, the probe provided the first direct sampling of Titan's atmosphere and the first detailed views of its surface, surviving for several hours after landing. Huygens now holds the record for a man-made machine landing the farthest away from the Earth. Together, Cassini and Huygens are leading to an explosion in our knowledge of the ringed planet and its huge, cloudy moon.

Cassini–Huygens was always a very ambitious mission, conceived as a collaboration between the United States and 17 European countries, working through their space agencies, NASA and ESA. Although the mission's objectives span the entire Saturnian system, for Cassini (as for Voyager before it) Titan was a priority target and the mission was designed to address our principal questions regarding the giant moon and its atmosphere. The spacecraft together are equipped with 18 science instruments (12 on the orbiter and 6 on the probe), gathering both remote sensing and *in situ* data. Cassini communicates through one high-gain and two low-gain antennas. Because of the great distance from the Sun, electricity to drive the spacecraft systems is provided by nuclear power, in the form of three radioisotope thermoelectric generators fuelled by plutonium.

Weighing about 6 tons (actually 5,712 kg), the Cassini–Huygens spacecraft was launched successfully on October 15, 1997, from the Kennedy Space Center at Cape Canaveral at 4:43 a.m. EDT. Because of its massive weight, Cassini could not be sent directly to Saturn but used the 'gravity assist' technique to gain the energy required by looping twice around the Sun. This allowed it to perform flybys of Venus (April 26, 1998 and June 24, 1999), Earth (August 18, 1999) and Jupiter (December 30, 2000). Cassini–Huygens reached Saturn and performed a flawless Saturn Orbit Insertion (SOI) at 10:30 p.m. EDT on June 30, 2004, becoming trapped forever in orbit like one of Saturn's natural moons. As well as imaging the atmosphere and surface, the Huygens probe took samples of the haze and atmosphere as it descended. These *in situ* measurements complement the remote-sensing measurements made from the orbiter.

During its four-year nominal mission, the Cassini Orbiter makes about 40 flybys of Titan, some as close as 1000 km (Voyager 1 flew by at 4400 km from the surface), and takes a huge number of measurements with the visible, infrared, and radar instruments. The instruments perform remote studies of Saturn, its atmosphere, moons, rings and magnetosphere, measuring temperatures in various locations, plasma levels, neutral and charged particles, compositions of surfaces, atmospheres and rings, solar wind, and even dust grains in the Saturn system. Some perform spectral mapping and obtain high-quality images of the ringed planet, its moons and rings. The final aim is to better understand the Saturnian system in a unique opportunity to look for answers to many fundamental questions about the physical processes that rule the origin and evolution of planets and moons, perhaps even — through the study of Titan — to the conditions that give rise to life.

4.2 The Spacecraft and its Systems

Cassini–Huygens is the most complex interplanetary spacecraft ever built, and the 18 instruments on board the orbiter and the probe represent the most advanced technological efforts of the countries involved in the endeavour. Cassini went into orbit around Saturn in a complex, multiple trajectory scheme that allows it to conduct nearly a decade of detailed studies of the Kronian system, while also carrying the Huygens probe and assuring the relay of its data to Earth. The descent module remained dormant for most of the trip to Saturn and only "woke" when, after release, it reached the top of Titan's atmosphere, deploying its parachutes and performing a total of more than 5 hours of intensive measurements in the atmosphere and on the surface.

Cassini carries three nuclear power sources or RTGs, which use the heat from the radioactive decay of 33 kg of plutonium dioxide to produce 885 Watts of power for the spacecraft and its payload. In addition, 117 smaller radioisotope heater units — like

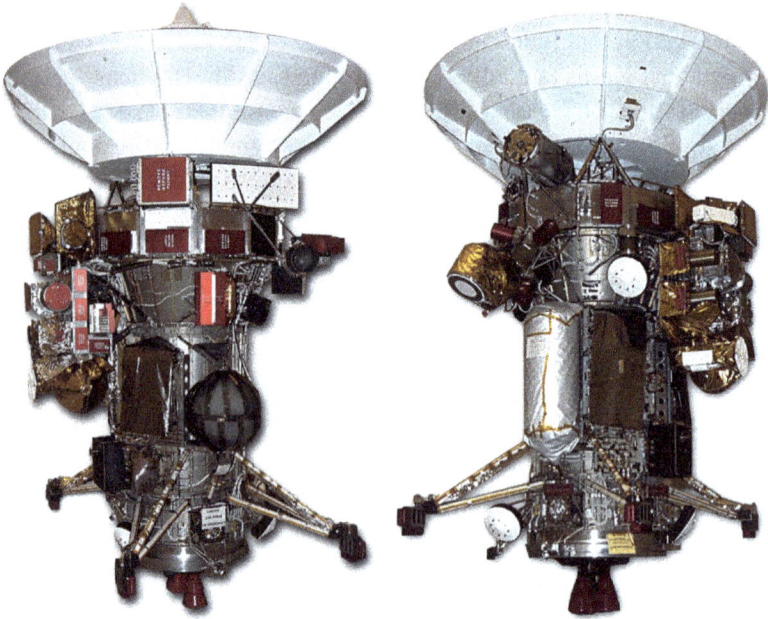

Figure 4.1   Detailed layout of the Cassini orbiter showing the main subsystems and the scientific instruments (NASA).

the ones recently used on the Mars Exploration Rovers — are used to keep the electronics at their operating temperatures. The generators and the heater units have had a long history of safe and reliable performance as part of the Voyager and Galileo missions.

While Cassini was still near the Earth, it used its two low-gain antennas to communicate with the operators on the ground. At the beginning of 2000, as the spacecraft entered the cooler outer regions of our Solar System, it turned its high-gained antenna toward the Earth and began to conduct all subsequent communications through it. The large communications dish is 4 metres across, and capable of relaying data from Saturn at rates as high as 140,000 bits per second. The on-board storage of data is by solid-state memory with a total capacity of 3.6 Gbytes. Telecommunications use a 20 W transmitter, operating at a frequency of 8.4 GHz (X-band).

Individually, the Cassini orbiter weighs 2,125 kg (4,685 pounds) and the Huygens probe 320 kg (705 pounds). The whole thing is 6.7 m (22 feet) tall, 4 m (13.1 feet) wide with a mass at launch of 5,712 kg (12,593 pounds), including the probe, 335 kg of scientific instruments, and 3,132 kg of propellant. Thus, more than half the spacecraft's total mass at launch was fuel, more than the propellant mass of the Voyager and Galileo spacecraft together. Half of it (about 1,500 kg) was required just for the orbit insertion at Saturn; the rest is used for manoeuvres during the course of the mission. Some of the propellant is hydrazine for the 16 small thruster jets, part of the orientation and flight path control system of the spacecraft. This keeps the communications dish pointed towards the Earth most of the time, with small excursions from this alignment when necessary to point the scientific instruments, which are bolted onto the spacecraft, at their targets. When major changes to Cassini's trajectory are required, propulsion is provided by one of the two main engines, which use monomethylhydrazine as the fuel and nitrogen tetroxide as the oxidizer.

Energy from the RTGs is distributed to the other subsystems, including the scientific instruments, as a 30 V supply. The 12 engineering subsystems in the orbiter control the spacecraft functions such as wiring, electrical power distribution, computers, telecommunications, orientation and propulsion. The subsystems distribute commands, and collect and format the data for transmission; they also control the propulsion devices, such as the orbit insertion motor and the attitude control jets; monitor and control the temperature everywhere on the spacecraft, and operate special devices like that which separates the probe from the orbiter. The spacecraft is controlled from the Earth through a sophisticated sequence of software commands. A given command sequence can operate on one of the computers on board for more than a month without interference from ground controllers.

The shape of the spacecraft was designed to accommodate the requirements of the various instruments and communication systems. The main body of the orbiter is cylindrical, in three parts or 'modules', two for equipment, with the propulsion module in the middle, capped by the high-gain antenna. Four booms protrude: the magnetometer is mounted on an 11 m long boom that extends outward from the spacecraft, as do three 10 m antenna booms. To protect Cassini against the extreme conditions in space and keep the computers, mechanical devices and electronic systems safe, the spacecraft and its instruments are covered with a thick shiny amber-coloured or black thermal blanket. Mylar is incorporated into this material to protect against micrometeorites that could hit the spacecraft. The onboard computers are designed to survive even solar flares, which can increase the interplanetary activity a thousand times. Cassini has some 22,000 wire connections and more than 12 km of cables linking the instruments and subsystems. Inside the spacecraft, insulating blankets are used, including kapton, to protect the instruments from the dust and to retain the heat.

The Huygens probe is 2.7 m (8.9 feet) in diameter and weighs 320 kg (705 pounds), including 43 kg of scientific instruments. For power, Huygens takes advantage of the orbiter while onboard via an umbilical cable. After separation, power was provided by five 23-cell lithium sulphite batteries, which were charged from the orbiter power supply before release of the probe. Data was radioed to the orbiter, for relay to Earth, at a rate of 8 kbps, using two separate transmitters, each with its own antenna. The link was one way only; all of the functions of the probe were automatic once it was released from the orbiter and the 'umbilical' connection severed. After separation from Cassini, Huygens flew for 22 days through the cold regions around Saturn, using 35 heater units to keep the equipment operational.

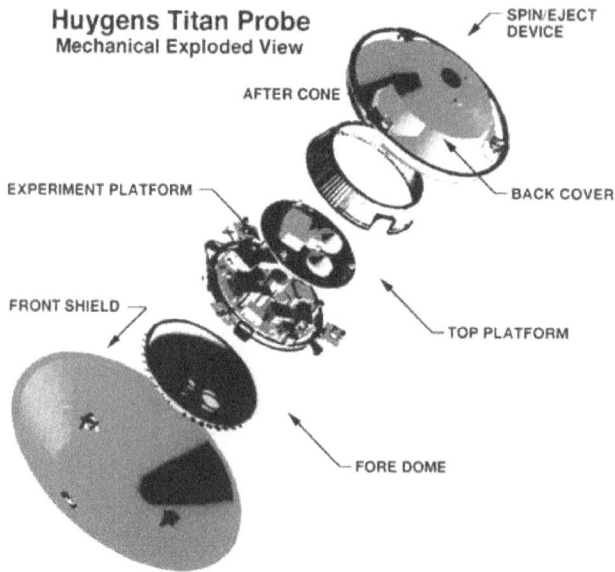

Figure 4.2   An exploded view of the Huygens probe showing the main subsystems and the experiment platform that carries the scientific instruments (ESA).

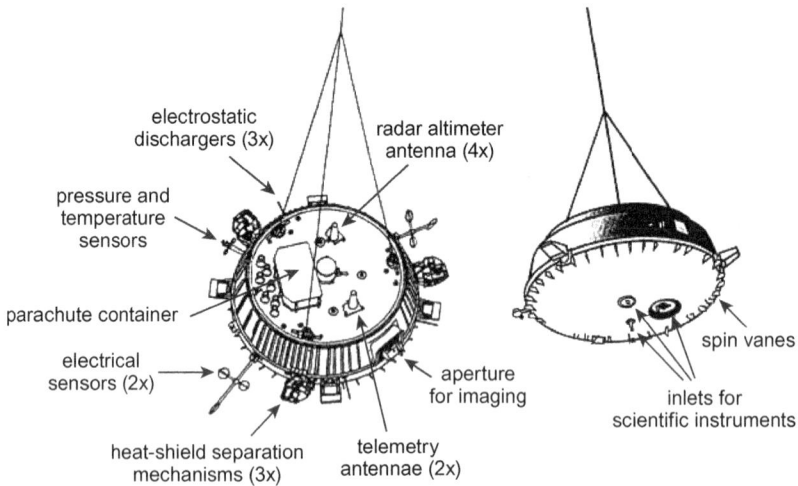

Figure 4.3   The Huygens probe under its parachute. The HASI booms are deployed, and the DISR sensor head can be seen. To the right, the disposition of the GCMS, ACP and SSP inlets is shown (ESA).

Huygens consists of the descent module, a front shield, an aft cover and a spin-eject device, the latter being part of the support structure on the orbiter which propelled the probe toward Titan and caused it to spin. The front shield, 2.7 m in diameter, consisted of a special thermal protective material called AQ60. This was also used in the aft cover, which ensures a slow and safe descent. During entry into Titan's atmosphere, the front shield had to sustain temperatures above 1,500 °C (2,700 °F), but thanks to the layers of insulation, the equipment inside the probe remained at temperatures below 50 °C (122 °F).

The probe's main body consists of two platforms and an aluminium shell. The central experiment platform also carries the electrical subsystems, which manage the data, keep time, switch on the probe prior to entry, assure telemetry, etc. The top platform above it is used to stow the parachute and carries the transmitter that sent the data to Cassini.

4.3 Scientific Objectives

Cassini–Huygens left Earth equipped with a set of interrelating instruments designed to address many of the most important questions about the complex Saturnian system. The general scientific objectives of the mission are concisely summed up in

Figure 4.4   Lift-off for Cassini on a Titan IVB/Centaur at 4:43 a.m., October 15, 1997 (NASA/JPL).

NASA's official statement of its purpose: to investigate the physical, chemical, and temporal characteristics of Titan and Saturn: its atmosphere, rings, icy satellites and magnetosphere. For Saturn, this includes discovering the sources and the distribution of lightning, by looking visually with television for flashes and listening at radio and other frequencies for electrostatic discharges and 'whistlers'. It also includes infrared remote sensing of the temperature field, cloud properties, and the composition of the atmosphere. None of these properties remains constant — Saturn's atmosphere is a very dynamic environment — so repeated measurements are necessary, to obtain not just the basic atmospheric structure parameters but also some indication of their variability. Of particular interest is the global wind field, including its wave and eddy components. Winds are hard to measure directly, but can be computed using observations of the movement of cloud features, and calculations based on pressure gradients inferred from temperature measurements. This gives the dynamics of the observable regions in the region at and above the cloud tops, so the internal structure and rotation of the deep atmosphere must be inferred by extrapolation from the layers above, while higher levels can be studied by ultraviolet spectroscopy and the fields and particles experiments to understand the diurnal variations and magnetic control of the charged particle concentrations in the low-density region.

A list of objectives for Titan was drawn up before the mission was launched. It will be many years after the observations end before we can finally say in detail how well they have all been addressed, but well into the mission new questions have already arisen and we already have some ideas about what still needs to be done by the extended Cassini or by new and even more sophisticated missions. The initial list reads:

- Determine the abundances of atmospheric constituents (including any noble gases).
- Establish isotope ratios for abundant elements, which will help constrain scenarios of formation and evolution of Titan and its atmosphere.
- Observe vertical and horizontal distributions of trace gases.
- Search for even more complex organic molecules.
- Investigate energy sources for atmospheric chemistry.
- Model the photochemistry of the stratosphere.
- Study the formation, composition and distribution of aerosols.
- Determine the winds and map the global temperatures, investigate cloud physics, general circulation, and seasonal effects in Titan's atmosphere.
- Search for lightning discharges.
- Determine the physical state, topography, and composition of the surface with *in situ* measurements at different locations of the disk.
- Infer the internal structure of the satellite.
- Investigate the upper atmosphere, its ionisation, and its role as a source of neutral and ionised material for the magnetosphere of Saturn.

4.4 The Long History of the Cassini–Huygens Mission

The outer planets have always been difficult targets for exploration, primarily because of their vast distances (Saturn is about 30 times as far from Earth as Mars, for example) that can only be spanned by using powerful launch vehicles. Another reason is the need to use nuclear power to provide electricity to run the spacecraft and its scientific instruments since, at Titan's distance, the Sun is just a big star and the use of solar cells is impractical: it would take an array the size of a tennis court. Finally, the journey takes a long time, and the operations costs (such as tracking station time, or keeping teams together who built and understand the spacecraft, and who could detect and solve problems as they arise) as well as the need for reliable components and redundant systems on board the spacecraft, all push the cost up until it is counted, not in millions, but in *billions* of dollars. In the future, when the initial surveys are over, exploration of the outer Solar System will be feasible with smaller spacecraft, with fewer, more focussed objectives, especially as new technology comes along so that small, inexpensive missions can do as much or more as big spacecraft do now. We may never see big missions to the outer Solar System like Cassini again.

Fortunately, Cassini promises a lot to be going on with. Before it arrived, only glimpses of Saturn and Titan had been obtained from the Pioneer and Voyager fast flybys of the 1970s and 1980s. As we saw in Chapter 2, these went tearing past Saturn in a couple of days, and past Titan in just a few hours. As an orbiter/probe mission, Cassini's heritage is not so much the Voyager missions, which recently ended their long odyssey in the Solar System, but the Galileo mission to the Jovian system. Galileo was the first artificial satellite of Jupiter, and it deployed the first probe into any outer solar system atmosphere.

From its initial conception to the completion of the mission in 2010, the Cassini–Huygens mission is an achievement that spans almost 30 years. Originally known as Saturn Orbiter–Double Probe (abbreviated SOPP), in the early days of its design when it was to have carried a probe for Saturn itself as well as one for Titan, Cassini was conceived as the Saturnian version of Galileo. Table 4.1 summarises the major milestones in the development and implementation of Cassini and Huygens. As early as 1982, a working group composed of the Space Science Board of the National Academy of Science in the U.S. and the Space Science Committee of the European Science Foundation was formed to study possible modes of cooperation between the U.S. and Europe for their mutual scientific, technological and industrial benefits. Among other possible projects, the group proposed a Saturn Orbiter–Titan Probe mission in which, it was suggested at this time, Europe would provide the orbiter while their colleagues in the USA would develop the probes.

This proposal was submitted both to the European Space Agency and to the National Aeronautics and Space Administration in the USA, and by 1985 the joint ESA/NASA assessment of the Saturn Orbiter–Titan Probe mission was completed and under development by both agencies in Phase A. This is the stage at which

Table 4.1   Milestones in the history of Cassini–Huygens and timeline for its trip to Saturn.

1982	Joint Working Group of European Science Foundation and US National Academy of Sciences proposes a Saturn Orbiter–Titan Probe mission
1983	US Solar System exploration Committee recommends a Titan Probe and Radar Mapper for NASA's core programme
1984–5	Joint NASA-ESA Assessment Study of Saturn Orbiter–Titan Probe mission
1987	ESA Science Programme Committee approves Cassini for a Phase A study
1987–88	NASA develops Mariner Mark 2 concept for Cassini and Comet Rendezvous-Asteroid Flyby (CRAF)
1988	ESA selects Cassini mission, names probe Huygens
1989	US Congress approves CRAF and Cassini
1992	Cassini restructured and rescheduled; CRAF cancelled
1995	US House Appropriations Committee targets Cassini for cancellation
1996	Spacecraft and instruments begin testing
April 1997	Cassini shipped to Cape Canaveral
October 15, 1997	Launch
April 1998	First Venus gravity assist flyby
June 1999	Second Venus gravity assist flyby
August 1999	Earth gravity assist flyby
December 2000	Jupiter gravity assist flyby
December 2001	First gravitational wave experiment
June 2004	Phoebe flyby at 52,000 km
July 2004	Spacecraft goes into orbit around Saturn
December 25, 2004	Release of Huygens Probe on a trajectory to enter Titan's atmosphere
January 14, 2005	Huygens returns data as it descends through Titan's atmosphere and reaches the surface; Orbiter begins tour of the Saturn system
July 1, 2008	Nominal end of mission
2010+	End of extended mission

detailed designs are worked out and a reliable budget established, prior to a final commitment to build the hardware. Cassini was developed along with another ambitious mission, the Comet Rendezvous/Asteroid Flyby (CRAF), both funded in 1989 by the U.S. Congress. The all-American SOPP had eventually evolved into Cassini/Huygens, now composed of an American orbiter and a European descent probe — sometimes referred to as 'Inissac' at the time, because it was the original Cassini concept backwards! This made it the first truly international planetary mission, in addition to its other breakthroughs. NASA and ESA launched a call for instruments for both missions, but, in 1992, CRAF was cancelled and Cassini almost was too. In the end, the international collaboration with its contractual obligations saved Cassini, but it had to be restructured to reduce the cost by simplifying the spacecraft capabilities. The articulated instrument platform was abandoned and the instruments hard-mounted to the spacecraft, which now had to turn in its entirety in order to point at the target. Other losses include tape recorders, mechanical gyroscopes and the steerable antenna on the probe, which was replaced with a cheaper fixed one. The total cost of the mission was then about \$3.26 billion, of which \$2.6 billion was to be the U.S. contribution while the rest, mainly for the probe and amounting to an estimated \$660 million, came from Europe. At about this time, ESA gave a separate name to the probe, and the combined mission was then called Cassini–Huygens.

Trouble brewed for Cassini at an early stage when NASA developed its "faster, better, cheaper" strategy, which sought to replace large expensive missions with small ones that did the same thing. Just how this miracle was to be brought about was never fully demonstrated, but suddenly Cassini/Huygens was no longer fashionable, in the US at least, as a terrestrial ambassador to the planets. The NASA Administrator took to referring to Cassini as 'Battlestar Galactica', and it was not meant as a compliment. However, it should not be forgotten that the research and development for Cassini–Huygens has brought forth new technologies that, while not satisfying NASA's new motto certainly provided innovations that found their way into the new low-cost 'Discovery' class missions, such as Mars Pathfinder. While the instruments on board Cassini and Huygens used the knowledge accrued from Pioneer, Voyager, and Galileo, they also introduced newer technology to address many of the known mysteries of the Saturn system much better than the earlier probes could, even if they had not gone by so quickly and so far away. For example, CIRS, the infrared spectrometer on Cassini, is a much more advanced version of IRIS on Voyager, with an optimal spectral resolution that, at 0.5 cm^{-1}, is an order of magnitude better than IRIS, and with better sensitivity through the use of cooled infrared detectors. It obtains much better spatial resolution, including an improvement of a factor of 2–3 over Voyager in the vertical (height) dimension, due to its better optics and 6 times closer approach to Titan than Voyager, and much better coverage from multiple encounters on different orbits of Saturn.

Somehow Cassini survived; the threat was always political rather than scientific or technical, and did not find much of an echo with NASA's European partner. Huygens of course is completely innovative: the first probe from Earth to actually enter Titan's atmosphere and land on the surface, and the most distant lander in our Solar System. We now have *in situ* temperature-pressure-density profiles, details of the atmospheric composition as a function of height, and direct data on the surface conditions and its enigmatic appearance.

While NASA produced the main Cassini orbiter spacecraft and ESA provided the Huygens probe, the Italian Space Agency (ASI) was responsible for the spacecraft's radio antenna and portions of three scientific instruments. In the United States, NASA's Jet Propulsion Laboratory in Pasadena, California, where the Cassini orbiter was designed, assembled, and tested, manages the mission. In Europe, an industrial team from all over the continent created the Huygens probe, with contributions

VENUS 1 FLYBY
26 APR 1998

VENUS 2 FLYBY
24 JUN 1999

SATURN ORBIT
INSERTION
1 JUL 2004

SATURN'S ORBIT
29.4 YEARS

JUPITER'S ORBIT
11.8 YEARS

VENUS
TARGETING
MANEUVER
3 DEC 1998

SUN

LAUNCH
15 OCT 1997

EARTH FLYBY
18 AUG 1999

JUPITER FLYBY
30 DEC 2000

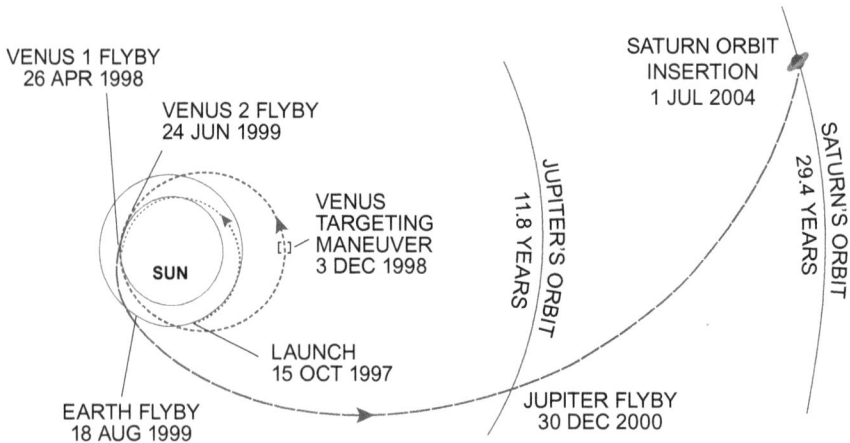

Figure 4.5   Cassini's trajectory between Earth and Saturn, showing the close encounters with Venus, Earth and Jupiter which were required to give 'gravity assisted' boosts to the spacecraft.

from France, Germany, Italy, UK, Ireland, Sweden, Spain, Denmark, Switzerland, Belgium, Austria, Finland, and Norway. Communications with Cassini are carried out through the stations of NASA's Deep Space Network in California, Spain and Australia, with the data from the Huygens probe forwarded from the DSN to an ESA operations complex ESOC in Darmstadt, Germany.

The scientific instruments aboard the mission were prepared in numerous European and American laboratories, in many cases with a strong collaboration between scientists and engineers on both sides of the Atlantic working on an individual instrument. European scientists lead two experiments on NASA's orbiter and participate in all of them, while US-led teams supplied two instruments in ESA's Huygens probe and American experts contribute to three others. The human involvement in this enterprise is vast: more than 5,000 people worked on some portion or other of the mission, and more than 260 scientists in the U.S. alone have contributed to the project. All in all, a fine example of international space science collaboration.

4.5 Departure for the Saturnian System

More than 200 scientists, engineers and others who had been involved in the development of this ambitious space mission gathered at Cape Canaveral Air Force Station, Florida to see it off on its seven year, 3.5 billion km trek to the realm of Saturn. They gathered first during the early morning of October 13, 1997 to watch from a location, as close as the safety officers would allow and giving a perfect view, 5.5 km away from the launching site. Unfortunately on that day there were strong winds blowing at altitudes of about 15 km, just where the detachment of the booster rockets from the spacecraft occurs. These have to fall to the ground in a predictable way, and land

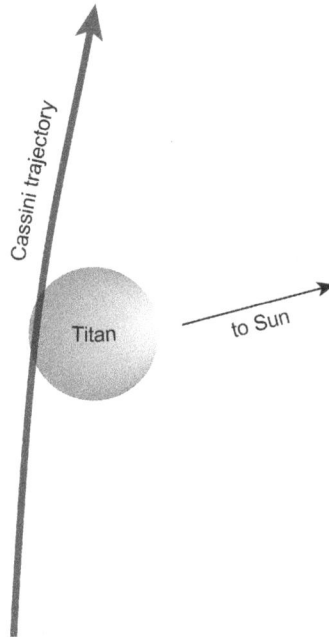

Figure 4.6 Cassini's tour of the Saturnian System is arranged to make it possible to have a close approach to Titan on most of the orbits. This view shows the fifth such encounter, which took place on March 31, 2005, to scale (NASA/JPL).

safely in the ocean, so the launch had to be put back to wait for better weather. Only two days later, a little before 5 o'clock on the morning of Wednesday, October 15, everyone gathered again and this time they watched in awe as the heavy spacecraft lifted off on its long journey.

The designated area for the launch-watchers is quite beautiful, part of a nature reserve, and on that warm Wednesday morning the sky was clear, with a huge full moon. For those who had worked and waited long years for this occasion, it was a magical moment. Screens set in front of the benches allowed a direct view into the control room, and loud speakers gave us all an opportunity to follow the procedures as all the systems went "GO" to indicate readiness for departure. The launch system was a powerful Titan IVB, with a Centaur upper stage. The coincidence turned out to be a good omen. The countdown had us all hypnotized and we held our breaths as we came to "lift-off". Many of us had never experienced a launch before and it was very much a personal affair, but few if any of the people gathered there managed to escape the rise in emotion.

There followed a loud roar and a bright flame along the ground, as the Cassini spacecraft, carrying the Huygens probe, began to rise slowly towards the sky. At 15 km, as planned, the boosters were discarded and the spacecraft crossed the dark sky like a bright comet and set off in its course to Saturn. The first measurements showed that the launch and the trajectory were precisely according to plan.

Holding hands or arms, many of us stood there and watched what soon became a splendid night show, as Cassini pushed its way through a cloud, a perfect image of power and phantasmagoria at the same time, offering a fireworks-like display. In spite of all the reassuring comments from the loudspeakers that were positioned around the site to relay the words from the mission controllers, it took the whole two hours before we could stop fretting that something might go wrong, dry our eyes and wish our "baby" an uneventful journey. On the way back to the hotels, we all agreed that this was a memorable event, something to cherish on our memories for the rest of our lifetimes.

4.6 Journey to Saturn and Orbit Insertion

Following the launch from Cape Canaveral, the combined orbiter and probe space-craft first travelled in a direction away from Saturn, into the inner Solar System, where it executed gravity-assist flybys of Venus in late April 1998 and again in June 1999, followed by one of Earth, in mid-August 1999 and one of Jupiter in late December 2000. These manoeuvres used the gravitational fields of each target planet to increase the spacecraft's velocity (the "slingshot" effect), giving it the boost needed to reach Saturn in "only" 6.4 years.

Without the assistance of Venus' gravity, such a large spacecraft as Cassini could not be injected into the interplanetary trajectory towards the outer Solar System at all, at least not without a much larger and more expensive launch vehicle. With the multiple flyby approach, the total mass that can get to Saturn is sufficient to include a large science payload, and enough fuel to make a number of manoeuvres while in orbit at Saturn. Without these in-orbit motor firings, Cassini would not be able to alter its path around Saturn to make repeated encounters with the satellites, including Titan.

The spacecraft completed nearly three-quarters of an orbit around the Sun prior to the first Venus flyby. The perihelion of this initial orbit was 0.676 AU. Only limited science data collection was allowed during the Venus flyby, in order not to put spacecraft and instruments at risk while they were so close to the Sun, since they were designed to survive this hot environment, but mostly not to operate in it. Some data collection for calibration was incorporated into the Earth flyby, but science observations did not begin in earnest until the time of the Jupiter flyby, the fourth and final gravity assist, in December 30, 2000. Cassini arrived at the Jovian system while the Galileo mission was still in place, and so could take advantage of the two different vantage points to gain improved knowledge on the shape of the Jovian magnetosphere and its interaction with the solar wind, to study the aurorae, and to check out the performance of all of the instruments.

Upon arrival at Saturn on June 30, 2004, nearly seven years after leaving Earth, the main rocket engine was turned to face the direction in which it was travelling, and fired to slow Cassini down. The spacecraft approached Saturn from below the ring plane, flying through the gap between the F and G rings, about 150,000 km from the centre of Saturn and just less than 2 hours before the spacecraft's closest approach to the giant planet. The spacecraft was oriented with the high-gain antenna used as a shield to provide protection from any small particles that could present a threat during the ring crossing. The main engine burn began shortly after the ring crossing and ended 97 minutes later. Had the motor failed, or not fired for long enough, the spacecraft would have hurtled past Saturn and eventually gone into a distant orbit around the Sun. Orbit insertion was cause for anxiety for many of us who lived through those 97 minutes watching and hoping that all would go well.

In fact, all did go well and the orbiter, its scientific instruments now operating, began its complex dance around the giant planet. The closest approach (periapsis) altitude during Saturn orbit insertion was 0.3 Saturn radii, which is the nearest it gets to the planet during the whole four-year tour, consisting of some 60 orbits around the planet. The arrival period also provides the first and best opportunity to observe Saturn's rings, and the images returned during this event were breathtaking and an enormous input for ring science.

4.7 Huygens Descends onto Titan

The orbits are designed so that most of them involve passing close to Titan, when observations could be made using the instruments onboard the orbiter. For the first two of these, Cassini was still carrying the Huygens probe. After the second encounter, which took place on December 13, 2004, the main rocket motor was fired again to raise the periapsis and to bring the orbiter and probe, still combined, into an orbit that passes close to Titan. On December 17, 2004, small onboard jets were fired to fine-tune Cassini's flight path to align it on a trajectory that would actually result in a collision with Titan. This probe targeting maneuver was essential since Huygens had no motors of its own and, once released, would simply follow the trajectory on which it had been placed by Cassini. Conditions were now right for the release of the probe from the mother ship that had brought it all the way to Saturn.

Final commands were sent to Huygens on December 21, 2004 including the setting of the Mission Timer Unit, a sophisticated alarm clock that would be activated about 5 hours before Huygens reached an altitude of 1270 km above Titan's surface. On December 25, 2004 at 02:00 UTC the Spin/Eject Device separated Huygens from Cassini with a relative speed of approximately 0.35 metres per second. This event included the firing of the small explosive pyrobolts, engagement of the separation push-off springs, ramps and rollers and the separation of the electrical connectors, all in the space of approximately 0.15 seconds. Wrapped in its heat shield, Huygens was set spinning at seven revolutions per minute in order to remain stable as it approached its target. The landing was targeted for a southern-latitude site on the

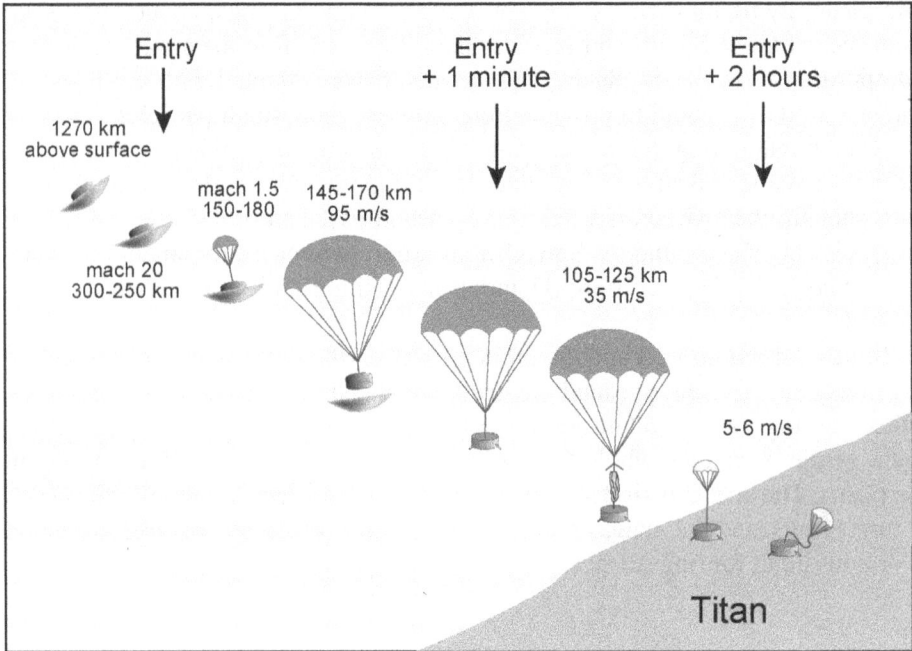

Figure 4.7   14 January 2005: The descent sequence followed by Huygens onto Titan. Using three parachutes, as well as the drag from the heat shield, the probe fell 1250 km in about two and a half hours, its velocity decreasing from 6 km per sec at the top of the atmosphere to about 5 metres per sec at impact (ESA).

dayside of Titan, near the large continental feature that had, at that time, not yet been named Xanadu. The probe entry angle into the atmosphere was controlled within a few degrees of a relatively steep 65° with respect to the vertical, this being the best compromise between a short descent and maximum deceleration by drag, designed to give Huygens the best opportunity to reach the surface alive.

The release of such a large mass had a considerable effect on the motion of the Cassini orbiter it left behind. Cassini also had to be re-targeted to avoid following Huygens into Titan's atmosphere, while remaining close to the probe's atmospheric entry point so it could achieve the radio relay link geometry. On December 28, an orbit deflection maneuver corrected the Cassini trajectory to fly past Titan at an altitude of 60,000 km and delayed the closest approach to until around two hours after Huygens reached the entry interface. The relative position of Cassini, Huygens and Titan enabled a theoretical maximum conversation between orbiter and probe of 4 hours 30 minutes.

The electronic alarm clock switched Huygens on, well before it reached Titan's upper atmosphere, on January 14, 2005 at 11:04 UTC. A few hours later, friction with the atmosphere acting on its heat shield began to cause the 300 kg probe to 'aerobrake' from its impact velocity of around 21,600 km/h (6 km/s) to less than

1,400 km/h at 180 km altitude, experiencing forces of up to 25 g and an entry temperature of more than 12,000 °C (21,600 °F). Once the speed had fallen to about Mach 1.4 (290 km/h at 170 km altitude), the shield was released and the main parachute, 8.3 m in diameter, was opened. The parachute was discarded 15 minutes later.

The people gathered in the ESA Control Centre at Darmstadt that January 14, 2006 received the first signal that Huygens is alive from an Earth-based radio telescope, as they waited for the first data and the first images to be transmitted by Cassini. Suspense was high because it takes about 67 minutes for the data from Cassini to get to the Earth, and the first stream of data, due to be transmitted through Channel A, failed because it was not turned on due to a programming error. However, after a few extra minutes, Channel B took over and the measurements began to come through, including the first images returned by the cameras of DISR while still falling through the atmosphere, soon stitched together to make the first panoramas of a fantastic and long-anticipated landscape.

The main parachute was discarded automatically after 15 minutes of descent to make sure that the orbiter would not pass below the horizon as seen from the landing site before all probe touched down, and had transmitted at least an hour's worth of surface data if it survived the impact. It was also necessary to find a compromise descent speed between shortening the duration, so Huygens and its payload did not freeze, on one hand, and achieving a sufficiently soft landing, so that it had a chance of surviving on the surface, on the other. Tests in which prototypes of Huygens were dropped through Earth's atmosphere allowed practical estimates to be added to theoretical calculations of the expected descent time; in fact, the descent from atmospheric entry to landing took 2 hr 27 m 50 s.

At around 125 km, falling at a speed of 360 km/h, Huygens deployed its smaller, 3 m diameter 'drogue' parachute, designed to stabilise the probe as it descended rather than to slow it too much more. Although a slower descent meant more scientific data on the atmosphere, it also meant a greater chance of technical failure, or loss of the communications relay, before the probe reached the surface. At about 60 km in altitude, the surface-sensing radar was turned on, and finally at an altitude of 700 m above the surface the descent lamp of the imaging instrument was activated. The purpose of this lamp was to enable scientists to accurately determine the reflectivity of the surface. For photography, the natural lighting by the Sun was used to illuminate the landing site. The light level on the surface of Titan was roughly 1,000 times less than we are used to on Earth by day, but 1,000 times stronger than the light of the full moon.

For those in Darmstadt, the most thrilling moment perhaps was the landing. Officially, the mission was planned to end when the probe hit the surface, with any further data being seen as a bonus should it survive. Huygens was a descent module and not a lander, as we were reminded several times during Cassini–Huygens meetings. At the very best, a few minutes of survival at the surface could be envisioned if "all went perfectly well and we didn't crash or sink".

As it happened, most things went more than just "perfectly well". Thanks to the parachutes, the surface impact speed was only about 20 km/hr (or 5 m/s). On the surface, the five batteries onboard the probe lasted much longer than expected, allowing Huygens to collect surface data for 1hr and 12 m. During its descent, the DISR camera returned more than 750 images and numerous spectra, while the probe's other three instruments (HASI, ACP and GCMS) sampled Titan's atmosphere to help determine its composition and structure. The Surface Science Package had plenty of time to acquire data after landing. The telemetry data from Huygens was relayed at a rate of 8 kilobits per second and stored in Cassini's solid-state memory while the latter was at an altitude of 60,000 km from Titan. Two nearly redundant channels, A and B, were planned, but in the event, as the result of a programming error, the only serious problem Huygens experienced led to loss of all data from channel A. The Doppler wind experiment was the main casualty, since unlike most of the instruments it did not have redundancy and unfortunately lost its data that were supposed to be transmitted only through Channel A. Happily, in the end all of the measurements were recovered because the weak signal from Huygens was captured by Earth-based radio telescopes, which became "Channel C", and the Doppler experiment was successful. The signal received via radio telescopes lasted 5 h 42 min, including 3 h 14 min from the surface. This was so much more than the few minutes that had been expected that it seems — even today — quite unbelievable.

The tremendous technological and scientific achievement of the Huygens mission will bear fruit for many years to come. But more than anything else, it once more proves the fantastic capabilities brought about by international collaboration. Landing on a new world 10 times farther from the Sun than our own planet stands

Figure 4.8   The Huygens Titan lander on Earth: a prototype in the snow of Northern Europe after a test drop from an aircraft (ESA).

with taking the first step on the Moon. Humanity has taken a huge step there towards broadening its horizons. In March 2007 the Huygens landing site was named the "Hubert Curien Memorial Station", in recognition of the former ESA director's contribution to European space.

4.8 Experiments and Payloads

Next we take a closer look at the all-important science payload, instrument by instrument. Remembering that Cassini was designed to explore the whole Saturn system, the emphasis here is on those devices that make measurements relevant to the problems and puzzles specific to Titan.

4.8.1 *The Scientific Instruments on the Orbiter*

The remote sensing instruments study their targets by making maps and pictures (imaging) them, by measuring at multiple wavelengths (spectroscopy), and by doing both at the same time (spectral imaging). The information sought comes from recognising structures in the images, and measuring how the reflection and emission of ultraviolet, visible and infrared radiation from the target varies with wavelength. These spectra also vary with position across the map of an inhomogeneous object like Saturn's atmosphere or Titan's surface and this gives us information about differences in composition or physical state (temperature, for example). With high-resolution imaging, extra insight comes from studying the morphology of the object and, in the case of clouds or other dynamic targets, its rate of movement or change.

Cassini carries four remote sensing instruments, one high spatial resolution television camera, with low spectral resolution filters, and three spectrometers covering the ultraviolet, visible/near infrared, and middle/far infrared ranges of the spectrum. The spectrometers are all designed to get spatial coverage, but not at nearly the high resolution of the television; this is usually called 'mapping' as opposed to 'imaging', to differentiate.

The Imaging Science Subsystem (ISS)

The Imaging Science Subsystem consists of a wide-angle television camera, with angular resolution of 60 microradians per pixel ($3.5° \times 3.5°$) and a narrow angle camera, with angular resolution of 6 microradians per pixel ($0.35° \times 0.35°$). The Wide Angle Camera is a 20 cm f/3.5 refractor with 18 filters in the 380–1100 nm spectral range, and a field of view of $3.5° \times 3.5°$. The Narrow Angle Camera is a 2 m f/10.5 reflector working from 2000–1100 nm in 24 filters; the field of view is an area 100 times smaller than that of the Wide Angle Camera ($0.35° \times 0.35°$).

The sensors are 1024×1024 CCD arrays. These devices for solid-state imaging are similar to the computerised digital cameras that are displacing film in domestic photography, and are specially developed to work out into the ultraviolet and

Table 4.2 Physical characteristics of instruments on the Cassini Orbiter, and some details of the people in charge of producing and operating them.

Name of Instrument	Web Site	Principal Investigator/ Institute	Mass (kg)	Mean Power Power (W)	Data Rate (kbps)
Imaging Science Subsystem (ISS)	http://ciclops.org/	C. Porco SSI Boulder	57.83	55.90	365.57
Composite Infrared Spectrometer (CIRS)	http://cirs.gsfc.nasa.gov/	M. Flasar NASA GSFC	39.24	26.37	6.000
Ultraviolet Imaging Spectrograph (UVIS)	http://lasp.colorado.edu/cassini/	L. Esposito U.Colorado	14.46	11.83	32.1
Visible-IR Mapping Spectrometer (VIMS)	http://wwwvims.lpl.arizona.edu/	R. Brown U. Arizona	37.14	21.83	182.78
Ion and Neutral Mass Spectrometer (INMS)	sprg.ssl.berkeley.edu/inms/	H.Waite SWRI	9.25	27.70	1.50
Cassini Plasma Spectrometer (CAPS)	http://caps.space.swri.edu/	D.T. Young SWRI	12.50	14.50	8.00
Cosmic Dust Analyser (CDA)	http://www.mpi-hd.mpg.de/dustgroup/	E. Grün MPI Heidelberg	16.36	11.38	0.524
Magnetometer (MAG)	http://www3.imperial.ac.uk/spat/research/missions/space_missions/cassini/mag_team	M. Dougherty U. London	3.00	3.10	3.60
Magnetospheric Imaging (MIMI)	http://sd-www.jhuapl.edu/CASSINI/	T. Krimigis APL	16.00	14.00	7.00
Radar (RADAR)	http://saturn.jpl.nasa.gov/spacecraft/instruments-cassini-radar.cfm	C. Elachi JPL	41.43	108.40	364.80
Radio Science (RSS)	http://saturn.jpl.nasa.gov/spacecraft/instruments-cassini-rss.cfm	A. Kliore JPL	14.38	80.70	N/A
Radio and Plasma Wave (RPWS)	http://www-pw.physics.uiowa.edu/plasma-wave/cassini/home.html	D. Gurnett U. Iowa	6.80	7.00	0.90

infrared, as well as at visible wavelengths. Compared to the vacuum-tube optical-only television used by Voyager, they offer much higher spectral and dynamic range, and are more linear. The extension of its coverage into the infrared means that ISS can image the surface of Titan in the spectral 'windows' at wavelengths near 0.94 and 1.1 μm. A special filter is provided for detecting lightning in the neutral nitrogen lines at 0.82 μm.

The ISS scientific objectives include investigating the composition, distribution, and physical properties of clouds and aerosols, including scattering, absorption and solar heating, as well as mapping the 3-dimensional structure and motions within Saturn's and Titan's atmospheres and searching for lightning, aurorae, airglow, and planetary oscillations as well. The ISS team is also engaged in mapping the surface of Titan in the near infrared windows to study the geology and get clues as to the nature and composition of the surface materials.

Composite Infrared Spectrometer (CIRS)

The Voyager IRIS instrument described in Chapter 2 demonstrated the value of infrared spectroscopy for studying the atmospheres of Saturn and Titan, but its relatively wide field-of-view and distant approach meant IRIS could not resolve the atmosphere at the limb for either Saturn or Titan. Limb measurements (i.e. tangentially viewing at the edge of the planetary disk) are particularly valuable because they are not complicated by emission from the surface, and because vertical resolution is largely determined by the field of view of the instrument rather than solely by properties of the atmosphere itself. Thus, an instrument with a sufficiently narrow field of view can obtain high vertical resolution by limb viewing.

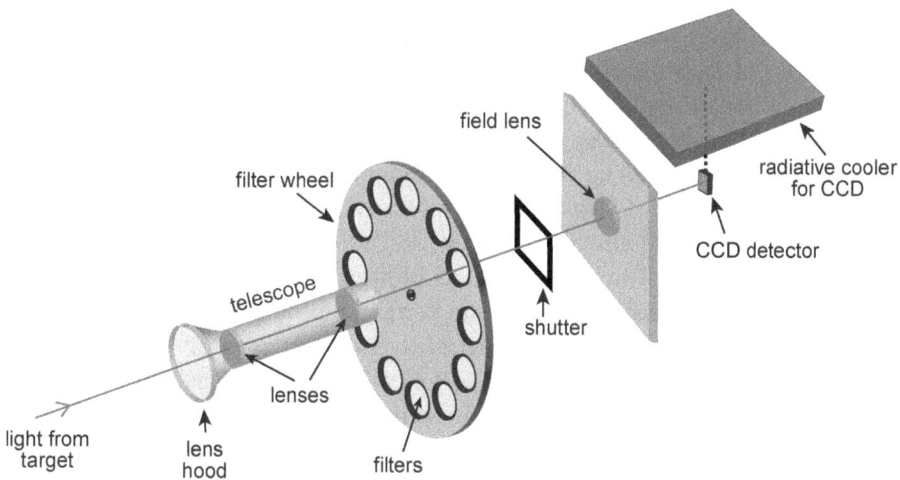

Figure 4.9 The Imaging Science Subsystem Wide Angle Camera. Team-Leader: C. Porco (NASA/JPL/SSI, University of Colorado).

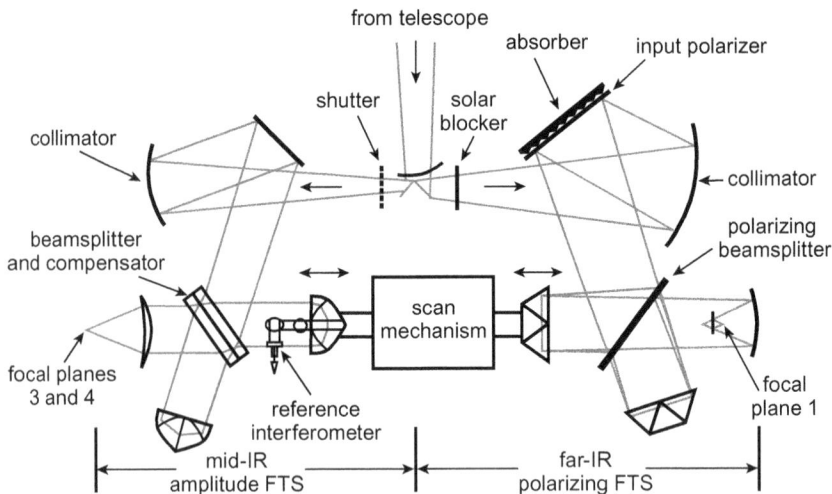

Figure 4.10   The Composite Infrared Spectrometer. PI: F. M. Flasar (NASA/JPL/GSFC).

The Cassini orbiter carries an improved version of IRIS called CIRS, the Composite Infrared Spectrometer. CIRS repeats the IRIS measurements at ten times higher spectral resolution ($0.5 \, cm^{-1}$), in an extended spectral range (7 to $1000 \, \mu m$, which is 10–$1400 \, cm^{-1}$), with a 4.3 mrad circular field of view and, of course, over a much longer period of time than Voyager, allowing the recovery of 40 times as many spectra. Built at Goddard Space Flight Center near Washington, DC, CIRS measures infrared emission from atmospheres, rings, and surfaces which can be used to map the temperature, hazes and clouds, and the chemical structure in Saturn's and Titan's atmospheres, including the global surface temperatures on Titan, to search for the spectral lines of new molecules, and to map the composition and thermal characteristics of Saturn's rings and other icy satellites.

As the 'composite' part of the name suggests, CIRS is made up of two interferometers in a single housing, served by a single 50 cm diameter telescope and contained in an overall mass of 39.24 kg. The far-infrared (10 to $600 \, cm^{-1}$) spectrometer has a 4.3 mrad circular field of view (FOV), which covers 600 to $1100 \, cm^{-1}$ with a 1×10 array of 0.273 mrad square fields of view, and a shorter-wavelength spectrometer covers the rest of the mid-infrared (1100 to $1400 \, cm^{-1}$) with a similar array.

The nature of the Cassini orbit leads to particular problems when trying to scan the limb of Saturn. The closer approaches to Titan occur at about 1000 km, or about 40% of Titan's radius, whereas a typical closest approach to Saturn is at 300,000 km, 5 times the planet's radius. To achieve a vertical resolution comparable with the atmospheric variability requires a much smaller field of view for Saturn measurements. CIRS has a spatial resolution of 0.3 mrad for wavelengths less than $17 \, \mu m$, compared to the 4.3 mrad of IRIS. The smaller field of view, combined with the use of ten element detector arrays at these wavelengths, allows extensive limb

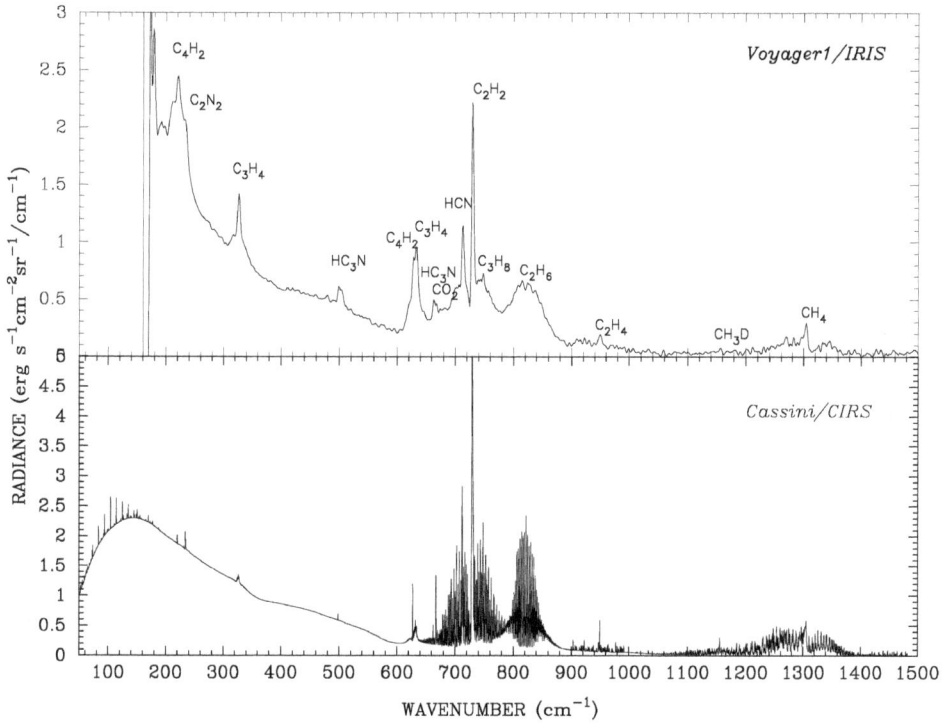

Figure 4.11   A composite CIRS spectrum taken by Cassini in 2005 compared to a spectrum taken by Voyager 1/IRIS in 1980. The difference in the two spectra resides both in the CIRS higher spectral resolution and in the larger spectral range.

observations of temperature and constituent abundances of both Titan and Saturn at vertical resolutions that are comparable with the vertical scales of processes within each atmosphere.

The combination of smaller field of view and higher spectral resolution is possible because of the use of cooled detectors. Team members at Oxford provided the radiative cooler and cold focal plane assembly incorporating mid-infrared detector supplied by collaborators in France and the US. When typical spectral measurements by CIRS are compared to IRIS data taken 25 years before, the CIRS spectrum provides both higher resolution and larger spectral coverage, with the inclusion of the $10\text{--}200\,\text{cm}^{-1}$ region. Regions of the spectrum where the noise (i.e. the noise equivalent spectral radiance) level is higher than the expected signal from Titan can be studied by increasing the duration of the observation, to give more signal.

In addition to improved sensitivity and vertical resolution, the long wavelengths accessible to CIRS allow sounding deeper into both atmospheres than was possible with IRIS. The high performance of CIRS enables several key objectives, including global mapping of the vertical distribution and temporal variation of gaseous species, temperature and clouds in both the tropospheric and stratospheric regions of Saturn and Titan, with latitude, longitude and height, as well as allowing for the possibility of discovering previously undetected chemical species in both atmospheres. New information on some species in Jupiter's atmosphere, particularly tropospheric NH_3 and PH_3 and stratospheric C_2H_2, and C_2H_6, were obtained during the Jupiter flyby in December 2000.

The scientific applications of the compositional data expected may be summarised under three broad headings. The first of these is evolution, or the attempt to infer from information on the relative elemental abundances and their isotopic ratios an improved understanding of the formation and evolution of the Saturnian system and the Solar System as a whole. Secondly, knowledge of the atmospheric composition and its vertical and horizontal variation is important for understanding the physical and chemical processes presently active on Saturn and Titan, in particular photochemical and radiative processes and those which give rise to the observed clouds and hazes. Finally, compositional data contributes to our understanding of the circulation and dynamics of the atmosphere, since transport between sources and sinks must occur and many species are useful tracers of the dynamical motions and timescales.

Ultraviolet Imaging Spectrograph (UVIS)

The UV Imaging Spectrograph has a set of detectors to measure the light reflected from atmospheres, surfaces and rings over wavelengths ranging from 55 to 190 nm, i.e. shorter than visible light, with a resolution of 0.2 to 0.5 nm. UVIS is actually made up of four instruments, a far ultraviolet (110 to 190 nm) spectrograph, an extreme ultraviolet (55.8 to 118 nm) spectrograph; a high speed photometer (115 to 185 nm) and a hydrogen-deuterium absorption cell operating at 121.5 nm wavelength. Standard stars are used to calibrate the spectra, and to obtain solar and stellar occultation profiles of the thermosphere as they set behind the planet.

The data that this instrument acquires are being used to investigate the composition, aerosol content, and temperature of the upper reaches of Titan's atmosphere. Species which can be detected in the UV include CH_4, C_2H_6, C_2H_2, H, H_2, N, N_2, Ar, CO, C_2N_2, and the D/H ratio can also be measured. These can be mapped to get the vertical/horizontal composition, the latitude-longitude variability, and any

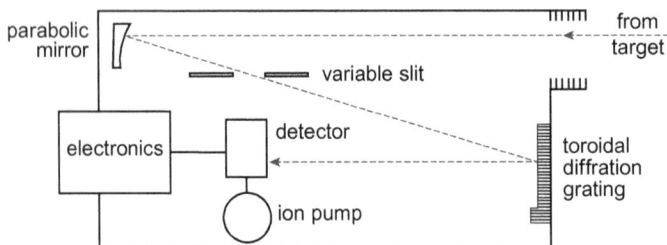

Figure 4.12   The Cassini Ultraviolet Imaging Spectrograph. PI: L. Esposito (NASA/JPL/University of Colorado).

interhemispheric differences, to elucidate the atmospheric chemistry occurring in Titan's atmosphere, as well as the distribution and scattering properties of aerosols, the nature and characteristics of the circulation, and the distribution of neutrals and ions.

UVIS also studies Saturn's atmosphere, again at the higher levels; the radial structure of the rings, and the surfaces and tenuous atmospheres of the icy satellites.

Visual and Infrared Mapping Spectrometer (VIMS)

The Visual and Infrared Mapping Spectrometer spans spectral windows between 0.6 and 5 μm, allowing the identification of surface materials with high ($\sim$500 m resolution) as well as resolved composition measurements. The VIMS instrument consists of a pair of imaging grating spectrometers designed to measure the reflected and emitted radiation from atmospheres, rings and surfaces over wavelengths from 0.35 to 5.1 μm, in two bands: 0.35 to 1.07 μm (called the visible subsystem, and having 96 spectral channels) and 0.85 to 5.1 μm (the infrared subsystem with 256 channels). All of these channels have 32 $\times$ 32 mrad fields-of-view.

VIMS is studying the composition and distribution of atmospheric and cloud species and the temperature structure on Titan and Saturn and on the icy satellites and rings. It helps to search for lightning and active volcanism on Titan, to observe the surface and map temporal behaviour of winds, eddies and other transient features on Saturn and to observe Titan's surface. Looking at Titan's nightside, the instrument can search for lightning and thermal emission from active cryovolcanoes.

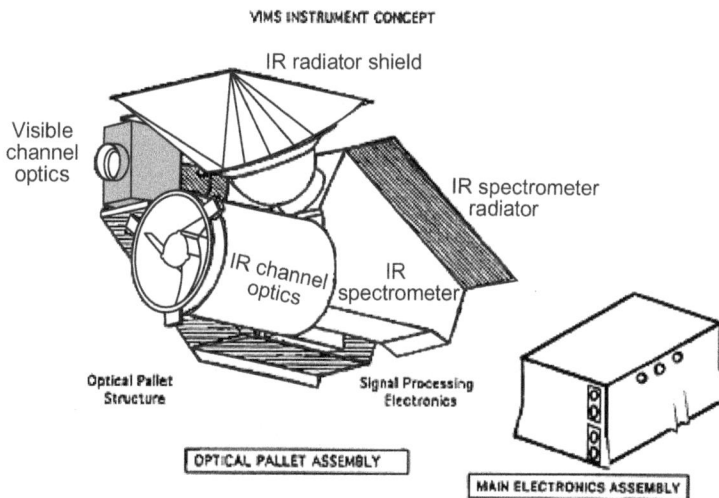

Figure 4.13   The Visible and Infrared Mapping Spectrometer. PI: R. Brown (NASA/JPL/University of Arizona).

Radar

The Cassini multimode radar, which completely penetrates the hazy atmosphere of Titan, operates as a radiometer (to measure surface temperature or emissivity), a scatterometer and altimeter (to measure the reflectivity and topography along the orbiter groundtrack) and has a synthetic aperture imager mode (SAR). It makes use of the five-beam feed assembly associated with the spacecraft high gain antenna to direct radar transmissions toward targets, and to capture blackbody radiation and reflected radar signals from targets.

The hardware, developed at the Jet Propulsion Laboratory, consists of a synthetic aperture radar imager (13.78 GHz Ku-band; 0.35 to 1.7 km resolution), an altimeter ((13.78 GHz Ku-band; 24 to 27 km horizontal, 90 to 150 m vertical resolution), and a radiometer (13.78 GHz passive Ku-band; 7 to 310 km resolution). This set of frequencies is used to investigate the geologic features and topography of the solid surface of Titan, and is providing evidence for whether extensive bodies of liquid exist on the surface or not. The radar uses the main data relay antenna for three functions: sending the radar pulse, receiving the echo, and transmitting the scientific data back to the Earth (of course, when the radar is operating, the data has to be stored for later transmission when the dish resumes its Earth-pointing mode).

The imaging mode, nearest closest approach, maps Titan's surface at 0.5 to 2 km resolution (which is not quite as good as the radar on Magellan achieved at Venus) over about 1% of Titan's surface for each flyby devoted to radar measurements. Thus perhaps 20% of Titan's surface will be imaged — in long thin strips — by the time the nominal mission ends. The imager on the orbiter carries filters at 940 nm, tuned to the windows between the methane bands, and so — like the HST — is able to measure surface contrasts. Additionally, polarizers are carried which remove most of the light scattered by the haze at near 90° phase angle, so these measurements also study the surface. The exact resolution achievable depends on the scene contrast and the haze optical depth at the time of the measurement, as well as the image motion compensation that can be achieved, but is likely to be better than 100 m. Other filters are able to probe different altitudes in the atmosphere.

Radio Science Subsystem (RSS)

Using a well-established technique developed on earlier planetary missions, the Cassini spacecraft's communication link is used to track the spacecraft, a task that can be done with remarkable precision. The signal is used to follow the bending of the radio beam as it passes through Titan's atmosphere; from this the refractivity of the medium can be deduced, and from that its molecular weight, temperature and pressure. The team is also engaged in the study of the solar corona, the radial structure and particle size distribution within Saturn's rings, the temperatures and electron densities with Saturn's and Titan's ionospheres and a search for gravitational waves coming from beyond the Solar System.

Cassini's capabilities for this sort of study are again much better than those available to Voyager. The transmitting and receiving frequencies available are S-band (around 2 GHz), X-band (around 8 GHz, and Ka-band (around 34 GHz), usable in non-coherent mode, where the frequency reference is the on-board ultrastable oscillator, or coherent mode where the Ka-band or X-band uplink is also used. The oscillator is stable to 1 part in 10^{13}; in coherent mode, 1 part in 10^{16} is achievable. Since the latter requires a two-way path to the Earth, it is not available during Titan occultations, hence the need for the oscillator.

A total of 15 Titan occultations are obtained during the course of the mission. With this data, it is possible to obtain accurate values for the lapse rate in the troposphere, thus further constraining the composition, in particular the argon abundance. The temperature-pressure field measurements allow zonal wind fields to be deduced, and yield temperatures and electron densities within Titan's ionosphere.

Because it allows very precise tracking of Cassini, the RSS also reveals how Titan affects the orbit of the spacecraft, and hence provides details of Titan's gravitational field, from which models for the shape of the solid body of Titan can be calculated. It will tend to be oblate (flattened) as a result of the effect of its rotation, as most large planetary bodies are; however, the tidal forces exerted by its large parent, Saturn, will tend to have the opposite effect, and induce a prolate shape. Exactly what the end result is will depend on the elasticity of the body, which in turn depends on factors such as the amount and distribution of volatile material.

Thus, we gradually obtain insights into questions such as whether Titan is in hydrostatic equilibrium, whether the interior is in fact differentiated, perhaps even whether there is a sub-surface ocean as some models predict. In conjunction with Earth-based Titan stellar observations, the results may also shed some light on the question of how Titan's orbit round Saturn comes to have such a large eccentricity, by permitting better estimates of the tidal dissipation which should be tending to circularise the orbit. If Titan were flexing a lot, as it would if it contains a lot of volatiles, then the most likely explanation would be the fairly recent impact of a large body. This could have left an observable record on the surface, which the radar can pick up as it covers the surface one region at a time.

Particle and Fields Experiments

This family of instruments is designed to investigate the magnetosphere of Saturn and its interaction with the solar wind, the rings, and the satellites, including of course Titan. This interaction is both a source and a sink for charged particles and molecules at Titan, and an understanding of the detailed processes involved is probably crucial for understanding the evolution of Titan's atmosphere on long time scales.

The magnetosphere of Saturn, discovered by Pioneer 11 and explored in more detail by the Voyager 1 and 2 spacecraft, has been shown to extend beyond the orbit of Titan at local noon, and to be extremely dynamic due to both external (solar wind) and internal drivers. The Cassini orbiter carries comprehensive instrumentation that has provided a comprehensive picture of the magnetic field, radio emissions, plasma, and energetic particle environments since orbit insertion on July 1, 2004. In the energetic particle and magnetic field data the latter increases close to the planet and then decreases with distance, as expected, but also exhibits what appear to be periodic variations. These correspond to plasma sheet encounters as marked by the vertical bars, and occur roughly at the planetary rotation period. The source of this periodicity, observed not only in particle and magnetic field intensities but also in planetary radio emissions is not fully understood at present.

The orbit of Titan is surrounded by a doughnut-shaped torus of charged hydrogen, nitrogen and other atoms that escaped from Titan. The Cassini payload is designed to characterise the ion flow around Saturn and Titan, measuring the composition of the torus, including heavy ions, and to investigate how they originate in the interaction between the planetary atmospheres and ionospheres with the magnetosphere. The principal loss process for particles leaving the torus may be particle precipitation into Titan's ionosphere; the instruments examine the processes responsible for the two-way traffic between the upper atmosphere and the torus.

The *Ion and Neutral Mass Spectrometer (INMS)* measures, by direct sampling, the distribution of positive ion and neutral species in the upper atmosphere of Titan, down to about 950 km, which is the closest Cassini is likely to approach to the surface of the satellite. These allow direct comparison with the density profile measured by the entry deceleration of the Huygens probe. INMS has recently allowed the detection of a rich mixture of hydrocarbons and nitriles which are found with mixing ratios that vary from 10^{-4} to 10^{-7}: acetylene, ethylene, ethane, benzene, toluene, cyanogen, propyne, propene, propane, and various nitriles.

The *Cassini Plasma Spectrometer (CAPS)* consists of an ion mass spectrometer, an ion beam spectrometer and an electron spectrometer. These measure the flux of ions as a function of mass per charge and the flux of ions and electrons as a function of energy per charge and angle of arrival at the instrument. Surprisingly, given the composition of the neutral atmosphere, CAPS has found that nitrogen ions are comparatively rare in Titan's magnetosphere, which instead is dominated by plasma composed almost entirely of ionized water and water products, including O^+, OH^+, H_2O^+ and H_3O^+, probably originating from the plumes on Enceladus.

The *Dual Technique Magnetometer (MAG)* consists of a combination of a vector/scalar helium magnetometer and a fluxgate magnetometer, used to determine the magnetic field of Saturn and hence its interactions with Titan and the icy satellites, the rings and the solar wind. Among other achievements, MAG provided the first hints on the water atmosphere around Enceladus.

The *Magnetospheric Imaging Instrument (MIMI)* maps the composition, charge state and energy distribution of energetic ions, electrons and fast neutral species in Saturn's magnetosphere. At Titan it monitors the composition and loss rate of particles from Titan's atmosphere due to ionisation and pickup, to determine the

MIMI: Magnetospheric Imaging Instrument

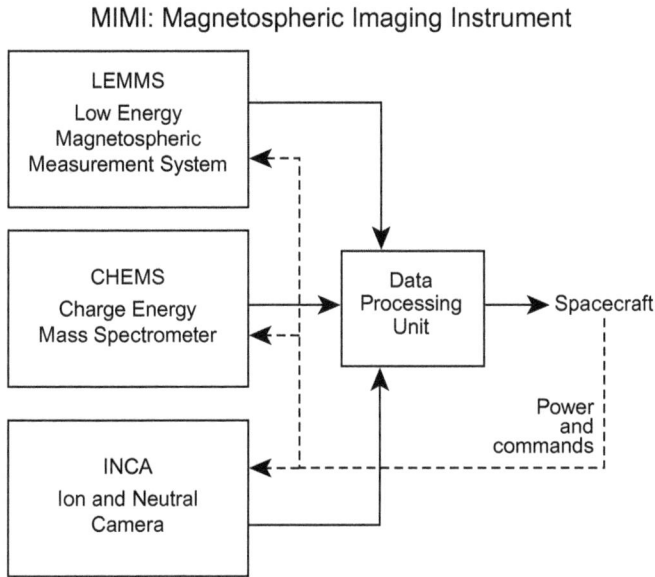

Figure 4.14   The Magnetospheric Imaging instrument. PI: T. Krimigis (NASA/JPL/APL).

importance of Titan's exosphere as a source for the atomic hydrogen torus in Saturn's magnetosphere.

The *Radio and Plasma Wave Science (RWPS)* instrument has electric and magnetic field sensors, a Langmuir probe, and high, medium and wide band receivers to measure the electric and magnetic fields and the electron density and temperature in the interplanetary medium, Saturn's magnetosphere, as well as Titan's induced magnetosphere and ionosphere, and the production, transport, and loss of plasma from Titan's upper atmosphere. It also searches for radio signals characteristic of lightning in Titan's atmosphere (although a similar search by Voyager failed to indicate any such emission).

Finally, Cassini carries an instrument to measure the number mass, velocity and composition of solid particles of dust in the environment around Saturn and Titan. The *Cosmic Dust Analyzer (CDA)* is investigating the physical, chemical, and dynamical properties of these particles, and their interactions with the rings, satellites, and magnetosphere of Saturn. This data tells us about the origin of oxygen compounds in Titan's reducing atmosphere, since they are probably due to photochemical reactions involving meteoric water.

4.8.2 The Scientific Instruments on the Probe

The Huygens probe carried six instruments through Titan's atmosphere down to the surface, to study the lower stratosphere and the conditions in the troposphere

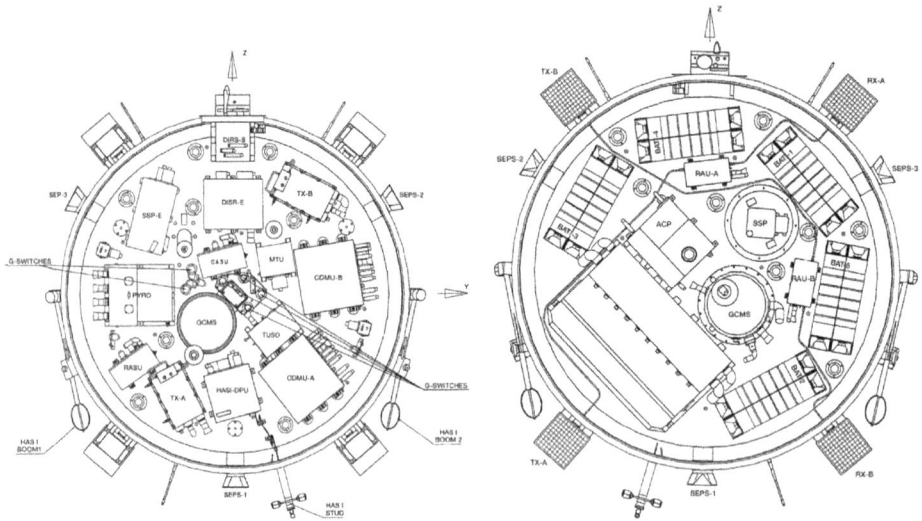

Figure 4.15   Scientific instruments on the Huygens Probe. Views of the payload accommodation on the top and bottom parts of the experiment platform (ESA).

during the descent and to send back information on the satellite's surface. The probe survived the landing and transmitted from the surface for a much longer time than expected because it touched down on a relatively soft solid surface, so the period of operation on the surface was limited by the uplink to the Cassini orbiter. Once that went over the horizon as seen from Huygens, the probe's useful life was over even though we know, from the continued reception of the signal by ground-based radio telescopes, that it continued to function for a further few hours. Data on the wind field on Titan were recovered by measuring the probe signal on Earth by radio telescopes.

Aerosol Collector Pyrolyser (ACP)

The Aerosol Collector Pyrolyser under the responsibility of Guy Israel (Service d'Aéronomie du CNRS) was designed to identify the composition of the aerosols in Titan's atmosphere by sampling them during the descent of the probe. Once inside the instrument, droplets were evaporated and thermally dissociated in an oven, and a chemical analysis of the products made with the gas chromatograph- mass spectrometer (GCMS) instrument which is also on board the probe. ACP sampled the aerosol from two atmospheric layers (150 to 45 km and 30 to 15 km from the surface), each followed by a 3-step pyrolysis at temperatures of 20 °C, 250 °C and 650 °C before injection of the products into the GCMS via a dedicated transfer line. The latter was performing its own analysis of Titan's gaseous atmosphere most of the time, as discussed below; the ACP experiment made use of its capability for approximately 20% of its operating life time.

Figure 4.16   The Aerosol Collector Pyrolyser instrument. PI: G. Israel (ESA).

Descent Imager/Spectral Radiometer (DISR)

The Descent Imager/Spectral Radiometer was built under the responsibility of M. Tomasko at the University of Arizona, with German and French collaborators. It measures solar radiation using silicon photodiodes, a two-dimensional silicon charge-coupled detector and two InGaAs near-infrared linear array detectors. Fibre optics collected light from upward and downward visible (480–960 nm) and infrared (0.87–1.7 μm) spectrometers with resolutions of 2.4 to 6.3 nm. For surface imaging, DISR used downward and side looking cameras with 0.06° to 0.20° fields-of-view. In the event, these were sensitive enough to photograph the surface, despite the fact that only 10% of the incident sunlight — already a hundred times weaker than on Earth — reaches the ground on Titan, without the help of the floodlight that was also carried as a precaution.

DISR measures at solar and near-infrared wavelengths in both the upward and downward direction to look at the scattering properties of the aerosols and to find out at which height levels most of the energy from the Sun is deposited in the atmosphere. The downward flux minus the upward flux gives the net flux, and the difference in the net flux at two altitudes gives the amount of solar energy absorbed by the intervening layer of atmosphere. This determines the rate at which the Sun is heating the atmosphere, as a function of height, which depends on the absorption properties of the gases and aerosols. The result is important for determining the temperature structure, and the gradients that drive atmospheric motions.

Measurements at 550 and 939 nm of small angle scattering in the solar aureole, the bright area around the Sun caused by forward scattering by suspended particles, of side and back scattering and polarisation, and of the extinction as a function of wavelength, all allow the optical properties of the haze particles (quantities such as optical depth, single scattering albedo, and the shape of the scattering phase function)

Figure 4.17   The Descent Imager/Spectral Radiometer, with the floodlight prominently featured. PI: M. Tomasko (ESA/NASA/JPL/University of Arizona).

DISR Imagers Approximate Fields-of-View

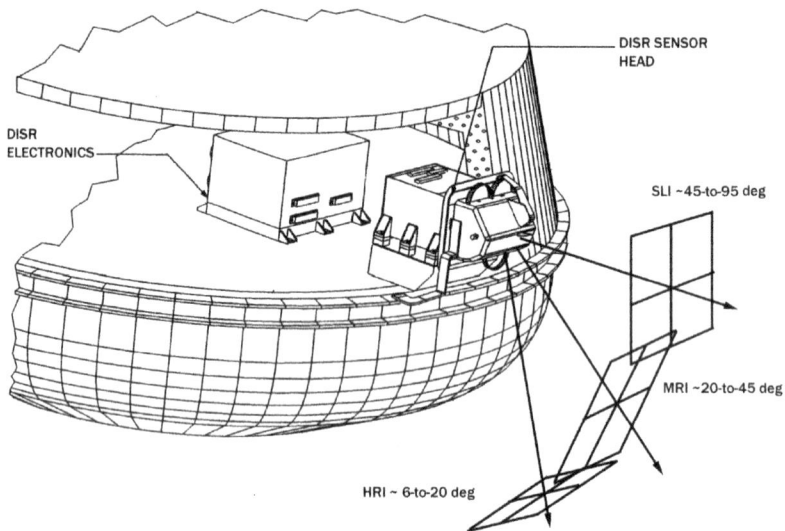

Figure 4.18   The Descent Imager/Spectral Radiometer, with the different directions of imaging (ESA/NASA/JPL/University of Arizona).

to be worked out. These properties together with determinations of size and shape can yield the imaginary part of the refractive index, and possibly constrain the real part also. Knowledge of the refractive index can be a strong, with luck, unique, clue to the composition of the particles. The DISR spectra of the downward streaming sunlight also showed the absorption bands of methane, allowing the determination of the profile of the mixing ratio of methane gas during the descent, analogous to a relative humidity profile on the Earth.

DISR also took spectacular pictures of the surface on the way down, observing in three different directions during its descent and on the ground to allow for the recovery of stereographic information. The many questions about the surface of Titan which are being addressed with the DISR pictures, including looking for channels, craters, lakes, glacial flows, frost and ice coverings, and active geysers and volcanoes, are addressed in later chapters. The descent imager pictures started at a height of 160 km, when the spatial resolution was about 1 km. They gave many clues about the topography of the surface, the reflection spectra of surface features, the composition of the different types of terrain observed, and the interactions of the surface and the atmosphere. They also provided the horizontal wind direction and speed calculated from images of the surface obtained every few kilometres in altitude from the drift of the probe over the surface of Titan as it descended.

Doppler Wind Experiment (DWE)

The primary scientific goal of the Doppler Wind Experiment led by M. Bird of the Univ. of Bonn in Germany, with Italian and American involvement, was to measure the direction and strength of the winds in the atmosphere of Titan, through Doppler tracking of the probe from the orbiter. The DWE used ultra stable oscillators within the radio transmitter on the probe and in the receiver on the orbiter to improve the stability of the relay link to make accurate Doppler measurements, the first time these devices have been used in deep space. They are compact, atomic resonance frequency-controlled oscillators whose output signal is obtained from a 10 MHz voltage controlled crystal oscillator, which in turn is frequency-locked to the atomic resonance frequency (6.834 GHz) of the ground-state hyperfine transition of the element ^{87}Rb.

The DWE data were lost due to the problem with Channel A in the communications link between Huygens and Cassini. Thankfully, the Huygens signal was captured and monitored from several large radio telescopes on Earth, and the experiment was saved. The Earth-based antennas that received the signal were the NRAO Robert C. Byrd Green Bank Telescope in West Virginia, USA, and the CSIRO Parkes Radio Telescope in Australia. From the detection of the incredibly weak Huygens radio signal, winds on Titan were deduced from the surface to a height of 160 km. The winds are weak near the surface and increase slowly with altitude up to about 60 km, and are from west to east at nearly all altitudes. Above 60 km large fluctuations sin the Doppler signal were observed, indicating a turbulent atmosphere with large values of wind shear. The science and engineering data recorded on board Huygens confirmed that the probe was buffeted by the winds in this region. A maximum wind speed of 120 m s^{-1} (430 km hr^{-1}) was found at an altitude of about

120 km. DWE also monitored the spin of the probe, and the amplitude of its swing under the parachute as it descended.

Gas Chromatograph/Mass Spectrometer (GCMS)

The Gas Chromatograph/Mass Spectrometer was developed by Hasso B. Niemann and his team from NASA Goddard Space Flight Center in Greenbelt, Maryland to measure the composition of Titan's atmosphere. The gaseous atmosphere was sampled directly during the descent, while the aerosol droplets were first vaporised by the pyrolysis experiment. The GCMS also had a heated inlet, so that the volatile component of the surface material at the landing site could be determined.

As Titan's atmosphere has so many components, these were separated in two dimensions by chromatography as well as mass spectroscopy. The instrument could handle an atomic mass range of 2 to 146 amu with a sensitivity of one part in 10^{12} and a mass resolution of about 10^{-6} at 60 amu. The gas chromatograph used H_2 as the carrier gas in 3 parallel columns. Its function is to increase the measurement capability of the instrument by time separation of species with different chemical properties for detection and identification and analyses by the mass spectrometer. The mass spectrometer system works by ionising the incoming gas using ion sources, then making species concentration measurements by passing them through a magnetic field which separates them by mass, and focuses them onto an ion detector.

Figure 4.19 The Gas Chromatograph/Mass Spectrometer instrument on the Huygens probe. PI: H. Niemann (ESA/GSFC).

The three gas chromatograph columns were optimised for individual objectives. The first was devoted to discriminating N_2 and CO, which have the same atomic mass and which therefore cannot be distinguished by mass spectroscopy alone. The second and third focussed on hydrocarbons and nitriles, and heavy hydrocarbons, respectively. The GCMS measured the major isotopes of carbon, nitrogen, hydrogen, oxygen and argon, and detected neon and the other noble gases to levels of 10–100 ppb. The GCMS collected data from an altitude of 146 km to ground impact. The Probe and the GCMS survived impact and collected data for 1 hour and 9 minutes on the surface. The major constituents of the lower atmosphere were confirmed to be N_2 and CH_4, and the vertical profile of the latter measured.

Huygens Atmospheric Structure Instrument (HASI)

The entry deceleration was measured by the Huygens Atmospheric Structure Instrument, built by the Italian Space Agency for a team led by M. Fulchignoni (LESIA, Paris Observatory). HASI was the only probe instrument to operate prior to parachute deployment on the probe. The deceleration is proportional to density, and from the density profile and the assumption of hydrostatic equilibrium, a temperature profile of the upper atmosphere was derived. HASI carried on making direct measurements throughout the atmosphere, obtaining atmospheric temperature, pressure, density and conductivity profiles from 170 km down to the surface as the probe descended by parachute. In addition to determining the atmospheric structure, these data help to identify the composition of the condensates in the regions where hazes or clouds form.

HASI also included a Permittivity, Wave, and Altimetry Experiment (PWA) which measured the electrical properties of the atmosphere (important in determining haze charging and coagulation physics) and searched for thunder and lightning.

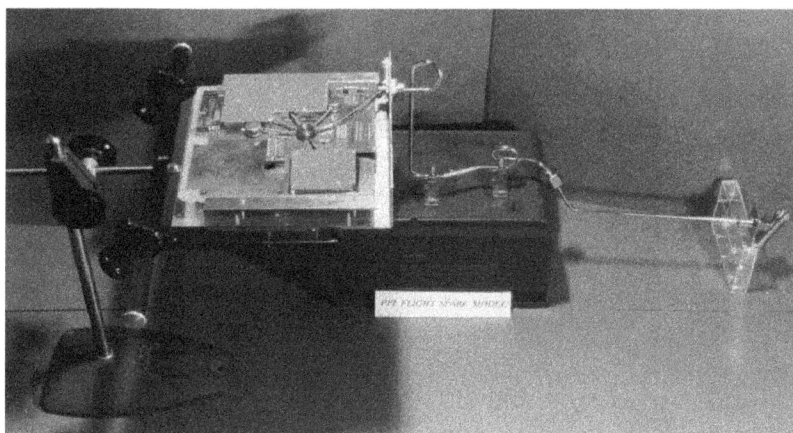

Figure 4.20 The pressure profiling sensor, part of the Huygens Atmospheric Structure Instrument. PI: M. Fulchignoni (ESA/Paris Observatory).

The probe also carried a radar altimeter, which estimated radar reflectivity and sur-face topography. The radar altimeter, part of the probe system itself, passed its signal to the PWA for science data processing.

On the surface, the conductivity was measured over a dynamic range from 10^{-15} to infinity, the relative permittivity in the range 1 to infinity, and acoustic density measurements in the 0–5 kHz range, with a sensitivity of 90 dB at 5 mPa. These data characterise the dielectric properties, conductivity and permittivity of the surface material.

Surface Science Package (SSP)

The material making up the surface of Titan at the landing site was directly inves-tigated by the Surface Science Package, which was developed at the University of Kent at Canterbury, England, by J. Zarnecki. SSP was equipped to measure the properties of a liquid surface, which was considered probable at the time the experiment was designed, although we now know that Huygens landed on a soft, solid surface like sand or snow. During descent, tilt sensors measured the probe's attitude, and an acoustic sounder measured the speed of sound in the atmo-sphere, which depends on temperature and molecular mass. Prior to impact, the sounder also probed the surface roughness at the landing site. During impact, SSP's accelerometer and penetrometer measured the mechanical properties of the surface material, from which particle size and stickiness can be estimated. The full complement of sensors used by the Surface Science Package is listed in Table 4.3.

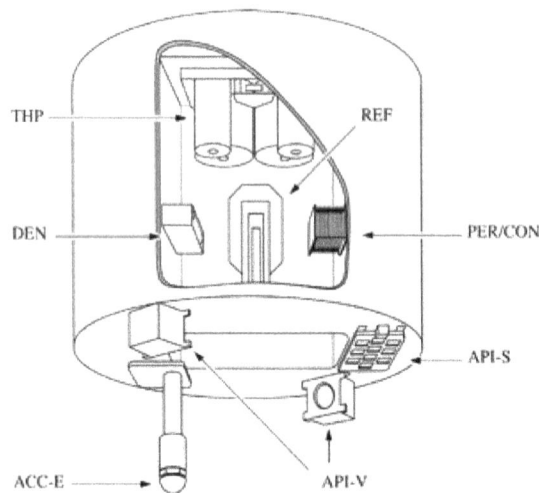

Figure 4.21   The Surface Science Package — see Table 4.3 for the meaning of the acronyms. PI: J. Zarnecki (ESA/Open University).

Table 4.3   The individual sensors that make up the Huygens Surface Science Package.

Acronym	Function	Description
ACC	Impact accelerometer	Piezoelectric type
TIL	Tilt sensor (X and Y axes)	Electrolytic type
THP	Thermal properties	Hot wire
API-V	Velocity of sound	Piezoelectric transducers
API-S	Acoustic sounder	Piezoelectric transmitter/receiver
PER	Fluid permittivity	Capacitance sensor
DEN	Density of fluid	Archimedes sensor
REF	Refractive index	Critical angle refractometer

4.9 Touring the Saturnian System

Once communication with Huygens was lost, the orbiter continued on its tour of the Saturnian system, returning to make additional close flybys to acquire more remote sensing data on Titan and its atmosphere. The geometry of each flyby is calculated to give the best compromise between Titan observations and getting the right gravity assist to send the spacecraft on the path to its next target. Huygens, of course, has been silent on these subsequent Titan encounters. In a typical timeline for one of the Titan encounters, remote sensing starts 12 hours before, and ends 4 hours after, encounter, with a concentrated burst of other measurements when the spacecraft is at its closest approach.

Besides Titan, flybys of selected icy satellites are used to determine their surface compositions and geologic histories, and some orbits are aligned in an anti-solar direction to permit studies of the magnetic tail (Table 4.4). Near the end of the four-year tour, the orbital inclination will be increased to approximately 85°, passing over the polar regions of Saturn. Then it can investigate the field, particle, and wave environment at high latitudes, and try to identify the source of very long radio waves (kilometric radiation) from the planet, a phenomenon unique to Saturn. High inclinations also permit high-latitude Saturn radio occultations, viewing of Saturn polar regions, and more nearly vertical viewing of Saturn's rings.

The nominal mission was scheduled to end on July 1, 2008, for a total mission duration since launch of 10.7 years, but it is to be extended since enough propellant is left to support the function of the orbiter's systems and science instruments until at least 2010.

4.9.1 Observations of Saturn

Measurements by Cassini have been essential for improving our understanding of the giant planet, which in turn bears important consequences in the studies of comparative planetology, key to our comprehension of the origin and evolution of the Solar System and, by extension, of the exoplanetary systems. Among the various

Table 4.4   Cassini flybys of Titan and the icy satellites in the nominal mission.

Orbit	Satellite	Flyby date	Altitude at closest approach
0	Phoebe	June 11, 2004	1,997 km (1,241 mi)
A	Titan	October 26, 2004	1,200 km (746 mi)
B	Titan	December 13, 2004	2,358 km (1,465 mi)
B	Probe Release	December 24, 2004	n/a km (n/a mi)
C	Iapetus	January 1, 2005	65,000 km (40,398 mi)
C	Titan	January 14, 2005	60,000 km (37,290 mi)
3	Titan	February 15, 2005	950 km (590 mi)
3	Enceladus	February 17, 2005	1,179 km (733 mi)
4	Enceladus	March 9, 2005	500 km (311 mi)
5	Titan	March 31, 2005	2,523 km (1,568 mi)
6	Titan	April 16, 2005	950 km (590 mi)
11	Enceladus	July 14, 2005	1,000 km (622 mi)
12	Mimas	August 2, 2005	45,100 km (28,030 mi)
13	Titan	August 22, 2005	4,015 km (2,495 mi)
14	Titan	September 7, 2005	950 km (590 mi)
15	Tethys	September 24, 2005	33,000 km (20,510 mi)
15	Hyperion	September 26, 2005	990 km (615 mi)
16	Dione	October 11, 2005	500 km (311 mi)
17	Titan	October 28, 2005	1,446 km (899 mi)
18	Rhea	November 26, 2005	500 km (311 mi)
19	Titan	December 26, 2005	10,429 km (6,482 mi)
20	Titan	January 15, 2006	2,042 km (1,269 mi)
21	Titan	February 27, 2006	1,812 km (1,126 mi)
22	Titan	March 18, 2006	1,947 km (1,210 mi)
23	Titan	April 30, 2006	1,853 km (1,152 mi)
24	Titan	May 20, 2006	1,879 km (1,168 mi)
25	Titan	July 2, 2006	1,911 km (1,188 mi)
26	Titan	July 22, 2006	950 km (590 mi)
28	Titan	September 7, 2006	950 km (590 mi)
29	Titan	September 23, 2006	950 km (590 mi)
30	Titan	October 9, 2006	950 km (590 mi)
31	Titan	October 25, 2006	950 km (590 mi)
35	Titan	December 12, 2006	950 km (590 mi)
36	Titan	December 28, 2006	1,500 km (932 mi)
37	Titan	January 13, 2007	950 km (590 mi)
38	Titan	January 29, 2007	2,776 km (1,725 mi)
39	Titan	February 22, 2007	953 km (592 mi)
40	Titan	March 10, 2007	956 km (594 mi)
41	Titan	March 26, 2007	953 km (592 mi)
42	Titan	April 10, 2007	951 km (591 mi)
43	Titan	April 26, 2007	951 km (591 mi)
44	Titan	May 12, 2007	950 km (590 mi)
45	Titan	May 28, 2007	2,425 km (1,507 mi)
46	Titan	June 13, 2007	950 km (590 mi)
47	Tethys	June 27, 2007	16,200 km (10,068 mi)

(Continued)

Table 4.4 (*Continued*)

Orbit	Satellite	Flyby date	Altitude at closest approach
47	Titan	June 29, 2007	1,942 km (1,207 mi)
48	Titan	July 19, 2007	1,302 km (809 mi)
49	Rhea	August 30, 2007	5,100 km (3,170 mi)
49	Titan	August 31, 2007	3,227 km (2,006 mi)
49	Iapetus	September 10, 2007	1,000 km (622 mi)
50	Titan	October 2, 2007	950 km (590 mi)
52	Titan	November 19, 2007	950 km (590 mi)
53	Titan	December 5, 2007	1,300 km (808 mi)
54	Titan	December 20, 2007	953 km (592 mi)
55	Titan	January 5, 2008	949 km (590 mi)
59	Titan	February 22, 2008	959 km (596 mi)
61	Enceladus	March 12, 2008	995 km (618 mi)
62	Titan	March 25, 2008	950 km (590 mi)
67	Titan	May 12, 2008	950 km (590 mi)
69	Titan	May 28, 2008	1,316 km (818 mi)

measurements that can contribute to constraining planetary formation models, the definition of the chemical composition of the giant planets and in particular the content in heavy elements in Saturn, is most critical. The budgetary restrictions that eliminated the Saturn probe that was initially part of the Cassini–Huygens mission, prevented *in situ* measurements similar to those performed by the Galileo probe in Jupiter in 2005, and thus the determination of the elemental composition in the well-mixed region in Saturn.

Some of the most recent equilibrium thermodynamics models developed for the giant planets predict a region where water is well-mixed on Saturn below 20 bars, for the nominal case where the condensable volatile abundances found on the planet are 10 times solar. However, because Saturn, like Jupiter, is a convective and turbulent planet, this level may, in fact, lie well below the 20 bar level, at higher pressures. Storms on Saturn have been observed by the Cassini ISS in the visible and by Cassini VIMS at $5\,\mu$m, locating them near the 6–8 bar pressure level, and consequently pushing estimates for the well-mixed regions for water to as low as 50–100 bars, where the temperatures are in the order of 400–500 K.

4.9.2 *The Icy Satellites, and Saturn's Rings*

It is of interest to look also at the lists of objectives for Saturn's rings, and for the Saturnian satellites other than Titan, usually referred to as the 'icy' satellites because they show icy surfaces to our telescopes and cameras (although there is plenty of ice on Titan as well, of course, hidden below the haze). As well as the indirect bearing that these have on understanding Titan, which formed as one of a family of satellites, and in an environment that also produced the rings, it has to be born in mind that

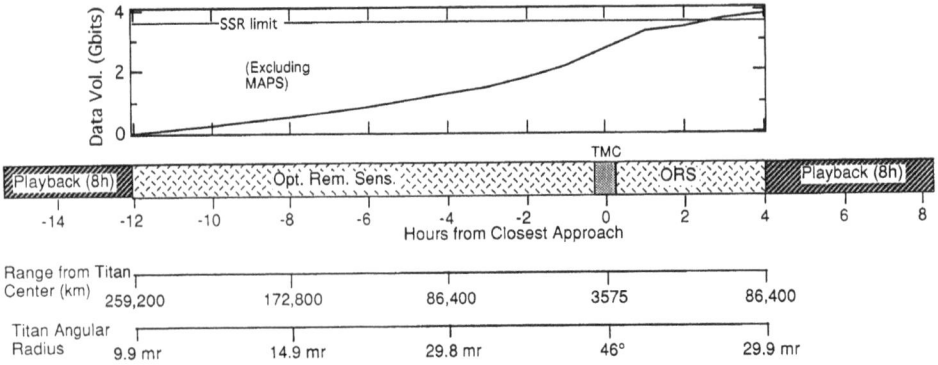

Figure 4.22   A typical Titan encounter sequence. The bar in the middle shows the time from closest approach to Titan, when the spacecraft is just 800 km above the cloud tops and the satellite's diameter subtends more than 90°, filling half the sky. The panel at the top of the figure shows how the data accumulates with time in the spacecraft memory, before the playback period starts four hours after closest approach. The bars at the bottom show the distance of Cassini from the centre of Titan, and the angular size of the satellite as seen from the spacecraft (NASA/JPL).

Figure 4.23   Saturn's rings photographed edge-on by Cassini, showing how thin they are. The bright spot is the moon Enceladus (NASA/JPL).

Titan is not the only objective of Cassini/Huygens. Some of the instrumentation was selected with these other objectives in mind, and even a big spacecraft can carry only a limited number of experiments.

High-resolution imaging can determine the thickness of Saturn's rings and gather information about the sizes, composition, and physical nature of the individual particles. With this data it is possible to determine the rate and nature of energy and momentum transfer within the rings, and study gravitational interactions between

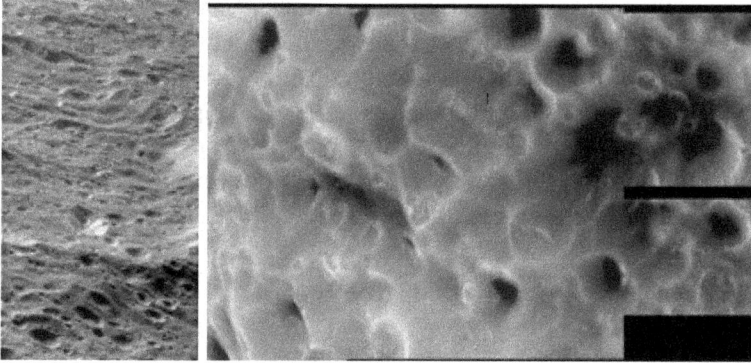

Figure 4.24 The surfaces of Rhea, left, and Hyperion, right, illustrating the extraordinary detail of the topography being captured by Cassini's cameras, exceeding anything achieved previously (NASA/JPL/SSI).

the rings and Saturn's satellites. Other issues being worked out now include:

- Studying the configuration of the rings and dynamical processes (gravitational, viscous, erosional, and electromagnetic) responsible for ring structure.
- Mapping the composition and size distribution of the ring material.
- Investigating the interrelation of rings and satellites, including imbedded satellites.
- Determining the dust and meteoroid distribution in the vicinity of the rings.
- Studying interactions between the rings and Saturn's magnetosphere, ionosphere, and atmosphere.

A similar set of goals is being worked on for the airless satellites (some of which, we now know, do in fact have very tenuous atmospheres). These include:

- Determining the general characteristics and geological histories of the satellites, from their cratering record for instance.
- Defining the mechanisms of crustal and surface modifications, both external and internal, particularly for major features like cliffs and rifts.
- Investigating the compositions and distributions of surface materials, particularly dark, organic rich materials and low melting point condensed volatiles, primarily by spectroscopy (which is much easier for the ices than it is for the complex, tarry organics; a complete understanding of the latter needs the analysis of samples in a laboratory, but some defining characteristics can be divined from spectra).
- Constraining models of the satellites' bulk compositions and internal structures, from their densities and the effects of their gravitational fields on the spacecraft as it passes close by, and also from their shape, where that departs from spherical.
- Studying the interactions of the satellites with the magnetosphere and ring systems of Saturn, including material injected into the magnetosphere from the satellites, as for example by the plumes on Enceladus.

Phoebe was the first satellite to be encountered by Cassini, from 2000 km away, a rather unsatisfactorily large distance that was dictated by the requirements for the rest of the trajectory. Among the most impressive observations by Cassini are the images of Hyperion with its heavily cratered surface and Iapetus with its unexplained dichotomy. Also, the first detection of water geysers on Enceladus, give a new dimension to our understanding of the "habitable zones" and in any event push back our beliefs of where liquid water could be found in the Solar System.

4.9.3 *Saturn's Magnetosphere and Titan*

Like Jupiter, Saturn has a strong internal magnetic field and hence an extensive magnetosphere. This contains neutral gas, some of which can become ionized and contribute to the charged particle populations of the radiation belts. The mechanism is probably sputtering by charged particle and meteoroid impact on the material in the rings and on the surfaces of Saturn's icy moons. This is believed to create an extensive neutral cloud of water molecules and water dissociation products, including ionized species, around the planet in the inner magnetosphere.

Titan and Saturn's magnetosphere interact: Titan's substantial atmosphere and ionosphere act as both a source and a sink of neutrals and ions to Saturn's outer magnetosphere. The primary processes involve energetic ions that enter Titan's atmosphere and sputter off neutral atoms or molecules. Certain photochemical reactions can achieve the same effect, if they energize the neutrals beyond the escape speed of Titan. These ejected neutrals will end up orbiting Saturn, producing a torus composed mainly of hydrogen and nitrogen. These neutral atoms and molecules can become ionized and picked up by the local magnetic field, producing a plasma torus around Saturn. Alternatively, neutrals in Titan's atmosphere can be ionized by impact with Saturn's magnetospheric electrons or by solar extreme ultraviolet photons. Ions created at lower altitudes may flow out of the atmosphere, down the wake, while ions created at higher altitudes are picked up by the external plasma and magnetic field, leading to a Titan plasma plume. This wraps all the way around Saturn, so as Titan orbits it can encounter the plasma again, leading to some complex interactions.

As is evident form the data taken by MIMI, energetic particle intensities in the vicinity of Titan's orbit are highly variable and fluctuate in response to changes in solar wind pressure and internal magnetospheric activity. In fact, Titan often finds itself outside the magnetopause near local noon, and exposed to the heated solar wind in the magnetosheath. This variability in energetic particle input has important consequences on Titan's ionosphere and upper atmosphere. The energetic neutral atoms around Titan that originate as fast ions trapped in the magnetic field can charge-exchange with the upper atmosphere. The maximum intensity occurs at the relatively high altitude of ~3000 km, surrounding the entire moon.

The Cassini teams are working to clarify this picture, with the overall goal being to understand what role these interactions have had, and may have in the future, for the evolution of Titan's atmosphere. The first task is to measure the configuration of the planetary magnetic field, globally and in the vicinity of Titan, and determine the composition and the sources and sinks of the various types of magnetosphere particles. These will eventually form the basis for improved models of the interactions of Titan's atmosphere and exosphere with the solar wind and Saturn's magnetospheric plasma.

4.10 Being Involved: Scientists and Instrument Providers

Missions like Cassini do not just happen. They are conceived and brought into being, not so much by the space agencies who manage them but by visionary people in the world scientific community. Acting through their institutions, learned societies and through committees of various scientific and research organisations, it is generally a committed few who take the initial steps to define and lobby for a project, often in the form of a very rudimentary concept, which may not be the same as that which finally evolves. We have already seen how the 'SOPP' version of Cassini, meaning a Saturn orbiter carrying two Probes, one for Saturn and one for Titan, was scaled down to save money.

Mission proposals gather momentum by attracting funds for studies, which define the scientific objectives and the details of what will be required to accomplish them. Usually, the anticipated cost, compared as objectively as possible with the new knowledge and other gains anticipated, is the crucial thing that emerges as the 'bottom line'. If a mission concept attracts enough support from the community, as sampled and evaluated by the agency's committees (themselves usually staffed by working scientists from universities and research centres), it can then get approved as a flight project and the requisite funds become part of the agency's budget. Cassini required this process to be undergone in the USA and in Europe in tandem.

Once approved, the work to build the spacecraft, acquire the launcher and all of the other mired requirements of such a complicated venture, are set in motion. Many of them require contracts with large industrial companies. Plans to track and control the spacecraft are put in place and time reserved on the giant antennae of the Deep Space Network, the only ones large enough to pick up the high-speed data link from Saturn. By the time Cassini becomes a memory, around the year 2020 or so, many thousands of people will have contributed to it in a variety of ways.

The really exciting role to play, for a scientist, is that of a scientific investigator. This means you have been selected by the space agency, advised by peer review groups, to carry out a key scientific role in the mission (and, incidentally, for a long mission like Cassini, have been given a job for life!). There are three main kinds of investigators: Principal Investigator (abbreviated to 'PI'), Co-Investigator, or Co-I, and Interdisciplinary Scientist, or IDS. These people take the lead in planning, guiding, analysing and reporting the scientific aspects of the mission, which of course underlie everything else. How does one get such an exciting and rewarding job on a mission, an adventure, like Cassini?

The answer is that you write a proposal, having first made enough of a reputation that the proposal will be taken seriously. The space agency requests proposals when it announces its plans for the mission, and makes arrangements to have them reviewed and evaluated. Principal Investigators lead teams which provide instruments, so their proposals have to show what their instrument would measure, what technique they would use, and how they would get it all done (and at what price!). Co-Investigators write part of the same proposal, showing how they would bring specific expertise to the design or building of the instrument, or perhaps to some aspect of planning the observations or analysing the data.

Table 4.5   The Interdisciplinary Scientists (IDSs) on Cassini–Huygens.

Interdisciplinary scientist	Tasks
Dr. Michel Blanc, Observatoire de Midi-Pyrenees, France	Transport of mass, linear and angular momentum, and energy in the magnetosphere/ionosphere/thermosphere system of Saturn
Dr. Jeffrey N. Cuzzi, NASA Ames Research Center, USA	Chemical composition, structure, particle sizes, origin, and evolution of Saturn's rings
Dr. Tamas I. Gombosi, University of Michigan, USA	Molecular composition, structure, and dynamical behaviour of the plasma within Saturn's magnetosphere
Dr. Tobias C. Owen, University of Hawaii, USA	Abundances and isotopic ratios of atmospheric constituents on Titan and Saturn, and their implications for their evolutionary paths
Dr. Laurence A. Soderblom, United States Geological Survey Flagstaff, USA	Geologic processes, evolutionary history, composition, and physical state of the surfaces of Saturn's icy satellites
Dr. Daniel Gautier, Observatoire de Paris-Meudon, France	Titan's atmospheric aerosols, photochemistry, general circulation, and upper atmosphere
Dr. Jonathan I. Lunine, University of Arizona, USA	Titan's atmospheric and surface evolution and the stability of the present atmosphere
Prof. François Raulin, LISA, Univ. Paris 12, France	Organic chemistry in Titan's environment and its implications for exobiology and the origins of life

IDSs usually write their own proposals, explaining how they have special knowledge or skills that would blend in with and enhance the efforts of the experiment teams. This might take the form of an advanced computer model of the dynamics of Titan's atmosphere, for example, which could be used to analyse the data from many instruments. Eight 'interdisciplinary' scientific investigations are part of the Cassini/Huygens mission (5 on Cassini and 3 on Huygens). In these, named scientists lead teams to look at specific problems using a combination of models and data from as many of the different instruments as apply.

Finally, the Cassini Project Scientist, D. Matson of the Jet Propulsion Laboratory and the Huygens Project Scientist, J.-P. Lebreton of the European Space Science and Technology Centre, are the overall spokespersons for the orbiter and probe parts of the mission respectively.

Once selected, the teams meet regularly and plan and supervise the development of their experiment. They also contribute members to teams that develop particular aspects of the mission, such as the observing sequences. This can be painful at times, especially for the PI whose instrument falls behind schedule, or goes over budget, with launch only a year or two away. The 'window' from Earth to Saturn is only open for a few weeks, when the two planets are aligned correctly, so the launch cannot wait (especially one with such a complicated trajectory as Cassini, with no less than four planets needed in the right place at the right time!). Teams then work all night, relax some of the requirements on their hardware, or any of a dozen other options to get back on track. It is a rare experiment, or a simple one, which does not have to make compromises at some stage, usually to save money when it runs short.

4.11 Reaping the Benefits

Even before launch, the Cassini project produced valuable technology, management tools, and knowledge, resulting in new applications for aerospace, business, communications, computing, and education. Many of Cassini's key technological innovations are now in use by other low-cost, high-efficiency space mission programmes.

As examples one can cite the use of solid-state data recorders with no moving parts, new electronic chips on the orbiter computer which directs the operations, a new integrated circuit on the computer system, an innovative solid-state power switch, a new X-band radio transponder, and a new hemispherical resonator gyroscope. The Cassini project also develops curricula tools and classroom supplements to enhance science teaching in schools. A European Network, called EuroPlanet (http://europlanet.cesr.fr/), has been activated recently and focuses on the promotion of planetary exploration from missions like Cassini. The presence of a Cassini signature disc, comprising more than 616,400 handwritten signatures from 81 countries all over the world, marks the interest of the public for this enterprise and is reminiscent of the souvenir that Voyager was carrying in the 1970s.

Cassini and Huygens have already provided a wealth of data. The analysis is still far from over, as we write, and the mission promises to unveil yet more of Titan's secrets in the months and years to come. The following chapters provide an account of the latest information we have on Titan's environment, from all available means of investigation, shortly before the end of Cassini's nominal mission.

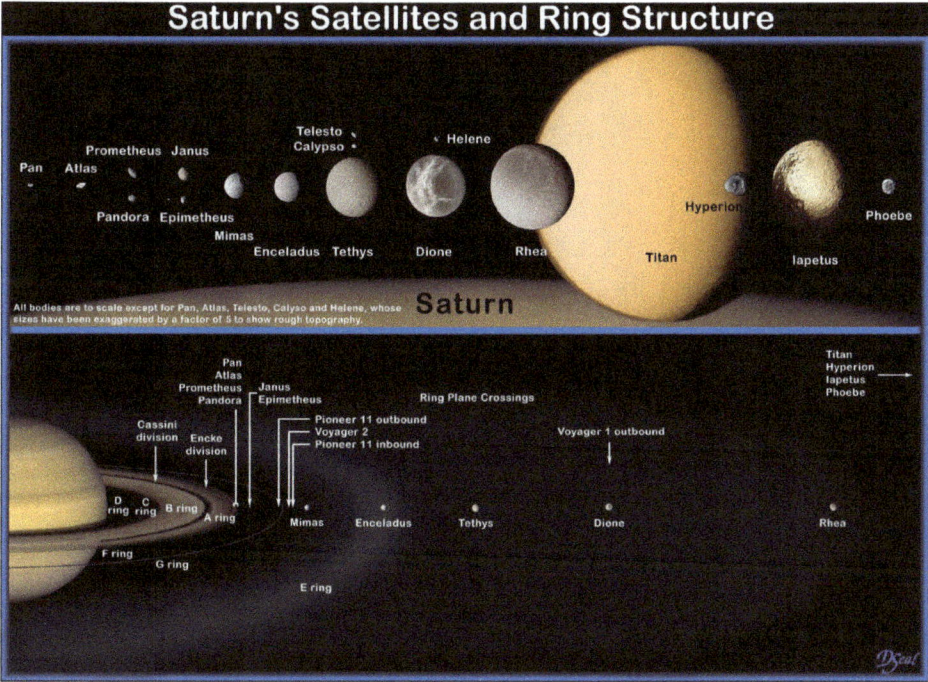

Plate 1   Saturn's satellite and ring system (NASA/JPL).

Plate 2   Two views of Titan's disk. Left: taken by the Voyager 2 cameras in Aug. 1981 and showing Titan's North to South asymmetry, with the southern hemisphere appearing lighter and a well-defined band near the pole. The extended haze is also clearly visible around the satellite's limb, but no glimpses of the surface were possible. Right: 23 years later, Titan's complex surface is at last revealed by Cassini/VIMS, by imaging at a longer wavelength in the near infrared (NASA/JPL).

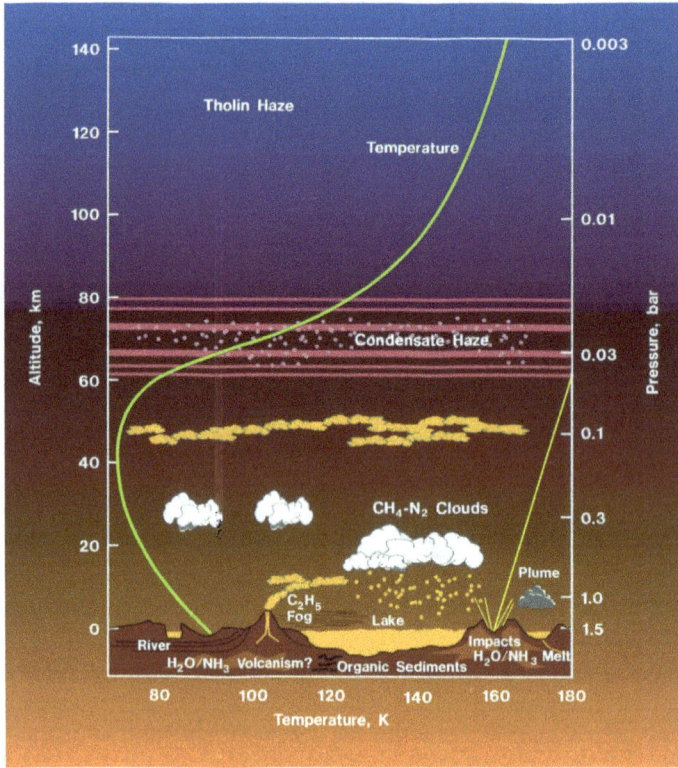

Plate 3   A schematic representation of the atmospheric structure and surface interactions on Titan.

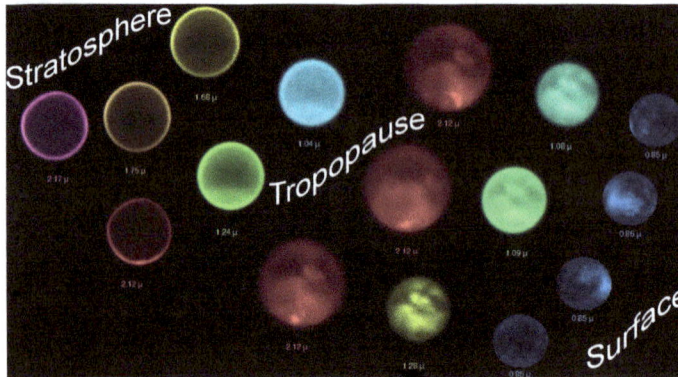

Plate 4   Views of Titan at various wavelengths, from Earth-based telescopes, in a montage by M. Hirtzig. From upper-left to bottom-right, the images are sorted in descending order of the altitude probed. The diameter of the image reflects the size of the telescope used: 10 m at the Keck (large), 8.2 m at the VLT (medium), and 2.4 m for the HST (small). The colour code corresponds to the wavelength, from blue (near IR around 1 μm) to red (longward of 2 μm). Several atmospheric features can be detected, including a bright northern limb in the upper stratosphere due to a local excess of aerosols, a bright polar collar in the lower stratosphere, and clouds above the south pole in the troposphere. Surface features are also clearly visible in the right-hand images, with high contrast and up to 300 km per pixel spatial resolution, while the four HST images show the rotation of Titan's surface, made obvious by the movement of the bright continent Xanadu.

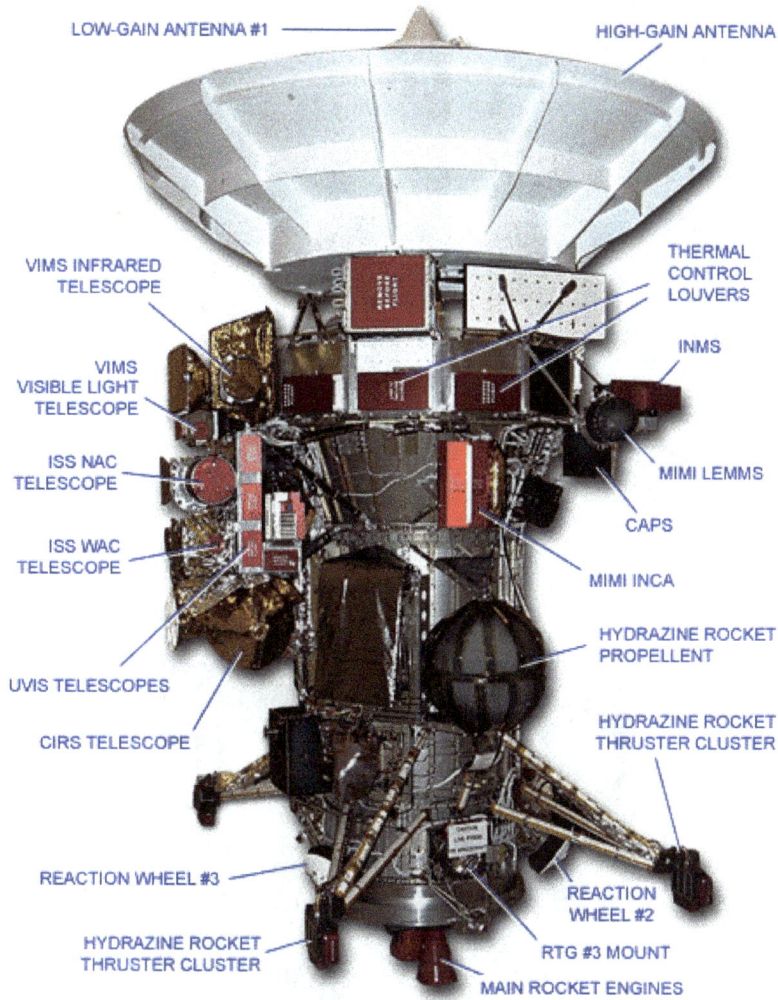

Plate 5   The Cassini spacecraft, showing the instruments and subsystems attached to the main structure (NASA/JPL).

Plate 6   Close-ups of the Huygens probe configuration (left) and the technical and mechanical structure of the combined Cassini–Huygens spacecraft (right) (ESA/NASA/JPL).

Plate 7   The launch of the Cassini–Huygens mission from Cape Canaveral on October 15, 1997, on
an appropriately named Titan rocket (NASA/JPL).

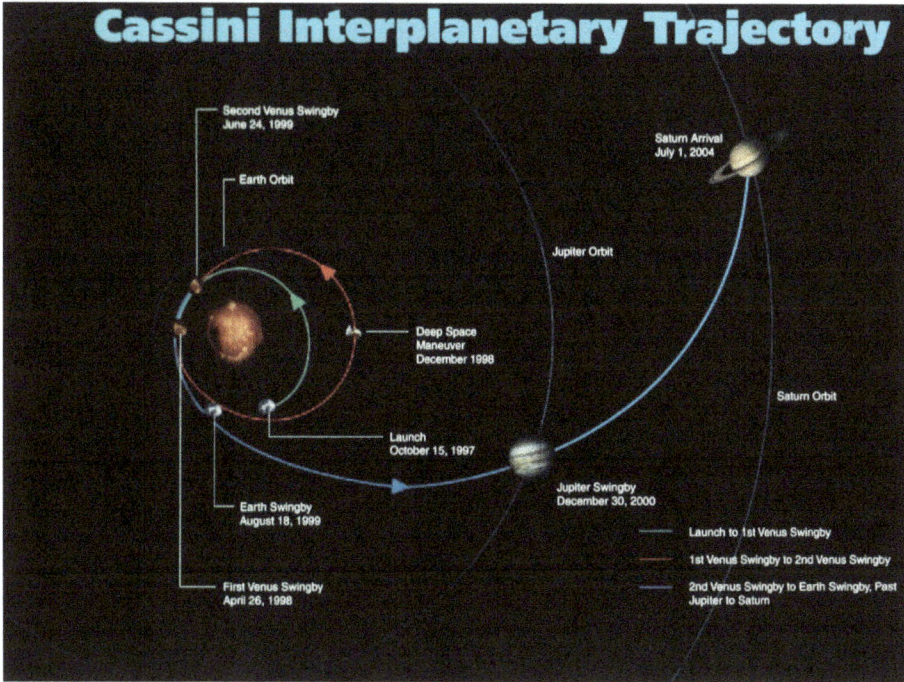

Plate 8   The trajectory followed by Cassini–Huygens through the Solar System (NASA/JPL).

Plate 9   An artist's concept of Cassini during the Saturn Orbit Insertion manoeuvre on July 1, 2004. The orbit insertion motor fired for 90 minutes, allowing Cassini to be captured by Saturn's gravity into a five-month orbit. The spacecraft's proximity to the planet at this time offered a unique opportunity to observe Saturn and its rings at extremely high resolution (NASA/JPL).

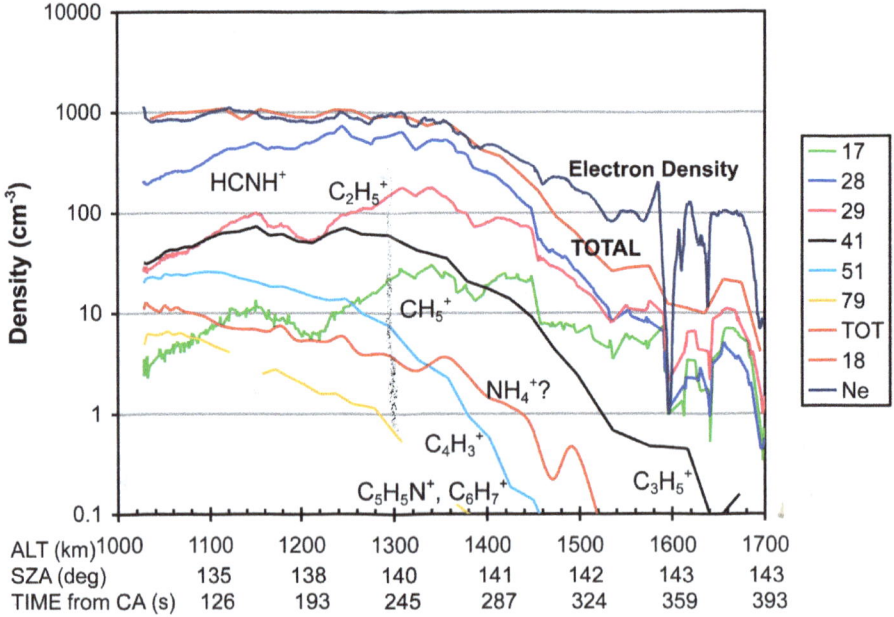

Plate 10   Density profiles of several important ion species measured by the INMS versus altitude, time from CA, and solar zenith angle for the outbound T5 encounter of Cassini with Titan. An electron density profile measured by the RPWS experiment is also shown (Cravens *et al.*, 2005).

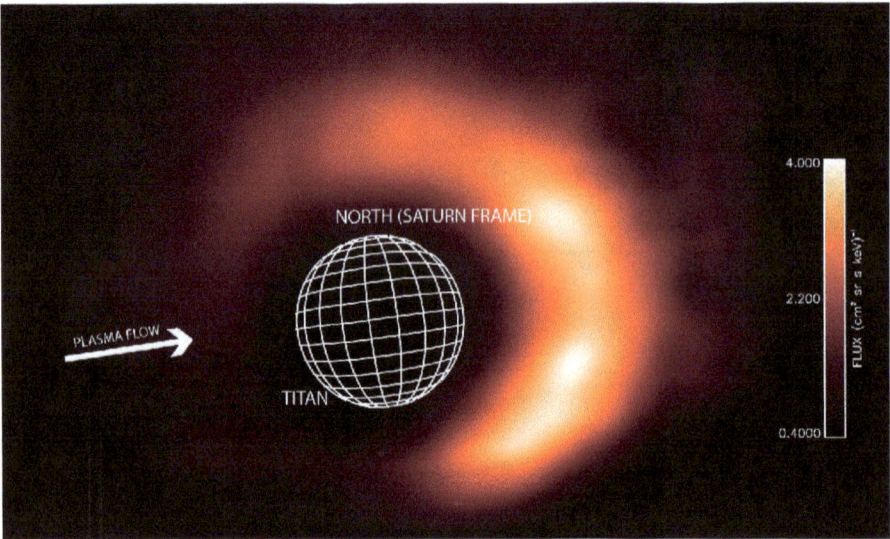

Plate 11   Cassini MIMI observations of magnetosphere dynamics near Titan, showing the emission from hydrogen atoms in Titan's exosphere, about 2500 km above the surface, stimulated by the energetic plasma flow in Saturn's magnetic field (NASA/APL).

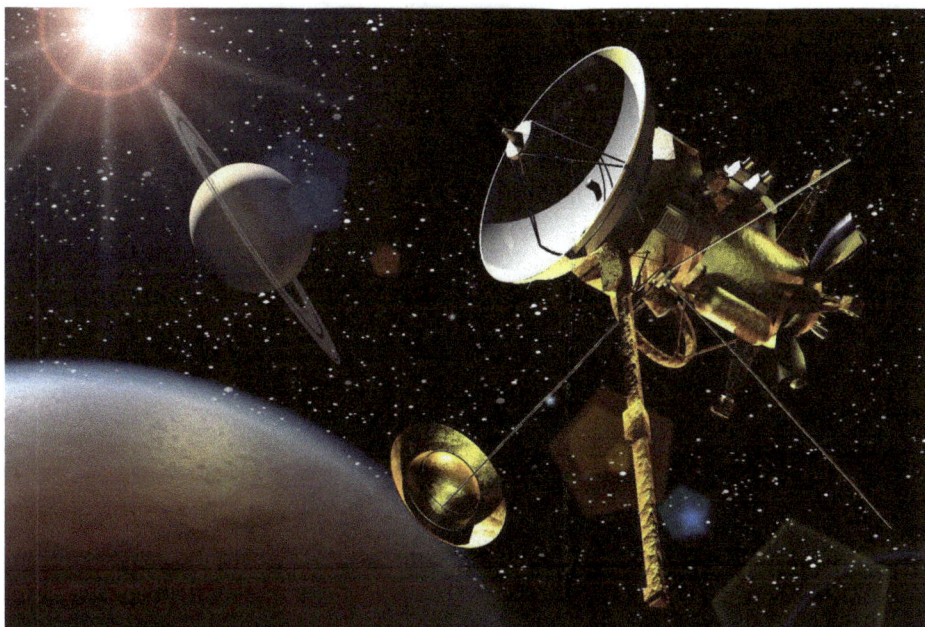

Plate 12 Christmas Day, 2004: Cassini releases the Huygens probe, seen inside its flat, conical heat shield, in the direction of Titan, seen below in the foreground with Saturn in the distance (NASA/JPL).

Plate 13   An artistic view of Huygens landed on the Titan surface (ESA).

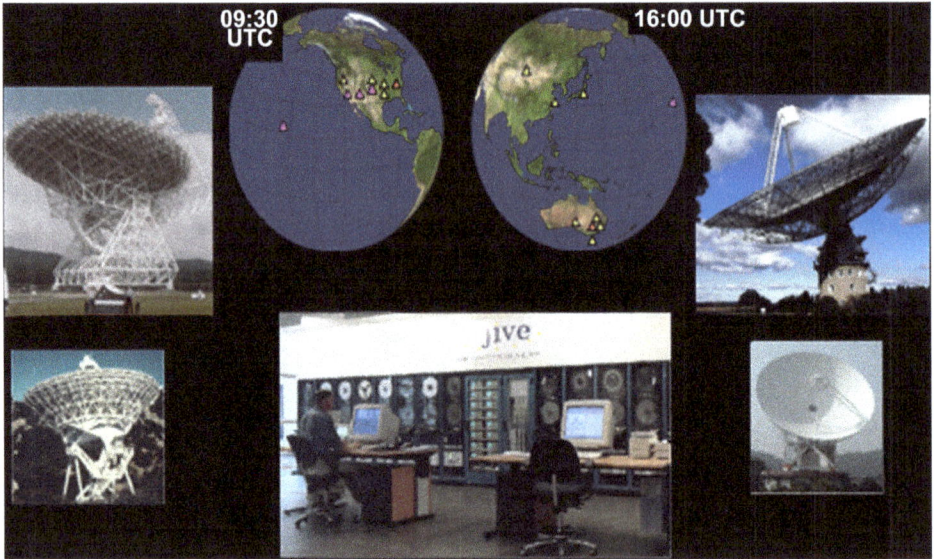

Plate 14   A montage illustrating the Doppler Wind Experiment, which used observations of the Huygens signal from several Earth-based radio telescopes.

Aerial Views of Titan around the Huygens Landing Site

Plate 15   Views of Titan's surface from the Huygens/DISR cameras during the descent through Titan's atmosphere on January 14, 2005. This montage shows flattened (Mercator) projections of the scene taken at four different altitudes (ESA/NASA/JPL/University of Arizona).

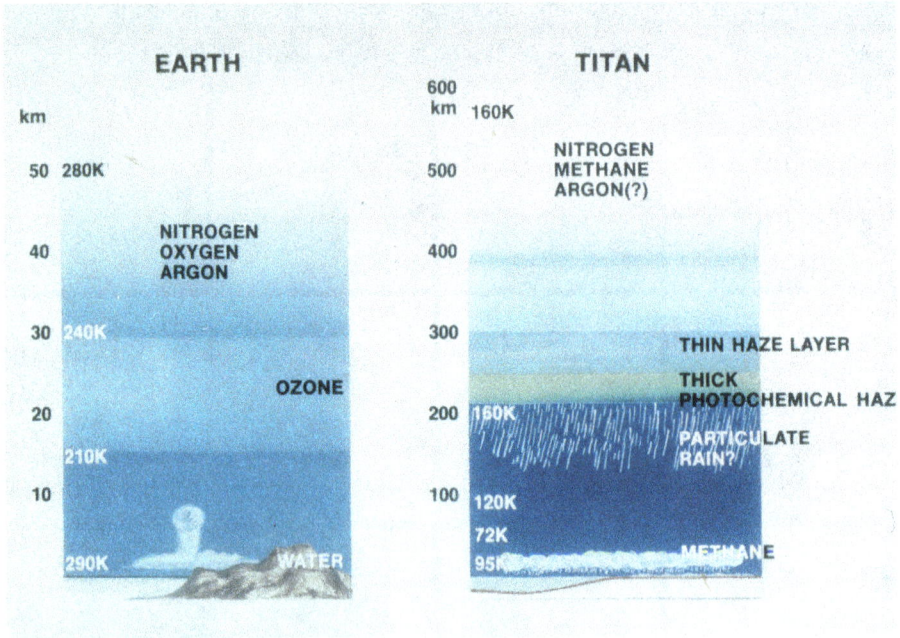

Plate 16   Titan and Earth atmospheres compared (NASA/JPL).

Plate 17   An imaginative view of sunrise over a hydrocarbon lake on Titan, with Saturn visible through a gap in the cloud (© Kees Veenenbos).

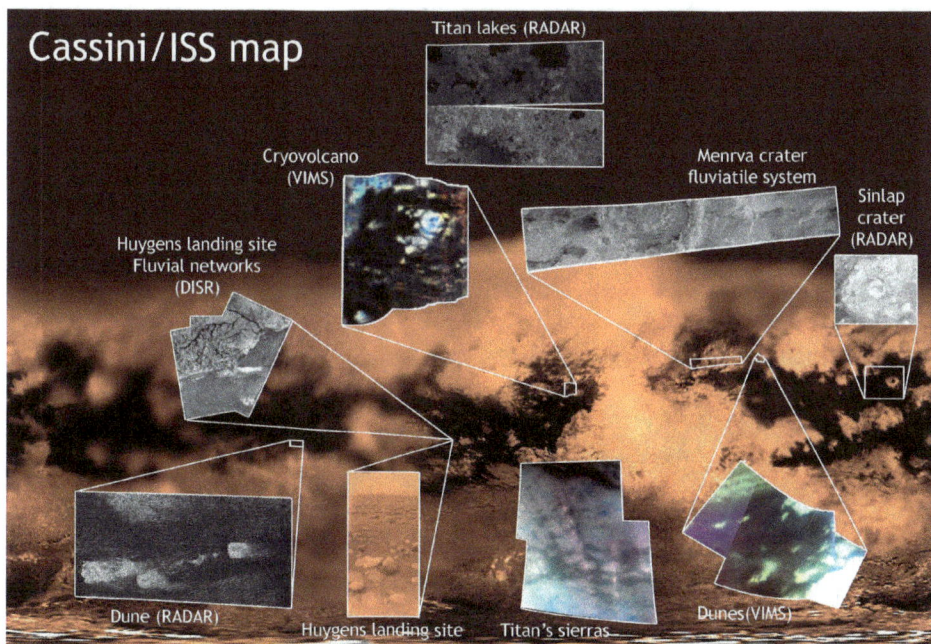

Plate 18   Synopsis of current observations for Titan's surface from Cassini and from the Huygens probe. The background map is from a mosaic of Cassini ISS images (NASA/ESA/JPL/University Arizona; LPG/Nantes & Sigal@LESIA).

Plate 19   Venus, Mars and Titan, roughly to scale.

Plate 20   The future exploration of Titan by instrumented platforms floating in the atmosphere: above, a Zeppelin-type dirigible over a varied landscape and below, a hot-air balloon over the shore to a hydrocarbon lake (artistic views by T. Balint).

CHAPTER 5

Titan's Atmosphere and Climate

Scientific materialism, which is commoner among lay followers of science than among scientists themselves, holds that what is composed of matter or energy and is measurable by the instruments of science is real. Anything else is unreal, or at least of no importance.

Robert M. Pirsig, *Zen and the Art of Motorcycle Maintenance*

5.1 The Climate on Titan

In planetary science, 'climate' usually refers to the structure of the atmosphere, especially the global and annual mean distribution of pressure and temperature as a function of height, and how that affects conditions at the surface. Apart from distance from the Sun, the most important factors determining the climate on any planetary body are the total mass of the atmosphere (determined from the surface pressure and acceleration due to gravity), and its composition. Of course, the average conditions are a very complex derivative of these basic parameters, in particular they are the result of some very dynamic meteorology and cloud chemistry. Both of these topics are complex enough to be the subjects of separate chapters later in the book. The long-term evolution of Titan's atmosphere, within the context of the Solar System as a whole, will also be the subject of a separate chapter.

The data from Cassini and Huygens resulted in an explosion of knowledge about Titan's atmosphere, particularly the dynamical processes mentioned above, and the near-surface environment. They have also permitted some refinement of the basic facts about Titan that affect its climate, including key atmospheric parameters, which we summarise in Table 5.1.

5.1.1 *Atmospheric Pressure Profile*

The accelerometer on Huygens first detected the aerodynamic drag forces due to Titan's atmosphere at a height of about 1,500 km, about ten times higher than would have been the case if its target had been the Earth. The pressure scale height (the vertical height in which the pressure falls by a factor of e^{-1}, or by about one-third of its value at some baseline) is proportional to temperature, so it varies with height,

Table 5.1   Titan's orbital and body parameters, and atmospheric properties from recent Cassini–Huygens measurements.

Surface radius	2575 km
Mass	1.35×10^{23} kg (= 0.022 × Earth)
Atmospheric Mass	6.25×10^{18} kg (= 1.19 × Earth)
Mean density	1880 kg m^{-3}
Surface gravity	1.354 m s^{-2}
Distance:	
from Saturn	1.23×10^{9} m (= 20 Saturn radii)
from Sun	9.546 AU
Orbital period	
around Saturn	15.95 days
around Sun	29.5 years
Obliquity	26.7°
Surface temperature	93.65 K
Surface pressure	1.467 bar
Atmospheric composition near surface:	
Nitrogen, N_2	95.1%
Methane, CH_4	4.9%
Argon, Ar	0.004%
Isotopic ratios:	
$^{14}N/^{15}N$	183 (0.67 × Earth)
$^{12}C/^{13}C$	82.3 (0.915 × Earth)
$^{36}Ar/^{40}Ar$	0.0065 (2 × Earth)

and inversely proportional to the product of the mean molecular weight, which is nearly the same on Earth and Titan (about 28) since the atmospheric compositions are similar, and the acceleration due to gravity, which of course is much smaller on Titan by almost an order of magnitude. The scale height is typically about 8 km in the troposphere of the Earth, 37 km on Saturn, with its low mean molecular weight, and 20 km in the lower atmosphere of Titan. In the stratosphere, Titan's scale height is closer to 40 km.

The atmospheric pressure as a function of height was measured by Huygens from the velocity as a function of time, determined by integrating the measured deceleration while it was descending under its parachutes, using the altitude obtained by integrating the vertical component of the velocity using tracking data obtained by the Cassini navigation team. Pressures were obtained from the density profile under the assumption of hydrostatic equilibrium using the known values for Titan's mass, radius and surface gravity. Once the heat shield had been discarded and the parachute deployed, the on-board pressure and temperature sensors were exposed to the atmosphere. Stable readings commenced at about 150 km altitude and continued down to, and on, the surface. The pressure at the landing site (at 15°S, 192°W) was $1,4767 \pm 1$ hPa, remaining steady within this range during the short time measurements were made on the surface.

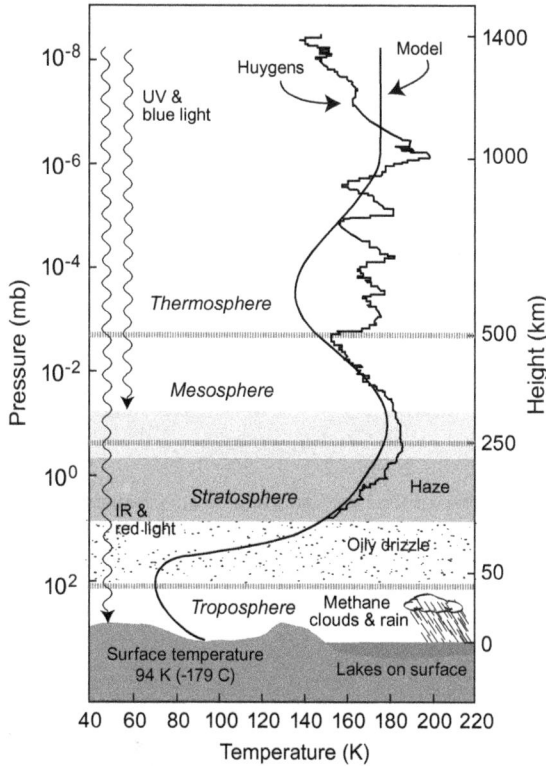

Figure 5.1   A model of Titan's atmosphere showing the temperature profile, clouds and haze layers, precipitation, and surface, and some of the important radiation fluxes. The jagged curve is the temperature profile derived from Huygens measurements (Fulchignoni *et al.*, 2005) and the smooth curve is a pre-Cassini (but post-Voyager) theoretical model. The two coincide closely below the haze layers.

5.1.2 *Atmospheric Thermal Structure*

Huygens also obtained a temperature profile in the upper atmosphere as a function of altitude from about 1,400 km down to 150 km, using the measured decelerations, the inferred pressures and densities, and the equation of state for a perfect gas, and assuming a value for the atmospheric mean molecular weight. Below 160 km, measurements with higher accuracy (better than 1 K) were obtained by direct measurement, using resistance thermometers exposed to the atmosphere.

By analogy with the Earth, Titan's atmosphere is subdivided into regions defined by the way temperature varies with height above the surface. On both planetary bodies, the temperature profile is characterised by inversions (locations above which the temperature profile switches from increasing with altitude to decreasing, or vice versa). This is usually produced by a localised region of heating, such as occurs in the middle atmosphere of the Earth due to the stratospheric ozone layer, and also in Titan's haze layers. Both of these absorb solar energy and heat the region, producing a local temperature maximum. The regions between the temperature

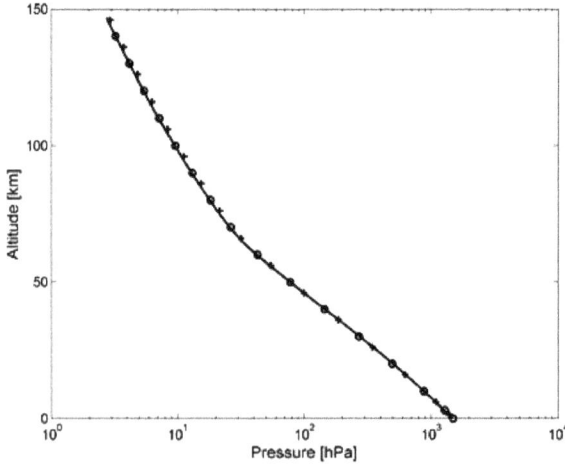

Figure 5.2   The pressure profile of the lower atmosphere as measured by the HASI experiment on Huygens (solid line), corrected for dynamical effects. These new data have an estimated uncertainty of 1% at all altitudes, and are not very different from the pressure values obtained by Voyager 1 radio occultation, which are shown as circles (ingress) and crosses (egress) (Fulchignoni *et al.*, 2005).

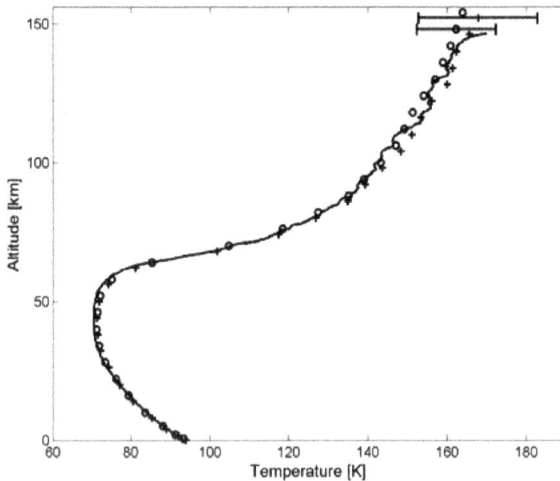

Figure 5.3   Temperature profile of the lower atmosphere as measured by HASI on Huygens. The estimated uncertainty is ±0.25 K in the range from 60 to 110 K, and ±1 K above 110 K. Like the pressure profile, they provide a good match to temperatures calculated from the Voyager radio occultation experiment assuming a pure nitrogen atmosphere, shown as circles (ingress) and crosses (egress) (Fulchignoni *et al.*, 2005).

maxima and minima are given the corresponding names in both atmospheres, while the boundaries between regions are identified by terms ending with -pause (from the Greek $\pi\alpha\upsilon\sigma\eta$ = end). However, the vertical extent of each region is different, with Titan's atmosphere generally being more extensive than the Earth's. In fact, the extent of Titan's atmosphere is comparable to the radius of the solid body (2,575 km), while that of the Earth is sometimes likened to the skin on a grape.

5.1.3 *Troposphere*

Much of the electromagnetic radiation emitted by the Sun and reaching a planet is at wavelengths in or near the visible part of the spectrum. The haze and cloud layers scatter the photons and reflect a portion of the radiation back to space, while some of it is absorbed by the aerosols and gases in the atmosphere. The rest of the energy, about 10% on Titan, reaches the ground where it is absorbed. The lower atmosphere is then heated by the ground, and becomes unstable against convection. This simply means that, if the temperature gradient is too steep, lower layers will be more buoyant than those above and they will tend to rise, forcing more powerful convection. If the gradient is too shallow, the motion will stop until the temperature gradient builds up again. The net result is that the gradient is usually close to a value called the adiabatic lapse rate, at which convection is just possible. Assuming that hydrostatic equilibrium applies, and that there is no net exchange of energy between a parcel of air and its surroundings, then the perfect gas law and some elementary thermodynamics tell us that the temperature gradient with height is just the acceleration due to gravity divided by the specific heat of the air. Thus it is constant for a given composition, and works out to be equal to approximately $10 \, \mathrm{K \, km^{-1}}$ for dry terrestrial air. For moist air, it is less (about $6.5 \, \mathrm{K \, km^{-1}}$ typically) and can be as small as $3 \, \mathrm{K \, km^{-1}}$. Since Titan has the same main constituent as on Earth, the difference in lapse rate is due mainly to the difference in gravity, and its value on Titan works out to be only about $1.4 \, \mathrm{K \, km^{-1}}$. Again, this varies depending on the condensable vapour present in the form of methane, although the temperature lapse rate measured by Huygens shows that the temperature gradient, or lapse rate, is shallower (more stable) than the dry adiabatic lapse rate, the rate for the buoyancy of a dry air parcel to just balance the vertical pressure gradient.

On the Earth, the region where convection dominates vertical heat transport is known as the troposphere ('turning-region'). The upper boundary occurs near the level where the overlying atmosphere is of such a low density that a substantial amount of radiative cooling to space can occur in the thermal infrared region of the spectrum. At the *tropopause*, radiation cools rising air so efficiently that the temperature tends to become constant with height and convection ceases. The Earth's tropopause varies in height over about 6 km with latitude, being highest (around 16 km) in the tropics, where solar heating is greatest, but it is generally quite a distinct feature in the temperature profile everywhere, and usually occurs only slightly below the temperature minimum.

On Titan, the temperature falls with height from the surface up to about 40 km, so this level is also referred to as the tropopause. At the Huygens entry site, the probe's sensors found that the temperature minimum, and hence the tropopause temperature, was 70.43 K at about 44 km where the pressure was 115 ± 1 hPa. This is about 1 K

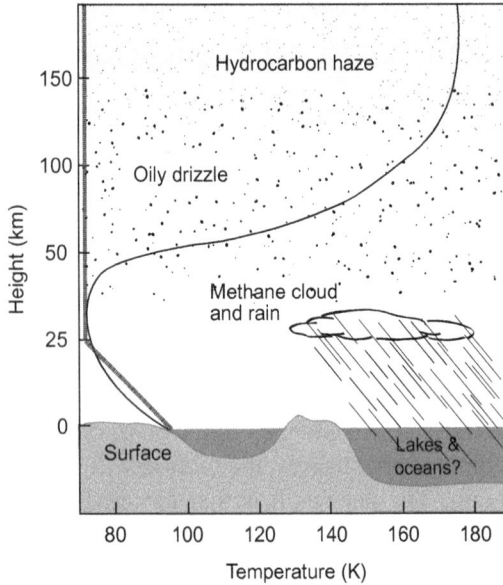

Figure 5.4 The mean temperature-height profile for Titan represented by a simple radiative-convective equilibrium model (heavy line), compared to the profile measured by Huygens. The model fails completely at higher levels because it takes no account of the heating due to aerosol, which greatly modifies the temperature. However, the profiles compare quite well in the middle to lower atmosphere, and at the surface.

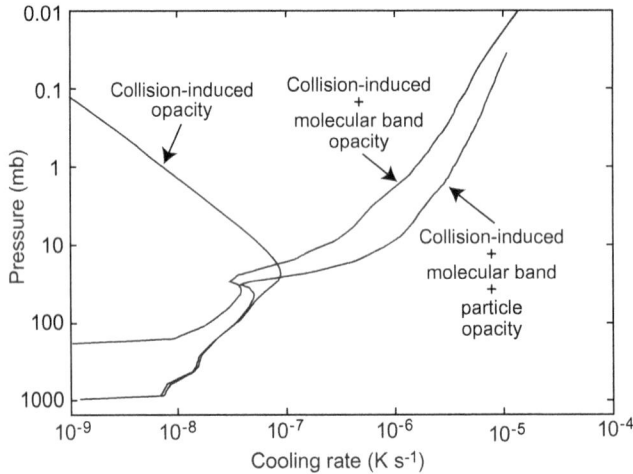

Figure 5.5 Calculated cooling rates in the atmosphere of Titan, showing the effects of considering the two different kinds of gaseous absorption, separately and together, and adding in the effects of infrared emission from aerosol particles (Bézard et al., 1995).

colder than Voyager radio occultation measurements, but most of this difference can be ascribed to the assumption by Voyager scientists of an atmospheric composition of pure N_2. The addition of 1.5% of methane brings the two measurements into agreement, a remarkable result considering they were made at different locations on Titan, and more than 20 years apart.

5.1.4 *Stratosphere*

By analogy with this philosophy for naming the troposphere, the lower stratosphere on Titan is then the shallow, quasi-isothermal region near 50 km, while the upper stratosphere is the much deeper region from about 50 to 300 km altitude where temperature increases with height, i.e. the region between the first temperature minimum and the first maximum above the surface.

The Earth's stratosphere originally got its name because it is the region where convection stops and the air forms layers that tend to stay put, i.e. the atmosphere is stratified. The absence of enough absorption above the tropopause to stop emitted photons from reaching space causes the lapse rate to tend to zero (i.e. to a constant temperature with height). Each layer is heated by radiation from the optically thick atmosphere below, and cooled by radiating to space, so height is no longer important to first order. The fact that the stratospheric temperature is soon observed to increase is, as we have seen, a consequence of the absorption of solar ultraviolet radiation by photochemically-produced ozone in the Earth's stratosphere. Ozone absorbs in the Hartley band, which forms a continuum from 0.2 to 0.3 μm. Below 70 km, virtually all of the energy absorbed is converted to heat. The ozone concentration in the stratosphere peaks near 25 km, but a calculation of the heating rate finds that this peaks at a height of about 50 km. The temperature is also a maximum at this level, which forms the *stratopause*.

On Titan, the corresponding effect is due to the absorption and thermalization of UV solar radiation by different gases and aerosols in the stratosphere, the temperature again rising with altitude starting from a minimum of 70.43 K at 44 km altitude (where the first temperature inversion occurs) up to 186 K at 250 km as found in the measurements by Huygens. Prior to Huygens entry measurements, M. Flasar and colleagues working with data from the Composite Infrared Spectrometer on Cassini provided information on the lower and upper stratosphere from roughly 70 to 400 km in altitude, indicating the presence of a stratopause at around 310 km of altitude with the same maximum temperature of 186 K.

If the lower stratosphere, just above the tropopause, is modelled as an optically thin slab, i.e. one with emissivity much less than one, its temperature can be related to the effective radiative, or equivalent black body, temperature of the planet by using well-known radiation laws due to Planck, Boltzmann and Stefan. The stratosphere is heated from below by a flux of infrared radiation from the planet, and cooled by identical emitted fluxes in the upward and downward directions, each proportional to the fourth power of its temperature. By balancing the two, we find that the mean stratospheric temperature is about 210 K for Earth and about 70 K for Titan, both values being close to what is observed.

5.1.5 Mesosphere

Above the stratopause, the temperature declines again, reaching a minimum at the *mesopause*, where the second temperature inversion occurs, signifying the end of the *mesosphere*. On Earth, the mesosphere begins at around 50 km altitude, whereas on Titan it runs through the 300–600 km altitude range. This is the coldest level anywhere in the Earth's atmosphere. Huygens observed an inversion region corresponding to the mesopause, less contrasted than that inferred from Voyager 1 data, at 490 km with a temperature of 152 K.

5.1.6 Thermosphere

The pressure at the mesopause is only a few microbars. With such low densities of gas above, solar photons in the extreme ultraviolet, and very energetic particles too, penetrate into the region, ionising and dissociating molecules and releasing kinetic energy. The heating thus produced causes the temperature to increase rapidly with height, leading to the name *thermosphere* ($\theta\varepsilon\rho\mu\acute{o}$ = warm), the most extensive part of the atmosphere, in which the energy is transported by thermal conduction. The Earth's thermosphere begins around 85 km altitude and, as might be expected, is much warmer at more than 1,000 K. On Titan, the thermospheric temperatures were expected to increase steadily with altitude from about 140 K at the mesopause at around 600 km, up to around 190 K above 1,200 km. Huygens found rather higher temperatures than the models predicted (average temperature of 170 K), with vertical waves of 10–20 K in amplitude, attributed by the science team to gravity waves and tidal phenomena, recorded above 500 km.

5.1.7 Exosphere

Between the tropopause, where large-scale vertical convection ceases, and the mesopause, the atmosphere remains fairly well mixed by turbulence produced by a variety of instabilities in wave motions and the mean flow. A fairly small distance up into the thermosphere, at around the 100 km altitude level on Earth, diffusion takes over as the dominant process and the atmosphere starts to separate into its lighter and heavier components. For many practical purposes, this level (the *homopause*) may be considered to be the effective top of the atmosphere. The very tenuous region above is often called the *exosphere* ($\dot{\varepsilon}\xi\omega$ = out) since here light elements escape the planet's grip and are lost to space. Titan's atmosphere is generally much more extended than Earth's, as we have seen, and the exosphere begins much higher above the surface on the lower-gravity world. According to estimates based on Voyager data, the homopause on Titan falls at an altitude of around 1,000 km; the Cassini mass spectrometer team recently refined this to 1,200 km.

5.2 Radiation in Titan's Atmosphere

5.2.1 *Solar and Thermal Radiation*

The most energetic solar photons at UV wavelengths and shorter are removed at various, mostly high, levels in the atmosphere, where they participate in the dissociation and ionisation of atmospheric gases. The longer solar infrared wavelengths (1 to 5 μm) are absorbed in some regions of the near infrared spectrum by molecular vibration-rotation bands, particularly those of water vapour and carbon dioxide on Earth and methane on Titan. The remainder, radiation at or near visible wavelengths, mainly propagates to the ground unless reflected or absorbed by clouds. On a global average, about 30% of the energy from the Sun falling on Titan is reflected back to space, 60% is absorbed in the atmosphere, and 10% at the ground; for the Earth, the corresponding numbers are 30%, 25% and 45%.

Because of its low temperature relative to that of the Sun, the planet's surface re-emits the absorbed energy at a much longer wavelength than it was received. Some of this thermal infrared radiation is transmitted directly to space by the atmosphere, but at most wavelengths — substantially longer than visible — the atmosphere is very opaque. The main source of this opacity is not the main constituent, nitrogen, and oxygen does not contribute much opacity on Earth, nor argon on Titan. Rather, it is the minor and trace constituents, especially water vapour, carbon dioxide, ozone, methane and nitrous oxide on Earth and methane on Titan which are the principal absorbing gases.

The reason for this is that molecules made up of identical atoms, like N_2, and monatomic ones like Ar, usually do not interact with infrared radiation. Molecules of mixed composition like methane and carbon dioxide, on the other hand, have very rich infrared spectra because their internal charge distribution produces a net dipole moment. They absorb photons at many wavelengths, leaving only a number of 'window' regions where clouds and haze are the dominant opacity sources. Nitrogen does contribute a significant opacity on Titan at longer wavelengths, through its collision-induced spectrum. During collisions, the molecule is distorted and a temporary dipole moment results. This effect is also present on Earth, but is much less important. The reason is the low temperature on Titan, the almost complete absence of water vapour, which dominates the long-wavelength spectrum on Earth, and the presence of a large abundance of methane, which turns out to be a particularly effective collision partner for nitrogen. In fact, N_2–CH_4 absorption is the main opacity source on Titan at wavelengths where most of the flux is emitted.

5.2.2 *Energy Balance and Surface Temperature*

What would we expect the temperature to be on Titan's surface? This important question can be addressed with some simple theory. Applying the Stefan–Boltzmann law, which states that the energy emitted as radiation by any body is proportional to the fourth power of its temperature, we can calculate how much energy the Sun is emitting, at its known temperature of approximately 5750 K. The answer is about 4×10^{26} W (400 yottawatts in SI units). Then it is easy to work out how much solar energy is falling on Titan from the inverse square law of distance from the Sun (just 228 megawatts, a solar constant of about $14\,\mathrm{W\,m^{-2}}$).

In order for Titan to be in equilibrium overall, the energy it intercepts, minus the amount reflected (about 30%), must be equal to the energy it emits as infrared heat radiation. If this were not the case, Titan would heat up, or cool down, until a balance was achieved. We assume this happened long ago on Titan, unlike its parent, Saturn, which is still emitting nearly twice as much heat as it receives, and so is gradually cooling. Gas giants, with their deep atmospheres, do this in general; Titan is smaller and its relatively thin atmosphere reaches equilibrium much faster, like that of the Earth.

Using the Stefan–Boltzmann law again, we find that the emitting temperature of Titan has to be approximately 82 K for that balance to be achieved (most of the uncertainty in this number, which is a few degrees, is due to the albedo). This is the temperature of a solid sphere the same size as Titan at the same distance from the Sun, but as a value for the surface temperature it is too low by 12 degrees or so. The reason for the discrepancy is the atmosphere. Its partial transparency to significant amounts of sunlight, and its high opacity to longer wavelengths, has the effect of trapping solar energy near the surface. Some of the energy from the Sun reaches the surface of Titan because solar radiation consists mostly of photons at near-visible wavelengths and the atmosphere is partially transparent at those wavelengths. The balancing, cooling radiation, however, is in the infrared part of the spectrum — because the source is so much cooler, near one hundred degrees instead of a few thousand. On Titan, 90% of the incoming solar radiation is absorbed and back-scattered in the higher atmosphere allowing only about 10% to reach the surface. Part of it is absorbed by the ground, which warms up and re-radiates infrared heat that is absorbed by the overlying atmosphere. Then it is the atmosphere, rather than the surface, which cools by radiating to space, and it is the atmosphere that has the characteristic temperature of 82 K. Of course, the atmosphere will radiate downwards as well as up, so the surface gets an extra contribution of radiated energy in addition to the input directly from the Sun, enough to increase the temperature by about twelve degrees, to 94 K ($82 + 12 = 94$), close to that observed by Huygens when it landed.

5.2.3 Model Temperature Profile

Using the surface temperature derived in the previous section, simple model profiles of temperature vs. height can be constructed for Titan by assuming convective equilibrium in the troposphere and radiative equilibrium in the stratosphere. The principal assumptions are that the temperature gradient is constant, and equal to the adiabatic lapse rate, in the troposphere, and constant, and equal to zero, in the stratosphere. Then, three parameters (surface temperature, tropospheric lapse rate and stratospheric temperature) provide a complete specification of the model climate as defined above. Despite their approximate nature, such models provide remarkably good approximation to the observed profile, in the lower atmosphere.

To formulate the model, the perfect gas law, the first law of thermodynamics and the hydrostatic equation are used to obtain the vertical gradient of temperature T with height z in the troposphere. For a 'dry' atmosphere, in which latent heat effects can be ignored, this is just $dT/dz = -g/c_p$, where c_p is the specific heat at constant pressure and g is the acceleration due to gravity. A more complicated but still straightforward expression can be obtained by introducing the Clausius–Clapeyron equation to allow for the latent heat effects of condensable species, primarily methane on Titan. Under the assumption of optically thin layers in radiative balance, the temperature T_s in the stratosphere is constant with height and given by $T_s = T_e/2^{1/4}$, where T_e is the equivalent blackbody temperature at which the planet cools to space. This, and the height of the tropopause at which the radiative and convective regimes change over, can be calculated either precisely by the detailed application of radiative transfer theory, or estimated by any of a range of approximate methods.

5.2.4 Radiative Equilibrium Temperature Profile

The exact solution of the problem of calculating how the temperature of the atmosphere is expected to vary with height involves proceeding from this very simple model to one that includes all of the complex physics of the inhomogeneous, variable atmosphere, with an assumed composition and other factors incorporated. The effects of clouds and hazes can be important in calculating the energy balance of individual atmospheric layers, so the number density, size distribution and composition of aerosols must also form an input to the calculation.

The calculated rate of cooling from Titan's atmosphere as a function of height, in units of $K\,s^{-1}$, shows that, when all of the cooling processes are included, the atmosphere at around the 1 mbar pressure level (for example) would cool down by one degree K in about a day, if there were no heating to compensate. The next step is to calculate how much solar energy is absorbed, how much is transmitted, and how much scattered into the forward and backward direction. The same is done for thermal energy, as a function of the temperature of the layer. The exchange of energy as radiation between every level in the atmosphere, and every other level, has to be taken into account, making the whole computation quite complex and time-consuming. Eventually, by iteration, we find the equilibrium temperature for every layer, i.e. the temperature when every layer emits as much energy as it absorbs.

This is still not the end of the computation, however, because some vertical temperature profiles, as we have seen, are unstable with respect to vertical motions. Whenever the calculated vertical gradient exceeds the adiabatic lapse rate, between 1 and 2 K per km, convection sets in which reduces the gradient and pulls it back to the adiabatic value. The temperature profile calculation has to include this adjustment. The result is called a *radiative-convective equilibrium model* of the atmosphere.

The set of such model calculations published by C. McKay and colleagues of the NASA Ames Research Center in 1989, include representations of what they deemed to be the most likely composition and aerosol properties, with different assumptions tried out where these properties are uncertain. They could then select the model that gives the closest agreement with the temperature profile measured by the Voyager radio occultation experiment. They could also see how varying one parameter at a time (for example the methane abundance), would be expected to affect the atmospheric temperature. They found that a 'nominal' model for Titan's atmospheric composition and haze properties, in other words, a model derived from other data such as the IRIS spectra, when used in the radiative-convective equilibrium model, produced a good match to the measured temperature profile, except that the model surface temperature comes out about 5 to 10 K cooler than observed. Interestingly, they also found that including methane or ethane condensation clouds in the model made very little difference to the temperature profile, when pressure-induced absorption by gaseous nitrogen, methane, and hydrogen are properly taken into account. The fraction of solar energy absorbed at the surface in the model is about 10%; this is the longer wavelength solar visible radiation, the UV component having been absorbed mostly in the stratosphere, giving rise to the positive temperature gradient above the minimum (70.5 K) at the tropopause (44 km).

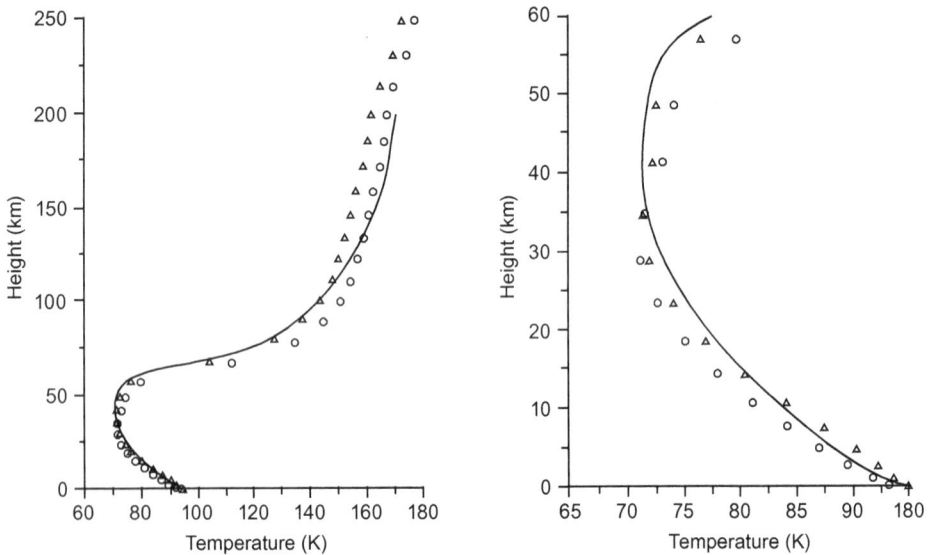

Figure 5.6   Titan's atmospheric temperature as a function of height. The solid line is the measured profile, from the Voyager radio-occultation measurements; the points are from a radiative-convective equilibrium model, with the different symbols representing different assumptions about the composition of the atmosphere. The curve on the right is an expanded version of the bottom part of the one on the left (McKay *et al.*, 1989).

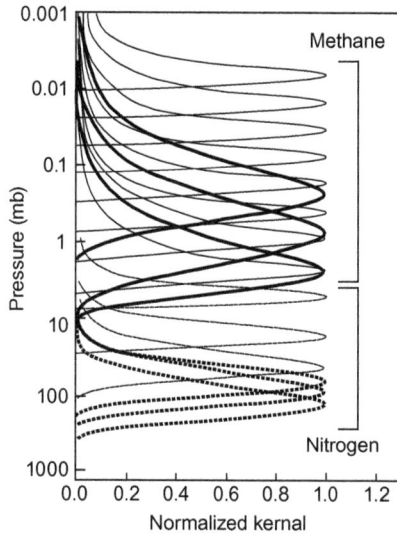

Figure 5.7   Contribution functions for atmospheric temperature sounding of the atmosphere of Titan by Cassini/CIRS. The curves show where in the atmosphere the measured radiation originates, calculated for nadir (downward) viewing (heavy lines) and limb (tangential) viewing of the atmosphere, in selected spectral intervals $0.5\,\mathrm{cm}^{-1}$ wide in the absorption band of methane near $7.7\,\mu\mathrm{m}$ and in the pressure-induced far-infrared band of nitrogen (After Flasar *et al.*, 2004).

5.3   Remote Atmospheric Temperature Sounding

Satellites and planetary probes employ infrared and microwave molecular spectroscopy techniques to obtain information about the physical properties of atmospheric gases. This is known as remote sensing or, where quantitative measurements of, for example, temperature profiles are concerned, *remote sounding*. Most of our current knowledge of Titan and its atmosphere was derived from remote sounding from Voyager or from the Earth, and the same approach is an important part of the Cassini mission.

The infrared flux emitted by Titan, and measured by instruments like IRIS and CIRS, depends on the profiles of temperature and absorbers in the atmosphere, so, in principle at least, measurements of the former can be analysed to estimate the temperature and composition profiles, since they are related by the radiative transfer equation. The radiance leaving the top of the atmosphere at a particular wavelength is given by a quantity called the contribution function, which has a maximum value at the height where the optical depth of the atmosphere is unity, which depends on the choice of wavelength. A selection of wavelengths in the wing of a strong absorption band of a species such as carbon dioxide for Earth, and methane for Titan, whose mixing ratio is known and approximately constant up to high levels, allows a family of contribution functions to be calculated which will cover a range of levels in the atmosphere.

A set of measurements of radiance can then be turned into a temperature profile by the inversion of a set of radiative transfer equations. The derived profile is a smoothed version of the actual atmospheric temperature profile, because of the averaging effect of the weighting functions. Some of the detailed structure can be extracted using many spectral channels corresponding to overlapping weighting functions; finding the optimum solution for the profile from a given set of measurements with their associated errors is therefore a complicated process.

The vertical resolution of remote sounding measurements is limited by the fact that, in order to get a measurement of radiance that is not dominated by instrument noise, narrow spectral intervals, each corresponding effectively to a single wavelength, cannot be used in practice. Instead, spectral bands containing many molecular vibration-rotation lines must be employed. This has the effect of further broadening the weighting function relative to that for an idealised, monochromatic observation. Typical widths at half maximum amplitude are in the range one to two atmospheric scale heights, or on Titan not less than about 30 km. To improve on this, and to allow higher levels to be sounded, has led to the development of limb scanning techniques, in which the instrument views the atmosphere tangential to the surface and images a thin slab of atmosphere onto the detector. This can improve the vertical resolution to a few km, depending on how close the spacecraft approaches to the limb. However, the gain is at the expense of poorer horizontal resolution, and the opacity of clouds and hazes is around a hundred times greater than when viewing vertically. This problem can be compensated for by using longer wavelengths, where aerosols tend to be less opaque, to probe the deeper levels.

The temperature profiles retrieved from infrared measurements made by CIRS include individual profiles and global maps. The spatial resolution of CIRS is such that it can observe Titan's atmosphere either vertically or tangentially, on the limb. In limb-viewing mode, where the line-of-sight extends through the atmosphere to deep space, the size of the individual detectors in the mid-infrared detector arrays, and the distance of the spacecraft from Titan, determine the altitude coverage and vertical resolution. On orbits offering a very close approach, the vertical resolution is better than 10 km. The coverage and repeatability of temperatures made by remote sensing allows the study of dynamical activity (see Chapter 8); the CIRS temperatures exhibit a high degree of spatial variability, including effects possibly due to the influence of vertically propagating waves.

The accuracy of the retrieved temperatures decreases below about 300 km, because of the uncertain effect of aerosol absorption on the measured radiances. Vertical profiles from about 10 mbar ($\sim$125 km) up to the 0.01 mbar level ($\sim$430 km) are feasible. A well-defined stratopause is evident near 0.07 mbar (310 km) with a temperature of 186 K. In this region, infrared emission to space by C_2H_6 is the dominant cooling mechanism.

Because the radiative relaxation time in the upper stratosphere is short ($\sim$1 yr) compared to Saturn's orbital period (29.5 yr), its temperatures and winds there should vary seasonally. Venus has little seasonal variation, because its spin axis is nearly

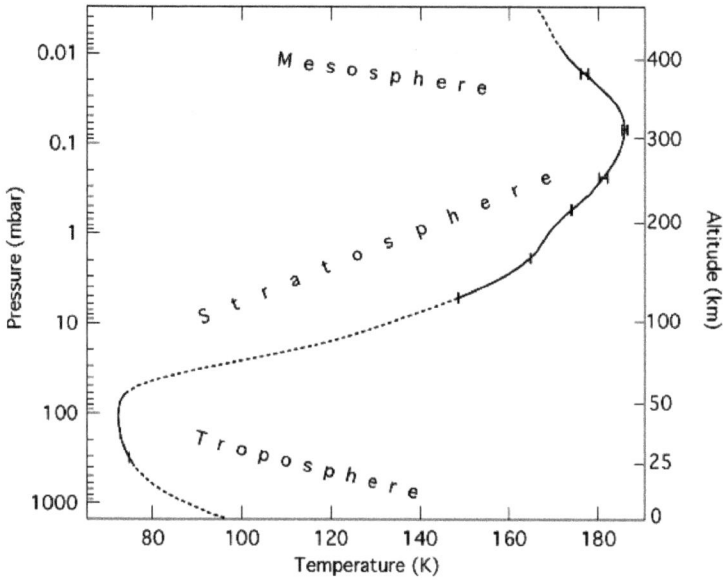

Figure 5.8   Vertical temperature profile in Titan's stratosphere near 15°S, retrieved from CIRS nadir- and limb-viewing spectra. The dashed portions of the curve represent the regions where temperatures are not well constrained by the spectra and are more influenced by the Voyager radio-occultation profiles and radiative mesospheric models used as the initial guess (Flasar *et al.*, 2005).

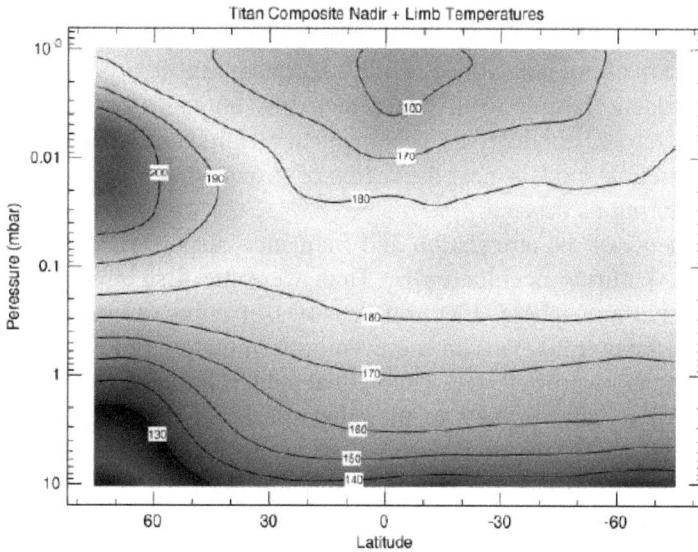

Figure 5.9   A meridional cross section, with latitude as the horizontal coordinate and pressure as the vertical coordinate, of zonally averaged stratospheric temperatures (K) from CIRS measurements (Achterberg *et al.*, 2008).

normal to its orbital plane. CIRS mapped stratospheric temperatures over much of Titan in the latter half of 2004, when it was early southern summer on Titan (solstice was in October 2002). The warmest temperatures are near the equator. Temperatures are moderately colder at high southern latitudes, by 4–5 K near 1 mbar, but they are coldest at high latitudes in the north, where it is winter.

These results can be compared to the predictions of three-dimensional general circulation models of the temperature structure, developed as part of studies of the global circulation since temperature and pressure gradients are associated with the large-scale motions. Radiative transfer calculations, based on an assumed composition and, where appropriate, its global variations, with the results constrained to be consistent with remote sensing and *in situ* measurements, are used to predict the temperature field and its seasonal variations.

5.4 Titan's Ionosphere and its Interaction with the Magnetosphere of Saturn

Bodies with atmospheres also have ionospheres — electrically conducting regions with a high concentration of ionized gases and charged particles. These are produced both by electromagnetic radiation from the Sun, primarily in the ultraviolet spectrum, and by energetic particle bombardment. The latter originate in the solar wind, in cosmic rays from outer space, and, in the case of Titan, as particles precipitated from the magnetosphere of Saturn. All of these act on the neutral atmosphere, resulting in ionization, charge exchange and secondary electron impact. The principal ionospheric layer on Titan, produced by precipitating electrons, lies between about 700 and 2700 km above the surface, while a secondary layer, produced by the more penetrating galactic cosmic rays, was detected by Huygens between 140 km and 40 km, with electrical conductivity peaking near 60 km, somewhat deeper than predicted by theoretical models, near the region of strong wind shear that was detected by Huygens during its descent.

Measurements of the attenuation and frequency dispersion of the radio signal from Voyager 1 during occultation by Titan were the first to confirm that Titan has an extensive ionosphere. Electron densities of approximately $3,000\,\mathrm{cm}^{-3}$ at an altitude of about 1,200 km, on the evening terminator, and $5,000\,\mathrm{cm}^{-3}$ on the morning terminator were detected. Theoretical models have matched these data, and predicted the abundance profiles of the other species expected to be present. The most abundant ion is H_2CN^+, followed by various hydrocarbons ($C_nH_m^+$) and nitriles ($C_nH_mNp^+$). These form as the products of chain reactions which begin with the ionization of molecular nitrogen to give N_2^+ and N^+, which reacts with both H_2 and CH_4, and methane to give CH_4^+. Numerous other ions form by charge exchange, atom/molecule-ion interchange, and other processes. The terminal ions are removed by electron recombination, leading to the formation of a range of neutral species, including the hydrogen cyanide (HCN) first detected by Voyager, as well as heavier molecules.

Some of the Cassini flybys of Titan have been at altitudes less than 1,000 km, close enough to be inside the ionosphere. The instruments on board have found that more than 10 percent of the ionosphere is made up of ionized hydrocarbon molecules chemically similar to compounds such as ethylene, propyne and diacetylene. These data are being used to put together a detailed picture of the composition and variability of the ionosphere by refining models until, eventually, a clearer understanding of the photochemistry and ion chemistry involved in the formation of Titan's orange haze will emerge.

Investigations of the properties of the ionosphere are also important for gaining an understanding of the processes by which gases are lost from the atmosphere of Titan over time. Because Titan has no strong intrinsic global magnetic field, charged particles in the rarified upper atmosphere are exposed to bombardment by the solar wind and by particles precipitated from Saturn's magnetosphere, as well as by cosmic rays from outer space. These interactions result in the ionosphere being drawn into a comet-like tail, with its population of ions and neutrals being lost to space at an unknown rate.

Determining, or at least modelling, these loss rates, is complicated because of the unique situation of Titan as the only satellite with a substantial atmosphere. Titan differs from, say, Venus, another planetary body with a thick atmosphere and no intrinsic magnetic field, in that it orbits Saturn, which has a very large field. Not only that, but at 20 Saturn radii from its parent, Titan orbits very close to Saturn's magnetopause, the boundary between the magnetosphere and the outer region where the solar wind flows around the planetary obstacle. Since changes in the solar wind dynamic pressure affect the position of the magnetopause, Titan can be inside or outside, depending on the level of solar activity. This leads to a widely varying particle flux onto Titan, both in intensity and direction, and to a variety of interaction scenarios.

Voyager 1 and Cassini observations established an upper limit of 4 nT at the surface and at the equator for Titan's intrinsic field, at least 10,000 times less than that of Saturn. As a result, when Titan is outside Saturn's magnetosphere, the solar wind interacts directly with the satellite's atmosphere. The incoming flow picks up the ions created from the ionization of Titan's exospheric neutrals and decelerates, as the magnetic field lines pile up in front of the satellite and drape around it. The heavy ions (N_2^+, H_2CN^+, etc.) end up in Titan's ionosphere between 700 and 2700 km above the surface, resulting in an asymmetric plasma flow, and a magnetotail with four lobes.

Below 700 km altitude, galactic cosmic rays penetrate down to the tropopause and cause some ionisation. At these altitudes, ionisation by meteorites also contributes. These sources of ions occur throughout the same altitude range as the methane photochemistry, and particle charge is known to strongly influence the coagulation of haze particles. Thus, Titan's atmosphere may be more closely linked with its space environment than on other planets. The exospheric ions are eventually picked up by the incoming magnetized plasma flow and carried away. The amount of atmospheric

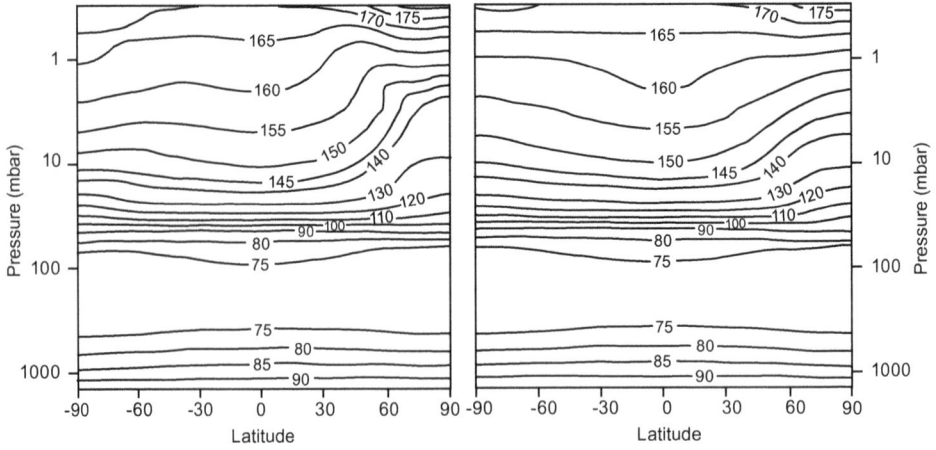

Figure 5.10   Model-calculated zonally averaged temperature in K, showing the seasonal variation in the rarefied upper atmosphere. The left frame shows northern winter solstice and the right the spring equinox (Hourdin *et al.*, 1995).

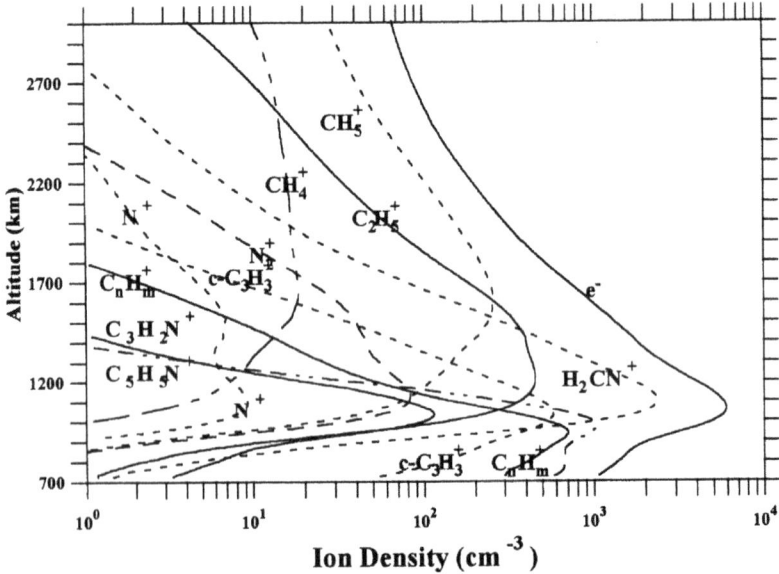

Figure 5.11   Major ion density profiles for Titan's ionosphere based on the neutral atmosphere from the photochemical model by Yung *et al.* (1984), as computed by Keller *et al.*, (1998).

constituents lost via this non-thermal escape process has not been estimated, but it is believed to be important.

The precipitation of electrons from the magnetosphere is energetic enough to excite nitrogen molecules in the upper atmosphere. They then emit light, the phenomenon which on the Earth is known as 'dayglow'. Some of the energy gets converted to heat, so that nitrogen atoms can have enough kinetic energy to escape from Titan. Hydrogen atoms are light enough to be lost at the normal temperatures that prevail in Titan's outer atmosphere. This means that the photochemical destruction of methane is irreversible, as the hydrogen produced is lost from Titan. The hydrogen does not have enough energy to escape from Saturn's gravity well, and instead it forms a toroidal cloud around Titan's orbit. Calculations of the loss rate of nitrogen suggest that, at the present rate, less than 1% of the present atmosphere will have escaped over the age of the Solar System, while that of hydrogen is, of course, much greater, with complete removal of all of the present atmospheric inventory of methane in just a few million years.

5.5 Climate Change on Titan

Like everything else in the Solar System, Titan has been subject to varying solar luminosity and may have experienced climate change in the past and face it in the future. Titan's present surface temperature is elevated above the radiative equilibrium value by about 12 K, due to the greenhouse effect from the combined effect of the collision-induced infrared opacity of nitrogen, methane and hydrogen and the radiative properties of the haze layers. Changes in atmospheric composition, surface pressure, haze and cloud cover, and solar input all have consequences for the surface temperature, with possible complex feedback processes introducing nonlinear effects.

On Titan, the main amplifying factor for climate change is the evaporation and condensation of volatiles, most notably methane. For large negative temperature excursions the main atmospheric constituent, nitrogen, can condense and, in the other, warmer, direction, significantly increased greenhouse gas contributions from higher hydrocarbons, ammonia, and even water vapour are possible. However, the most likely fate for the planet-sized moon is an eventual decline in the emission of methane into the atmosphere from surface lakes or from the interior. Then the greenhouse will collapse in a few million years and most of the atmosphere will freeze on the surface, as already happened on Neptune's large satellite Triton. If the absorptivity of the surface subsequently increases with time due to the photochemical conversion of material in the ice or accumulation of dark material from space, perhaps organics similar to those seen on the surfaces of some of the other Saturnian moons, the surface temperature will increase, perhaps to the point where nitrogen and methane could re-evaporate and build up the atmosphere again.

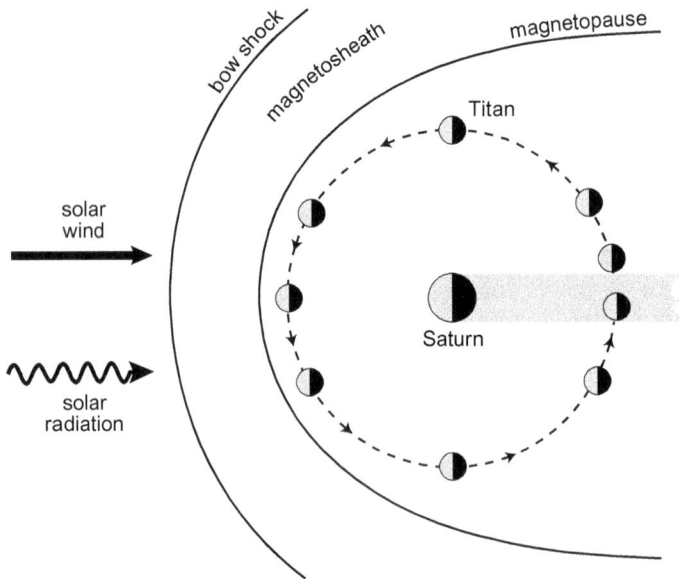

Figure 5.12   During its orbital motion around Saturn, the angle between the solar photon radiation and the magnetospheric flow direction at Titan varies through 360°. At times, the magnetopause moves inside Titan's orbit, removing the magnetospheric particle flux, while around the equinoxes there are times when the solar flux is eclipsed by Saturn.

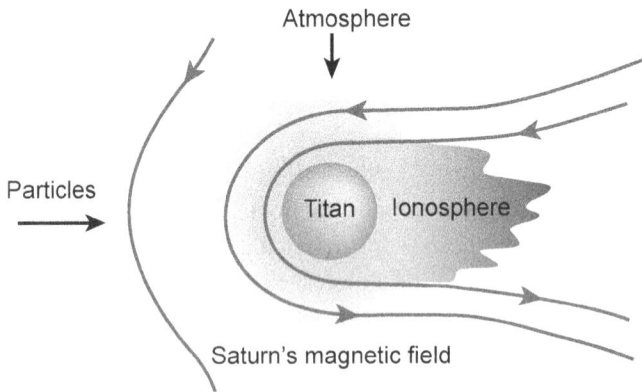

Figure 5.13   Titan's atmosphere obstructs the flow of charged particles in Saturn's magnetic field, producing a bow shock and distorting the ionosphere, from which charged particles can escape into space.

Conversely, if the methane supply is abundant, a perturbation such as a small increase in insolation over the present value or a large cryovolcanic episode adding greenhouse gases to the atmosphere could produce a dramatic warming of Titan's climate. For example, if the methane reservoir referred to above was in the form of liquid methane seas on the surface, as seems possible from recent Cassini observations, small increases in heating would raise the atmospheric mixing ratio of methane rapidly and produce an enhanced greenhouse effect that would further increase the temperature of the surface, leading to more evaporation. The greenhouse effect makes Titan fairly sensitive to increases in tidal heating, for example, such that a Titan-like moon in orbit around a giant planet might be rather warmer than if it were itself a planet in a heliocentric orbit. If Titan were $\sim$80 K warmer, its surface and near subsurface might be liquid (the ammonia-water peritectic melting point is $\sim$176 K). The co-existence of liquid water and the organics produced by photolysis allows the easy synthesis of prebiotic molecules such as amino acids.

As the Sun becomes a red giant, its luminosity will increase by over an order of magnitude for several hundred million years. However, as the atmosphere is warmed, it expands. Production of aerosol is tied to the absorption of UV at a given pressure (typically the aerosol production altitude in scattering models is 0.1 mbar) so that the altitude of haze production increases as the solar luminosity increases. Thus, for a fixed production rate, since the haze has much further to fall, the column optical depth increases. This largely compensates for the increased luminosity. A small offsetting effect is the change in the solar spectrum — as the solar surface cools, more red light is produced which penetrates deeper into the atmosphere than the present Sun. Some models suggest that, while Titan may get somewhat warmer when the Sun becomes a red giant and destroys the terrestrial planets, it still does not offer a strong prospect as a suitable abode for the survival of humanity.

Suddenly I was aware of something new. The air in front of me had lost its crystal clearness. .... I was aware of a faint taste of oil upon my lips, and there was a greasy scum upon the woodwork of the machine. There was no life there. It was inchoate and diffuse, extending for many square acres and then fringing off into void. No, it was not life. But might it not be the remains of life? Above all, might it not be the food of life, a monstrous life, even as the humble grease of the ocean is the food for the mighty whale?

Arthur Conan Doyle, *The Horror of the Heights*

6.1 Titan's Chemical Composition

What we knew about Titan's chemical composition until about a decade ago was based primarily on the data recovered by Voyager in 1980. More recently, measurements from the ground and by the ISO space observatory have complemented this knowledge with a number of additional atmospheric compounds and more refined abundance values. Finally, since 2004, our understanding of the atmospheric composition has been greatly enhanced using the data returned by the Cassini and Huygens instruments described in Chapter 4, not least by the mass spectrometer on the probe.

The main constituent in Titan's atmosphere is molecular nitrogen, just like the Earth. Nitrogen is difficult to observe by remote sensing at the usual infrared and microwave wavelengths; as a homopolar molecule (both atoms the same), it has no permanent dipole moment and cannot interact with photons in the normal way. A similar problem exists with argon, which eluded detection until *in situ* Cassini–Huygens data showed the presence of small amounts of ^{36}Ar and ^{40}Ar. Nitrogen and argon do register in the ultraviolet spectrum, and nitrogen was hence first firmly detected by the Voyager 1 UV spectrometer. However, because the UV spectrum originates in the rarefied upper atmosphere, it does not give a reliable indication of the bulk composition of the atmosphere. After nitrogen, methane, which has a rich infrared spectrum, is the most abundant molecule to have been definitely detected, followed by traces of H_2 and other more complex organics.

For a long time, our knowledge of the bulk composition relied heavily on the determination of the mean molecular weight m, recovered by Voyager data through a combination of the radio-occultation method and infrared spectroscopy. The occultation method is enabled when a spacecraft flies behind the planet so that its radio communication beam passes through the atmosphere and the effect on the signal can be used to derive a T/m profile. The temperature was thus retrieved in 1980 from the analysis of the emission of the methane band at 7.7 μm, recorded by the IRIS spectrometer, and the combined data, with their associated uncertainties, yielded a range of possible mean molecular weights for the atmosphere. In a preliminary study, which assumed the perfect gas law (which is not quite correct), m fell in the range from 27.8 to 29.3 amu. Since the value for nitrogen is 28, this was consistent with its role as the predominant molecule, but allowed for a few percent of a heavier molecule like argon (36 amu), or a smaller amount of a lighter molecule like methane (16 amu). None of the other noble gases has been detected on Titan to date, and the amount of argon present is limited to a fraction of one part per million.

The methane abundance is extremely important for its effect on the atmospheric thermal structure and contribution to the material on the surface, but this too remained uncertain until recently. The stratospheric abundance of methane is a balance between destruction by photochemical processes and vertical transport through the cold trap at the tropopause. Photolysis of methane takes place much higher ($\sim$700 km) and catalytic destruction by radicals — such as C_2H — are the main depletion mechanisms in the stratosphere. Prior to Cassini, the value was thought to lie between 1.7 and 3.0%, depending on the argon abundance assumed. The mole fraction of CH_4 near the equatorial surface was estimated to be higher, with probable values in the range 4.5–8.5%, the uncertainty depending primarily on the degree of supersaturation present. The near-surface abundance of N_2 was then likely to be between 95% (for zero argon and 5% methane) and 85% (7% argon, 8% methane). Studies based on the Voyager occultation data combined with the IRIS spectra between 200 and 600 cm^{-1}, tended to support the idea of methane supersaturation, where the species remains gaseous even though it is below the temperature where it can condense, in the troposphere. On the other hand, methane can condense to form clouds but it was not known until quite recently whether or not this actually happened. The evidence for clouds and aerosols of methane and other condensates, and the possible sizes and distributions of suspended particles, are discussed in Chapter 7. The exact altitude and frequency of occurrence of the low-level clouds is not precisely known yet, but the fact of their formation, which may depend on the availability of condensation nuclei for the droplets, means supersaturation at the same time is less likely.

Before Cassini, beyond the compositional uncertainties, corresponding uncertainties on the thermal profile also existed, since the measured spectrum depended on both in ways that cannot be completely separated. The state of play even after ISO, at the beginning of the 21st century, left room for uncertainties of about 2 K at the tropopause and 4 K between 150 and 200 km altitude. The surface temperature at the occultation point near the equator was estimated to lie between 92.5 and 95.5 K; of course, this is likely to vary with location and time.

Table 6.1   Constituents detected in Titan's atmosphere and first-time references. V1 stands for Voyager 1, IRIS for the Infrared Radiometer Interferometer Spectrometer, GCMS is the Huygens Gas Chromatograph Mass Spectrometer and ISO/SWS is the Infrared Space Observatory Short Wavelength Spectrometer.

Constituent	First detection/range/means	Refs. of first detection
Major		
Molecular nitrogen, N_2	Voyager radio occultation; UV	1,2
Nitrogen, N	Voyager, 1134 Å multiplet	2
Methane, CH_4	Ground-based, UV and IR: 6190 & 7250 Å, 1.1 & 7.7 μm	3, 4, 5
	Ionosphere with Cassini/INMS	6
Monodeuterated methane, CH_3D	Ground-based at 1.65 and 8.6 μm	7, 8
Hydrogen , H	V1, 1216 Å	2
Hydrogen, H_2	Ground-based, 3–0 S(1)	4
	Ionosphere, Cassini/INMS	6
Argon (Ar^{36}, Ar^{40})	Cassini–Huygens/GCMS	9
Minor		
Ethane, C_2H_6	Ground-based, 822 cm^{-1}	10, 11
Acetylene, C_2H_2	Ground-based, 729 cm^{-1}	7, 12
	Ionosphere, Cassini/INMS	6
Monodeuterated acetylene, C_2HD	Cassini/CIRS, 678 cm^{-1}	13
Propane, C_3H_8	V1/IRIS, 748 cm^{-1}	5, 14
Ethylene, C_2H_4	Ground-based, 950 cm^{-1}	7
Methylacetylene, CH_3C_2H	V1/IRIS, 328, 633 cm^{-1}	5, 14
Diacetylene, C_4H_2	V1/IRIS, 220, 628 cm^{-1}	15
Benzene, C_6H_6	ISO and Cassini/CIRS, 674 cm^{-1}	9, 13, 16
	Huygens/GCMS	
Hydrogen cyanide, HCN	V1/IRIS, 712 cm^{-1}	5
Cyanoacetylene, HC_3N	V1/IRIS, 500, 663 cm^{-1}	15
Cyanogen, C_2N_2	V1/IRIS, 233 cm^{-1}	15
Dicyanogen, C_4N_2	V1/IRIS, solid form at 474 cm^{-1}	17
Acetonitrile, CH_3CN	220.7 GHz multiplet	18
Carbon monoxide, CO	Ground-based, mm, submm, microwave, infrared	19
Carbon dioxide, CO_2	V1, 667 cm^{-1}	20
Water, H_2O	ISO/SWS, 237, 243 cm^{-1}	21
Ammonia, NH_3, C_2H_3CN, C_2H_5CN, CH_2NH	Suggested indirectly by modelling Cassini/INMS ionospheric data	22

[1]Lindal *et al.* (1983); [2]Broadfoot *et al.* (1981a); [3]Kuiper (1944); [4]Trafton (1972); [5]Hanel *et al.* (1981); [6]Waite *et al.* (2005); [7]Gillett (1975); [8]Lutz *et al.* (1981); [9]Niemann *et al.* (2005); [10]Gillett *et al.* (1973); [11]Danielson *et al.* (1973); [12]Caldwell *et al.* (1977); [13]Coustenis *et al.* (2007); [14]Maguire *et al.* (1981); [15]Kunde *et al.* (1981); [16]Coustenis *et al.* (2003); [17]Samuelson *et al.* (1997); [18]Bézard *et al.* (1993); [19]Lutz *et al.* (1983); [20]Samuelson *et al.* (1983); [21]Coustenis *et al.* (1998); [22]Vuitton *et al.* (2006).

The chemical processes in Titan's present-day atmosphere are dominated by reactions between molecules containing H, C, and N, giving rise to a suite of hydrocarbons and nitriles in the stratosphere, in varying abundances. The known oxygen chemistry involves small amounts of H_2O, CO and CO_2. The neutral species are produced through photochemical reactions, whereas ions are formed primarily as a result of charged particle precipitation, although these can form neutral species later in the reaction chain. Note however that solar extreme ultraviolet (EUV) photons are responsible for the photoionization of neutral species that produce the observed ionosphere in the upper atmosphere. Photoelectrons produced by the photoionization of neutrals, enhance the ion production, while cosmic rays and other external energetic particles produce the low altitude observed ions. The energetic particles from Saturn's magnetosphere have a smaller contribution than EUV photons and their contribution is mainly in the mesosphere. We have a rudimentary understanding of these processes, as they apply to Titan, and models have been developed to fill in the gaps where measurements are still needed. The organics combine in the atmosphere to produce aggregates of the materials called tholins, producing aerosols that grow chemically until a certain size beyond which they start to coagulate. At this point their growth depends on the collision rate. When they are large enough, gravitational settling will bring them to the surface; on the way, they may act as condensation nuclei for condensable species such as methane and lead to the formation of clouds and rain.

Titan's gaseous composition is undoubtedly even more complex than the currently measured inventory of gases, summarised in the following tables, suggests. Laboratory simulations and chemical models imply that Earth- and spacecraft-based spectroscopy of the neutral atmosphere is picking up only the simplest molecules, and may not be revealing the full chemical complexity of the satellite's atmosphere. Very large or 'macro' molecules, especially more complex hydrocarbons and nitriles than those observed from Earth or by Voyager, could be present in significant amounts on Titan. The fact that we did not have evidence for these species prior to Cassini–Huygens was attributed to them being present only in very small quantities, or having spectral bands that are weak or obscured by more abundant species, or all three reasons. However, in very recent Cassini observations from a combination of mass/charge and energy/charge spectrometers performed and analysed by Hunter Waite and his group, evidence was found for tholin material (as negatively charged massive organic molecules) and their formation at high altitudes (about 1,000 km) in Titan's atmosphere.

The formation and growth of solid and liquid particles in the atmosphere must, unless they evaporate in the relatively warm lower atmosphere, which is unlikely at least for the organics, eventually result in precipitation as rain or drizzle on Titan's surface, now explored by Cassini–Huygens. However, the possibilities for the composition of the surface on Titan remain numerous: water and other ices, rocky material, and accumulations of hydrocarbon snow and rain may all be present. Preliminary analyses of the properties of any liquid areas on Titan's surface, discovered

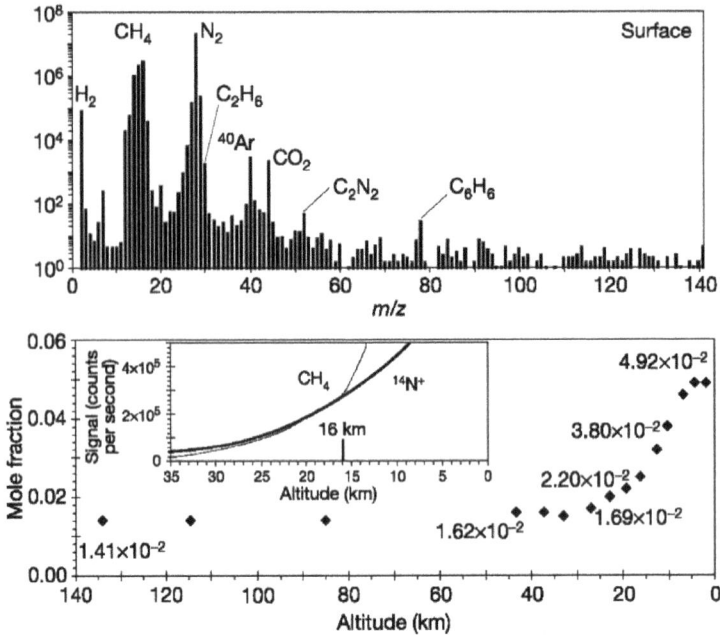

Figure 6.1   Above: the atmospheric composition at Titan's surface from the Huygens GCMS. Below: the GCMS report on the mole fraction of methane with respect to nitrogen versus altitude in Titan's atmosphere (Niemann *et al.*, 2005).

mainly in the north polar region, suggest that they could be mostly composed of liquid ethane and liquid methane, in unknown proportions, with a few percent of dissolved nitrogen (2 to 6%). If the liquid is warm, relatively speaking, it could be an efficient CO reservoir. More recent studies show however that if the surface and the atmosphere are in equilibrium (which is not certain), any liquid on the surface is likely to be about 60% CH_4 near the equator and 30% near the poles, with the remainder mostly liquid C_2H_6. A full discussion and speculation about the surface properties is deferred to Chapter 9.

6.2 The Bulk Composition of the Atmosphere

Interestingly enough, the proportions of the two main constituent of Titan's atmosphere were more difficult to determine than the abundances of the trace constituents. The methane stratospheric abundance, for instance, was not well constrained at the time of Voyager or ISO, but varied significantly among different estimates, while the thermospheric abundance was overestimated. The precise nitrogen content, although we now know that it is more than 90%, proved very elusive for many years.

Cassini/Huygens finally produced firm determinations for the major constituents, nitrogen, methane and hydrogen. The CH_4 mole fraction is 1.41×10^{-2} in the stratosphere, confirming methane as the second most abundant molecule on Titan.

It begins increasing below 32 km, until at about 8 km, where it reaches a plateau of about 4.9×10^{-2}. The data shows an increase of methane at 16 m/z, when compared to nitrogen (in this case $^{14}N^+$) at $m/z = 14$, near 16 km. This is probably due to condensates evaporating in the inlet system of the mass spectrometer as the Huygens probe passed through a layer of methane haze. The Huygens GCMS measurements in the lower atmosphere are in good agreement with the stratospheric CH_4 value inferred by CIRS on the Cassini orbiter ($1.6 \pm 0.5 \times 10^{-2}$) and the surface estimate given by the Huygens DISR spectra (also about 5% at the surface).

The GCMS also witnessed a rapid increase of the methane signal after the probe's landing heated the surface, which suggests that liquid methane exists on the surface or right underneath, together with other trace organic species, including cyanogen, benzene, ethane and carbon dioxide. The methane abundance measurement in the upper atmosphere by Cassini ended a long controversy over the value measured, and later revised, by the Voyager UVS team. The analysis by H. Waite and colleagues of the Cassini mass spectrometer measurements showed that methane's mole fraction is $2.71 \pm 0.1\%$ at 1,174 km, which led to significantly improved constraints on the whole methane profile.

Nitrogen and methane, the most abundant constituents in Titan's atmosphere, are both photodissociated by solar ultraviolet radiation, energetic particles from Saturn's magnetosphere and galactic cosmic rays, leading to the initiation of a complex organic photochemistry, which finally produces the haze. In the atmosphere, atomic hydrogen is transformed to molecular hydrogen in the presence of aerosol particles. The abundance of H_2 is estimated from Cassini and Voyager data to be in the 0.1–0.2% range. With the exception of trace amounts of hydrocarbons, nitriles and a few oxygen species, the rest of the atmosphere is almost entirely dominated by molecular nitrogen, more than about 97% in the stratosphere.

Argon was also expected to be a major atmospheric constituent, based on cosmogonical considerations. However, although it is indeed the only noble gas detected to date, it is found in small amounts only, mostly in the form of primordial ^{36}Ar (2.8×10^{-7}) or its radiogenic isotope ^{40}Ar (4.32×10^{-5}). The low abundance of primordial noble gases on Titan has been interpreted as meaning that nitrogen was originally captured in a relatively volatile form, i.e. as NH_3 rather than N_2. Subsequent photolysis could then have created the N_2 atmosphere we see today.

6.3 Ionospheric Chemistry

Prior to Cassini–Huygens, the most reliable information we had about Titan's atmosphere related to the middle atmosphere, the part at altitudes between roughly 100 and 500 km from the surface, because this is the region probed by the infrared spectroscopy measurements conducted by space missions and observatories like Voyager and ISO. The lowest part of the atmosphere, the troposphere, remained largely unexplored, except for a few long-wavelength ground-based measurements

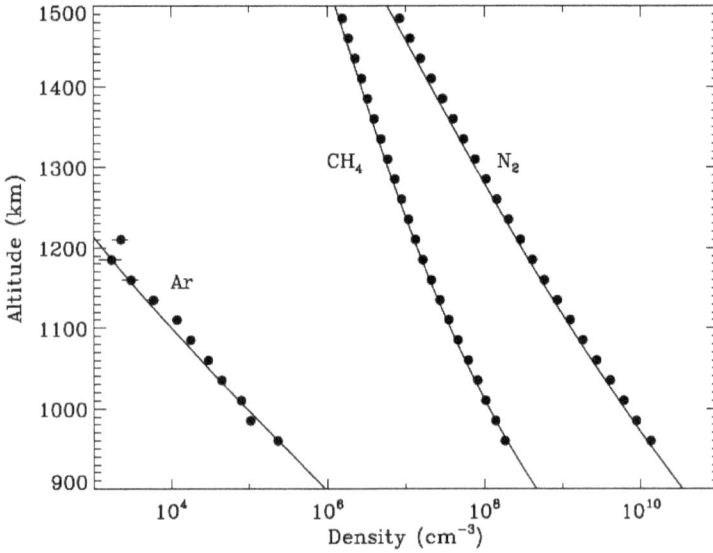

Figure 6.2   The Cassini/INMS-derived densities of methane, molecular nitrogen, and argon. The solid black lines represent the best-fit densities, which correspond to a 149 K isothermal temperature profile. The methane mixing ratio from this fit to the data is 2.7% at 1174 km (Waite *et al.*, 2005).

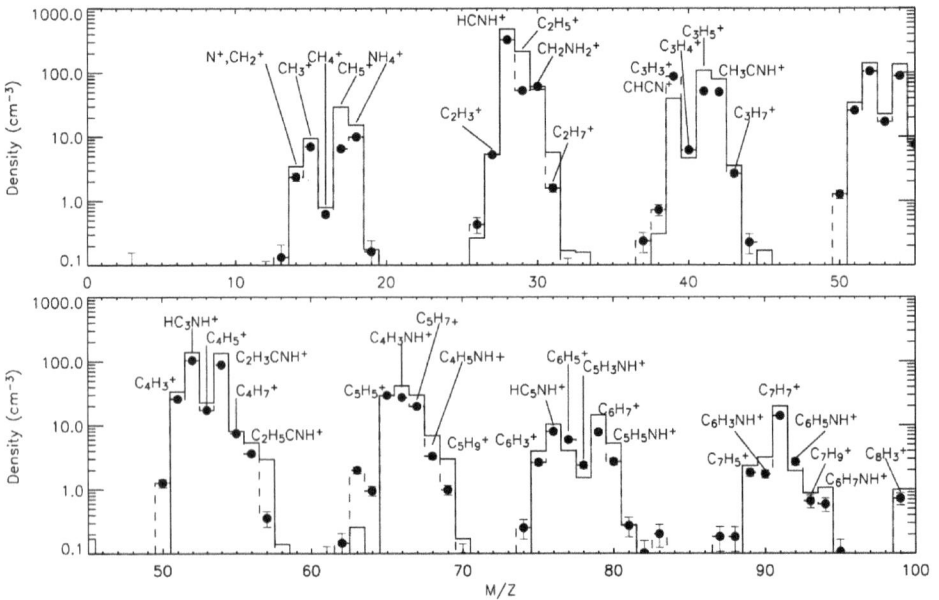

Figure 6.3   INMS measurements of April 2005 (dots) and modelled spectrum (lines). The prominence of NH_3, CH_2NH, CH_3CN, C_2H_3CN, C_2H_5CN was not predicted by pre-Cassini models (Vuitton *et al.*, 2006).

Table 6.2 Neutral mole fractions at an altitude of 1100 km, used in the model by Vuitton *et al.* (2006) for the species detected in the ionosphere of Titan by Cassini/INMS. Adapted from Vuitton *et al.* (2006).

Ionospheric species detected by INMS	Neutral mole fractions
H_2	4×10^{-3}
CH_4	3×10^{-2}
C_2H_2	3×10^{-4}
C_2H_4	6×10^{-3}
C_2H_6	1×10^{-4}
C_4H_2	6×10^{-5}
HCN	2×10^{-4}
HC_3N	2×10^{-5}
CH_3CN	1×10^{-5}
C_2H_3CN	1×10^{-5}
C_2H_5CN	5×10^{-7}
NH_3	7×10^{-6}
CH_2NH	$<1 \times 10^{-5}$

and the Voyager radio-occultation and IRIS data. The higher part of the atmosphere, including the ionosphere, was also little explored, although there were some ultra-violet observations, as was briefly discussed in Chapter 3.

Cassini/INMS recently determined the ionospheric abundance of a host of nitrogen-bearing molecules in Titan's upper atmosphere. Dissociation of N_2 and CH_4, the primary atmospheric constituents on Titan, by energetic photons and electrons, provides H, C, and N atoms in the upper atmosphere, which then combine to produce complex organic molecules. The density of ions in Titan's upper atmosphere depends closely on the composition of the neutral atmosphere and, for many species, measurements of associated ions coupled with simple chemical models provides a sensitive determination of their abundances. This technique proved useful to obtain the densities of C_2H_4, C_4H_2, HCN, HC_3N, CHCN, NH_3, C_2H_3CN, C_2H_5CN, and CH_2NH. The latter four species had not previously been detected in the gas phase on Titan, and none had been accurately measured in the upper atmosphere. The presence of these organics implies that nitrogen chemistry on Titan is more extensive than previously realized.

6.4 Trace Constituents in the Neutral Atmosphere

As we have seen in Chapter 3, the complexity and diversity of Titan's atmosphere was first intimated by ground-based observations in the 1970s and, in 1980, by the Voyager missions with the first close-up measurements. The spectrum of Titan in the thermal infrared (200–$1,500\,cm^{-1}$) was first acquired by the Infrared Radiometer Spectrometer (IRIS) aboard the Voyager 1 mission in November 1980. The spectral resolution was a constant $4.3\,cm^{-1}$ and the spatial resolution provided disk-resolved measurements of Titan's stratosphere at various latitudes and longitudes during the flyby. The thermal and compositional structure of Titan's stratosphere was retrieved from these spectra, identifying numerous hydrocarbons, such as methane, acetylene, ethylene, ethane, diacetylene, methylacetylene, propane and monodeuterated methane. Also, the signatures of nitriles, such as cyanoacetylene, cyanogen and dicyanoacetylene (the last as ice in emission at $478\,cm^{-1}$; it has no equivalent gaseous band), as well as CO_2, were found in the IRIS spectra. These species were discovered in a special Voyager IRIS observational sequence, consisting of $\sim$30 spectra recorded at grazing incidence over Titan's north polar region. This allowed the retrieval of meridional variations of these constituents, based on spectral averages from 53°S to 70°N. In 1981, Titan was again observed by Voyager 2, but from a much greater range.

Later on, in 1997 the Infrared Space Observatory (ISO) offered the possibility to take a high-resolution snapshot of Titan's disk a little over two Titanian seasons after the Voyager encounters. ISO was put into orbit around the Earth in November 1995 and lasted for 28 months. It carried a 60 cm diameter He-cooled telescope and operated in the 2–200 μm spectral region through 4 instruments: 2 spectrometers, a photometer and a camera. The Titan spectra, acquired on January 10 and December 27, 1997, at a spectral resolution about 10 times higher than Voyager/IRIS, were disk-averages. Those recorded by the Short Wavelength Spectrometer in the grating mode cover the range from roughly 2 to 50 μm and show emission signatures of all of the expected minor constituents in Titan's stratosphere (hydrocarbons, nitriles and CO_2), with a higher resolution (0.3 to $0.8\,cm^{-1}$) allowing the resolving of the different bands and the separation of the various contributions and so a better determination of the abundances and vertical distributions of the constituents. The ISO/SWS spectra provided the first detection of water vapour in Titan's atmosphere from 2 emission lines around 40 μm (the mole fraction derived at 400 km of altitude is about 10^{-8}), as well as the first hint of the presence of benzene (C_6H_6) at $674\,cm^{-1}$ for a mole fraction on the order of a few 10^{-10}. Since then, the benzene detection has been confirmed by Cassini at several locations on Titan's disk.

However, apart from the dedicated ISO/SWS observation around 40 μm ($250\,cm^{-1}$) where water vapour was detected, the spectral range at wavenumbers shorter than $200\,cm^{-1}$ escaped detection. The Cassini/CIRS instrument fills the gap, with spectra in the 10–$1,500\,cm^{-1}$ range at resolutions ranging from 0.5 or $2.5\,cm^{-1}$ depending on the objective, and this has confirmed the presence of the species observed by IRIS (except for solid C_4N_2 at the moment), while adding those found in the FP1 sub-mm region: CO, H_2O, and CH_4. In addition, the CIRS spectra have so far allowed the detection of several new isotopes, new vertical distributions, and a better characterisation of the haze. Pre-Cassini knowledge of the chemical composition of Titan's stratosphere as derived by Voyager and ISO data is summarised in Table 6.3.

Table 6.3 Atmospheric gaseous abundances for Titan from Voyager 1 equatorial data and ISO disk-average spectra. The observations were taken more than 2 Titanian seasons apart.

Molecule	Voyager (1980)	ISO (1997)
C_2H_6	1.3×10^{-5}	1.3×10^{-5}
C_2H_2	3.0×10^{-6}	2×10^{-6}
C_3H_8	5.0×10^{-7}	5×10^{-7}
C_2H_4	1.5×10^{-7}	8×10^{-8}
C_3H_4	5.0×10^{-9}	8×10^{-9}
C_4H_2	1.4×10^{-9}	1.5×10^{-9}
C_3H_4 (allene)		$<5 \times 10^{-9}$
C_6H_6 (benzene)		$\sim 5 \times 10^{-10}$
HCN	1.7×10^{-7}	1.5×10^{-7}
HC_3N	$<1 \times 10^{-9}$	
C_2N_2	$<1.5 \times 10^{-9}$	
CO_2	1.4×10^{-8}	1.5×10^{-8}
H_2O at 400 km		8×10^{-9}
CH_3D	1.5×10^{-5}	8×10^{-6}

While the Voyager missions and ISO had done wonders in providing a fairly complete general view of Titan's stratosphere in terms of mean chemical composition and temperature, there was still a shortage of information on vertical profiles. These are very important for understanding the link between composition, chemistry, and the atmospheric circulation. The horizontal-viewing north pole sequence taken by IRIS allowed vertical distributions to be inferred for some of the species. Progress was also made from Earth-based observatories, due to new technologies becoming available at telescopes much sooner than they could be deployed in space. In addition, the use of certain long-wave spectral domains such as the millimetre, radio and microwave ranges can be easier with Earth-based receivers. Thus, prior to the arrival of Cassini at Titan in 2004, ground-based observations provided not only new detections, including CO and CH_3CN, but also most of the vertical distribution information that we had.

Carbon monoxide (CO) was discovered from the Earth in the near-infrared range, as was monodeuterated methane (CH_3D). In 1992, thanks to the IRAM radio telescope, we witnessed the first detection of a more complex nitrile than any of those observed by Voyager, in the form of CH_3CN, acetonitrile, in the millimetre range around 220.7 GHz at 0.1 MHz resolution. The proportion of acetonitrile is found to strongly increase with height in the stratosphere, by about one order of magnitude between 200 and 300 km, reaching about ten parts per billion. Since then, it has been detected in Titan's ionosphere by the VIMS spectrometer on Cassini, but still not in the neutral atmosphere.

Ground-based high-resolution heterodyne millimetre observations of Titan also offered the opportunity to determine vertical profiles and partial mapping in some cases of HCN, CO, HC_3N, and CH_3CN, which showed that the nitrile abundances increase with altitude, in agreement with predictions by photochemical models which place the production zone above 300 km in the mesosphere — thermosphere. Various sinks, including condensation with subsequent precipitation, cause abundances to decrease in the lower stratosphere. CH_3CN and CO were not observed on Titan with Voyager or ISO. The vertical profile of CO was very recently investigated again both in the stratosphere and in the troposphere by use of infrared, millimetre and microwave observations from the Earth. These two molecules are discussed more extensively below.

The question of the stratospheric profile of HCN on Titan was also revisited recently using the IRAM 30 m radio telescope at Pico-Veleta in Spain. The vertical distribution, obtained from an analysis of the first rotational transition at 88.6 GHz, was found to increase with altitude. The HCN abundance increases from 50 ppb at 100 km, to 7 times as much near 200 km, through a mean value of about 160 ppb around the 160 km level, where the scale height is about 47 km. These results are consistent with the analyses of Voyager infrared measurements. On the other hand, the inferred vertical concentration gradient is much steeper, and the abundance in the lower stratosphere smaller, than photochemical models predict. Further ground-based work allowed for spectrally resolved observations of methane lines and a first detection of the mesosphere of Titan in the infrared, as well as high-resolution measurements of propane and ethylene on Titan. These have now largely been overtaken by the recent findings in Titan's neutral atmosphere reported by Cassini instruments.

6.4.1 *Stratospheric Composition Measurements with Cassini*

Even before the Cassini spacecraft arrived in the Saturnian system on July 1, 2004 after a 6.7 year trek, the Composite Infrared Spectrometer aboard Cassini had been active returning valuable information on Jupiter, Saturn and Titan. CIRS is a Fourier transform spectrometer consisting of two interferometers scanning the far-infrared (10–$600\,cm^{-1}$) and mid-infrared (600–$1,500\,cm^{-1}$) ranges with an apodized spectral resolution varying from 15.5 to $0.5\,cm^{-1}$ (see Chapter 4). This represents an order of magnitude improvement over the IRIS resolution, and is similar to that achieved with ISO in the mid-infrared range. The advantage offered by CIRS over ISO is obviously to be found in the disk-resolved high-resolution measurements, and the long period of study, possibly over a period of about 6 years until 2010 (end of the extended mission) or beyond, covering a large part of a Titanian season. With respect to IRIS, in addition to the higher spectral resolution and the increase in linear spatial resolution by a factor as high as an order of magnitude, there is the superior sensitivity and the higher limb-viewing resolution and time-coverage afforded by CIRS. As was the case with ISO, the spectral resolution of CIRS allows us to separate the contribution of species such as C_3H_4 and C_4H_2 (around $630\,cm^{-1}$), or

CO_2, C_6H_6 and HC_3N (around $670\,cm^{-1}$), which presented blended bands in the Voyager data, and to disentangle the abundances of HCN (at $713\,cm^{-1}$) and C_3H_8 (at $748\,cm^{-1}$) from that of C_2H_2 (at $730\,cm^{-1}$). Furthermore, the spectral range of CIRS includes the interval between 10 and $200\,cm^{-1}$, which was not available to IRIS.

After the data are received on the ground, the interferograms are edited and transformed into calibrated spectra. Steps in this process include eliminating bad scans, removing predicted noise patterns, calibrating, apodizing, and Fourier transforming. Algorithms remove interferograms that have anomalous intensities and adjust interferogram lengths for uniformity. Calibration is performed using untransformed interferograms and complex spectra of deep space and internal shutter reference targets, so that phase correction is an intrinsic part of the process.

The analysis of the CIRS data involves complex radiative calculations and modelling. CIRS Focal Plane 1 encompasses the $10–600\,cm^{-1}$ range in which the rotational and vibrational signatures of CO, HCN, CH_4, H_2O, and their isotopes have so far been detected. Focal Plane 3 includes the spectral bands which yield the molecular abundances for C_2H_2, its deuterated isotope (C_2HD), C_2H_4, C_2H_6, C_3H_4, C_3H_8, C_4H_2, HCN, HC_3N and CO_2, in the $600–1,000\,cm^{-1}$ spectral region. Methane and its monodeuterated isotope CH_3D, as well as propane, ethane and their isotopes, are observed in Focal Plane 4 between 1,000 and $1,400\,cm^{-1}$. CH_3D is important because it allows us to obtain a value of the D/H ratio in methane, which is a key parameter in cosmological models, since deuterium is destroyed in stars. The ν_4 CH_4 band at $1,304\,cm^{-1}$ serves as an atmospheric thermometer, particularly for the stratosphere, and analysis of its emission gives the temperature profile at different locations of Titan's disk. Once the thermal profile is known from the inversion of radiances measured in the methane ν_4 fundamental, the radiative transfer equation can be used again to derive the mixing ratios of the other gases whose features can be seen in the spectra. An iterative procedure is used, which can be described in outline as follows.

A line-by-line radiative transfer program, incorporating the temperature profile and using the abundances of the absorbers as parameters, generates synthetic spectra in the desired spectral range. These spectra are compared to the observations until the best agreement is obtained by trial and error. This gives the mixing ratios of the molecules exhibiting emission bands in the spectra at a given latitude; the procedure is repeated for measurements at different locations and times. The atmospheric opacity in the model must be correctly calculated including all of the important contributions from molecular sources and from the haze aerosols. Because the pressure on Titan is quite high, the radiative transfer calculations have to include the collision-induced absorption in the troposphere (with some contribution in the lower stratosphere at wavenumbers lower than $250\,cm^{-1}$) between the more abundant molecules, i.e. $N_2–N_2$, $N_2–H_2$, $N_2–CH_4$, $CH_4–CH_4$, as well as the vibration-rotation bands from the molecular species with permanent dipole moments. The most important of these are CH_4, CH_3D, C_2H_2, HCN, C_3H_4, HC_3N, C_4H_2, C_2H_4,

C_3H_8, C_2H_6, C_2N_2 and CO_2. The clouds and aerosols contribute continuum opacity, that is absorption which varies slowly over a wide spectral range. Since the composition is unknown, but the opacity tends to vary slowly with wavelength for solid and liquid absorbers, their effect can be simulated by a cloud model, fitted to the measurements outside the bands of the gaseous species. This tends to be the least reliable part of the procedure.

The synthetic spectra, generated by the radiative transfer program, are convolved with the properties of the relevant instrument to give the appropriate spectral resolution of 2.5 or 0.5 cm^{-1} prior to comparison with the measured spectra. When the instrument was viewing in the nadir direction, i.e. vertically downwards, or approximately so, the derived abundances are relative to some stratospheric level on Titan and contain relatively little information as to the vertical distribution of the component. The gas bands of the most abundant hydrocarbons like C_2H_2, C_2H_4, C_3H_8 and C_2H_6 probe levels in the atmosphere peaking around 3–5 mbar. The main emission observed in the bands of higher order and less abundant hydrocarbons (C_3H_4, C_4H_2) comes from lower altitudes (around 9 mbar), whereas the other molecules mainly probe intermediate pressure levels (5–6 mbar). Calculations indicate that the regions probed are similar for southern and mid-latitudes, but shift to lower altitudes (higher pressures) for northern regions.

However, when limb viewing (approximately tangential to the surface) is possible — as for example in a special north pole sequence obtained by Voyager — then vertical distributions can be derived, as will be explained below. Voyager limb data yielded vertical distributions for some of the hydrocarbons and nitriles. The vertical distributions generally showed an increase with altitude, confirming the prediction of photochemical models that these species form in the upper atmosphere and then diffuse downwards in the stratosphere. Below the condensation level of each gas, the distributions were assumed to decrease following the respective vapour saturation law.

6.4.1.1 *Hydrocarbons*

After methane, the most abundant hydrocarbons in Titan's stratosphere are ethane, acetylene and propane, in that order. Ethane is one of the major trace gases in Titan's stratosphere (the most abundant hydrocarbon, along with CH_3D, after CH_4) and showed no significant variation as a function of latitude in the Voyager spectra. From CIRS data we also find little variation in the C_2H_6 abundance from north to south (Table 6.4), with the most common value at lower latitudes around $1.3 \pm 0.3 \times 10^{-5}$, although some subtle increase to the north is suggested by the fits. This ethane abundance is in excellent agreement with the $1.3 \pm 0.1 \times 10^{-5}$ mean equatorial value inferred from IRIS.

The best fit for acetylene, using a constant-with-height stratospheric mixing ratio, requires a mole fraction of about $3.7 \pm 0.8 \times 10^{-6}$ near the equator, which holds throughout all latitudes within error bars. When this value is used the calculated synthetic spectrum satisfies the emission observed in the centre of the Q-branch at $729\,\text{cm}^{-1}$ for 5°S, but does less well in the right and left wings of the band, for both the medium- and the high-resolution selections. Given the CIRS nominal temperature profile used for 5°S, the contribution function for the $C_2H_2\nu_5$ Q-branch at $729\,\text{cm}^{-1}$ probes a large pressure range between 0.3 and 10 mbar, peaking at around 4 mbar, the wings of the band probing generally (except for the hot bands) somewhat lower atmospheric levels (around 5 mbar at $738\,\text{cm}^{-1}$ and 10 mbar at $675\,\text{cm}^{-1}$), so information can be inferred on the acetylene

Table 6.4   Chemical composition of Titan's neutral atmosphere, as found in ground-based observations or by spatially-resolved Cassini–Huygens measurements. The species are listed in decreasing abundance within a family. The stratospheric values pertain to pressure levels in the 3–9 mbar range. The 'North pole' values correspond to about 50°N, and can be higher at higher latitudes.

Gas		Mole fraction		Comments (Refs.)
Major components				
Nitrogen	N_2	0.97		Inferred indirectly
Methane	CH_4	1.4×10^{-2}		Stratosphere (1,2)
		4.9×10^{-2}		Surface (2,3)
Monodeuterated				
methane	CH_3D	8×10^{-6}		(4)
Hydrogen	H_2	0.0011		(5)
Argon	^{36}Ar	2.8×10^{-7}		(2)
	^{40}Ar	4.32×10^{-5}		(2)
		Equator	North pole	
Hydrocarbons				
Ethane	C_2H_6	7×10^{-6}	1.1×10^{-5}	(4)
Acetylene	C_2H_2	2.5×10^{-6}	3×10^{-6}	(4)
Monodeuterated				
acetylene	C_2HD	6×10^{-10}	2×10^{-9}	(4)
Propane	C_3H_8	3.5×10^{-7}	6×10^{-7}	(4)
Ethylene	C_2H_4	1.5×10^{-7}	5×10^{-7}	(4)
Methylacetylene	C_3H_4	5.2×10^{-9}	2×10^{-8}	(4)
Diacetylene	C_4H_2	1.1×10^{-9}	2×10^{-8}	(4)
Benzene	C_6H_6	2.0×10^{-10}	3.8×10^{-9}	(4)
Nitriles				
Hydrogen cyanide	HCN	7.7×10^{-8}	7.8×10^{-7}	(4, 6)
Cyanoacetylene	HC_3N	3.0×10^{-10}	4.4×10^{-8}	(4)
Cyanogen	C_2N_2	5×10^{-10}	9×10^{-10}	(6)
Dicyanogen	C_4N_2			Solid form only (7)
Acetonitrile	CH_3CN	1.5×10^{-9}		(8)

(Continued)

Table 6.4 (*Continued*)

Gas		Mole fraction		Comments (Refs.)
Oxygen compounds				
Water vapour	H_2O	8×10^{-9}		(9) at 400 km
Carbon dioxide	CO_2	1.1×10^{-8}	1.3×10^{-8}	(4)
Carbon monoxide	CO	$(2-4) \times 10^{-5}$		Troposphere (10, 11)
		$(2-6) \times 10^{-5}$		Stratosphere (1, 12, 13)
Isotopic ratios				
$^{13}C/^{14}C$		82.3 ± 1		(2)
$^{14}N/^{15}N$ in HCN		67		(11)
in N_2		183 ± 5		(2)
D/H in CH_3D		1.2×10^{-4}		(4)
in HD		2.3×10^{-4}		(2)
in C_2HD		$1-3 \times 10^{-4}$		(4)

1. Flasar *et al.* (2005) from Cassini/CIRS data.
2. Niemann *et al.* 2005 from Huygens/GCMS data.
3. Tomasko *et al.* (2005) from Huygens/DISR data.
4. Coustenis *et al.* (2007, 2008a) from Cassini/CIRS data.
5. Samuelson *et al.* (1997a) from V1/IRIS data.
6. Teanby *et al.* (2006) from Cassini/CIRS data.
7. Samuelson *et al.* (1997b) from V1/IRIS data.
8. Bézard *et al.* (1993) from disk-averaged ground-based heterodyne mm observations.
9. Coustenis *et al.* (1998) from ISO/SWS data.
10. Lellouch *et al.* (2003) from ground-based VLT data at 5 micron.
11. Marten *et al.* (2002) from disk-averaged ground-based heterodyne mm observations.
12. Gurwell and Muhleman (2000) from disk-averaged mm heterodyne data.
13. Baines *et al.* (2006) from Cassini/VIMS data.

abundance as a function of altitude from the study of this band at roughly two pressure levels: one around 3 mbar and one around 7 mbar.

The abundance of propane is, in general, difficult to extract from the Titan spectrum because the propane ν_{21} band at $748\,cm^{-1}$ overlaps the R-branch of a strong acetylene band. The best fits are obtained for values around $6.0 \pm 1.8 \times 10^{-7}$ near the equator, which, despite the uncertainty is in good agreement with the results from ground-based observations made at higher resolution, which allows easier separation of the lines of the two species.

The diacetylene ν_8 and the methylacetylene ν_9 emission bands near $630\,cm^{-1}$, which appeared blended at the IRIS spectral resolution, have clearly resolved contributions in the CIRS data because of the better spectral resolution, up to 9 times higher. This allows their abundances to be inferred with better precision from CIRS than from IRIS. These two constituents show the highest increase towards the north among all the hydrocarbons, starting in the south with mole fractions of about 1×10^{-9} for C_4H_2 and 4.5×10^{-9} for C_3H_4 up to about 40°N. These molecules then increase by factors of 10 and 3 respectively by 70°N. The derived ethylene mole

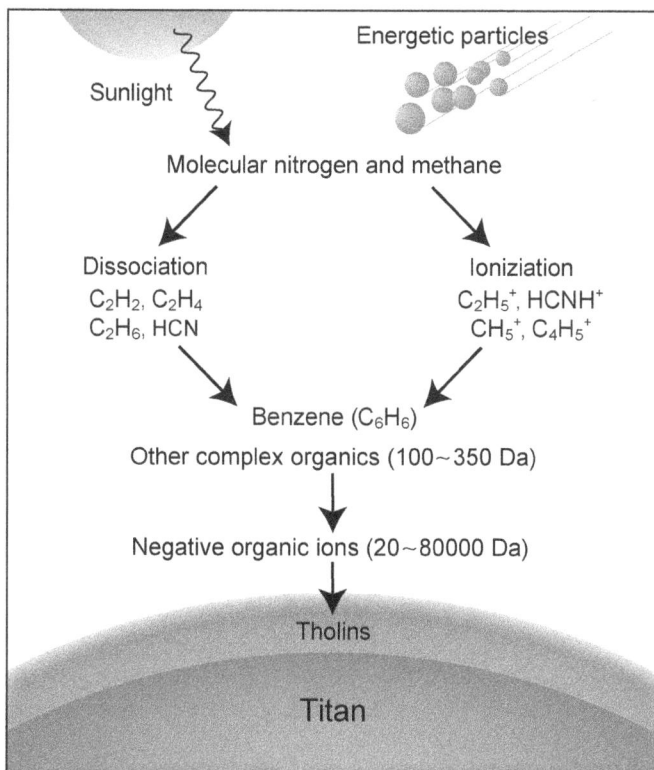

Figure 6.4   Ion chemistry photochemical scheme leading to the production of tholin material and negative organic ions in the upper atmosphere of Titan (After Waite *et al.*, 2007).

fractions also show considerable variability with latitude, from 1.5×10^{-7} in the south to 5×10^{-7} in the north.

No contribution by benzene (C_6H_6) was found in the IRIS spectra in 1980 at a resolution of $4.3\,cm^{-1}$, but its presence was first suggested in ISO data. Benzene had been predicted to exist in Titan's atmosphere from laboratory simulations and photochemical models. In the models, the primary mechanism for the production of benzene on Titan involves the recombination of propargyl radicals (C_3H_3) and aromatic chemistry that may hold considerable significance in the formation of hazes. Since that first indication, benzene has been regularly detected on Titan, in particular from Cassini–Huygens data, both in the atmosphere and on the surface. CIRS spectra near $674\,cm^{-1}$ show an emission feature that increases with latitude north of the equator, at the 10-σ level at 50°N and present at levels of a few 10^{-10} at lower latitudes. The benzene feature is affected, however, by the presence of a nearby additional emission observed at about $678\,cm^{-1}$ and not reproduced by the

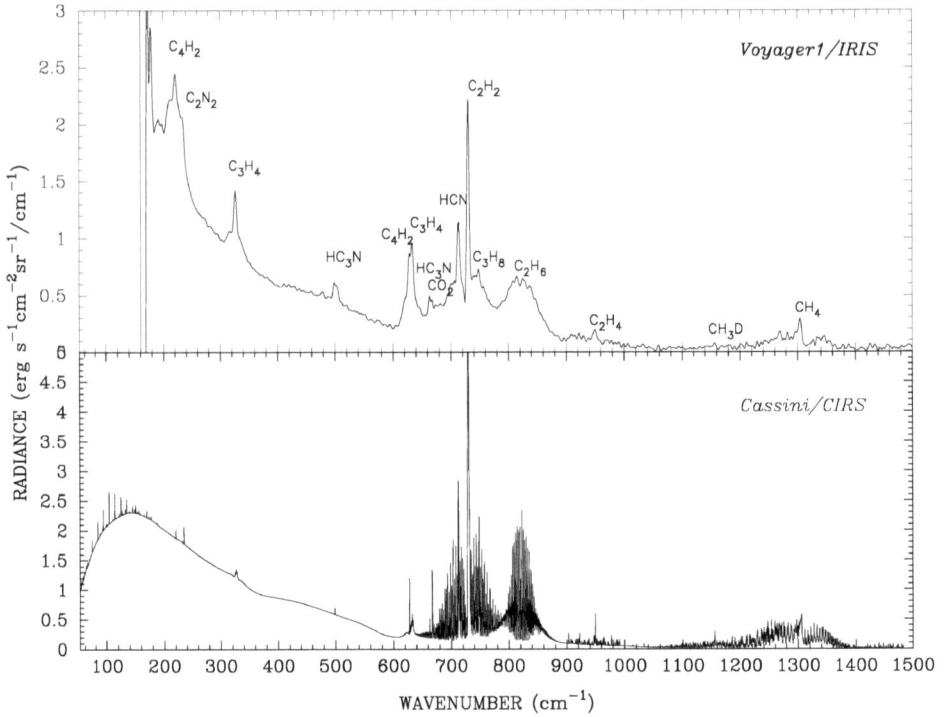

Figure 6.5   Comparison of spectral ranges and resolutions of a Voyager/IRIS spectrum (above) taken in 1980 near Titan's equator and its counter-part from Cassini/CIRS as observed in 2005 (below).

initial theoretical model. It turned out, after further modelling, that this feature is due to C_2HD, an acetylene isotopomer, present in amounts of a few ppb.

Benzene was also reported in Titan's higher atmosphere as a result of Cassini mass spectrometer measurements. The CIRS data yielded C_6H_6 abundances, up to 3.8×10^{-9} at 70°N, corresponding to atmospheric levels between 0.5 and 20 mbar (the C_6H_6 contribution function peaks near 6 mbar at the equator and 7.5 mbar in the north). This constant abundance assumed above the 30 mbar level yields a column density for Titan of about 4×10^{15} molecules cm^{-2}.

6.4.1.2 Nitriles

Nitrile detections are limited at present to HCN, HC_3N, C_2N_2, CH_3CN and C_4N_2, the last only in its solid form. The constant-with-height mixing ratio of HCN that best matches the CIRS data increases from around 5.7×10^{-8} in the south to an order of magnitude or so higher in the north. The value of 7.7×10^{-8} inferred from the emission observed in the v_2 713 cm^{-1} band near the equator is relevant to altitudes around 130 km (5–6 mbar).

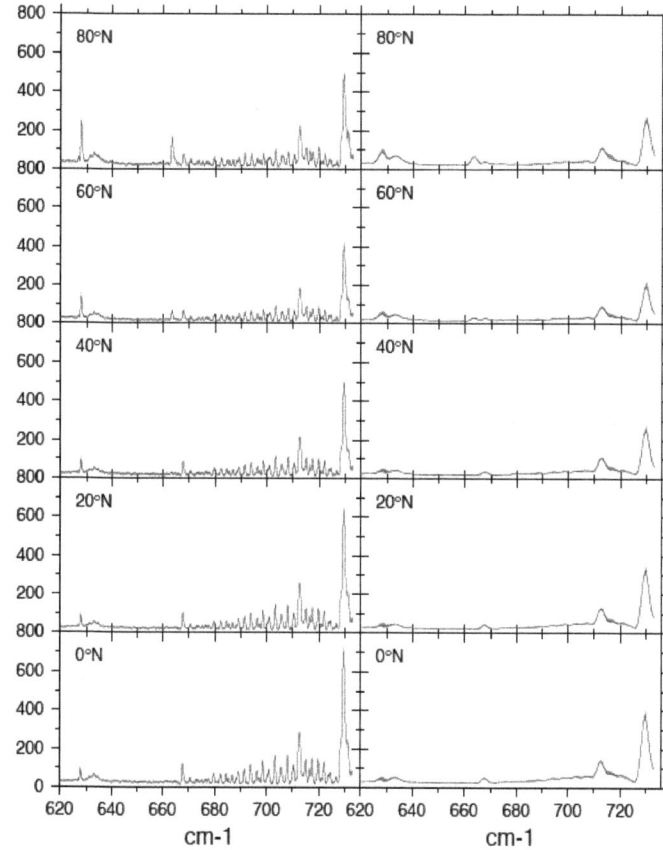

Figure 6.6   Example of fitting the CIRS spectra in part of the FP$_3$ focal plane where various emission features of gases appear (C$_2$H$_2$ at 730, HCN at 713, CO$_2$ at 667, C$_3$H$_4$ at 633 and C$_4$H$_2$ at 628 cm^{-1}). The observations (grey envelope) are compared to theoretical spectra (black lines); the spectral resolution is 0.5 cm^{-1} (left) and 2.5 cm^{-1} (right) (Teanby *et al.*, 2007).

Voyager IRIS spectra had already detected cyanoacetylene, HC$_3$N, near 500 cm^{-1} and also through its stronger ν_5 band at 663 cm^{-1}, albeit only at high northern latitudes (>50°N) at levels of a few 10^{-8}. No HC$_3$N emission was observed in Titan's thermal infrared spectrum at other locations, where only an upper limit of $\sim 10^{-9}$ was obtained. The second band of HC$_3$N was blended with CO$_2$ in the case of the IRIS resolution. More recently, the excess in emission observed at 663 cm^{-1} in the ISO spectra was attributed to HC$_3$N for an averaged abundance of $5.0 \pm 3.5 \times 10^{-10}$. The HC$_3$N contribution functions peak around 6 mbar for mid latitudes and around 9 mbar for high northern latitudes, always assuming a constant-with-height mixing ratio.

HC$_3$N emission is clearly visible in the CIRS nadir spectra taken at northern latitudes and shows variations from south to north. It can be detected even in the low- and mid-latitude CIRS nadir spectra at 663 cm^{-1} (where the 3-σ noise level is about

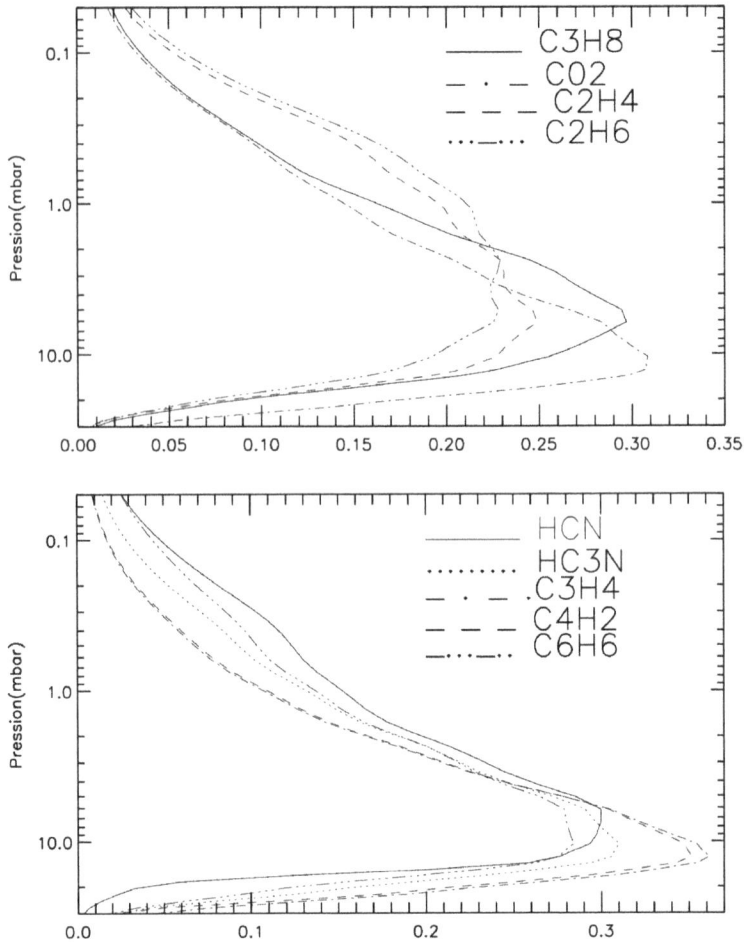

Figure 6.7   CIRS contribution functions for some species in Titan's stratosphere (Coustenis *et al.*, 2007).

2×10^{-9} Wcm^{-2}sr^{-1}/cm^{-1}), with associated abundances around 3×10^{-10}, but is more easily discernible in the higher-northern latitude spectra, where the mixing ratio can reach 4.4×10^{-8}.

C_2N_2 and C_4N_2 were found in the Voyager IRIS data, the first as a gaseous emission at 234 cm^{-1}, the second only in its solid form at 478 cm^{-1}. The first has been identified in CIRS spectra also, only at higher northern latitudes, but so far the second has not. C_2N_2 and HC_3N have lifetimes of less than a year in Titan's atmosphere. This leads to a large variation in abundance with altitude. In turn, any subsidence becomes extremely evident as a sharp increase in mixing ratio at lower altitudes. Short lifetimes mean that these molecules probe short time scale atmospheric motions.

6.4.1.3 Oxygen-Bearing Molecules: CO, CO$_2$ and H$_2$O

Carbon monoxide (CO) and carbon dioxide (CO$_2$) were for some time the only oxygen-bearing gases known in Titan's atmosphere, until the 1998 discovery of water vapour by ISO. CO$_2$ was the first molecule containing oxygen to be detected on Titan, when it was identified in the Voyager IRIS spectra from its emission in the ν_2 fundamental band at 667 cm^{-1}. In those data, the mean mixing ratio of carbon dioxide in Titan's atmosphere above the 110 mbar level was found to be about 1.3×10^{-8}, with some uncertainty depending on the vertical distribution assumed. IRIS did not resolve the emission bands found in the spectra and therefore contained little information on the vertical distribution of CO$_2$. It also did not cover any spectral region containing carbon monoxide bands, and so was unable to search for that species. Cassini/CIRS observed the same CO$_2$ band and the best fit was obtained with 1.3×10^{-8} as the mole fraction near the equator, remaining constant within the error bars from the south to the north. The CO$_2$ emission observed in the CIRS spectrum originates from pressure levels near 6 mbar for the low-latitude regions.

Prior to its detection, the presence of CO in Titan's atmosphere had been expected because it is a common species in the Universe, and because the Voyager IRIS detection of CO$_2$ suggested it should be there. Where carbon dioxide is present, the

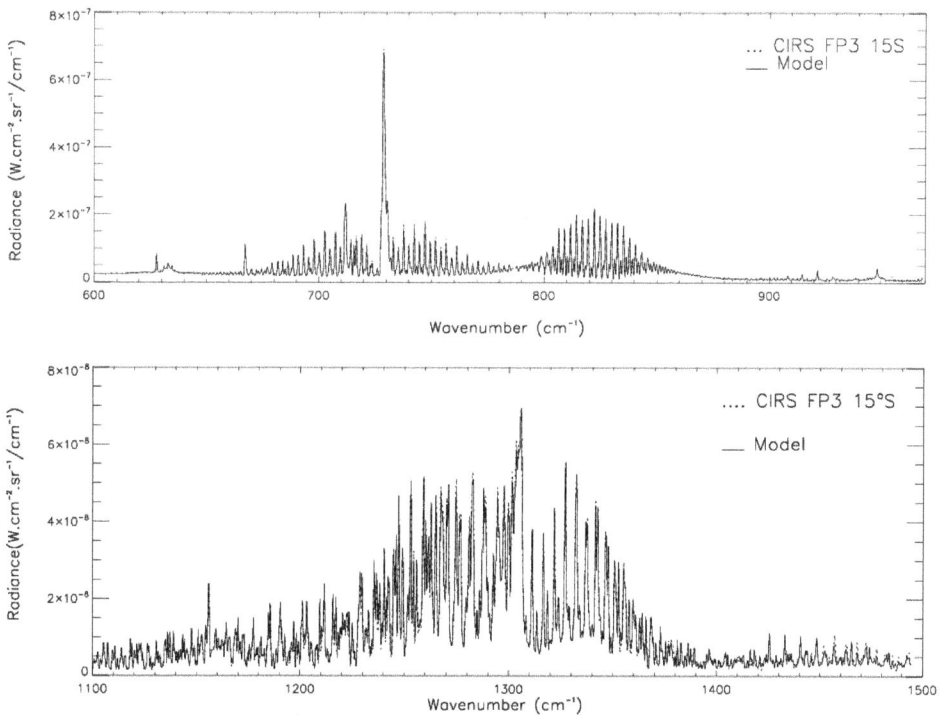

Figure 6.8   CIRS spectra and simulations from a radiative transfer model at Titan's 15°S (Coustenis *et al.*, 2007).

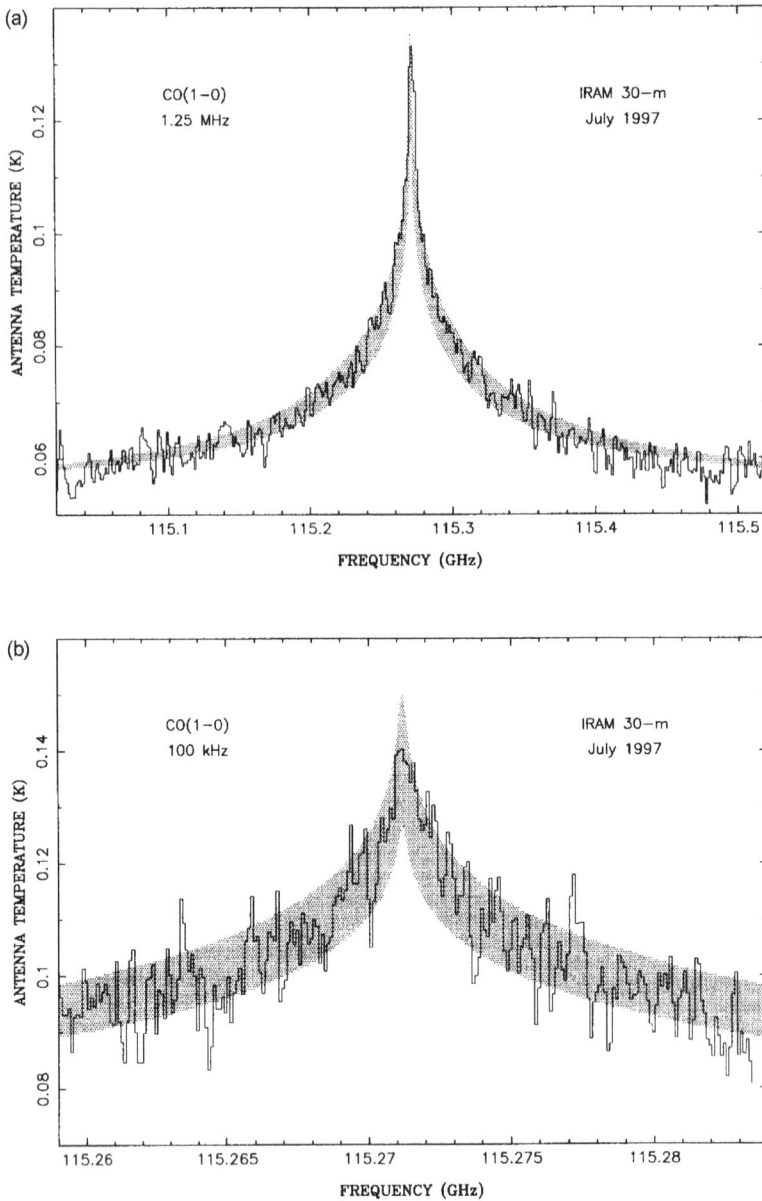

Figure 6.9   CO spectra in the millimetre range (Muhleman *et al.*, 1984).

monoxide is expected to be produced from photolysis of CO_2 by solar UV radiation in the upper atmosphere. CO is also produced from the reaction of the products resulting from the dissociation of incoming water and of the indigenous methane molecules. Once formed, CO is relatively stable and can accumulate to produce significant amounts. Also, it is possible that CO is, or has been, delivered directly

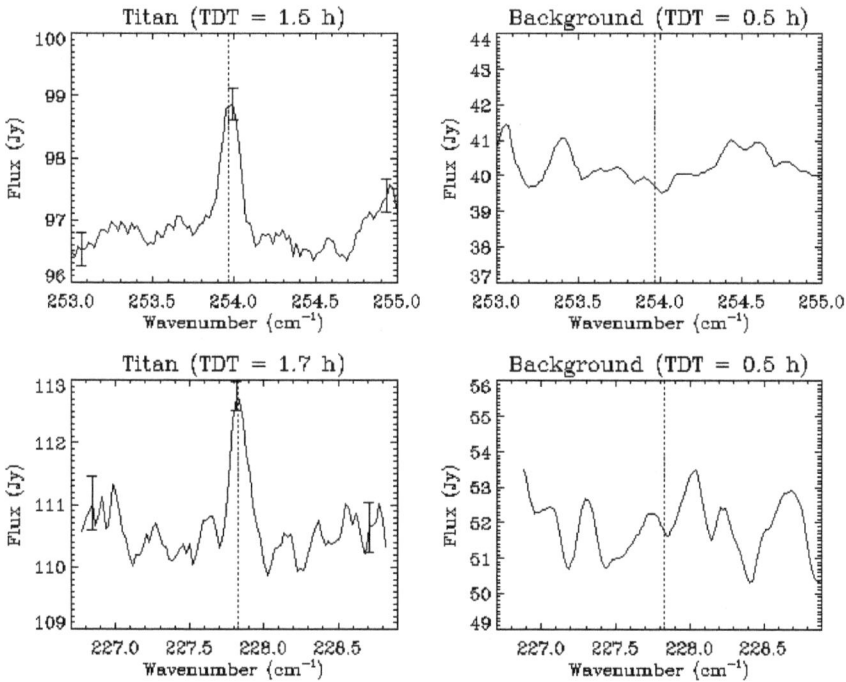

Figure 6.10   The two water vapour lines detected in Titan's atmosphere in 1998 with the ISO Short Wavelength Spectrometer. The on-off continuum flux is about 55–60 Jy, and the line peaks rise at about 1.5–2.0 Jy.

into Titan's atmosphere as a component of the icy material in comets, as discussed further below.

CO was finally detected in 1983 from ground-based observations in the near infrared at around $6,350 \, \text{cm}^{-1}$. The measured abundance of about 6×10^{-5} was fairly consistent with the prediction of the photochemical models then in existence, which was of the order of 1.1×10^{-4}. These near infrared measurements were mainly sensitive to the amount of CO in the lower part of Titan's atmosphere, and so again did not give any vertical profile information. Millimetre measurements of CO rotational lines, on the other hand, are more sensitive to the stratospheric mixing ratio, so when the J(1-0) transition near 115 GHz was observed, yielding a mixing ratio of $6 \pm 4 \times 10^{-5}$, this appeared to be convincing evidence for a uniformly mixed profile for CO.

This finding was challenged in 1987, however, when fresh observations of the 115 GHz CO line found only 4×10^{-6} for the stratospheric CO/N_2 mixing ratio, more than an order of magnitude lower than the tropospheric value, thus suggesting that CO is depleted in the stratosphere. The 115 GHz measurements were repeated in 1995 with improved sensitivity, finding $[CO] = 5 \pm 1 \times 10^{-5}$, confirming the 1984 value, and consistent with uniform vertical mixing. However, other recent microwave results, and new near-infrared observations, both found smaller values

$(2.5 \pm 0.5 \times 10^{-5})$ below 1 mbar, in agreement with 5 μm tropospheric soundings from the ground, which yielded $3.2 \pm 1 \times 10^{-5}$. These confusing results began to converge towards a constant concentration in the atmosphere somewhere in the $(2-6) \times 10^{-5}$ range (Table 6.4) after the Cassini/CIRS found the CO mole fraction in the stratosphere to be about 4.5×10^{-5} and the Cassini/VIMS observations reported a value of 3.2 ± 1.510^{-5} in the lower stratosphere. It then seems that CO is uniformly mixed, to an accuracy of about 50%, throughout the atmosphere (from the surface up to about 300 km in altitude). This implies that no input from episodic commentary impacts is required to explain the CO abundance, which is consistent with known chemical pathways. If the stratospheric abundance were indeed higher than in the troposphere, some mechanism — other than the classical photochemical concepts — must be found to produce a satisfactory explanation. It has been suggested that aerosols might alter the vertical distribution of carbon monoxide by adsorbing CO molecules and transporting them in the troposphere in sufficient quantities so as to generate the enrichment observed. This hypothesis requires massive condensation of methane and therefore, if all the observations are correct, there may be implication for large amounts of methane ice in the Titanian atmosphere. On the other hand, if the CO profile is constant, Cassini investigators are proposing rather that CO is being delivered into Titan's atmosphere by outgassing from the interior, where it contributes also to the formation of methane via the release of hydrogen through the serpentinization process, followed by Fischer-Tropsch catalysis.

The detection of water in Titan's atmosphere came after the detection of CO and CO_2 and was based on observations of two lines in the rotational spectrum at 37.4 and 43.9 μm. This ended a long search, because it was hard to explain the absence of H_2O in the presence of CO and, especially, CO_2. The H_2O abundance is of the order of a few tens of a part per billion at an altitude of 400 km. Although seeming small, this water vapour abundance implies a water influx on Titan significantly greater than what might be expected based on local and interplanetary sources alone. With the detection of water vapour, the source for the oxygen in the atmospheric carbon monoxide and carbon dioxide observed in Titan was at last found. The Cassini CIRS data have confirmed the ISO water vapour detection, and they are compatible with the same H_2O profile. Water quickly dissociates into OH which combines with methane photolysis products (such as CH_2, CH_3, etc.) and produces CO and CO_2. The source for the water itself may be the rings of Saturn, or meteorites or comets. CO, as well as H_2O, could be delivered to the atmosphere directly; if the incoming material was typical of comets, the mixture of ices would be roughly 90% H_2O and 10% CO, but all comets are not the same and the latter could be as high as 20%. These numbers suggest that any direct injection of CO (or CO_2) is a small source of CO when compared to photochemistry involving OH and dissociation products of methane. The implications of the water detection for Titan's photochemistry are discussed in Section 6.5.

An atmosphere of water vapour has recently been discovered around another of Saturn's moons: Enceladus. Its origin in this case seems however to be different from that on Titan and most probably internal (coming from the interior) rather than external.

6.4.1.4 Deuterium: CH₃D and D/H Ratio

Ground-based observations of the $3\nu_2$ monodeuterated methane band at $1.6\,\mu$m confirmed the value of $\sim 1.5 \times 10^{-4}$ found from Voyager analyses of the ν_6 CH_3D band at $8.6\,\mu$m. Both therefore support the evidence for a deuterium enrichment in Titan's atmosphere with respect to the protosolar value and that of the giant planets, both of which are 5 to 10 times smaller at D/H ~ 2–3.4×10^{-5}.

The D/H value on Titan is related to the ways in which hydrogen-containing compounds were incorporated into the satellite during its formation, and later. The preliminary value derived from the CH_3D ν_6 band observed in emission in the ISO SWS spectrum (9.5×10^{-5}) is close to the Voyager 1 value (1.5×10^{-4}), and to the determinations from ground based observations in 1983 and again in 1992, namely D/H $= 7.75 \times 10^{-5}$. Using the same nominal methane mixing ratio (1.4%) at all latitudes, a CH_3D abundance of 8×10^{-6} provides a good fit to all the data and in particular to the high-resolution mid-latitude spectra. This yields a D/H ratio on Titan of about 1.2×10^{-4}.

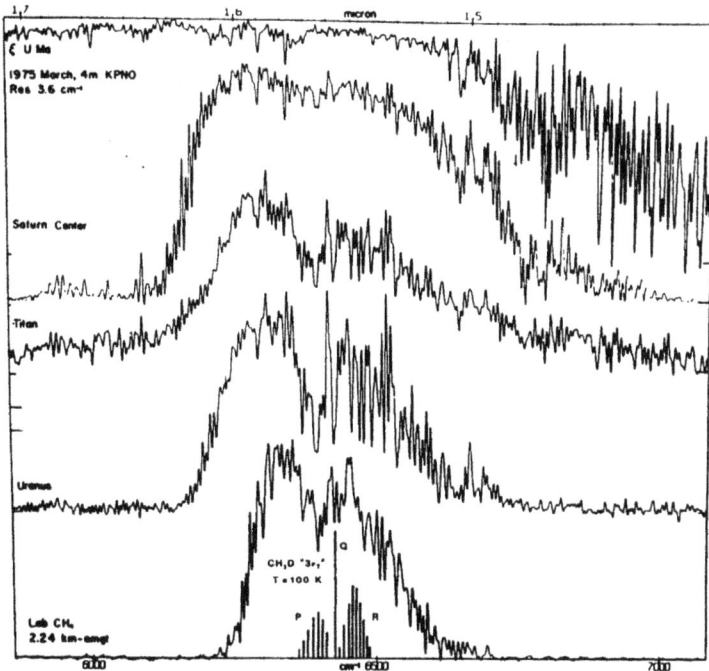

Figure 6.11   Spectra of Saturn, Titan and Uranus in the $1.6\,\mu$m spectral window region. The resolution for Titan is $3.6\,\mathrm{cm}^{-1}$. The top spectrum is a reference star, and the bottom two spectra show a laboratory spectrum of methane and the line positions for deuterated methane, CH_3D (Fink and Larson, 1979).

Although higher than the giant planet values, the D/H ratio in Titan is lower than those measured (as D_2O/H_2O) in the comets Halley, Hyakutake and Hale-Bopp, by a factor of 2 to 4. If, as is currently thought, the ices in comets acquired their enrichment in deuterium from hydrogen in the pre-solar cloud, the D/H ratio in water and in methane ices should have been near their equilibrium values at the temperature of the cloud. Theory predicts that the equilibrium value of D/H in methane is substantially higher than it is in water. Even if interstellar grains were partly re-equilibrated with hydrogen in the nebula prior to forming comets, the difference in deuterium enrichment between the two species should have remained approximately the same, since the isotopic exchange coefficient between CH_4 and HD is not very different from that between H_2O and HD. In other words, in comets, D/H in CH_4 should be, if anything, higher than D/H in H_2O. Therefore, the difference in the other direction between the D/H ratio in water in comets on one hand and the D/H in methane in Titan on the other, is significant, and tends to rule out the formation of the atmosphere of the satellite by cometary impacts, contrary to a scenario advocated by some authors.

More likely, the D/H ratio on Titan is consistent with the formation of the atmosphere by outgassing from the interior of the satellite. Most of the carbon was probably in the form of CO in the early sub-nebula of Saturn. When CO was converted to CH_4, later in the history of the sub-nebula but prior to the formation of the satellite, the methane was enriched in deuterium by isotopic exchange with hydrogen. The coefficient of isotopic exchange between HD and CH_3D vanishes to zero at low temperature, so the enrichment factor for D in methane with respect to the protosolar value in H_2 is limited to a low value, one which depends on temperature-pressure conditions in the sub-nebula. Recent estimates of this factor are about 1.4; this is probably an upper limit since it does not take into account diffusive processes that redistribute the deuterium enrichment throughout the sub-nebula.

The D/H ratio in methane present in the grains which came together to form Titan has been estimated to be in the range $3.3-3.7 \times 10^{-5}$ using arguments based on measurements of the $^3He/^4He$ ratio in the solar wind, and in Jupiter as measured by the mass spectrometer on the Galileo entry probe. Some fractionation, mainly due to the loss of the lighter H isotope following the photodissociation of methane, will have occurred during Titan's history. The corresponding fractionation factor is estimated to be in the range 1.7–2.2, resulting in a D/H ratio in methane today in the range 5.6 to 8.1×10^{-5}, consistent with the most recent measured values.

However, the estimate of deuterium enrichment by fractionation depends on the ratio of the total initial methane reservoir in Titan to the total present reservoir. The factor given above assumes that the atmospheric methane is being continuously replenished. Another scenario in which the atmospheric CH_4 is randomly and infrequently replenished from the interior, by cryovolcanism for instance, might result in a different estimate of the deuterium enrichment factor. For the moment, there is only preliminary evidence that such volcanism exists, and the high abundance of methane, in an environment where it is being continuously destroyed, tends to argue for continuous replenishment.

The possibility remains that Titan's atmosphere did after all result from volatile degassing of cometary grains, with some poorly-understood fractionation or other mechanisms acting during or after the formation of the satellite causing the deuterium enrichment now observed. Heterogeneous fractionation processes, those involving materials in different phases, are particularly difficult to evaluate, but may be very efficient. Among those which have been suggested are exchanges of deuterium between methane gas and cloud particles, the ethane-methane lakes on the surface, or the solid surface itself. A further possibility could involve isotopic exchange catalysed in the presence of metallic grains in the Saturnian nebula. Clearly, with so many unknowns, we do not have the measurements at present which are needed to distinguish between the different theories which seek to explain the distribution of deuterium on Titan and elsewhere in the Solar System. We will return to the discussion of the implications of the D/H ratio for Titan's origin and evolution in Chapter 10.

6.4.2 *Vertical Distributions*

As confirmed by Cassino–Huygens CIRS, limb data are very important in sounding the chemical and thermal structure of Titan. Already in the 80s, an observational sequence of 30 spectra recorded over Titan's north polar region turned out to be the source of some of the most significant data obtained by Voyager IRIS. Nine of the spectra were taken at grazing incidence over the satellite's surface, corresponding to lines of sight that do not intercept the surface but follow optical paths entirely situated in the atmosphere (limb-viewing). The advantage of such observations is that they can be associated with definite altitude levels and therefore yield height-dependent information on temperature and composition with better vertical resolution than nadir (downward) viewing.

Moving to lower altitudes in the atmosphere, the optical paths become much longer than the corresponding vertical path, and so the emission features in the spectra are enhanced. This aspect has allowed the detection of cyanoacetylene and cyanogen in the spectra taken near the north pole. Also, two emission bands of diacetylene and methylacetylene at 220 and 328 cm^{-1}, respectively, absent in the equatorial spectra, were observed near the north pole.

The vertical concentration profiles of all the gases detected above Titan's north pole at the time of the Voyager encounter were derived assuming that the distributions decrease below the condensation level following the relevant vapour saturation law. They show a general tendency towards an increase of the mixing ratio with altitude, confirming in general terms the prediction of theoretical photochemical models that these species form in the upper atmosphere and then diffuse downwards in the stratosphere. However, the actual gradients in abundance represented by these profiles were not always in accordance with the models, for acetylene (C_2H_2) for

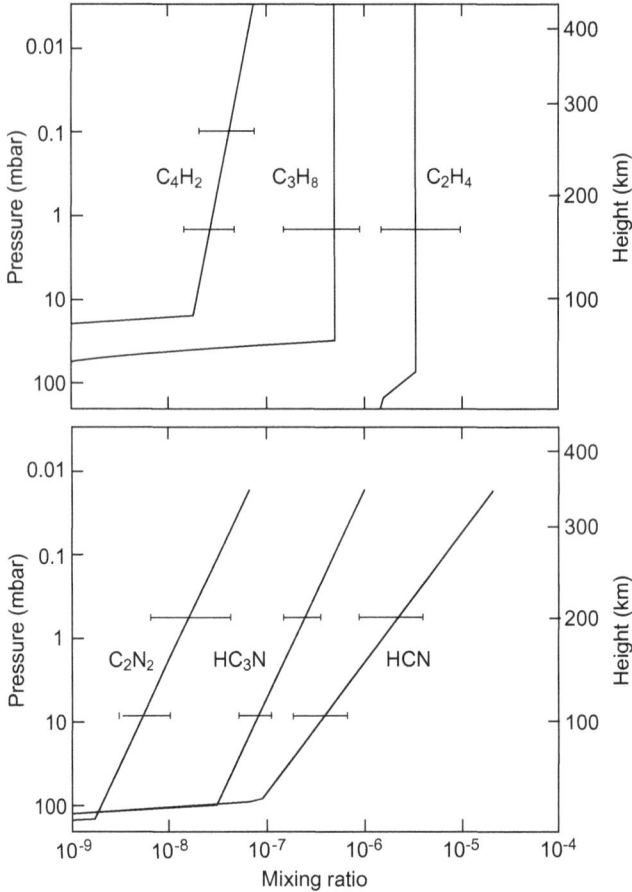

Figure 6.12   Vertical distributions for some hydrocarbons and nitriles over Titan's north pole from Voyager observations (Coustenis *et al.*, 1991).

instance. The vertical distribution obtained from ISO/SWS increases with height and has a slope $-\mathrm{dln}q/\mathrm{dln}P = 0.32 \pm 0.12$, similar to that obtained by V1/IRIS for Titan's north pole. The Cassini mission, and in particular the CIRS instrument, offered the opportunity to obtain more and better-constrained vertical distributions of the species from the fit of the emission observed in the band at $729\,\mathrm{cm}^{-1}$ in limb spectra with measured vertical distributions instead of an assumed constant-with-height profile.

Vertical abundance distributions for C_2H_2, C_2H_4, C_2H_6, CH_3C_2H, C_3H_8, C_4H_2, C_6H_6, HCN, HC_3N and CO_2 were thus were retrieved from Cassini CIRS limb data at $15°S$ (from the first flyby) and at $80°N$ (from a later flyby). The temperature profiles in Titan's stratosphere were also retrieved, in the altitude range 100–460 km for Tb and 170–495 km for T3, and these show a well-defined stratopause at around 310 km (0.07 mbar) and 183 K at $13°S$, moving up to 380 km (0.01 mbar) with

207 K at 80°N. Near the north pole, stratospheric temperatures are colder, and mesospheric temperatures are warmer, than near the equator. C_2H_2, C_2H_6, C_3H_8 and HCN amounts increase with height at 15°S and 80°N, consistent with their formation in the upper atmosphere, diffusion downwards and condensation in the lower stratosphere, as expected from photochemical models. The CH_3C_2H and C_4H_2 mixing ratios also increase with height at 15°S, but near the north pole their profiles present an unexpected minimum around 300 km, observed for the first time in the CIRS limb data. C_2H_4 is the only molecule showing a vertical abundance profile decreasing with height at 15°S. At 80°N, it also displays a minimum mixing ratio at around the 0.1 mbar level. More data taken at grazing incidence over Titan's disc covering more latitudes recently became available and are currently being analyzed.

For benzene, C_6H_6, an upper limit of 1.1 ppb (in the 0.3–10 mbar range) was derived from CIRS limb data at 15°S, whereas a constant mixing ratio profile of about 3 ppb was inferred near the north pole. At 15°S, the vertical profile of HCN exhibits a steeper gradient than other molecules suggesting that a sink for this molecule exists in the stratosphere. All molecules display a more or less pronounced enrichment towards the north pole, probably due to dynamical subsidence at the pole in winter that brings air enriched in photochemical compounds from the upper atmosphere to lower levels.

6.4.3 Spatial Variations

Trace constituents with finite atmospheric lifetimes can act as tracers for the photochemical and dynamical physical processes in Titan's atmosphere. However while, as we have seen, latitudinal variations are very prominent on the satellite, Cassini measurements have confirmed that there are no significant variations as a function of longitude in Titan's stratospheric composition. Variations as a function of longitude are also found to be virtually absent in the temperature field. The strong zonal winds plays an important part in smearing out latitudinal variations in temperature and in composition.

The meridional variations in Titan's trace constituents retrieved from Cassini CIRS observations have higher precision than their precursors from Voyager and ISO. Acetylene, ethane, propane and carbon dioxide, the most abundant trace constituents after methane, were found by CIRS to exhibit no significant compositional variations in latitude, but to be homogeneously mixed from pole to pole. C_3H_4, C_4H_2, HCN, HC_3N, and C_6H_6 show an enrichment in the north, by as much as an order of magnitude for HC_3N, and C_6H_6, which only become apparent (in emission at 663 and 674 cm^{-1}, respectively) at latitudes higher than 50°N. C_2H_4, on the other hand, exhibits a possible twofold decrease in mixing ratio from south to north and CO_2, C_2H_6, C_2H_2 and C_3H_8, remain fairly constant with latitude. Previous measurements of the latitude variation of nitriles and hydrocarbons displayed similar trends, but with greater contrast between low and high latitudes. This indicates that as winter progresses, north polar subsidence may lead to further enhancements for these gases.

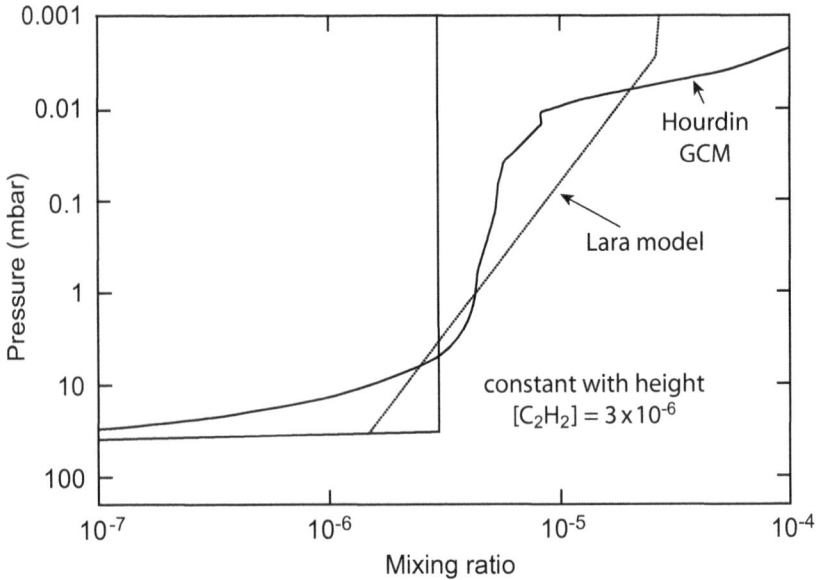

Figure 6.13   Acetylene CIRS profiles that satisfy the CIRS nadir data: Lara *et al.* profile, a constant-with-height above the condensation level abundance, and an intermediate case below 0.01 mbar from an updated GCM model by Hourdin *et al.* (2005).

We know only in very general terms what causes these latitudinal variations. The nitrile enhancement near the north pole has been tentatively attributed to an accumulation of these light-sensitive species in that region during the winter, when the polar region is in semi-permanent shadow. At the time of the Voyager 1 flyby, the north polar region was just coming out of winter, when it would, according to this explanation, have close to maximum nitrile abundances. The Voyager 2 encounter occurred about 8 months after northern spring equinox, and the chemical time constants may be long enough that the UV flux had not completed the process of depleting the strong nitrile concentrations that were observed.

Model studies suggest that purely radiative processes, including photochemistry, are not capable of causing the latitudinal variations of mixing ratio observed in Titan's stratosphere. However, the main latitudinal dependences do appear, at least qualitatively, in a realistic 2-D (latitude–altitude) combined photochemical and meridional transport model which uses a season-dependent wind field that features a subsidence zone in the winter polar stratosphere. The model tends to show latitudinal variations that are comparable to those observed in IRIS data, which pertain to conditions near spring equinox. Models for northern winter solstice (calculated at $Ls = 300$) show that there is reasonable agreement in the mid-latitude values for all the constituents, and a fair representation of the northern trends of C_4H_2, HC_3N and HCN. However, the data are not compatible with the enrichment found at the south pole in the model, in particular for diacetylene, ethylene and methylacetylene

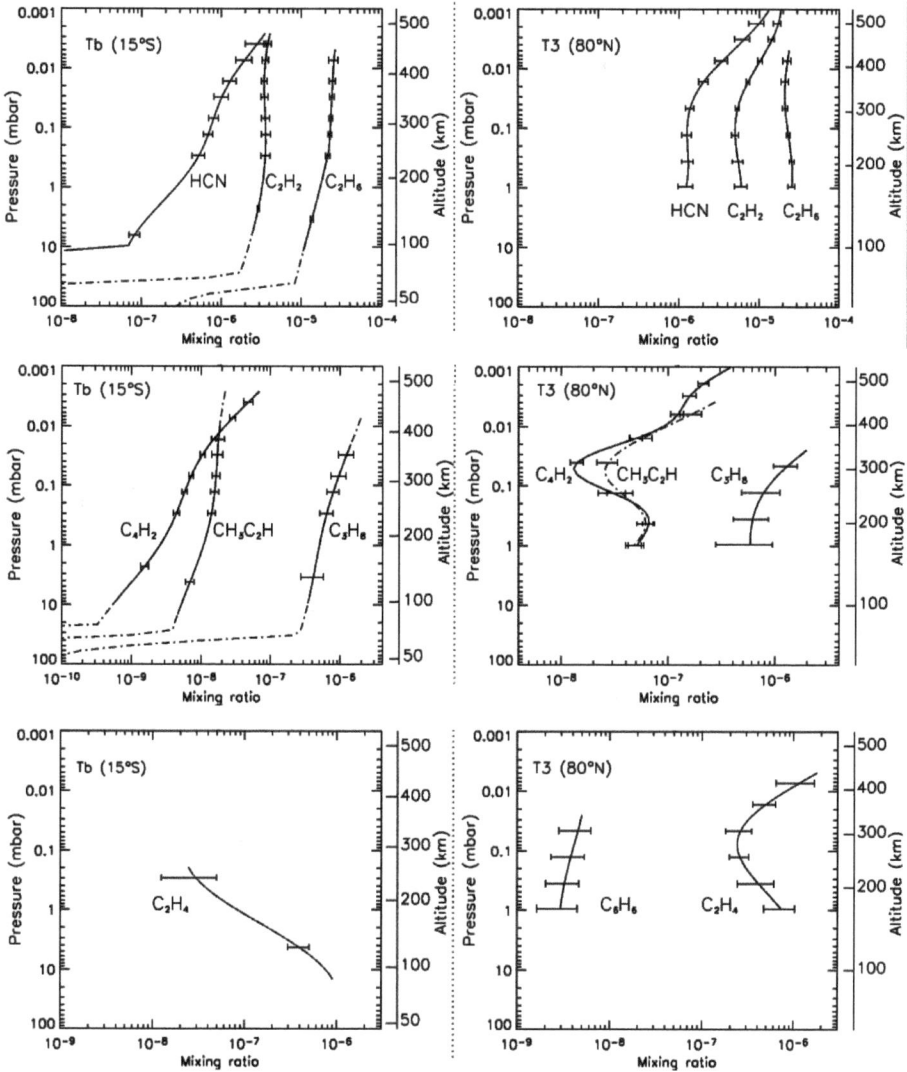

Figure 6.14 Vertical distributions for several hydrcarbons and nitriles in Titan's stratosphere, derived from Cassini/CIRS spectra at 15°S and 80°N (Vinatier *et al.*, 2007).

(C_4H_2, C_2H_4 and C_3H_4), nor with the depletion the model predicts at 60°S. The modellers attribute these discrepancies to the way the horizontal mixing is treated, and the existence of a residual circulation cell in the summer polar stratosphere. As more data is acquired and analyzed across the changing seasons, the need for a sound theory that can provide a satisfactory explanation for the distribution of nitriles and hydrocarbons becomes more urgent. The solution will require a full 3-D treatment of the problem, including seasonal dynamical effects, along with improved values for the various reaction rates.

6.4.4 *Temporal Variations of the Trace Constituents*

When comparing the disk-resolved Cassini/CIRS or V1/IRIS latitude-dependent results with the ISO disk-averaged abundances to look for seasonal changes, care is required because, besides the difference in the area of Titan involved in this comparison, the altitude range involved may also vary due to the differences in the geometry of the observations. Caution must also be exercised when comparing abundances of components showing significant variation with latitude or with height. Modelling disk-averaged observations with a single "average" temperature profile derived from the 7.7 μm methane band (as is the case for ISO), may affect the derived abundances because of the non-linear dependence of the Planck function with temperature.

The ISO observations were made in 1997, 16.5 years after Voyager 1, during northern autumn, a little over two Titanian seasons after the Voyager data taken in early northern spring, and one full season before the CIRS observations in early northern winter. The compositional polar enhancement observed during the Voyager encounter could be a seasonal effect which reverses every two seasons, if the gas content contributes along with the haze to the north-south asymmetry inversion observed in spatial images of Titan taken at these two times. All other things being

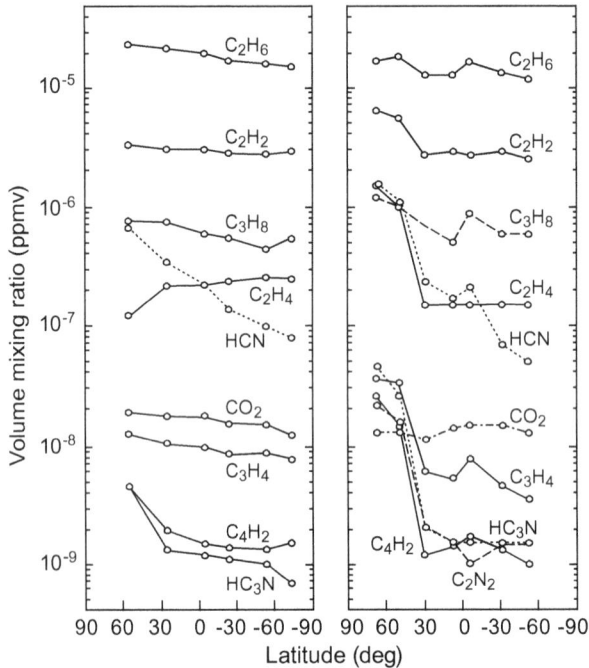

Figure 6.15 Meridional variations in Titan's stratospheric trace constituents almost a year apart in season from Voyager 1/IRIS measurements and Cassini/CIRS. Adapted from Coustenis and Bézard (1995) and Coustenis *et al.* (2007).

constant, a considerable difference in composition on a disk-average basis (mainly mid-latitudes) is then not expected between CIRS, IRIS and ISO, but cannot be excluded at the 20% level, reflecting the possible error in including the exact limb contribution in the ISO disk-averaged calculations. It may be more obvious in the abundances of trace constituents in Titan's stratosphere, in particular those of C_3H_4, C_4H_2, and HCN which showed significant variation with latitude at the time of the Voyager 1 encounter and some enhancement in 1997 with respect to 1980 for these three molecules, with factors up to 3–4. If these enhancements are real, they could be attributed to seasonal effects associated with circulation patterns.

A more meaningful comparison can be applied to the Cassini CIRS and Voyager IRIS data alone, exhibiting the differences within one season on Titan. The Voyager data had shown considerable enhancement in the abundances of some nitriles and hydrocarbons (within a factor of 2 for the most abundant molecules after CH_4: C_2H_2, C_2H_6 and C_3H_8; factors of 15–30 for the nitriles HCN, HC_3N and C_2N_2; about 10 for C_3H_4 and C_2H_4, and 30 for C_4H_2). Associated with these compositional variations, the temperature field distributed about the equator showed a temperature decrease of approximately 20 K at the 0.1 mbar level at 70°N. The variations in composition could explain the temperature field through cooling rate effects, although dynamical phenomena most probably also have a part to play. The comparison between results from 2007 and 1980 shows that for most of the species observed the abundance differences between CIRS and IRIS are small, except for the larger northern enrichments. The main difference is that the variations of mixing ratio with latitude found by CIRS, though qualitatively similar, tend to be quantitatively smaller. The difference in magnitude between the Voyager 1 and the Cassini eras must be due, at least in part, to the difference in seasons, and it will be exciting to await the arrival of northern spring equinox towards the end of the Cassini mission to measure the meridional variations and then to see if Titan returns to the IRIS values.

6.5 Photochemistry

The photochemistry which occurs in Titan's atmosphere begins with the absorption of solar radiation and the consequent ionization and dissociation of the main atmospheric molecules, in particular, N_2 and CH_4. The photolysis rates for these two species have their maximum efficiency in the vicinity of the homopause and the fragments produced react together and with the background gases, to produce the first set of hydrocarbons and nitriles. These are in turn photolyzed leading to the formation of other fragments and the same process continues, yielding increasingly more complex species. At these high levels, gas phase chemical reactions and vertical transport primarily determine the distribution of minor species. When they are convected or diffused to lower levels, below 100 km altitude, the low temperatures which prevail in the lower stratosphere result in strong condensation sinks for most of the compounds.

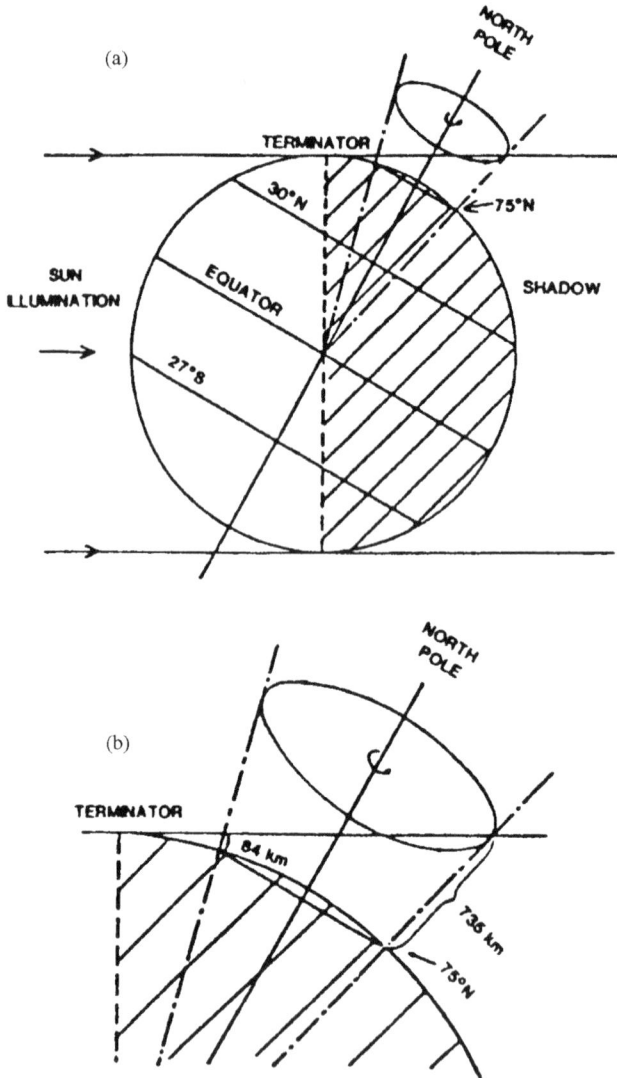

Figure 6.16   Titan geometry at the time of the Voyager 1 encounter, showing the conditions leading to polar enhancement (Yung, 1987).

The neutral photochemistry in Titan's atmosphere was investigated even before the Voyager era, but the models were restricted to hydrocarbons, the only species whose presence was certain at the time. After the Voyager encounters, when the dominating presence of N_2 was established, photochemical models began to take into consideration the formation of nitriles. With reactions between carbon, nitrogen, hydrogen and oxygen compounds, the chemistry becomes quite complex.

In spite of all the improvements, the photochemical models we have at present fall far short of precisely describing all of the likely gas phase chemical pathways, not to mention those involving condensation processes, analogues of which have recently been found to be crucial for the understanding of ozone chemistry in the Earth's stratosphere. Nor are they capable of reproducing the observed atmospheric gaseous composition. The current models do not include all the species detected by INMS, for example, not just because they are only recently discovered or because the pathways are complex, but because there is a significant lack of reaction rates, absorption cross section and photolysis yields which are necessary for the simulation of these species. If these laboratory data existed, they would greatly improve the models.

A common approach to modelling the photochemistry has been to generate the vertical temperature distribution, from the surface to the thermosphere, by combining measurements from Voyager and Cassini and model results for the temperature structure at different altitudes. In order to include the effects of the aerosols in the radiation transfer calculations, vertical profiles of haze opacity were either specified by a simple exponential decrease with altitude or in more recent work generated by microphysical models using a specified vertical haze production rate. Microphysical haze models, concerned with the growth rates and the size and number density distribution of the stratospheric aerosols as a function of altitude have made fairly detailed predictions of the aerosol vertical distribution (see next chapter), but again knowledge of the associated chemistry lags behind. Using this approach, photochemical models have managed to fit most of the atmospheric species concentrations available from observations before the Cassini/Huygens mission. An understanding of the interactions between photochemistry and microphysics, the formation of small organic nuclei from the condensation and polymerisation of complex hydrocarbons and nitriles, remains lacking, although clues on the mechanisms at work may be inferred from laboratory simulations, again as described in the next chapter.

By concentrating on the main reactions controlling the production and loss of the principal atmospheric gases, it is possible to achieve a theoretical view of the chemistry on Titan, in terms of a limited number of chemical steps, following the photolysis processes outlined above. The first photochemical model developed by Yung and colleagues made use of a fairly complete set of chemical reactions, based on the compilation of previous investigations, and was enriched by adding the photochemistry of oxygen compounds in a mildly reducing atmosphere. Vertical distributions were derived for most of the atmospheric gases and their average mixing ratios were compared with Voyager results at the relevant altitude levels.

In the light of new observational constraints, improved reaction rates, updated kinetics laboratory data, and other useful measurements, one-dimensional steady-state photochemical models were developed in 1995–1996, essentially pertaining to Titan's equatorial region for moderate solar activity. D. Toublanc and colleagues used an elaborate Monte Carlo description for solar radiation transfer within the atmosphere to investigate the possible production of oxygen-containing species arising from an influx of water vapour at the top of the atmosphere. A more complete model of this type, with an extensive list of chemical reactions and associated rates,

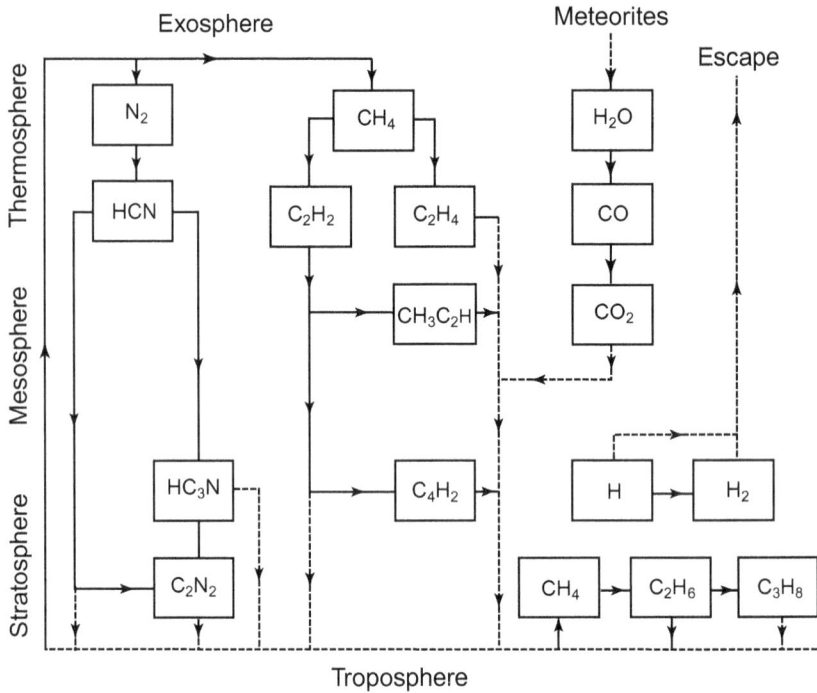

Figure 6.17 A model scheme for the principal photochemical processes occurring in Titan's atmosphere by Yung *et al.* (1984).

was published by L. Lara and colleagues in 1996. In this model boundary conditions were chosen as follows: near the tropopause (40 km altitude), the abundance of the condensable species was assumed to follow the saturation law. For the non-condensable, long-lived compounds (such as CH_4, H_2, CO and C_2H_4), the mixing ratios were established from their lower-stratospheric value. At the upper boundary, zero flux was assumed for all species except H, H_2, $O(^3P)$, and $N(^4S)$, which were allowed to escape into space. Lara *et al.* used an ablation profile for the water vapour influx, included the effects of GCR and presented a physical description of the condensation processes taking place in Titan's lower stratosphere. In 2001, S. Lebonnois and co-workers investigated the seasonal variation of the composition in Titan's stratosphere using a 2-D (latitude-altitude) model. Beyond neutral species chemistry, models have included ionospheric chemistry, as in the 2004 work of E. Wilson and S. Atreya where the contributions of energetic electrons and photoelectrons were included. The more recent models by P. Lavvas, I. Vardavas and colleagues, present a complete radiative/convective-photochemical-microphysical description of Titan's atmosphere through a model which generates the thermal structure, the atmospheric composition, and the haze structure in a self-consistent manner. The haze is produced from polymer production governed by the photochemistry, which is determined by and determines both the radiation field and atmospheric temperature structure. The photochemical processes expected to dominate in Titan's atmosphere are then incorporated within this framework, as described in the following sections.

6.5.1 Hydrocarbons

The major photodecomposition products of CH_4 are $^1CH_2$ (excited methylene), CH_3 (methyl radical)and CH (methylidyne), which subsequently undergo other reactions, leading to the formation of C_2H_6, C_2H_4, and C_2H_2. The photolysis of these initial hydrocarbons produces more and new fragments (radicals) that eventually lead to more complex compounds. Since the rates of the reactions which control the efficiency of each species production and loss, in principle, depend on the background temperature and pressure conditions, the net-production of each species can exhibit a strong vertical variability. Some examples of the basic reactions (or, in some cases, the net reaction of certain chemical schemes, which can include two or more chemical reactions) are given hereafter.

A first cycle, initiated by the production of CH, leads to the formation of ethylene in the mesosphere and thermosphere:

$$CH + CH_4 \rightarrow C_2H_4 + H.$$

Once ethylene is formed, its photolysis leads to acetylene:

$$C_2H_4 + h\nu(\lambda < 1700\,\text{Å}) \rightarrow C_2H_2 + H_2,$$
$$CH_2 + CH_2 \rightarrow C_2H_2 + 2H.$$

The acetylene (C_2H_2) produced in these reactions diffuses down to the stratosphere, where it photolyses, resulting in the formation of C_2H. The latter reacts preferentially with C_2H_2 again to give diacetylene, C_4H_2, i.e.,

$$C_2H + C_2H_2 \rightarrow C_4H_2 + H.$$

C_4H_2 can further react with C_2H to give rise to other polyacetylenes, representing a chemical sink for acetylene and diacetylene in the mesosphere, i.e.,

$$C_4H_2 \rightarrow C_6H_2 \rightarrow C_8H_2 \rightarrow \text{etc.}$$

Diacetylene can also be formed by

$$4CH_4 \rightarrow C_4H_2 + 10H + 2H_2.$$

The direct production of the methyl (CH_3) radical, from the photolysis of CH_4, allows, through a different cycle, the formation of ethane in the mesosphere and lower thermosphere (below 800 km), in addition to its formation in the stratosphere through catalytic processes:

$$CH_3 + CH_3 + N_2 \rightarrow C_2H_6 + N_2.$$

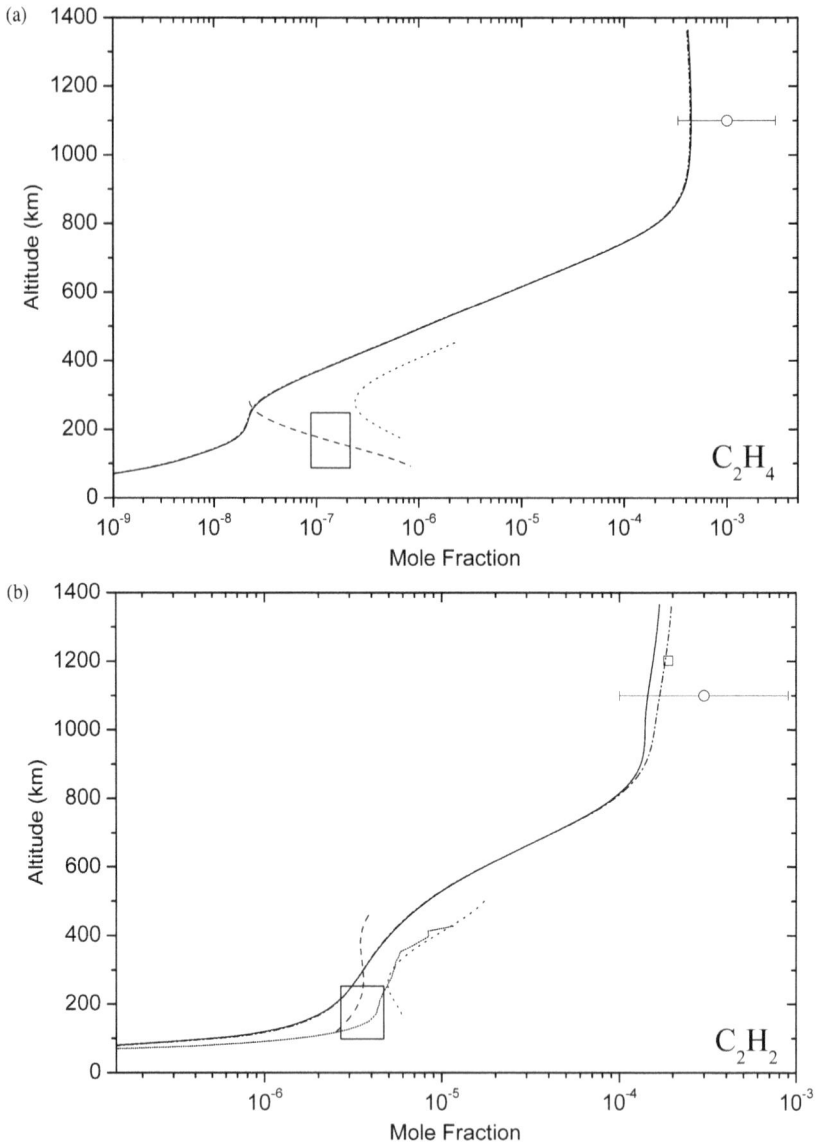

Figure 6.18   Calculated vertical profiles for ethylene (a) and acetylene (b) from the photochemical model of Lavvas *et al.* (2007), compared to Cassini observations. Solid lines correspond to model results, dashed and dotted lines correspond to the CIRS retrieved profiles for 15°S and 80°N, respectively (Vinatier *et al.*, 2006) based on limb spectra, while open squares and circles represent the INMS measurements for the upper atmosphere from Waite *et al.* (2005) and Vuitton *et al.* (2006a, b), respectively. CIRS abundances from nadir spectra at 33°N are also shown in boxes (Coustenis *et al.*, 2007) and the short-dotted line corresponds to the Hourdin *et al.* (2004) acetylene profile generated by a 2-D photochemical/dynamical model that was found to provide a good fit to the CIRS nadir spectra for the equator.

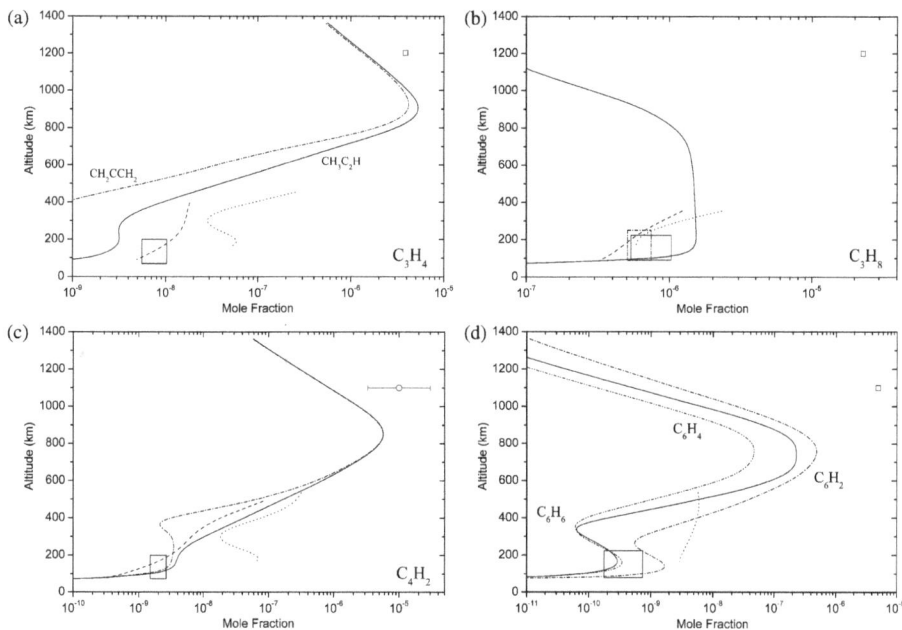

Figure 6.19   As for 6.18 but for methylacetylene (a), propane (b), diacetylene (c), and benzene (d), from the photochemical model of Lavvas *et al.* (2007) compared to Cassini observations.

The C_2H_6 production rate is much greater than the C_2H_2 production rate because of higher photolysis rates above 1450 Å (which produce primarily C_2H_6) than below 1450 Å (which produce C_2H_2). Photolysis of C_2H_6 and subsequent reactions with CH_3 result in the formation of propane (C_3H_8):

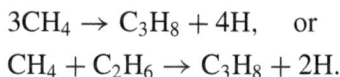

$$3CH_4 \rightarrow C_3H_8 + 4H, \quad \text{or}$$
$$CH_4 + C_2H_6 \rightarrow C_3H_8 + 2H.$$

The transport of C_2H_6 to the tropopause leads to its condensation, followed by rain-out. CH_4 is recycled mainly by reaction of CH_3 with H.

Yet another step introduces methylacetylene in the mesosphere according to:

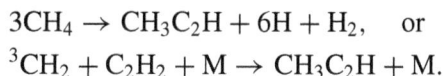

$$3CH_4 \rightarrow CH_3C_2H + 6H + H_2, \quad \text{or}$$
$$^3CH_2 + C_2H_2 + M \rightarrow CH_3C_2H + M.$$

The main loss of acetylene in the upper atmosphere is due to its reaction with methylene which leads to the formation of propargyl radicals (C_3H_3):

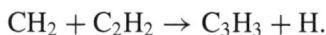

$$CH_2 + C_2H_2 \rightarrow C_3H_3 + H.$$

Two propargyls can then combine under high pressure:

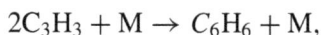

$$2C_3H_3 + M \rightarrow C_6H_6 + M,$$

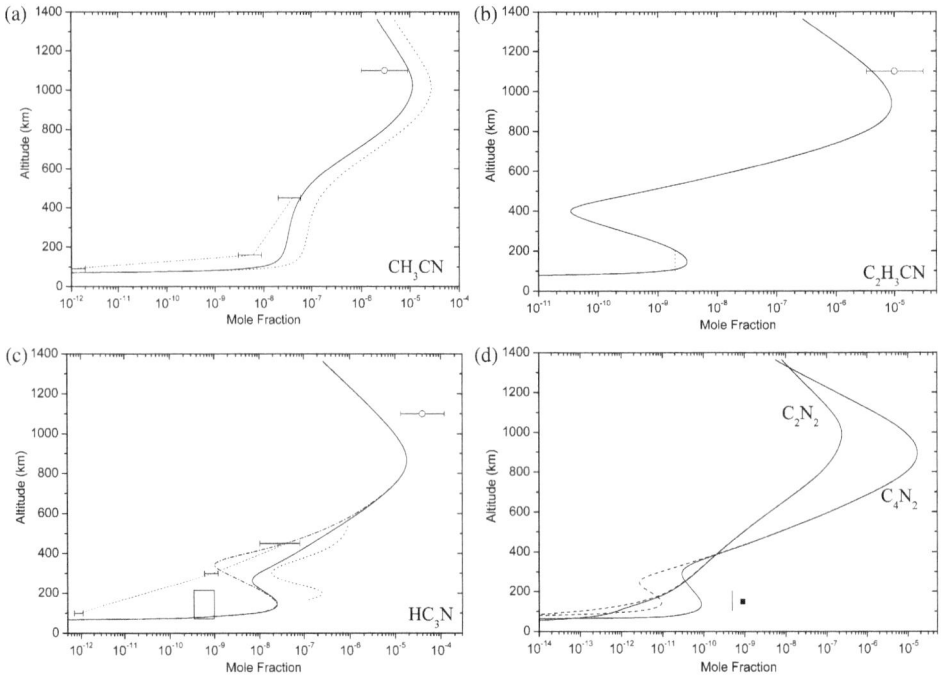

Figure 6.20   Model vertical profiles for the main nitrile species observed in Titan's atmosphere. Open circles correspond to the INMS derived abundances from Vuitton *et al.* (2006a, b), horizontal error bars connected with dashed lines present the Marten *et al.* (2002) ground-based observations and the box represents the CIRS observations from Coustenis *et al.* (2007). The vertical line and the filled square correspond to the upper limit-low latitude and high northern latitude retrieved abundance, respectively, for C_2N_2 from CIRS measurements by Teanby *et al.* (2006). From Lavvas *et al.* (2007).

to produce benzene, with a secondary contribution from

$$C_2H_2 + C_4H_5 + M \rightarrow C_6H_7 + M \rightarrow C_6H_6 + H$$

where C_6H_7 is produced first and eventually provides benzene. Finally a small contribution is provided by the addition of hydrogen to the phenyl radical (C_6H_5):

$$H + C_6H_5 + M \rightarrow C_6H_6 + M$$

with the chemical formation of the phenyl based on diacetylene.

6.5.2 Nitriles

N_2 is highly inert, and it does not react directly with the hydrocarbons in Titan's atmosphere. On the other hand, atomic nitrogen reacts readily with CH_4. N_2 is dissociated by solar UV photons at wavelengths below 1000 Å, by magnetospheric electrons, and by galactic cosmic rays, the first mechanism being by far the most efficient. The nitrogen atoms thus formed play a crucial role in the production of nitriles.

HCN production is the first step here, summarised by:

$$2CH_4 + N_2 \rightarrow 2HCN + 2H_2 + 2H.$$

Hydrogen cyanide then leads to the formation of HC_3N, with a net result as follows:

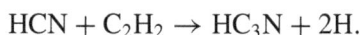

$$HCN + C_2H_2 \rightarrow HC_3N + 2H.$$

The production of cyanogen is controlled by the sequence:

$$N_2 + C_2H_2 \rightarrow C_2N_2 + 2H$$

while that of dicyanogen is preferentially:

$$N_2 + 2C_2H_2 \rightarrow C_4N_2 + 2H + H_2.$$

Following the recent discovery of CH_3CN, a mechanism was proposed for its formation, viz.:

$$HCN + 2CH_4 \rightarrow CH_3CN + CH_3 + 3H.$$

6.5.3 Oxygen Compounds

The traces of water found in the upper atmosphere of Titan by ISO probably originate from the influx of chondritic or icy meteorites. It photodissociates to produce hydroxyl (OH) radicals in the high atmosphere, which react with CH_2 and CH_3 to form CO, subsequently diffusing downwards. Some of the CO may be destroyed near 500 km, again by the action of OH, forming CO_2. The CO_2 is destroyed in turn by photolysis and various chemical reactions, but some probably reaches the lower atmosphere where it condenses and precipitates on the surface. There is no known sink for CO on Titan, other than OH, and so its lifetime is expected to be very long (comparable to the age of Titan), for those molecules that reach the middle and lower atmosphere. As a consequence, CO should be uniformly mixed with N_2 in the stratosphere and troposphere, while the CO_2 abundance there is expected to be small, as observed.

CO is produced in reactions between OH and CH_3 and CH_2. The formation of carbon dioxide from water vapour and carbon monoxide may be written as:

$$H_2O + CO \rightarrow CO_2 + 2H,$$

where the intermediate steps involve the production of OH through water photolysis and

$$CO + OH \rightarrow CO_2 + H.$$

A simplified version of the oxygen chemistry can then be expressed in terms of production and loss in a very crude form, where only leading terms are kept and where the equations are valid only when integrated over a column, as follows:

Carbon monoxide:

$$P(CO) = k_3 \times [OH][CH_3],$$
$$L(CO) = k_1 \times [OH][CO]$$

Carbon dioxide:

$$P(CO_2) = k_1 \times [OH][CO],$$
$$L(CO_2) = \text{Condensation and Precipitation}$$

Water:

$$P(OH) = J \times [H_2O],$$
$$L(OH) = k_1 \times [OH][CO] + k_2 \times [OH][CO_2],$$

where J is the photodissociation coefficient related to water, and k_1 and k_3 are reaction rates. Near carbon monoxide equilibrium, $P(CO) = L(CO)$, and hence $k_3 \times [CH_3] = k_1 \times [CO]$.

It follows that the production of carbon dioxide,

$$P(CO_2) = J \times [H_2O]/(1 + (k_3 \times [CH_3])/(k_1 \times [CO])),$$

is in fact equal to $J/2 \times [H_2O]$, and therefore independent of CO, because CO has a long photochemical lifetime at equilibrium.

In their 1996 photochemical model, Lara and colleagues were able to reconcile at least some of the observations of both CO and CO_2. They assumed a continuous supply of water into Titan's upper atmosphere of about $6 \times 10^6 \, \text{cm}^{-2} \, \text{s}^{-1}$, proposing that this could come from external sources such as the rings of Saturn, meteorites or comets, They also predicted the expected vertical profile of H_2O in the present-day Titan atmosphere in equilibrium with the assumed supply and chemical depletion. This prediction is roughly consistent with the interpretation of the ISO observations.

The ultimate sink of the oxygen, coming in at the top of the atmosphere as H_2O, is at the surface of Titan, where it should precipitate out as frozen CO_2. Thus, another material is probably depositing itself on top of the water ice that makes up the upper part or crust of the solid body of Titan, in addition to the hydrocarbon-based liquid and solid drizzle expected from the haze layers, and rain from any clouds present. If so, it is debatable whether water ice is in direct contact with the lower atmosphere anywhere on Titan, since this will depend on the nature and thickness of the condensate coating. One possibility is that low-lying regions are filled with accumulating precipitates of all kinds, while high ground is washed clean of solid or tarry deposits by methane rain.

If there is good contact between ice and air, at least in some regions, then the troposphere could be saturated with water, albeit at temperatures so low that this corresponds to a very small mixing ratio. The relatively warm upper atmosphere,

Icy particles
from meteors
and comets

H_2O

Sunlight

OH

CH_2, CH_3

CO

CO_2

condensation

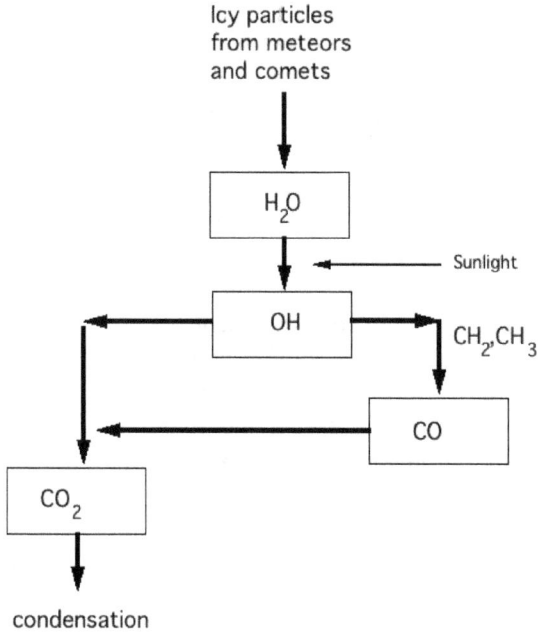

Figure 6.21   Basic water photochemistry on Titan.

which could hold proportionally more water, is prevented from obtaining this from
below by the cold-trap formed by the tropopause. This is the same reason that the
stratosphere of the Earth is very dry relative to the tropopause below, which of
course is in contact with the water in the oceans. The net effect is that the water

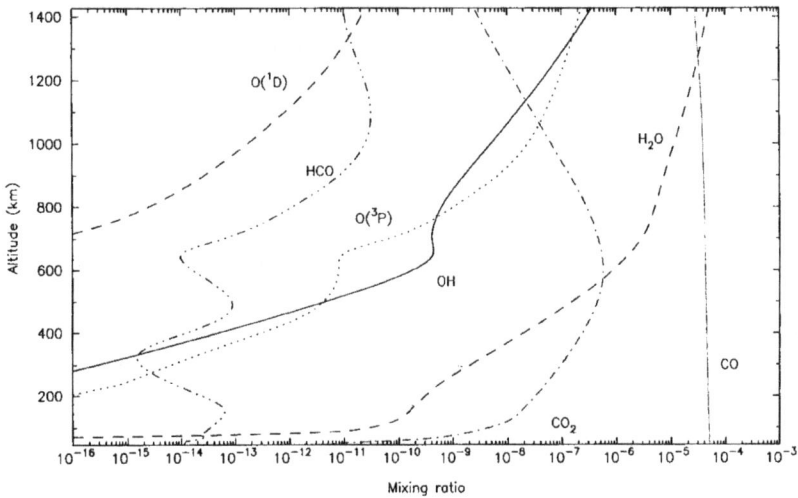

Figure 6.22   Calculated vertical profiles of oxygen components in Titan's atmosphere, from the model
by Lara *et al.* (1996).

ice on the surface of Titan cannot contribute significantly to the photochemistry in the upper atmosphere, even if it is exposed to the air. If Titan was moved closer to the Sun, this would change. Warmer surface and atmospheric temperatures would cause the water to sublime into the atmosphere and propagate to higher levels. The oxygen provided in this way would rapidly combine with the methane and other hydrocarbons and form large amounts of carbon dioxide, similar to what has happened on Venus and Mars. The character of Titan's atmosphere would radically change, but not necessarily towards something closer to the Earth.

Abundances which depend on the statistical influx of meteoritic and cometary material might be expected to be time-dependent. To investigate this, the amount of CO_2 measured in Titan's atmosphere from Voyager has been compared with that from the ISO and CIRS observations almost a Titan year apart. The agreement between the two is excellent; it appears therefore that a mixing ratio of around 10^{-8} of CO_2 may be stable in Titan's stratosphere. However, the water influx on Titan required to produce this much CO_2 is higher than recent estimates from water detection on Saturn by ISO ($5 \times 10^6 \, \mathrm{cm}^{-2} \, \mathrm{s}^{-1}$). Theoretical work, involving calculations of the gravitational focussing effect of the large mass of Saturn, predicts 7 times or less water influx on Titan than on Saturn from interplanetary dust. The local sources (the icy satellites and rings) cannot make up the difference, because gravitational effects favour Saturn as the greater recipient of water coming from these sources, too, except possibly for material from the outermost satellites (Hyperion, Iapetus and Phoebe). Further progress on the whole question of oxidised molecules in Titan's atmosphere awaits new and more precise values for their abundances and fluxes.

The most recent models produce vertical profiles for the Titan atmospheric components, some of which match the observations. New investigators are now interested in the subject of Titan's photochemistry, trying to combine such processes as radiative transfer and convection mechanisms with GCMs, and more complete schemes for Titan will no doubt see the light of day in the near future.

6.5.4 Condensation Efficiencies

The vapour pressures and condensation rates for the most abundant minor species on Titan have been studied to see which could produce condensation products under Titanian conditions. Models show that H_2O, HCN, HC_3N, C_4H_2 and C_2N_2 are expected to condense easily at pressures of 10–20 mbar (75–100 km altitude), whereas methane, ethane and ethylene condense at low altitudes (under 60 km) or not at all. C_3H_4, C_3H_8 and C_2H_2 are intermediate. The more abundant species, and those which condense at higher altitudes, are obviously the favourites when it comes to selecting candidates for the observed features of condensed material in measured spectra. The spectra of condensates differ from the gas phase of the same material, in general, by having more extensive features (i.e., extending across a wider range of wavelengths) and a lack of fine structure (continuum absorption, rather than sharp spectral lines).

These species have different condensation rates, which, with their abundances and the physical conditions such as temperature, control the probability that they will appear in large quantities in Titan's lower stratosphere. H_2O, CH_3CN and C_3H_4 have low condensation fluxes (on the order of 10^4 cm^{-2} s^{-1}) compared to C_4N_2, C_4H_2 and C_2H_4 (by 2 orders of magnitude), to HC_3N or C_3H_8 (by 3 orders of magnitude), and to C_2H_2, HCN and C_2H_6 (by 5 orders of magnitude). So, since solid C_4N_2 has been already observed in Titan's stratosphere, it should also be possible to find solid HC_3N, C_3H_8, HCN, C_2H_2, C_4H_2, and C_4N_2. Condensed C_2H_6 and H_2O are also possible, if less likely, while C_3H_4, C_2H_4 and CH_4 are more or less ruled out in this model. Acetonitrile, acrylonitrile and propanenitrile are expected to condense near the same location as HCN, but in such small amounts that they, and indeed many other of the more complex photochemical products, could be present and still remain undetected for the foreseeable future.

6.5.5 Aerosol Production

The preceding discussion of the condensation of photochemically-produced molecules, and their subsequent condensation to form particles, leads naturally into a discussion of the aerosols we see on Titan, and which dominate the visual appearance of the satellite. Can we work out, with a combination of the present theoretical understanding of Titan's atmospheric chemistry, as outlined in this chapter, plus spectroscopic measurements of Titan, what is the composition of the visible layers in the stratosphere and above? Another approach that is available is to try to make synthetic Titan haze material in the laboratory, such as the reddish-brown 'tholins' produced by C. Sagan and colleagues, beginning in 1989. Earlier, F. Cerceau, F. Raulin and colleagues had produced and measured the infrared spectra of various gaseous nitriles like those in the atmosphere of Titan by a similar approach. Of course, these types of experiments cannot faithfully reproduce the planetary conditions and require careful interpretation. But they do support the idea that Titan is harbouring the necessary conditions for the formation of numerous highly complex aerosol molecules. How these issues tie in with current knowledge of the properties of Titan's aerosols is the subject of the next chapter.

The interior of Rama was completely blanketed with clouds, and nowhere was a break visible in the overcast. The top of the layer was quite sharply defined; it formed a smaller cylinder inside the larger one of this spinning world, leaving a central core, five or six kilometres wide, quite clear except for a few stray wisps of cirrus.

Arthur C. Clarke, *Rendezvous with Rama*

7.1 Introduction and Overview

In the last chapter, we discussed the structure and composition of the atmosphere of Titan in terms of the gases present. However, this description is obviously incomplete because we know from various observations that the atmosphere also contains a fine dispersion of small particles or aerosols, which form extensive layers of haze surrounding the entire globe. The presence of aerosols does not come as a surprise, since the complex chemistry and low temperatures on Titan virtually guarantee that some of the species present will form solid or liquid particles, by a combination of chemical synthesis and condensation. Exactly what forms, where, and what eventually happens to it, is one of the most complicated and puzzling aspects of the satellite's mysterious atmosphere.

The haze on Titan is often described as a photochemically-produced smog, and it extends very high in the atmosphere. The effect of this is to increase the apparent diameter of Titan by several hundred kilometres, when it is viewed through a telescope from the Earth, and is why it was long thought to be the largest satellite in the Solar System, whereas in fact Ganymede wins by a small amount if only the solid bodies are compared. We saw in the previous chapter how the products of the photo-dissociation of nitrogen and methane are expected to recombine and produce larger molecules, which mix downwards into cooler regions and form the stratospheric haze layers.

If this process continues for long enough, some of the higher-order products will become sufficiently abundant that sooner or later they are bound to form larger particles that fall out of the stratosphere. The expectation is that at least some of these will consist of refractory materials, that is, oily or solid substances which, once formed, do not evaporate under the conditions that prevail on Titan. Once in the troposphere, they drizzle down to the surface, and apparently contribute to the level of atmospheric turbidity detected by the Huygens probe.

These droplets also provide possible nucleation sites for the growth of larger droplets where the conditions of temperature and relative humidity are right for the condensation of abundant species such as methane and ethane. Whether or not the particles of photochemically-produced haze have a role in the formation of condensate clouds in the lower atmosphere, the process by which they reach the surface and are removed, probably permanently, from the atmosphere, will be accelerated in the regions where they end up 'rained out' by being entrained inside millimetre-sized drops of liquid methane.

While the existence of the thin, ubiquitous, photochemical haze covering Titan was inferred even before the Voyager missions from measurements of Titan's spectral albedo, it was only relatively recently confirmed that relatively dense, localised, cumulus-type clouds are also present. These form below the haze in the lower atmosphere, and are inferred to be composed primarily of methane, since calculations show that this very abundant species will condense and evaporate at the temperatures in Titan's troposphere where the clouds are seen to form, much as water does on Earth. The clouds are limited in horizontal extent and frequency of occurrence, and were detected, first indirectly from ground-based telescopic observations and then in the Cassini images at near-infrared wavelengths that also show features on the surface.

The size and shape of the haze particles in Titan's atmosphere has been the subject of debate for a long time. Instruments on the Pioneer 11 and Voyager spacecraft obtained the first photopolarimetry measurements of scattered light from Titan at high phase angles (the angle between the Sun–planet and observer–planet directions), since this angle is always small when viewing Titan from Earth. The interpretation of this data by the Pioneer team, assuming the aerosols are made up of spherical particles, constrained the particle radius to about 0.1 μm. However, the high phase angle brightness measurements made from Voyager some years later required the particle sizes to be between 0.2 and 0.5 μm, with a higher likelihood for the upper limit. Attempts to reconcile the two included the suggestion of a multimodal distribution for the particle sizes, and the possibility that the particles are irregularly shaped, possibly aggregates constructed from smaller units. In any case, a model in which the particles are all spherical and of nearly the same size (which works quite well for the upper cloud layers on Venus) seems to be too simplistic for Titan.

Theoretical work on the possible size and shape of the particles in Titan's photochemical haze layers has been focussed in recent years on microphysical models involving fractal geometries. Fractals are patterns that occur in nature over a range of size scales and produce irregular shapes and surfaces. If the Titan aerosols are assemblages of many small solid particles, then fractal theory can produce models of them that can be tested against observations of the spectral dependence of the brightness and polarization of radiation reflected from Titan at different wavelengths and phase angles. In order to include the effects of the aerosols in the radiation transfer calculations that make up the model, vertical profiles of haze opacity have to be specified as well as the microphysical properties of the particles. The simplest assumption is a homogeneous slab with a constant number density of particles, or a simple exponential decrease in number density with altitude, such as would occur if the particles are uniformly mixed with the atmospheric gases. In more recent work, more complicated profiles are generated by microphysical models using a calculated or pre-specified production rate of haze material as a function of height.

In addition to the shape and size, the refractive index of the haze particles is an important parameter in the model calculations. The refractive index is based on laboratory measurements of candidate materials, produced in experiments that simulate the production of the haze material from methane and nitrogen. The optical properties of the organic matter (tholin) produced in these experiments provide a reasonable match to the general features of the observed spectrum of Titan, supporting the inference that the haze is composed of refractory organics of photochemical origin.

The model predictions can be tested against observations of Titan's spectral geometric albedo from the ultraviolet to the near infrared, as obtained from ground-based and spacecraft instruments. As was first shown by C. P. McKay and colleagues in 1989, the fit to the spectral geometric albedo is most sensitive to three parameters; the optical properties (refractive index, size, shape and amount) of the haze particles, the methane profile, and Titan's surface reflectivity. Using observations over a range of spectral domains, from ultraviolet to far infrared, constraints can be set on the values of these parameters.

The main achievement of the fractal approach has been that it provides, in general, a good fit to the geometric albedo both in the UV, visible and near-IR regions while at the same time matching the polarization data. However, fractal models have been unable to provide as good a match to the methane absorption feature at $0.62\ \mu$m as the spherical particle models do. The fit to the data with fractals was improved by applying a haze cut-off below 100 km, as suggested by observers seeking to interpret early HST measurements. A haze clearing was also found to be necessary to improve the fit with observations when using spherical particle models, although in this case at lower altitudes (below 30 km). The first in-situ measurements, from the DISR instrument on the Huygens probe showed, however, that the haze opacity extends down to the surface, introducing a conundrum which, like many other questions relative to the properties of the hazes and clouds in Titan's atmosphere, remains unsolved.

In the following sections we review the terminology used and a few basic facts about clouds and precipitation on the Earth, as background for considering related behaviour in the atmosphere of Titan. We discuss the observations we have of clouds and condensates on Titan and of hazes in the middle and upper atmosphere: visual images from Voyager and recently from Cassini have identified their extent and layering, while Earth-based observations, especially of the albedo and its hemispherical asymmetry, limb darkening, and behaviour during occultations of stars by Titan, continue to set valuable constraints on cloud and haze models. Thermal and dynamical coupling with the haze layers are considered next, then the work which has been done to infer the vertical distribution of different particle sizes, and the spectroscopic evidence for their composition. We go on to describe the laboratory simulations that give clues as to the chemical origin of the haze material and that, in spite of their limitations, have proved extremely useful in understanding the elemental composition, solubility, and refractive indices of possible aerosol particles. Finally, we discuss modelling and computer simulations of the haze formation that have been used to retrieve information on the physical properties of the particles and on the rate of haze production, approaches that are central to current efforts to analyze the Cassini–Huygens data.

7.2 Terrestrial Clouds and Precipitation

A brief discussion of the properties of the more familiar clouds we experience on the Earth will help to introduce the discussion of the corresponding phenomena on Titan. The latter is necessarily highly theoretical, given the limited information we have at present, and terrestrial analogues are usually helpful. For instance, the emphasis in scientific work on Titan clouds on soluble condensation nuclei as a prerequisite for droplet formation, even when supersaturation is present, derives directly from terrestrial experience.

On the Earth, substantial clouds of water droplets are usually confined to the lowest scale height of the atmosphere (0 to 8 km), although very thin 'noctilucent' clouds have been observed as high as the mesopause (85 km). There are many different classifications of clouds, nearly all formed by some version of the same basic process: convection upwards of moist air and subsequent condensation of water droplets or ice crystals around dust nuclei. In the usual definition, clouds differ from hazes primarily in that they have larger particles, often referred to as 'droplets', although they are technically still aerosols. The particles are more numerous, so clouds are usually optically thick over quite short distances, whereas haze or aerosol layers are usually optically thin unless they are physically very deep.

A typical terrestrial cloud droplet is a few microns in diameter, while a typical raindrop is one hundred times larger, i.e. contains as much water as a million cloud particles. The process by which cloud droplets grow is basically by diffusion and condensation of water vapour on a nucleation site, until they reach about 30 μm in diameter, then by collision and coalescence until they eventually rain out.

In addition to their obvious importance in meteorology and the terrestrial water cycle, clouds play an important role in controlling the global climate by scattering, absorbing and emitting radiation. They interact strongly with both incoming solar and outgoing thermal (planetary) radiation, and so are an important component of the atmospheric greenhouse effect. Because the radiative transfer properties of clouds depend on variable and relatively inaccessible quantities such as the number density and size distribution of particles, their role is a particularly difficult one to quantify in climate models.

Figure 7.1 Titan image taken one day after Cassini's first flyby on July 2, 2004 and showing a thin, detached haze layer that appears to float above the main atmospheric haze over Titan's limb. The Voyager spacecraft also detected such detached haze layers on Titan during their flybys in the early 1980s (NASA/JPL).

7.3 Visible Aspects of Titan's Haze

Half a century ago, Earth-based observations of Titan revealed the presence of methane absorption bands in the infrared spectrum and indicated the presence of a dark-orange or brown aerosol haze layer in the upper atmosphere. The presence of CH_4, which was known to be easily photolysed (i.e. dissociated by radiation, especially the energetic ultraviolet photons from the Sun), and the colour of the cloud deck, led to the suggestion at an early stage that organic aerosols were being produced in the atmosphere.

The highest spatial resolution images of Titan obtained from Voyager in 1990, especially photographs of the limb, confirmed this basic picture, showing an optically thick haze deck composed of several distinct layers in the stratosphere completely hiding the surface at visible wavelengths. The uppermost of these is very tenuous and shows up primarily in the ultraviolet region of the spectrum. In Voyager television observations of the high phase angle brightness, i.e. those in which the incident and scattered rays are almost in a straight line, particles are observed at altitudes as high as 500 km above the surface, with an extensive, detached haze layer occurring from 300 to 350 km. The visible limb of the planet, where the vertical haze optical depth is 0.1, is about 220 km above the surface. The haze controls the propagation of sunlight in Titan's atmosphere and thereby significantly influences the temperature of the atmosphere and of the surface. It was previously assumed that its base would probably be located between 100 and 200 km altitude, where condensation and coalescence may go on more rapidly, forming larger and more numerous particles so that the haze condenses into denser conglomerations. In view of recent Huygens results, however, it would seem that at least some haze extends all the way down to the surface; this may be particles formed at higher levels falling out as drizzle.

Attempts were made to infer some properties of the haze aerosols, such as the radius and the refractive index, from early Earth-based observations. The reflectivity was known to vary with a period of half a Titan year (15 Earth years) even before the arrival of the first space mission in the late 1970s. A large effort was devoted, starting in the early 1980s, to understand the physics and chemistry of the haze system, using polarimetric observations, large phase angle photometry, Earth-based observations of geometric albedo, near-infrared telescopic observations, theoretical modelling,

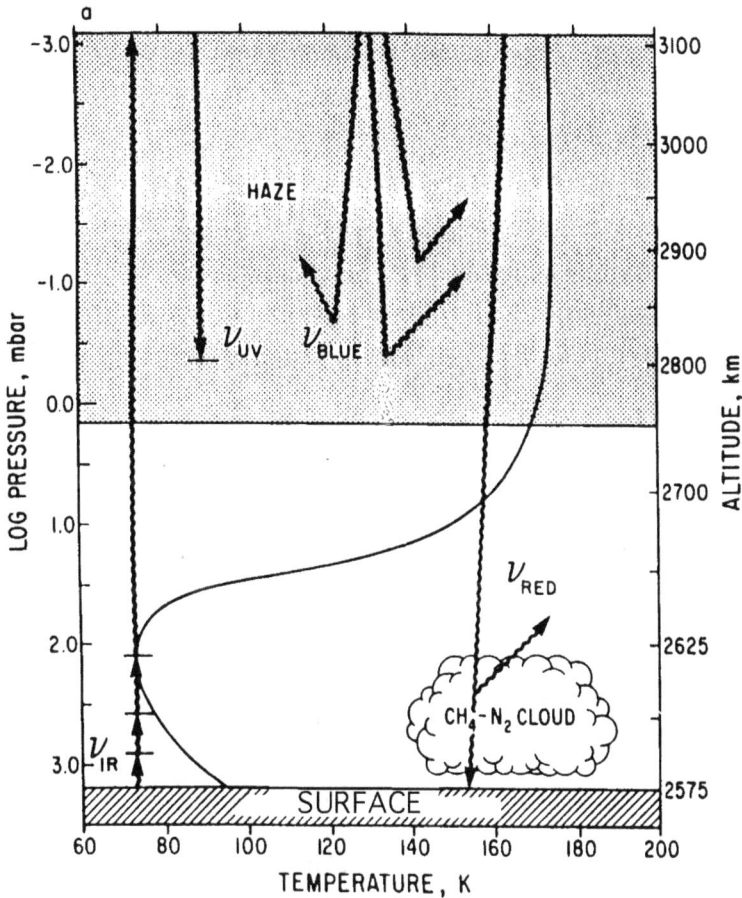

Figure 7.2   The approximate locations of haze production and of the occurrence of sparser but thicker methane condensation clouds, relative to the vertical temperature profile. The arrows show the penetration of radiation in different wavelength ranges.

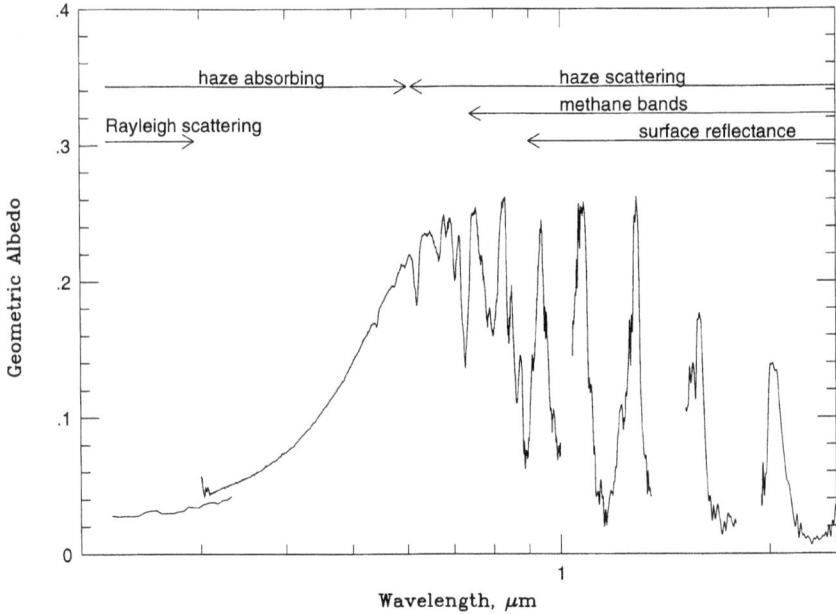

Figure 7.3   The geometric albedo of Titan from International Ultraviolet Explorer (0.2–0.3 μm) and ground-based observations, with an indication of the processes that determine the albedo in each wavelength range.

and laboratory experiments. This has culminated in the current close monitoring of Titan's haze with Cassini, while observations from the ground continue to contribute thanks to recent improvements in Earth-based instrumentation, and with the advent of techniques such as adaptive optics, speckle interferometry and space telescopes like Hubble.

A north-south haze asymmetry of the Titan albedo was observed by the Voyager spacecraft in the visible region of the spectrum, the southern hemisphere being brighter by about 20% than the northern one. This is related to seasonal variations in the stratospheric haze and appears to operate over a 30-year cycle. J. Caldwell and colleagues reported the first observational evidence for these seasonal changes in 1992, using observations from the Hubble Space Telescope. The asymmetry reaches a maximum during the equinoxes, and reverses in the periods after the equinoxes. Later work suggested a dependence on wavelength; the effect is more evident at blue wavelengths with a peak around 450 nm, weaker in the ultraviolet, and reverses at red and near-infrared wavelengths longward of 700 nm. As we discuss later, a dynamical origin seems to be the most plausible way to explain the asymmetry and its dependence on wavelength: since the seasonal haze changes are altitude dependent, different haze asymmetries would be observed, depending on the wavelength used and hence on the altitude probed.

The Cassini ISS camera recently observed a faint thin haze layer that encircles the denser stratospheric haze, rather like the detached haze layer observed by Voyager

25 years ago. However, there is a difference in altitudes: the thin current haze layer is located 150–200 km higher than the one seen by Voyager. Dynamical models are unable to render the complexity of seasonal phenomena or circulation patterns on Titan that could be responsible for such an upward shift. Cassini images also show a multi-layer structure in the north polar hood region and at lower latitudes in some cases, possibly due to gravity waves similar to those that have been detected on Titan at lower altitudes. Some of these layers may be related to the two global inversion layers observed in stellar occultations of Titan above 400 km in altitude. The long-wavelength filter in the Cassini camera, centred at 0.94 μm, which is just beyond the range of the human eye, can see through the haze and image the surface.

7.4 Size and Vertical Distribution of the Haze Particles

By achieving the high phase angle coverage that is possible only from spacecraft, Pioneer 11 and Voyager photopolarimetry data provided the first detailed information on the scattering properties of the haze particles. If the haze particles were spherical, the large polarisation of light scattered by Titan's atmosphere implied that they be no larger than 0.1 μm. On the other hand, studies of the Voyager observations of Titan led K. Rages and colleagues to propose particles that are at least 0.2 μm and probably as big as 0.5 μm in radius. There are ways in which these two sets of measurements can be reconciled. One is by invoking a double-peaked distribution in the atmosphere, by placing the detached haze layer with larger particles above 300 km, assuming it was put there by rising air currents. Another is by realising that we probably should not expect Titan's haze particles to be simple spheres. It is now believed that they may have quite a complex shape, possibly complicated aggregates of spherical or non-spherical particles stuck together forming shapes that are described by fractal geometry.

In August 1995, the Earth crossed the plane of Saturn's rings, thus diminishing their interference with observations of Titan and Saturn, and allowing a number of interesting eclipses and other events to be observed by ground-based telescopes, including Titan's shadow cast onto Saturn. These measurements are analogous to the high-phase angle Voyager images, with the additional advantage that the rings could act as a fiducial marker to determine the location of Titan's centre, whereas no such marker was available in the spacecraft observations. The data show that the altitude of the optical limb (where the atmosphere appears to end above the surface) is between 180 and 360 km, depending on the wavelength, with an aerosol scale height of 45 km. If the aerosol particles were spherical, this would mean sizes of 0.3 μm in radius for those in the north, while smaller ones, about 0.1 μm, would exist in the south. This model fails to match limb observations in the UV, which is probably another indication for fractal rather than spherical particles. Nevertheless, the larger northern and smaller southern particles probably again reflect the existence of two separate aerosol layers — with the upper 'detached' layer only present in the north. The timing of this — late northern summer — is consistent with the seasonally dependent interpretation of the haze layering from Voyager and stellar occultations if the higher haze is assumed to build up when the solar irradiance is strongest.

Measurements taken with the Voyager IRIS instrument in the far-infrared range showed that the stratospheric continuum opacity increased monotonically between 250 and 600 cm^{-1} in a way that is similar to that observed in the tholins produced in laboratory simulations. The effect is more and more prominent with decreasing altitude, interpreted as indicating aerosol particles that are fairly small at high altitudes in the stratosphere, with an additional component of larger particles (possibly ethane condensation clouds with particles of radius $\sim$10 μm) lower down. An aerosol mass abundance of $\sim$6 $\times$ 10^{-8} was derived at altitudes around 160 km, with associated scale heights 1.5 times larger than the atmospheric density scale height for the minor aerosol component, and somewhat less for the large-particle component. The opacity was found to be roughly independent of latitude, except for an apparent fall-off at high southern latitudes. At shorter wavelengths in the mid-infrared, where the opacity in Titan's spectrum is due mainly to the presence of particles in the stratosphere and is less sensitive to those in the troposphere, there was an increase in stratospheric opacity by a factor of 2 or more at latitudes above 50°N, compared to mid or southern latitudes, at the time of the Voyager encounters.

The nature of the haze aerosols measured by the Huygens Descent Imaging System (DISR) during its passage through Titan's lower atmosphere came as a surprise to scientists recalling the results from Pioneer and Voyager and the predictions by the most advanced cloud physics models available at the time. The new observations, the first to be obtained actually inside the atmosphere of Titan, yielded an

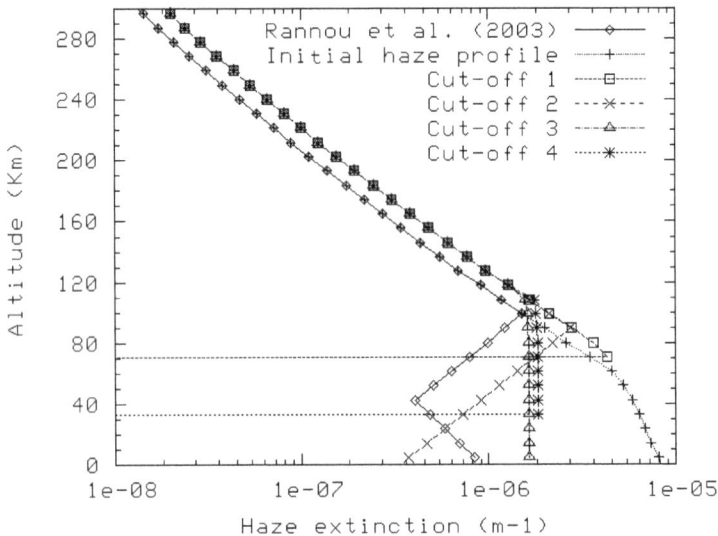

Figure 7.4   Haze extinction profiles that fit the Titan data from Rannou *et al.* (2003) and from Coustenis *et al.* (2006). Note that all of the retrieved profiles, as opposed to the initial or 'a priori' profile, have a sharp drop in the haze opacity below about 70 to 100 km altitude.

estimate for the radius of the monomers, the individual particles making up the fractal aggregates, to be around 0.05 μm, which is in good agreement with previous values. However, the DISR data also showed that the haze optical depth varies from about 2 at 0.935 μm to only about 4.5 at 0.531 μm, and to fit this the number of monomers N in, a haze particle needs to be as much as ten times larger than previously assumed. With $N = 512$, the equivalent sphere with the same projected area has a radius of 0.9 μm, nearly three times the favoured size in most of the models.

7.4.1 *Haze Vertical Profiles*

Measurements by the DISR also showed that the number density of the haze particles does not increase with depth nearly as dramatically as predicted by the older cloud physics models. In fact, the number density increases by only a factor of a few over the altitude range from 150 km to the surface. This implies that vertical mixing is much less than had been assumed in the models, where the particles were distributed approximately as the gas density with altitude, i.e. assumed to be well mixed. The new evidence for weaker mixing has implications for the haze extinction by making the flow towards the surface (where the particles are considered to be irreversibly lost) smaller. However, particle loss could be dominated by other processes, such as coalescence and rainout.

Another key finding by the probe was the absence of a clear region at low altitudes, in contrast to the conclusions of the first modellers of Titan's haze, based on the analysis of ground-based data. They had found it necessary to assume a low concentration of haze below a "cut-off" altitude, generally located between 70 and 100 km, to account for the geometric albedo observations in the methane band wings. A profile with suppression of haze at lower altitudes had been applied regularly since the first retrievals of Titan's haze profiles, and was commonly used in past radiative transfer studies using 1-D models. P. Rannou and colleagues found that improved 2-D models yield haze vertical profiles which are very different from the 1-D ones, due to the more complete representation of the effect of the circulation of Titan's atmosphere. However, the 2-D model did not intrinsically predict a haze cut-off at lower altitudes, and it was still necessary to introduce this *ad hoc* in order to match the observations. A recent upgrade to a full general circulation model finally did produce haze suppression at lower altitudes, through the inclusion of methane clouds in the troposphere with haze particles acting as condensation nuclei. In other modelling work by P. Lavvas and colleagues, the particle flux towards the surface is enhanced by an increased eddy coefficient, another way to simulate the loss of particles at low levels and avoid the need to apply an artificial cut-off to the haze profile.

Having reconciled the models with the haze "cut-off" at lower altitudes that several different authors had required in the analysis of Earth-based observations, in January 2005 researchers suddenly had to explain why this seems to be incompatible with the results of the Huygens DISR instrument, which measured a continuously-increasing opacity with depth for three wavelengths from 140 km altitude down to the surface.

If both remote and in-situ techniques are assumed to be equally reliable, the answer to this apparently contradictory result might lie in the principal aspect that distinguishes the two approaches: the geometry of observation. It seems reasonable to assume the coexistence of haze particles with solid methane crystals at higher altitudes and with liquid drops of methane at lower altitudes. Although the aerosol particles are very conservative scatterers, with values of the single scattering albedo close to 100% for wavelengths around 1 μm, their asymmetry factor values are only around 0.5 at these wavelengths, implying that a significant part of the scattered light is directed back into space. The hypothetical solid particles and liquid drops, on the other hand, would scatter mostly in the forward direction and would be harder to detect by an observer on the Earth. Such an observer outside Titan's atmosphere, detecting a decrease of backscattered light at lower altitudes, has to assume a reduced concentration of scatterers at lower altitudes by introducing some kind of haze "cut-off".

However, the transparent solid crystals or liquid drops could coexist with the haze particles without being detected by the Earth-based telescopes since they would mostly scatter light in the forward direction. A probe inside Titan's atmosphere observing upwards, away from the surface, like the Upward-Looking Visible Spectrometer (ULVS) and the Upward-Looking Infrared Spectrometer (ULIS) instruments on the Huygens probe, would however detect not only the light scattered by the haze particles but also the light forward-scattered by the solid and/or liquid methane particles. The probe would observe a halo around the Sun, as on Earth when a cloud or fog covers the Sun in the sky, which would add to the haze opacity and yield a perceived total opacity near the ground higher than that inferred from the Earth-based observations. This hypothesis is the most plausible one at present and is supported by some recent findings of a relative humidity consistent with the coexistence of

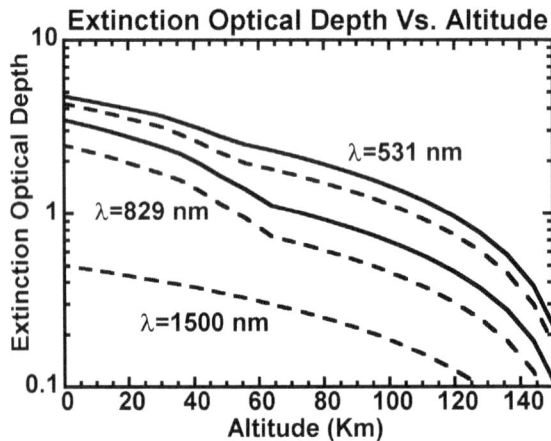

Figure 7.5   Extinction optical depth vs. altitude, as measured by DISR, for three different wavelengths: 0.531 m (top), 0.829 μm (middle) and 1.5 μm (bottom). The dashed curves correspond to the 256 monomers per aggregate particle case, and the solid curves correspond to the 512 monomers per aggregate particle case. No haze suppression is found at lower altitudes (Tomasko *et al.* 2005).

liquid or, depending on the temperature and hence the altitude, solid methane with the gaseous methane in Titan's atmosphere at the Huygens landing site.

7.4.2 *Haze Opacity Spatial Variations*

Since 1996, measurements of Titan's haze optical depth at different regions on the disk have been available to provide information on the global variation of the haze. In that time the solar longitude L_s has varied from about 190° to over 300°. Observations by Gibbard and colleagues found that the haze optical depth in the equatorial region at 2 μm dropped from 0.18 to 0.12 between 1996 and 1997, and then increased to 0.15 in 1998, remaining constant until 2003, and then finally increasing to 0.2 in 2004 (at Ls ~290°). These values may be compared to calculations by A. Negrão *et al*, using data obtained at the CFHT and at the VLT, of the integrated haze opacity at 2.04 μm,

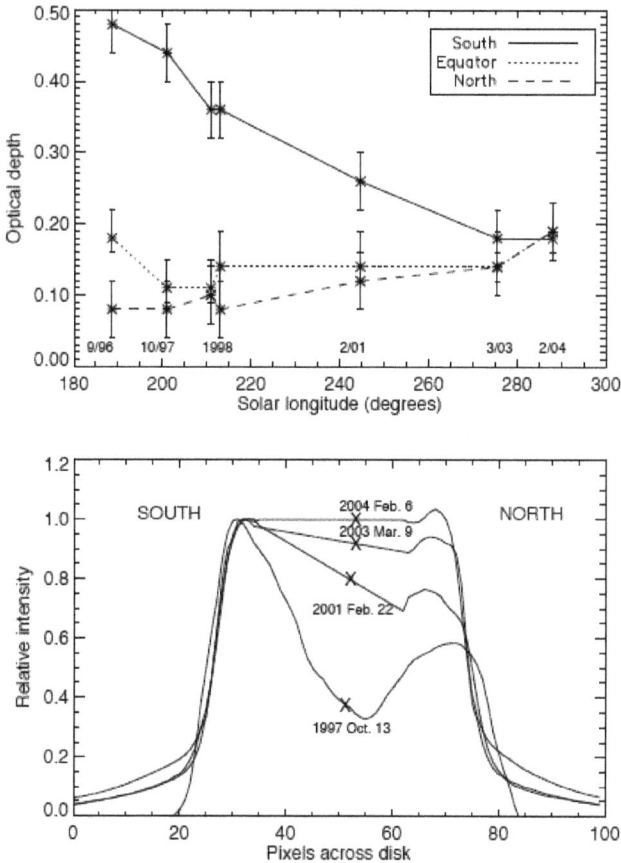

Figure 7.6 (Top) 2 μm haze optical depth values from September 1996 to February 2004. (Bottom) north/south scans across the centre of Titan images from 1997–2004 showing how the haze asymmetry has changed during this time. Surface features have been removed (except for the 1997 data) and replaced with a linear fit. The scans are normalized to the southern limb brightness. Location of the equator is marked by 'X' (Gibbard *et al.* 2004).

which yielded a value of 0.264 for the Huygens landing site, 0.250 for a dark site located at the equator, and 0.258 for a bright neighbouring site. The solar longitude at the time of the Huygens landing was around 302°. The probe value of the haze optical depth near the equator is reasonably compatible with the remotely observed increase of the equatorial haze optical depth between 2003 and 2004, with the slope of the curve about the same between 2003 and 2004 and between 2004 and 2005.

Scans across the disc of Titan on various dates reveal the progression with time of the north-south asymmetry in Titan's albedo. Since, at the time of the Voyager encounters, the northern hemisphere was brighter than the southern one at near-infrared wavelengths, the 30-year cycle in the asymmetry means that we are currently, close to 30 years after the Voyager observations, seeing a migration of the haze from the southern to northern latitudes, passing by but not accumulating at the equator.

7.5 Tropospheric Condensate Clouds

Information on localised clouds and condensates in Titan's lower atmosphere was, until very recently, not only scarce, but even more indirect and speculative than that on the global haze layers. Spatially-resolved ground-based and HST observations in the past decade, as well as the recent Cassini–Huygens measurements, have changed all of this. We expect clouds to occur most readily in updrafts where moist air is lofted by vertical motions, cooling as it expands and causing condensation to occur. On Titan the most likely material to behave in this way is methane, because of its large abundance in the atmosphere, and because Titan atmospheric temperatures are close to methane condensation values. Studies of moist convection in Titan's atmosphere had suggested that the maximum area of the globe expected to be covered by convective plumes would be very small, in which case intense but short lived "storm clouds" on Titan would be rare. Then came the Voyager occultation profiles for the temperature structure of the lower atmosphere, which suggested that the CH_4 abundance is high enough to exceed the saturation vapour pressure in the upper troposphere, leading to a tendency for methane clouds, and methane rain, to occur on Titan even in the absence of major updrafts.

After the Voyager 1 encounter, it also became possible, using IRIS spectra coupled with radio occultation temperature profiles, to estimate the tropospheric opacity in the infrared at wavenumbers between 200 and 600 cm^{-1}. This sparked a controversy over whether the gaseous opacity of the atmosphere, which at these wavelengths is primarily the collision induced absorption from the various combinations of N_2, CH_4 and H_2, was sufficient to account for all the tropospheric opacity. Early studies suggested that it was not, and it became fashionable to assume that liquid or solid methane clouds might provide the remainder, particularly since the temperatures observed by Voyager had revealed that methane condensation was to be expected. A further, indirect piece of evidence for the possibility of such clouds was the increased scintillation in the radio occultation data at the level where CH_4 cloud formation was expected, possibly signalling a turbulent region in the atmosphere.

However, for a long time convincing observational evidence for extensive methane clouds on Titan was not forthcoming. Studies showed that, if the missing opacity comes from clouds, they would need to be made of very small (submicrometer radius) particles, with very large cross-section to volume ratios in the visible and in the infrared, in order to fit the observations. If they had a large particle density in order to provide enough infrared opacity, they would also prevent the penetration of sunlight down to the surface, modifying the greenhouse effect and making it difficult to explain the observed surface temperature on Titan. In fact, all quantitative analyses of the thermal balance in Titan's lower atmosphere show that any cloud layer, if present at all, could not be extensive or persistent. There was also the evidence from near-infrared images of Titan obtained using HST and ADONIS that any tropospheric clouds, if present, must be patchy or thin. For all these reasons, clouds were not the preferred solution to the interpretation of the IRIS spectra.

The only obvious way remaining to solve the problem of the mid-infrared opacity was to accept that a higher amount of methane must be present, together with the fact that the amount required corresponded to a high degree of supersaturation in the troposphere. Supersaturation did not seem an unreasonable solution to the problem of the missing spectral opacity, if it can be argued that there is likely to be a shortage of seed nuclei on which cloud droplets could start to grow. This would lead instead to the eventual formation of large (at least 30 μm in radius, and possibly much larger) methane droplets, which would quickly precipitate through the atmosphere. Rain would fall, without necessarily forming much in the way of clouds first, and after the shower, supersaturation would build up again.

These arguments are not conclusive, however. It is not simple to reconcile the proposed paucity of seed nuclei in the lower atmosphere with the idea, now quite well established, that the continuous build-up of the upper atmospheric hazes must lead to a steady drizzle of small particles down to the surface, probably those observed by Huygens. Perhaps these particles are themselves quite large, since they would be expected to coalesce into relatively small numbers of quite large particles before they became massive enough to fall into the troposphere. They might, therefore, be too scarce to be effective as condensation nuclei and therefore consistent with the conjectured level of methane supersaturation.

In addition to photochemical hazes at high levels and possible methane clouds in the troposphere, a variety of condensate zones may exist at intermediate levels in the stratosphere. As discussed in the previous chapter, many of the CH_4–N_2 photochemistry by-products are expected to condense in the colder parts of Titan's middle atmosphere, with the exceptions of CO, H_2 and perhaps C_2H_4. Furthermore, the condensation of ethane varies significantly over latitude due to the temperature gradient, which also explains why C_2H_6 clouds form mostly at high latitudes. Thus, clouds of various compositions may be found in the stratosphere, again provided suitable seed nuclei are present to facilitate condensation. If condensation is not promoted for some reason, for example if haze material is insoluble in liquid ethane, which is the case with tholins made in laboratory experiments, then some or all of these potential cloud layers may not exist, even if the condensate itself is present.

Something of this kind is thought to be the situation on Neptune, where no condensate layers have been detected at low temperatures, where they might be expected. Voyager infrared spectra of Titan do, however, include broad emission features that have recently been attributed to condensates like C_4N_2, C_2H_2, HCN and perhaps C_2H_6 and C_2H_5CN (see below, Section 7.7).

One night in 1995, observations with UKIRT by C. Griffith and colleagues at wavelengths from 1.5 to 3 μm led to a report of variable behaviour of the infrared spectrum of Titan, which was shown to be consistent with the presence of methane clouds at altitudes of 15 km covering about 10% of the disk. The same team found evidence for more clouds in 1997 and 1999, using the same method, but the data concerned Titan's disk as a whole and no information could be retrieved on the spatial distribution of the clouds. Reports on detections of clouds followed by scientists using the Hubble Space Telescope, and the UKIRT and Keck ground-based telescopes in Hawaii, and since the turn of the century direct imaging of clouds on Titan's resolved disk has been achieved from Earth-based observatories by several other groups. Most of the clouds seen are located in Titan's southern hemisphere, as expected since it was southern summer on Titan, which means that solar heating was concentrated there, driving rising motions. The same images of Titan in the near-infrared methane windows that provided conclusive evidence that Titan is not permanently covered with thick tropospheric clouds, have, in a recent careful reanalysis, indicated the possibility of small transient features. Observers using these and other large telescopes such as the VLT reported a large cloud complex evolving over Titan's south pole, which has been visible consistently since 1999 in the near-infrared (at 2.12 μm for instance), although no previous indication of it was ever reported. It was extremely bright in 2001–2002, but Cassini images have shown that it has been slowly dissipating since the few first Titan flybys in 2005 and is quite marginal now. Its shape is indeed irregular and changing with time, resembling more a cluster of smaller-scale clouds than a large compact field. Should it prove that this system's life was indeed on the order of 5–6 years (fairly close to a Titan season), constraints can be placed on the seasonal circulation patterns on Titan. Other discrete clouds detected at mid-latitudes are infrequent, small and short-lived. Keck and Gemini data by Roe and colleagues indicate that they tend to cluster near 350°W and 40°S. They may be related to some surface-atmosphere exchange (such as geysering or cryovolcanism) because they are not easily explained by a shift in global circulation. A dozen or so large-scale zonal streaks have also been observed by Cassini preferentially at low southern latitudes and mostly between 50–200°W. In December 2006 and January 2007 and Cassini VIMS and ISS detected the presence of a large northern polar cloud or cap, which may be composed of ethane particles suspended between 30 and 60 km in altitude.

The various types of clouds that have now been identified on Titan can be classified as:

(i) widespread permanent ethane clouds, or mist, formed above the tropopause and in the troposphere everywhere in the polar regions (beyond about 60° latitude);

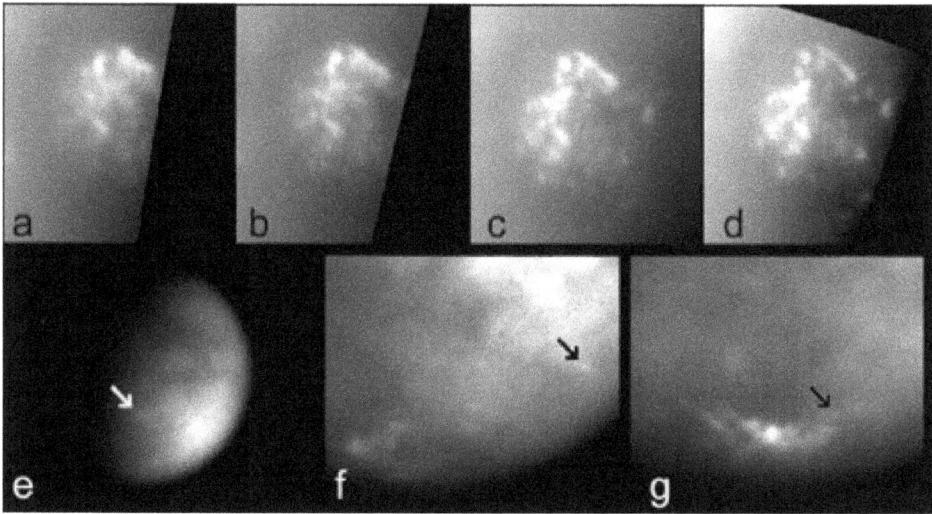

Figure 7.7 Titan's meteorology observed with the Cassini imaging system. The top row is a sequence of four images showing the temporal evolution over the period 05:05–09:38 hrs of the Titan south polar cloud field on July 2, 2004. In the lower row, three examples of discrete mid-latitude clouds for which motions could be tracked in successive images are marked with arrows (NASA/JPL).

(ii) sporadic methane clouds, localized in space and time, at altitudes between 15 and 30 km at tropical latitudes;

(iii) frequent and thick sporadic methane clouds at 15 km altitude around 40° latitude in the summer hemisphere (south in 2005); and

(iv) thick and frequent methane clouds at both poles below or around the tropopause.

Cassini–Huygens has provided new information on the role of methane and the methane cycle in Titan's atmosphere. The relative humidity of methane (about 50%) at the surface found by DISR and the evaporation witnessed by the GCMS show that there has been and will probably again exist fluid flows on the surface, implying precipitation of methane through the atmosphere. Although some argument took place for a considerable time as to whether Titan's lower atmosphere could support convection or not and as to whether methane was supersaturated or not, there is clear evidence today that clouds exist in Titan's troposphere, although in general they are less common and tend to appear higher than expected after Voyager. While the DISR team reported no definite detection of clouds during its descent through Titan's atmosphere, they did see something that might have been a thin haze layer at around 21 km of altitude, possibly due to methane condensation.

Spectra from Cassini's Visual and Infrared Mapping Spectrometer reveal that the horizontal structure, height, and optical depth of Titan's clouds are highly dynamic. Vigorous cloud centres have been seen to rise from the middle to the upper troposphere within 30 minutes and dissipate within the next hour. Their development indicates that Titan's clouds evolve convectively; dissipate through rain; and, over the

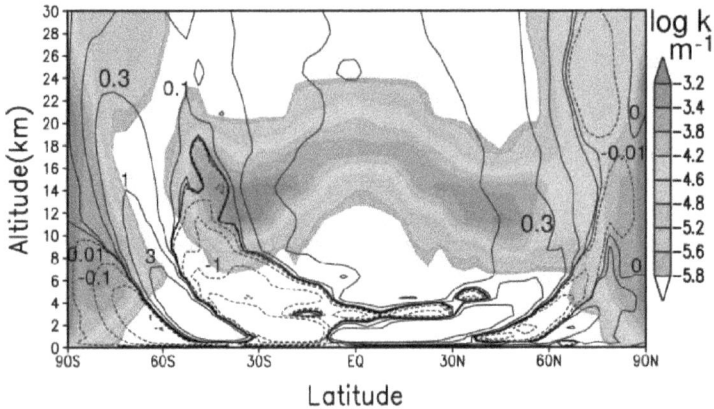

Figure 7.8   A Titan General Circulation Model plot of cloud opacity and wind (Rannou *et al.*, 2004).

next several hours, stretch downwind to achieve their great extents in longitude. The observed characteristics suggest that temperate clouds originate from circulation-induced convergence, resulting from forcing associated with Saturn's tides, and variations in Titan's surface topography or even its composition. Ethane clouds were also detected at northern latitudes between 51°N and 68°N, on December 13, 2004, August 21 and 22, 2005, and September 7, 2005. It is thought that they are formed as a result of stratospheric subsidence and the particularly cool conditions near the north pole, where winter conditions prevailed at the time.

More recent observations between September and December 2006 found a markedly reduced level of cloud activity. Only about 0.15% of the globe was covered, compared to more than three times as much to a few years earlier. Whether or not this is a seasonal variation will not become clear until we have a much longer sequence of observations, including the period either side of the autumn equinox in 2009.

7.6 Thermal and Dynamical Interactions with the Haze

In the upper part of the atmosphere, the haze is responsible for the absorption of sunlight, which accounts for the relatively high temperatures prevailing in Titan's stratosphere compared to the surface. According to models, the haze reflects about a third of the incoming solar radiation and absorbs an approximately equal amount, the other third being transmitted to the lower regions. Being optically thin in the thermal infrared, the haze allows the heat from the lower atmosphere to escape readily into space, while it partially blocks the penetration of sunlight. This effect is the opposite of the greenhouse effect, and is sometimes referred to as an 'anti-greenhouse effect'. Calculations suggest that it reduces the surface temperature by about 9 K, relative to a haze-free model, whereas the greenhouse effect due to methane warms the surface by about 20 K, so that the net effect on the surface is an excess of around 11 K over the radiative equilibrium temperature of Titan, i.e. the temperature it would have with the same albedo but no atmosphere, which is about 82 K.

Changes in the asymmetry between the two hemispheres appear to lag behind the solar forcing by about 90° of orbital phase. In other words, when the sunlight on each hemisphere was about the same (at equinoxes in 1980 and 1995), the albedo contrast between the north and south hemispheres was at its strongest. No spatially-resolved data exists for Titan near solstice, although the disk-integrated albedo (which varies sinusoidally with a period of about 14 years) indicates that the hemispheres were about the same brightness during solstice. The phase lag is also apparent in the difference in brightness temperature between the two hemispheres, as recorded in the emission from the methane band near 7.7 μm by Voyager IRIS. The radiative time constant, the characteristic time for the atmosphere to cool if the sunlight is removed, is small compared to the length of the season for the atmosphere at the height levels sounded by this band. Since the insolation on the two hemispheres was the same, another process has to be invoked to explain the difference. The difference in temperature may have a purely radiative origin, due to the enhanced concentration of gas and haze opacity in the high northern latitudes at the time of the Voyager encounter. The haze albedo is expected to be temperature, as well as gas composition, dependent, and this in turn affects the energy balance and so the temperature, so some kind of feedback loop is probably responsible for the rapid, large-scale changes observed.

A different interpretation is that the phase lag between the stratospheric temperatures and the solar forcing is caused by dynamical, rather than radiative, inertia. This implies in turn that the lower levels of the atmosphere are coupled to the upper layers through vertical motions. This possibility can be investigated with general circulation models, full 3-D models that include all of the relevant physics, which for Titan must include unique phenomena such as the tidal effects of the gravitational field of Saturn. Radiative, dynamical and chemical effects are all involved at the same time and are coupled to each other. The haze absorbs sunlight and warms the stratosphere, and this warming is in part responsible for driving the circulation. Photolysis obviously depends on sunlight, and the subsequent reactions are temperature, as well as species concentration, dependent. Concentrations of reactants are moved around by the winds. Titan clearly is a system where understanding even the few observations, mainly of temperature and albedo, that we have so far, would benefit from using a full radiative transfer and completely coupled dynamical model, including the interactions between the haze, the condensation and the circulation, to try to reproduce them.

In 2006, P. Rannou and colleagues started work to couple a cloud microphysical model to their general circulation model. The results follow a simulation of 25 Titan years, which is the time required to achieve a converged state. The cloud extinction from this model along with the wind stream function at the season of Cassini arrival, two terrestrial years after the northern winter solstice, shows the haze is strongly scavenged (removed by cloud sedimentation) below 50 km at latitudes between 40°S and 40°N, falling to around 10 km in polar regions. The scavenging produces a haze inversion layer (a zone where extinction locally increases with altitude) similar to that inferred from telescope observations. The circulation moves the haze, allowing particles to grow larger where there is upwelling, and accelerating its removal by rainout where there is downwelling. Additionally, lateral transport by winds will have a role in controlling the haze abundance.

In the 1980 Voyager images, the haze appears to be detached most if not all the way around the globe of Titan. Only the north polar hood appears undetached, possibly because cloud formation is occurring there and is filling in the region between the detached portion and the main body of the haze. If upward motions are responsible for the detached haze, upwelling would have to occur almost everywhere, with the corresponding downwelling taking place over the north polar hood. Dynamical models indicate that during the period around solstice there is a pole-to-pole Hadley-type circulation in Titan's stratosphere, rising in the summer hemisphere. Such a circulation might push small aerosols to the winter pole. It seems plausible that the strength of the circulation, and perhaps also the amount of transported aerosol, would correlate with the intensity of the solar flux on the illuminated pole. At solstice, the opposite pole is in darkness for the entire Titan season.

It has been suggested that the amplitude of the north-south albedo contrast cycle may be asymmetric — i.e. the bright/dark albedo ratio at northern spring equinox (i.e. the time of the Voyager encounter, after southern summer) may be larger than at northern autumn equinox. If so, this additional asymmetry may be an effect of Saturn's eccentric orbit around the Sun, and the fact that perihelion occurs close to southern summer solstice. The daily-averaged peak insolation is higher at the south pole ($7.4\,\mathrm{W\,m^{-2}}$) than at the north pole (about $6\,\mathrm{W\,m^{-2}}$) during their respective summers.

Rannou's model was applied to the time of the Cassini arrival, when the ascending and subsiding motions at the summer and winter hemispheres respectively still exist in the stratosphere. The winter downwelling brings gases and haze down to the troposphere, where they condense producing a thick polar cloud below 60 km and possibly droplets. The pole-to-pole cell changes direction every 12 terrestrial years (0.4 Titan years). A secondary cell, which turns in the opposite direction in the summer hemisphere, is predicted by this model at altitudes between 50 and 200 km but is as yet unobserved. Similar processes occur at the summer pole due to the descending branch of this secondary cell. In the troposphere, a more symmetrical situation is predicted due to the longer dynamical and radiative time scales. The observed methane clouds appear at locations where ascending motions are predicted, in the ascending branch of the tropospheric Hadley cell near 40°S, but the model predicts more clouds at other locations that have not yet been observed.

The model predicts that both methane and ethane are cycled between equatorial and polar latitudes, by evaporation, transport, and precipitation. Ethane rainfall is estimated to amount to ∼0.2 mm per Titan year in polar regions, while methane can reach as much as 1 m per year. Rannou and colleagues point out that the methane cycle on Titan has some similarities with water cycles on Earth and on Mars. For instance, in all three cases the source is at the surface, and the circulation is dominated by equator-to-pole overturning in a Hadley-type circulation. The Hadley cell system produces a cloud belt in the intertropical convergence zone on Earth and on Mars and also produces the clouds seen frequently at 40°S on Titan. In the two-dimensional model of Titan, slantwise convection polewards of the Hadley cell produces clouds because methane-rich air is transported toward cold regions. This mechanism is probably the cause of the south polar cloud seen in the current season on Titan.

7.7 Observational Evidence on the Aerosol Composition

Most of the features seen in emission in Titan spectra are due to the various gases in the atmosphere. However, some broad features appeared that cannot be attributed to any known gaseous component. Since broad, relatively featureless absorption is characteristic of solids or liquids, rather than gases, these probably arise from the condensed species in the atmospheric hazes.

It is natural to suspect that they are the signatures of some of the organic molecules formed in the photochemical chains discussed earlier, those which should solidify at

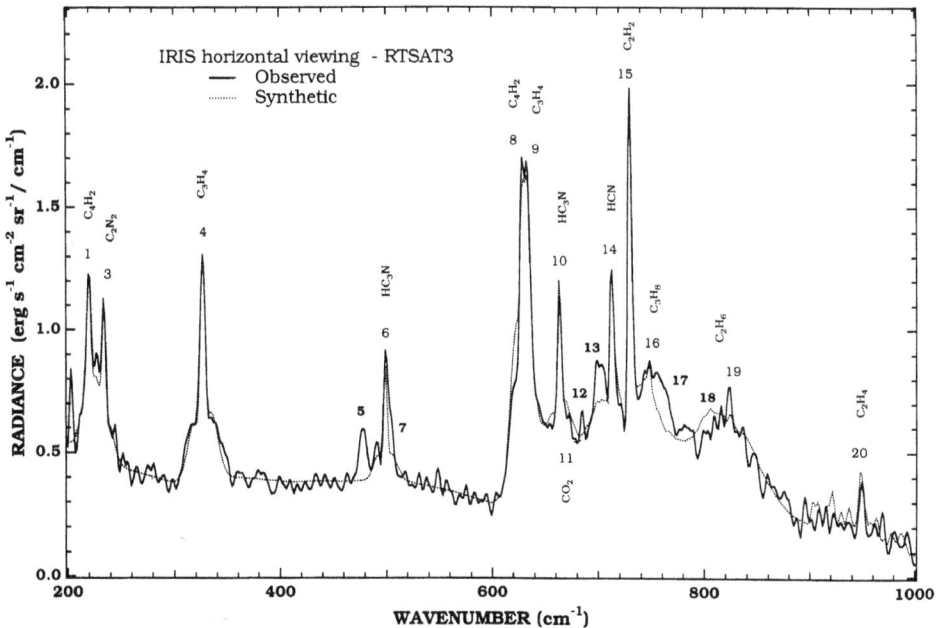

Figure 7.9 Voyager/IRIS spectrum taken in 1980 at grazing incidence over Titan's north polar region and the calculated synthetic spectrum including gas contributions. The remaining broad features are unaccounted for, but thought to be due to solids in the stratosphere of Titan (Coustenis *et al.*, 1999).

low temperatures if they are present in sufficient quantities. Studies in the laboratory suggest that the material in the photochemically-produced tholin is probably not soluble in condensed non-polar hydrocarbons, such as CH_4 and C_2H_6, and therefore the particles in the main aerosol layer are not likely to act as an effective source of seed nuclei for these condensates. The tholin might, however, be slightly soluble in condensed nitriles, and so might be the source of seed nuclei for the condensed C_4N_2, which has been definitely identified in the north polar hood from IRIS data.

The C_4N_2 band lends itself well to analysis due to its isolation from other features in the spectrum. Scattering and radiative transfer modelling performed by R. Samuelson and colleagues suggest a C_4N_2 cloud optical thickness between 0.04 and 0.15, depending on the vertical distribution of material, with a mean particle radius about $5\,\mu$m. Further analysis suggests that the abundance of condensed C_4N_2 is perhaps two orders of magnitude larger than could be sustained by the gas phase in a steady state, as follows. An upper limit to the abundance of C_4N_2 vapour (4×10^{-10}) above the cloud top at 90 km was inferred from the absence of its gas phase band at $471\ cm^{-1}$ in the spectrum. When this abundance is extrapolated to obtain an upper limit to the total column abundance of vapour that condensed to form the cloud, it accounts for only about one percent of the amount of cloud actually inferred from

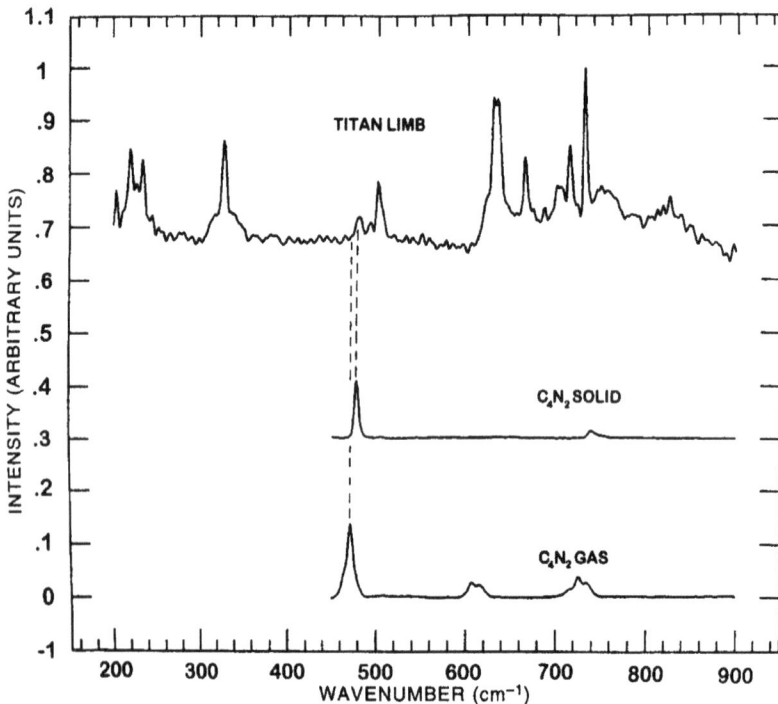

Figure 7.10   Identification of solid C_4N_2 in Titan's stratosphere (Samuelson *et al.*, 1997).

the analysis. Thus, a time-independent downward vapour diffusion process from above — coupled with condensation at the cloud top and steady precipitation from the cloud bottom — is not consistent with the observed C_4N_2 cloud/vapour abundance ratio. Part of the reason for this apparent discrepancy may be that the C_4N_2 was restricted to a very narrow region, in the north polar hood, and at one particular time, when Voyager made its observations. These two facts taken together suggest that a time-dependent process — possibly associated with seasonal variations of temperature, chemistry, and dynamics — is responsible.

One possible scenario is that the abundances of C_4N_2 vapour and condensate both vary with the season. There may be a build-up of vapour during the polar winter, when it is protected from photolytic decomposition while in the long polar night. As spring approaches, two things occur: 1) the stratosphere becomes colder, which causes increased condensation, and 2) the polar shadow recedes, which enhances photolysis and hence the decomposition of dicyanogen vapour above the shadow. The combination of these two processes may lead to an excess of condensate relative to vapour, compared to steady state conditions. Particles of the size inferred from the data — about 5 μm in radius — will precipitate out from the stratosphere in less than a year, which is short compared with the duration of spring (about 7 years). Thus precipitation may be confined to a short time during the early part of polar spring.

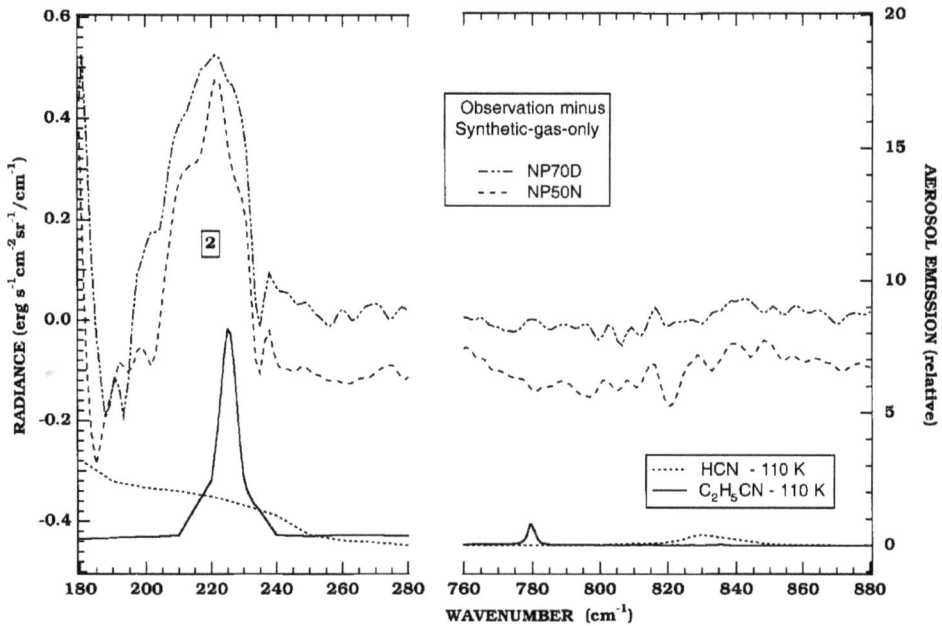

Figure 7.11   Comparison of laboratory and IRIS data showing the possible presence of C_2H_5CN in Titan's stratosphere (Coustenis *et al.*, 1999).

If such a process operates for dicyanoacetylene, it may operate for other volatiles as well, including water, which has been detected on Titan in vapour and solid form by ISO and from the ground. In recent re-analyses of such data, Samuelson and colleagues have studied the spectral continuum between 200 and 600 cm^{-1} in the Voyager IRIS limb spectra and found indication of a tenuous but relatively uniform cloud of small condensed organics (1–5 micron in size) permeating the lower stratosphere (60–90 km in altitude). Perhaps the dicyanogen (or dicyanoacetylene) particles act as seed nuclei for the ethane haze which is suspected of being the large particle component of the north polar stratospheric continuum between 200 cm^{-1} and 600 cm^{-1}. Two independent studies of the spectral continuum between 200 and 600 cm^{-1} suggest the presence of condensed ethane near or slightly above the tropopause. In one, IRIS limb spectra continua of the north polar hood were analysed, yielding the wavenumber-dependence of the aerosol continuum in the stratosphere. The ratio of opacity at high wavenumbers to that at low wavenumbers increases with decreasing depth, implying that condensates contribute to the aerosol continuum in the lower stratosphere or upper troposphere. Later, a reanalysis of these data in the context of scattering theory inferred that an ethane haze consisting of particles with mean radii about 10 μm could explain the data. Condensates other than ethane could also explain the wavenumber-dependence of opacity, but are less likely from relative abundance considerations. Also, vapour pressure arguments imply condensed nitriles near 90 km, the most likely being HCN. A thorough study of the IRIS spectra in comparison with laboratory data found hints of the possible presence of C_2H_2, C_2H_5CN, HCN, HC_3N, and C_2H_6 as condensates in Titan's stratosphere, but not water ice.

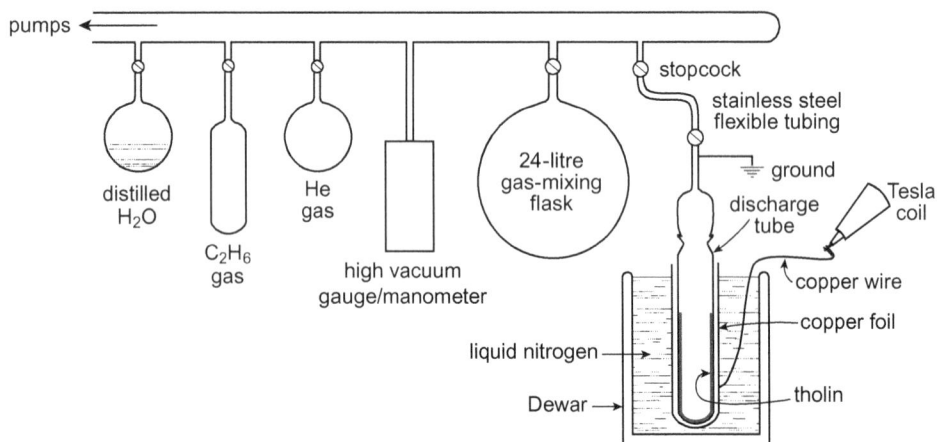

Figure 7.12   Laboratory experiment for producing organic solids in reducing atmospheres. Tesla coils supply a discharge of about 50 kV for a period of about one month; the resulting 'tholins' are collected on a cold finger in the vessel on the right, and analysed by pyrolysis gas chromatography and mass spectroscopy (Khare *et al.*, 1981).

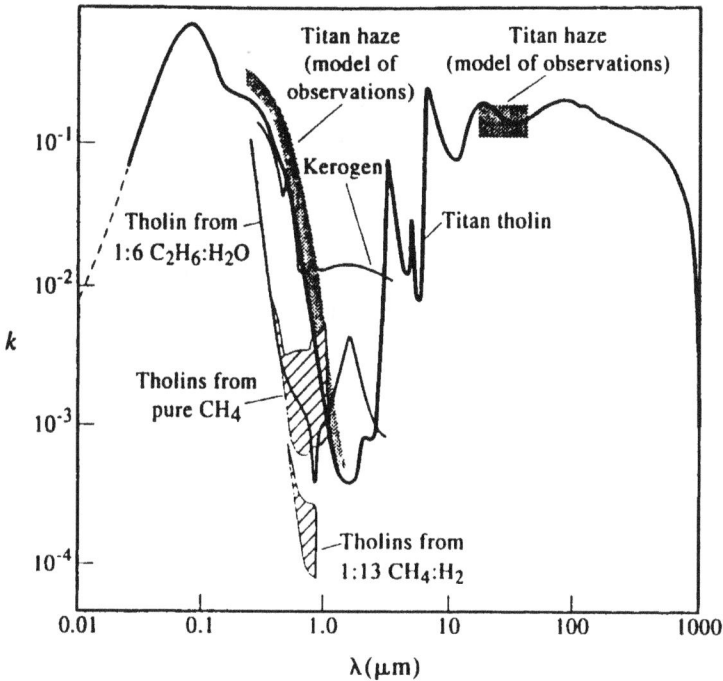

Figure 7.13 The imaginary part of the refractive index of laboratory tholins as a function of wavelength, for different methods of production (Sagan *et al.*, 1992).

7.8 Laboratory Simulations of Haze Materials

With so little direct information on the chemical composition of the aerosols in Titan's haze layers, we have to rely on indirect inferences from theory or from laboratory simulations. In this section we will look more closely at efforts made over the past 30 years to manufacture haze material in the laboratory under simulated Titan conditions, and the evidence that the resulting substances are consistent with photochemical models and such spectroscopic evidence as we have.

At high altitudes, high energy particles from Saturn's magnetosphere, cosmic rays, and UV radiation from the Sun all contribute to the partial decomposition of molecular nitrogen and methane, the fragments of which will recombine into simple hydrocarbons and nitriles. These, in turn, are subject to further chemistry and are expected to build up into quite complex molecules. The exact composition, however, cannot be predicted theoretically, and our best course at present is to resort to laboratory simulations of the process. These, of course, cannot reproduce the conditions present on Titan exactly (in particular, the long time periods involved) but they are still our best hope in elucidating the nature of the haze material and they have already provided useful constraints on the elemental composition, solubility, and optical properties of the particles.

7.8.1 *Chemical Composition of Tholins*

Since the early work of B. Khare and C. Sagan in 1973, many laboratory simulations
of Titan's atmosphere have been conducted by this team and others. Khare and his
colleagues at Cornell University set up a synthetic Titan atmosphere consisting of
molecular nitrogen (90%), with 10% of methane added to make a total pressure
of 0.2 mbar, representative of Titan's stratosphere at 250 km. They then passed a
cold plasma electric discharge through the gas at standard temperature for up to
several months. At the end of this time, a brown film of organic material had been
formed on the inside of the reaction vessel. It was then that this type of mixture of
substances was given the general name of 'tholins'. When subsequently analysed
by pyrolysis gas chromatography/mass spectroscopy, the film was found to contain over 75 different compounds, mainly hydrocarbons and nitriles, plus traces of
some oxygen-bearing species that apparently resulted from contamination within the
system.

Considerable work has been directed towards investigating organic synthesis
from mixtures that represent the products of N_2–CH_4 photochemistry, such as
C_2H_2 and HCN (Table 7.1). Using electron microscopy observation of photochemically produced particles in Ar–C_2H_2 mixtures, A. Bar-Nun and colleagues observed
that either spherical particles can be formed, or aggregates built from spherical
monomers. The latter implies that the growing aerosols may not behave as liquid
drops and that, at a moment of their growth, they can stick together to form clusters.
Such particles were also observed in mixtures involving He or N_2, and C_2H_2, HC_3N
and C_2H_2.

Table 7.1   A summary of tholin synthesis experiments in the laboratory.

Pressure (Pa)	Temperature (K)	Initial gas mixture	Spectral range (μm)	References
20	300	N_2 (90%), CH_4 (10%)	0.02–920	Khare *et al.* (1984)
~200	100–300	N_2 (98%), CH_4 (2%)	2.35–40	Coll *et al.* (1999)
~100	300	N_2 (98%), CH_4 (2%)	0.2–0.9	Ramirez *et al.* (2002)
13300	297	N_2 (98%), CH_4 (1.8%) H_2 (0.2%)	0.2–16	Tran *et al.* (2003)
93320 13-2300	~300	N_2 (90%), CH_4 (10%)	0.3–1	Imanaka *et al.* (2004)
			0.2–2.5	Coll *et al.* (2004)
90		N_2 (98%), CH_4 (2%)	2.5–20	Bernard *et al.* (2006)

In general, these studies may be telling us something but they do not really correspond to the mixture of gases on Titan, and it is hard to be sure how to interpret them. More recent simulations may have got closer: P. Coll, F. Raulin and colleagues used CH_4 and N_2 at the temperatures and pressures of Titan's stratosphere, and analysed the products without exposing them to air, thus avoiding oxygen contamination, hoping to obtain the most accurate simulation of Titan's haze formation to date. Future simulations and models may incorporate small amounts of H_2O, CO and CO_2 deliberately, to provide traces of oxygen-bearing molecules corresponding to their observed abundance on Titan.

Recent detailed analyses of the organic compounds contained in laboratory tholins, produced under different conditions found a complex organic mix of simple alkanes, aromatic compounds, heteropolymers, and amino acid precursors. In addition to the organic structure, there is interest in the simple elemental composition of the possible haze analogues in order to determine their potential importance as a sink for photochemical products. Table 7.2 compares this type of result for different experiments, and shows that the results vary widely, probably owing to differences in the temperature and pressure of the simulations and the energy source used. Because of the importance of temperature in the condensation process, it is likely that the low temperature results of Coll and colleagues represent the most accurate simulation; those at higher temperatures apparently underestimate the incorporation of carbon containing compounds into the tholin material by a factor of 2 or more.

Estimates of the photolysis rate of CH_4, the dissociation rate of N_2, and the production rate of tholin in terms of total mass, carbon atoms, and nitrogen atoms are compared in Table 7.3. The elemental composition is based upon laboratory tholin data summarised in Table 7.2 using an average value of the C/N ratio of 4, while the haze production rates have typical values given in Table 7.4 below. These comparisons suggest that the production of tholin on Titan is a minor sink for C, while about 12% of the N produced is lost as solid organic material. This sink of N atoms is large enough that it must be accounted for in a self-consistent way in photochemical models of Titan's atmosphere, particularly if the N incorporated into the haze derives from HCN. This seems likely since the rate of HCN production in current photochemical models is comparable to the sink of N in haze material listed in Table 7.3.

Table 7.2 Comparison of the relative abundances of carbon, hydrogen and nitrogen in laboratory tholins produced by different experimental groups.

Stoichiometry	C/N ratio	Conditions	Energy	References
$C_8H_{13}N_4$	1.9	Low P	Tesla coil	Sagan *et al.* (1984)
$C_{11}H_{11}N$	11	Low T	Tesla coil	Coll *et al.* (1995)
$C_{11}H_{11}N_2$	5.5	High T, P	Tesla coil	McKay (1996)
$C_{11}H_4N_4$	2.8	Low T, P	Tesla coil	Coll *et al.* (1999)

Table 7.3   Production rates of mass, N, and C (from McKay, 1996).

Process	Mass $(g\,cm^{-2}\,s^{-1})$	N $(cm^{-2}\,s^{-1})$	C $(cm^{-2}\,s^{-1})$	References
C from CH_4 photolysis	2.6×10^{-13}	—	1.3×10^{10}	Toublanc *et al.* (1995)
N from N_2 dissociation	2.0×10^{-14}	8.4×10^8	—	Strobel *et al.* (1992)
Tholin production	10^{-14}	1×10^8	4×10^8	Toublanc *et al.* (1995) Assuming C_4H_4N (C/N = 4)

It is likely that the computed profiles of HCN will be altered when this additional sink of N is included. This could be important because the HCN profile determined from observations is used to infer the eddy diffusion coefficient. Photochemical modelling indicates that the eddy diffusion coefficient derived from the HCN values is too large to be consistent with the observed distribution of hydrocarbons. The Cassini CIRS team found the same problem when attempting to use a single mixing profile to fit the vertical distributions of HCN and C_2H_6. Including an important additional sink for HCN could resolve this discrepancy. Recent modelling has shown that attributing the HCN (and other species) chemical loss to haze formation, can solve this problem. Regarding the loss of carbon in forming haze material, current photochemical models already include the loss of C to haze in the reaction scheme through the chemistry of C_2H_2, thought to be the main hydrocarbon species from which the solid haze material derives.

Another of the factors that can alter the optical properties of tholin produced in the laboratory is the C/N ratio of the starting mixture. With better measurements of the composition of the material on Titan, it may be possible to use the elemental ratios to deduce the chemical mechanisms of tholin formation and hence be able to distinguish between, say, HCN and C_2H_2 polymerisation. Furthermore, the reactions responsible for incorporating N-containing compounds into the haze may operate at different altitudes from those for C-containing compounds. Thus the C/N ratio, and therefore presumably the optical properties of the haze, may vary with altitude. The Voyager data show that the ratio of C_2H_2 to HCN, the species most likely to be the principal starting points for C and N incorporation into the haze, also vary with latitude. It follows that the C/N ratio and optical properties of the haze vary with season, which may help to explain the seasonal brightness changes and play a role, along with particle density and size variations forming in the hemispherical asymmetry in albedo.

We return now to the question of the solubility of tholin in possible cloud and rain-forming liquids on Titan. Recent studies showed that tholin was completely insoluble in liquid ethane down to the level of measurement (<0.03% by mass).

Tests in a variety of other solvents confirmed theoretical predictions that tholin material is much more soluble in polar solvents (like water, ethanol, methanol, glycol, and dimethylsulfoxide) than in non-polar solvents (ethane, hexane, benzene). The insolubility of haze material in non-polar compounds has implications for the condensation of ethane and methane in Titan's atmosphere. Condensation may be inhibited or completely suppressed by the insolubility of the tholin and the resulting large contact angles with condensing ethane and methane. This could allow supersaturation of ethane by a factor of over 500 in Titan's stratosphere, and supersaturation of methane by a factor of over 2 in the troposphere.

7.8.2 *Optical Properties of Tholins*

The experiments discussed above produce in general a solid organic material, on which studies can be conducted to obtain not only its elemental composition and organic structure, but also its optical properties, in particular the real and the imaginary refractive indexes. The index of refraction depends in general on the electric conductivity σ, the dielectric constant ε, and the magnetic permeability μ of the medium. The real part, n, of the index is sometimes called the propagation constant (although both parts depend on wavelength) while the imaginary part, k, is responsible for the absorption of radiation and is proportional to the absorption coefficient of the material.

The experimenters measured the optical properties of the tholin mixture, obtaining the real and imaginary parts of its refractive index from 250 Å to 1,000 μm. These have been widely exploited since then in atmospheric radiative transfer studies of Titan, for example by C. McKay and colleagues in 1989, when they determined

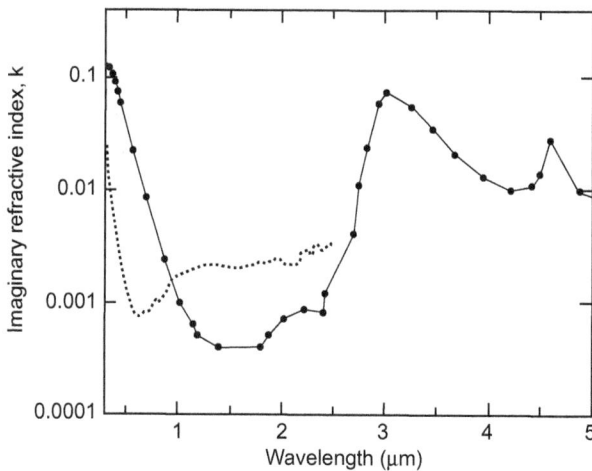

Figure 7.14 Tholins imaginary refractive index from Khare *et al.* (1984) and Coll *et al.* (2004), in the visible and near-infrared spectral region.

the absorption required to fit the geometric albedo in the atmospheric windows across Titan's spectrum. Their theoretical calculations of the scattering properties of model Titan clouds, assuming the optical properties of laboratory tholin, matches the geometric albedo spectrum of Titan quite well, except that they needed to apply a scaling factor of 4/3 to the imaginary part of the tholin refractive index. This gives quantitative support to the idea that tholin could account for the colour of Titan, albeit with some manipulation. From this analysis, it appears that the laboratory tholin is too bright in the visible and too dark in the near-infrared. When McKay and colleagues tried to fit the Voyager IRIS far-infrared spectrum from 400–600 cm^{-1}, they found that this time they required the imaginary refractive index of laboratory tholin to be scaled by a factor of 1/2.

It was always expecting rather a lot to assume that the laboratory discharge would come up with exactly the right mix of complex chemicals. The basic idea still seemed useful, and until recently, most modellers used these first laboratory measurements for the refractive index of Titan haze-type analogues, scaled by a factor which depends on the wavelength and the type of particles used: 4/3 in the shortwave region of the geometric albedo for spherical particles, while for fractal particles it is 3 in the UV and 1.5 in the visible. More recent measurements have shown that the optical properties of the laboratory haze analogues depend significantly on the experimental conditions, with factors such as the pressure under which the analogues are made defining their chemical structure and hence their radiative properties, further reason to think that photochemically-produced particles at different altitudes in Titan's atmosphere could exhibit different optical properties.

The imaginary refractive index values obtained by the Khare team in the 0.3–5 μm region show a strong decrease between 0.7 and 2.8 μm, being optically thin in this spectral region (the limits correspond to $k < 0.01$). The absorption due to aerosols is stronger below 0.7 μm and above 2.8 μm. The high absorption of the tholins below 0.7 μm agrees with the spectral behaviour of Titan's aerosols, which are optically thick in the visible region of the spectrum, giving rise to the reddish veil that hides Titan's surface, giving further support to the idea that laboratory tholins can be regarded as reasonable analogues of Titan's aerosols. Above 2.8 μm the absorption due to atmospheric methane becomes more important than that of the haze and dominates Titan's infrared spectrum.

Coll *et al.* performed elemental analysis, pyrolysis, solubility studies and infrared spectroscopy at relatively high pressures and only above 2.5 μm, while Ramirez *et al.* performed measurements at lower pressures but only covered the visible range. Tran *et al.* used a photochemical flow reactor and measured imaginary refractive indices from the ultraviolet to the infrared. Their measurements are affected by the uncertainty in the thickness of the film produced, but they were able to retrieve a value of around 0.03 in the 0.8–2.5 μm range. Imanaka *et al.* performed laboratory experiments on tholins formed in cold plasma at various pressures, but their data do not cover the near-infrared region where many of the most important observations of Titan are made. The values by Coll *et al.* are, on average, 5 times larger than the

Khare *et al.* values in the 0.9–2.5 μm range, but both sets show similar slopes in the visible. Finally the recent measurements by Bernard *et al.* produced two types of tholins with different spectral behaviours.

7.9 Microphysical Models of Titan's Haze

A variety of theoretical models have been developed to simulate the microphysics of the haze on Titan. Microphysical models take into account the detailed structure of the haze, in particular the size and shape of the particles, and their number density. The models are of two types: those derived from a fit to data, usually observations of the intensity and polarisation, as a function of direction, of light at various wavelengths scattered from the haze, and those which consider the physics of the processes by which haze droplets form and dissipate. However, most of them do not try to explain the full evolution from methane photolysis to the final aerosol particle and, in fact, this path is not yet well understood.

7.9.1 Organic Haze Production

Table 7.4 is a summary of the estimates of the production rate of haze material that exist in the literature. The estimates made before the Voyager encounter, were based on the rate of CH_4 photolysis and on the observed albedo of Titan. These suggested a production rate of 3.5×10^{-14} g cm^{-2} s^{-1}. A detailed microphysical model, on the other hand, found that a production rate ten times higher (3.5×10^{-13} g cm^{-2} s^{-1}) was needed to fit the geometric albedo, although this assumed a pure CH_4 atmosphere.

After the Voyager encounter showed that the atmosphere of Titan was primarily N_2, and the detection of hydrocarbon gases in the stratosphere confirmed that the haze was likely to be organic material, photochemical models of that time, while having only a perfunctory treatment of the chemistry of higher order species (C_nH_m with $n > 4$), did include a series of reactions that produced hydrocarbon polymer. The net production of haze material in these reactions is given as $\sim 10^8$ molecules cm^{-2} s^{-1}, corresponding to a mass production rate of 2×10^{-14} g cm^{-2} s^{-1}. A simple monodisperse haze model (one in which there is only one particle size at each altitude) based on the Voyager atmospheric profile, suggests that a haze production rate of 1.2×10^{-14} g cm^{-2} s^{-1} would fit the geometric albedo from 0.3 μm to 2 μm. An estimate for the column mass of aerosol required to match the longer-wavelength, thermal infrared spectrum of Titan, with some reasonable assumptions about the sedimentation velocity, yields a production rate of 0.8×10^{-14} g cm^{-2} s^{-1}. A detailed aerosol model confirmed that a similar haze production rate of 1.2×10^{-14} g cm^{-2} s^{-1} would fit the geometric albedo.

Thompson and colleagues used a low-pressure (0.24 hPa) plasma discharge to simulate radiation chemistry on Titan. Their estimate for the production of solid organic material was from 0.3 to 3×10^{-14} g cm^{-2} s^{-1}, later revised to 0.5 to 4×10^{-14} g cm^{-2} s^{-1}; the value listed in Table 7.4. Although the dynamics of the haze

Table 7.4   Estimates of Titan haze production rates.

Description of model	Rate $(10^{-14} \, \mathrm{g\,cm^{-2}\,s^{-1}})$	Work by
C_2H_4, C_2H_2, and CH_4, photolysis	3.5	Podolak and Bar-Nun (1979)
Microphysical model in CH_4 matching albedo	35	Toon et al. (1980)
Photochemical model, hydrocarbons only	~2	Yung et al. (1984)
Simple microphysical model matching albedo	1.2	McKay et al. (1989)
Voyager thermal infrared spectra	0.8	Samuelson and Mayo (1991)
Wind and detailed microphysics model matching albedo	1.2	Toon et al. (1992)
Laboratory simulations	0.5–4	Thompson et al. (1994)
Photochemical model, hydrocarbons only	1.7	Lara et al. (1994)
Fractal microphysics matching albedo	1.4–3.5	Rannou et al. (1995)
Fractal microphysics matching high phase angle photometry	0.6	Rannou et al. (1997)
Spherical microphysical model	0.4	Tomasko et al. (1997)
Fractal microphysics model	1.5	Cabane and Chassefière (1995)
Photochemical model	4	Lebonnois et al. (2002)
Photochemical model	3.2	Wilson and Atreya (2003)
Photochemical / Microphysical (spherical) / model matching albedo, temperature and composition	1.3	Lavvas et al. (2007)

is quite different with particles that are complex aggregations, the production rates inferred for fractals are similar to those previously obtained for spheres.

All the post-1989 haze models that included radiation transfer calculations have achieved a good fit to the spectral dependence of Titan's geometric albedo by using the optical constants of tholin material produced in the laboratory, with the multiplying factor discussed earlier. The haze particle production rate is assumed to be a symmetrical distribution (usually Gaussian) centred at some chosen altitude, usually between 350 and 600 km depending on the model, and the total column production rate varied to fit the geometric albedo. A problem is that vertical haze production profiles generated from photochemical models are significantly different from these simple profiles adopted in current haze microphysical models to match albedo observations. Also, some models calculate the radiation field but do not calculate the temperature profile that results from the model haze structure, in which case the two are not necessarily compatible.

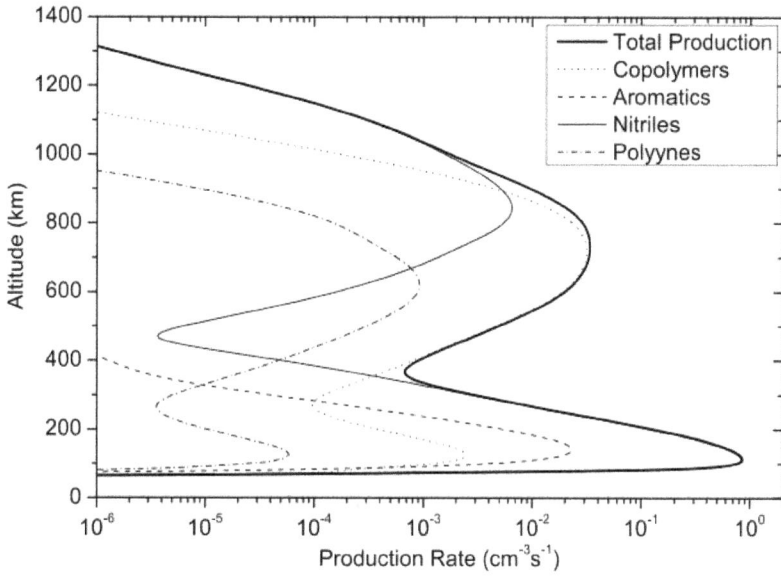

Figure 7.15   Vertical production profile of haze particles with different photochemical composition (Lavvas *et al.*, 2007).

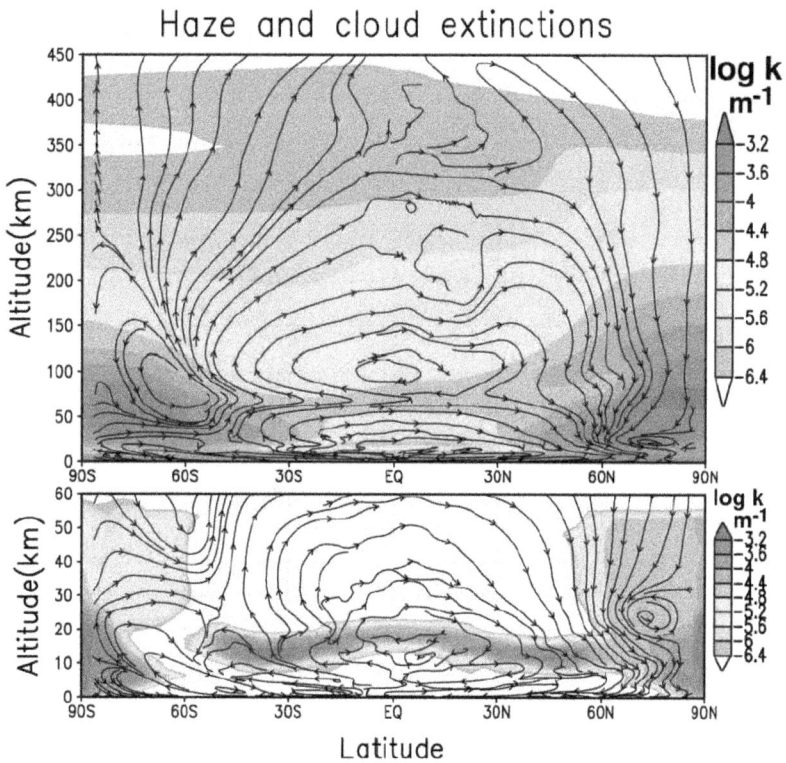

Figure 7.16   Haze and cloud extinctions from model calculations (Rannou *et al.*, 2006).

In the 2007 model by Lavvas *et al.*, based on reaction pathways implied by the laboratory experiments discussed above, and including polymer structures of different chemical composition, the haze production profile was obtained from the photochemistry and then used to produce the haze vertical structure and its radiative properties self-consistently. From the total vertical production rate, along with the contribution of each family of pathways, it is evident that there are specific altitude regions where the contribution of each family is important. In the lower atmosphere, below 300 km, the production is controlled by the nitrile family along with the aromatics. At higher altitudes, the aliphatic copolymers dominate the production up to 1,000 km, while at even higher levels the nitrile contribution dominates again.

Nitriles are produced at very high altitudes due to their dependence on N_2 destruction, which is possible only by high energy UV photons, and they have their maximum deposition above 1,000 km. Hydrocarbon production, which is initiated by methane photolysis, has its maximum at lower altitudes due to the absorption of these species at longer wavelengths. Hence, above 1,000 km, nitrile pathways dominate haze production, while at lower altitudes the contribution of hydrocarbons (copolymers) increases; CH_4 photolysis has its maximum at ~800 km. In the stratosphere where galactic cosmic rays have their greatest impact, increased N_2 dissociation leads to the enhancement of nitrile production that leads to a further large contribution by this family to the total haze production profile at these relatively low levels.

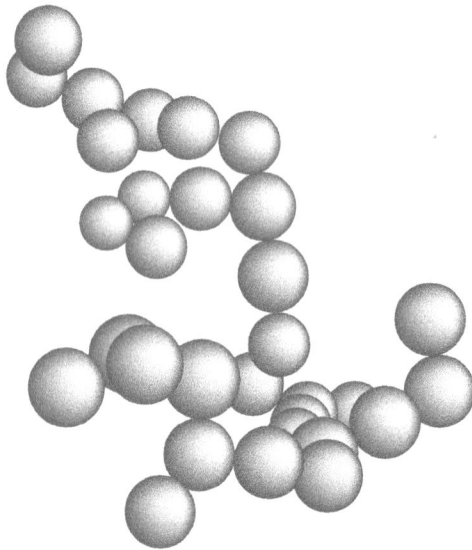

Figure 7.17   Aerosol aggregates, produced in the laboratory, which may be analogues of Titan's haze particles (Cabane and Chassefière, 1995).

7.9.2 *Fractal Models and Scattering Properties of the Haze*

The basis of the fractal models is the observation that when liquid particles combine, they form a new compact sphere corresponding to the combined mass, but when solid particles combine they can form a range of shapes from compact spheres to long strings of particles. In this case the aerosols grow up as spheres until a certain altitude level, below which the monomers sediment and coagulate to form aggregate structures. These structures interact differently with light of different wavelengths. Monomers interact with the short wavelength radiation and determine the polarization, explaining the ultraviolet and polarization data. For the longer wavelengths, the aggregates act like large particles and can explain the geometric albedo values and the forward scattering properties of the aerosols. Fractal objects have the same structure when observed at different scales, although in cases like that of Titan's aerosols only from a statistical point of view. The shape is characterised by the *fractal dimension*, which gives a measure of the space occupied by the object. When monomers combine directly onto aggregates, building the aggregates one monomer at a time, the fractal dimension is 2.5 if the mean free path is small and the monomers diffuse into the aggregates. When the mean free path is large and the monomers follow ballistic trajectories, the fractal dimension is 3 and compact spheres result, similar to liquid particles. If growth of the aerosols is dominated by aggregates combining with other aggregates, the fractal dimension is 1.75 if the mean free path is smaller than the particles, and 2 when it is larger than the particles.

The microphysical models of Titan's main haze suggest that it can be physically divided into two regions, a high altitude region in which the fractal dimension is 3 (monomer growth) and a lower altitude region in which the fractal dimension is 2 (cluster-cluster ballistic growth). The haze is constantly replenished with newly produced macromolecules and larger, older, particles. Since the coagulation process is favoured by the different particle sizes, this leads to permanent development of the largest particles in the photochemical zone. These growing particles, roughly spherical and compact evolve into the monomers which later build up the aggregate. As soon as they are sufficiently heavy, they begin to fall out of the photochemical zone. Estimates of the altitude of the photochemically active zone and the photo-chemical production rate may be used to compute the size of the monomers. A rather low production altitude, below 450 km, leads to monomer radii consistent with the optical properties of Titan's atmosphere. Below the photochemical zone, collisions occur without the presence of smaller particles, since these are incorporated into the monomers faster than they settle.

As the particles fall to lower altitudes, the coagulation process occurs between clusters formed from the aggregation of monomers. If condensation of hydrocarbons begins to occur on the haze particles at about 80 km, then the average size of an aggregate increases as each monomer grows. The aggregates grow not because of the bigger monomers but due to diffusion and condensation of some atmospheric gases. Microphysical models that include settling, coagulation, etc., are able to compute the distribution of fractal particles at any altitude level, from which the average number of monomers in the aggregates at a given level in the atmosphere can be deduced.

As we saw above, it had been concluded early in the study of Titan's atmosphere that a bimodal distribution was necessary to explain the optical properties of the aerosol. Models of the haze composed of monomers in the upper altitudes and fractals in the lower layer effectively respond like a bimodal distribution of particles. The small particles, both the free monomers and the monomers within aggregates, interact with short wavelength radiation and thus the UV and polarisation data can be explained. At longer wavelengths, the aggregates behave as large particles and can reproduce the geometric albedo and forward scattering radiation properties observed for Titan. In addition, as discussed above, fractal particles seem to better satisfy the limb data derived from Titan's shadow on Saturn, which indicated that the absorption at short wavelength (0.337 μm) was too large to be consistent with spherical particles of one size. This study showed that the fractal dimension of aggregates has to be 2, which is the same as that predicted by microphysical arguments, and the implied production altitude is around 385 km, corresponding to a monomer radius of 0.066 μm.

The optical properties of a fractal particle differ from a sphere of equivalent mass or area to the extent that, in the fractal case, one has to deal with the interactions of fields radiated by each of the monomers of the aggregate. P. Rannou and colleagues developed a treatment that considers the monomers within each aggregate with the simplifying assumption that the radiation field within the aggregate itself is uniform (the mean field approximation). Since the monomers are identical, they all scatter the light in the same way, and their interaction with the incoming radiation can be treated using Mie theory. Using spherical harmonic sum functions it is possible to obtain the scattering matrix of the monomers which constitute the fractal, and so to derive the field radiated by the aggregate.

Once the optical properties (phase function, cross sections, etc.) of a given particle are known, the transfer of radiation in Titan's atmosphere may be computed, and can be used in comparisons with observations to determine the production rate and associated altitude, as well as the optical properties of the haze material. A comparison of the geometric albedo computed with the fractal model and with the spherical model shows that the low UV albedo produced by many small monomers is fairly well reproduced in the fractal model. The values of the absorption index of the haze obtained from the fractal model are comparable, but higher, than the experimental values obtained with terrestrial analogues by factors of ~3 in the UV and ~1.5 in the visible and they follow the same variation with wavelength.

The fractal model of the aerosols has also been applied to reproduce the limb profiles at phase angles of 140° and 155°, as observed by Voyager; here again the method leads to a good fit with the observations. The intensity profiles are consistent with production at an altitude $z \leq 385$ km, while the production rate and absorption index of the haze are consistent with the previous values.

7.10 Discussion and Conclusion

The subject of clouds and haze on Titan and, generally speaking, of what happens in the lower atmosphere and on the surface, is clearly quite complex and still poorly understood. We must struggle with the bits and pieces of information offered by observations, laboratory simulations, and computer modelling and try to put together a jigsaw puzzle that will reveal the nature and existence of a distant and well-concealed planetary body. Perhaps the most useful product of such an exercise is the list of key questions that it generates, pointing the direction for future research. In the following list, we summarise first the key points in our present understanding of Titan's organic haze and condensates, then some outstanding questions.

- Titan's haze is optically thick at visible wavelengths and has a detached upper layer. The haze varies between hemispheres in a seasonal pattern, and which hemisphere is brighter depends on wavelength, with visible and near-infrared data showing opposite effects.
- Microphysical models, photochemical models and laboratory simulations all suggest that the production rate of the haze is in the order of 10^{-14} g cm^{-2} s^{-1} and that the total mass loading of the haze is $\sim 2.5 \times 10^{-5}$ g cm^{-2}.
- The optical depth at 0.5 μm of the sum of all of the layers is about 3 and thus about 10% of the incoming sunlight reaches the surface.
- At wavelengths above about 1 μm, the haze becomes increasingly transparent both as a result of the change in the ratio of particle size to wavelength and because the absorption coefficient of the haze material becomes smaller.
- The shape of the particles in the main haze deck is probably fractal (with fractal dimension 2) with an equivalent volume of a sphere of radius 0.2 μm.
- The haze is probably composed of organic material with a C/N ratio of between 2–4, and a C/H ratio between 0.5–1.
- Absorption of sunlight by the haze, which is transparent in the thermal infrared, results in a negative greenhouse effect that cools the surface by 9 K. Since the gas in the atmosphere warms the surface by about 20 K, a final surface temperature 11 K higher than the effective temperature is found.
- Condensate clouds are present at low levels (around 20 km above the surface) but are patchy and transient.

Key questions about Titan's haze and clouds that remain to be clarified are:

- What are the chemical pathways that lead to the formation of the haze particles?
- What produces the detached haze at high altitude?
- What is the cause of the seasonal variation, and how is it related to dynamics?
- Is the haze material at high altitude, particularly the detached haze layer, of different composition than lower in the atmosphere?
- Are there seasonal changes in haze composition?

- How closely does the tholin produced in the laboratory resemble Titan aerosol, optically and chemically?
- What is the exact nature of the condensed phases, and where are they located in the atmosphere?
- What is the efficiency of haze particles as condensation nuclei for stratospheric hydrocarbons and tropospheric methane and ethane?
- Is the lower atmosphere cleared locally of haze by rainout?
- What happens to the haze particles and condensates after they reach the surface? What is their contribution to the surface features observed (e.g. the bright and dark regions) and their role in the methane cycle?
- What meteorological phenomena occur on Titan and what is their contribution in the geological features observed, for instance by erosion?

The Cassini/Huygens mission to Saturn and Titan has started to recover some answers to these and other questions about the nature of the aerosols and of the surface on Titan. The Aerosol Collector Pyrolyser has sampled the aerosols during the descent of the Huygens Probe and identified ammonia in their composition using the gas chromatograph-mass spectrometer instrument also on board the probe. The Descent Imager/Spectral Radiometer made optical measurements at solar wavelengths from the Probe to look at the scattering properties of the aerosols. The Cassini orbiter continues to make observations in the near-infrared with VIMS, in the thermal infrared with CIRS, and with ISS in the visible windows, to provide data on cloud properties and compositional variations, as part of an increasingly more comprehensive view of Titan's surface and atmosphere that will not be complete for many years, or even decades, to come.

CHAPTER 8

Atmospheric Dynamics and Meteorology

To the Sea, to the Sea! The white gulls are crying,
The wind is blowing, and the white foam is flying.
West. West away, the round sun is falling.
Grey ship, grey ship, do you hear them calling,
The voices of my people who have gone before me?

J.R.R. Tolkien, *The Return of the King*

8.1 Introduction

Titan meteorology is a very recent subject: before Huygens, no man-made object had ever entered the atmosphere of Titan to provide direct measurements of winds or storms or any of the dynamical aspects of meteorological systems as we know them on Earth. Before the extensive mapping from Cassini, with its advanced multispectral cameras and spectrometers, there had been precious little seen through telescopes or television cameras, either, by way of clear evidence of the weather activity on Titan or clues as to its nature.

In fact, both clouds and haze occur on Titan, at different levels in the atmosphere and produced by the condensation of completely different types of material. By 'clouds', we mean physically localised clumps of droplets or crystals of some condensed or frozen species, and by haze, some more extensive, generally horizontal layer of similar aerosols in a much lower concentration. Hazes usually have smaller particle sizes, as well as concentrations, compared to clouds, and often can be penetrated optically or in the near infrared. Clouds are seldom transparent except at very long (e.g. microwave) wavelengths.

Even before Cassini and Huygens arrived, it was clear that Titan's atmosphere is far from static. There was plenty of indirect evidence from analogies with other planetary atmospheres, measurements of temperature contrasts, and from the subtle markings in the haze structure, to suggest that rapid motions are present. The winds around the equator at high altitudes seem to far exceed the speed at which the surface rotates below, a condition which Titan shares most notably with Venus, and which is known as zonal super-rotation. The property of slowly-rotating planets with deep, hazy atmospheres to exhibit particularly rapid superrotation is probably the most

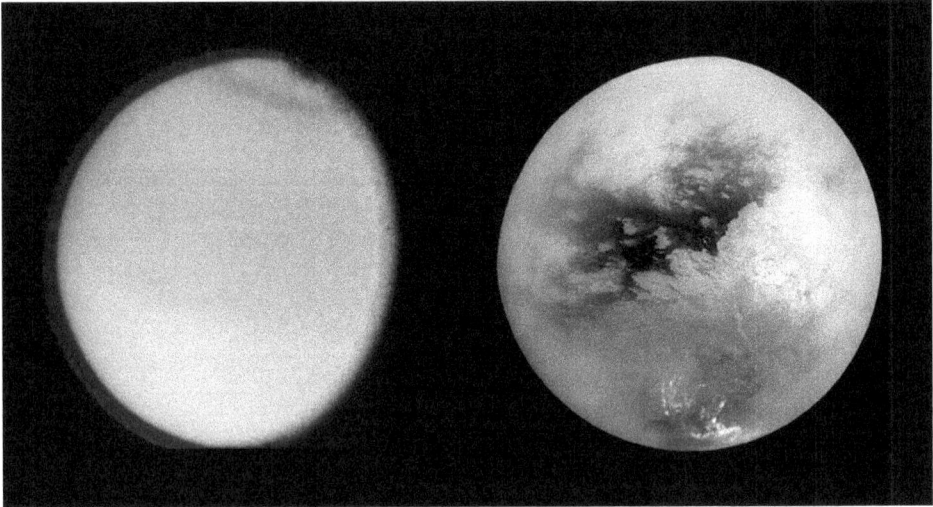

Figure 8.1   Left: a Voyager image of Titan obtained on November 9, 1980 from a distance of 4.5 million km, clearly showing the north-south difference and the North Polar hood, but little else. The Cassini/ISS image (right) was taken on October 26, 2004, from about 500,000 km using a near-infrared filter centred at 0.938 μm. Centred at 15°S and 156°W, the image shows Xanadu (the bright equatorial region) and bright clouds in Titan's southern hemisphere (NASA/JPL).

studied aspect of Venus' and Titan's dynamics, but is not fully understood for either of them.

The early results returned by the new missions, not yet fully analysed, suggest that the global-scale circulation on Titan is accompanied by waves, storms, fronts, and all of the paraphernalia of weather, although not exactly as we know it on Earth. A day on Titan is 16 Earth-days and a year is close to 30 terrestrial years. It will be many years then, including the gathering of more *in situ* data, before we know about the general circulation and dynamics of the atmosphere in all of its regions. A start has been made on modelling Titan's atmosphere, based on the physics that it has in common with the Earth and using the computational framework of terrestrial general circulation models.

8.2 Dynamics of Planetary Atmospheres

As with many other aspects of planetary science, our thinking about dynamics is conditioned by centuries of trying to understand the Earth's atmosphere, and descriptions of Titan's dynamics use much of the same terminology. It is also particularly interesting to examine contrasts between the behaviours of the two atmospheres, and to see if we can understand the physical basis for major differences. Because very little is actually known about the circulation of Titan's atmosphere, we rely very much at present on analogies and extrapolations in any description that we attempt, such as the expectation that Titan has important basic features in common with Venus, as well as with Earth.

The general circulation of an Earthlike atmosphere is driven principally by the pressure gradients that arise as a result of the gradient with latitude in the solar heating rate. Near the equator, where the incoming energy per unit area is greatest, the heated air tends to rise, due to convection, and to move polewards in such a way as to reduce the temperature gradient between equator and pole. The net result is, on the average, a general overturning of the troposphere, with rising motions at low and descending motions at high latitudes, known as the *Hadley cell*. This cell does not extend all the way to the pole on Earth (although it may on Venus, see below), primarily because the pressure gradient force generated by the unequal solar fluxes at low and high latitudes is opposed by the rapid rotation of the planet, and the resulting Coriolis forces, tending to so-called *geostrophic* balance. This limits the latitudinal extent of the Hadley cell, forming a second cell (the *Ferrel cell*) at mid to high latitudes, and a third near the poles. The horizontal transport of energy at middle and high latitudes by this kind of general overturning is complemented by an equal or larger transport by turbulence and large-scale wave motions. The relationship between large and small-scale cells, waves and flows in atmospheres is an extremely complicated research field to which many complete books are devoted.

Venus resembles the Earth physically, but differs in that it rotates much more slowly (1 Venus sidereal day — the period of rotation relative to the fixed stars — is equal to 243 Earth days). Coriolis forces are much less important on Venus and in consequence the Hadley-type circulation is more stable than it is on Earth. Cloud motions photographed from spacecraft show the equator to pole motions, with velocities of a few metres per second on average, quite clearly; what is less certain is the horizontal extent of the cell, and whether or not there are several Hadley cells stacked on top of each other to cover the whole vertical range in Venus' deep atmosphere. There is some inconclusive evidence for this in the meridional wind data from the Pioneer Venus entry probes.

The Venusian atmosphere, as a whole, rotates much more rapidly than the solid planet. This 'superrotation' is greatest at heights near about 60 km, roughly where the observable cloud patterns lie. At that level, the pressure is about 100 mbar, and the zonal wind (parallel to the equator) is around $100 \, \mathrm{m \, s^{-1}}$, about fifty times faster than the rotation of the surface beneath. Super-rotation also occurs in the Earth's upper atmosphere, and on the outer planets, driven by the transfer of momentum upwards by wave motions. The combination of equator-to-pole overturning in the Hadley cell, and the fast zonal winds, leads to huge amounts of angular momentum being transported polewards in the upper branch of the cell. The result on Venus is a giant vortex over each pole, and they are observed to have a curious double structure, which led to the coining of the term 'polar dipole'. Something comparable may exist on Titan: the distribution of minor constituents observed by Cassini suggests a special circulation pattern over the poles, as discussed further below.

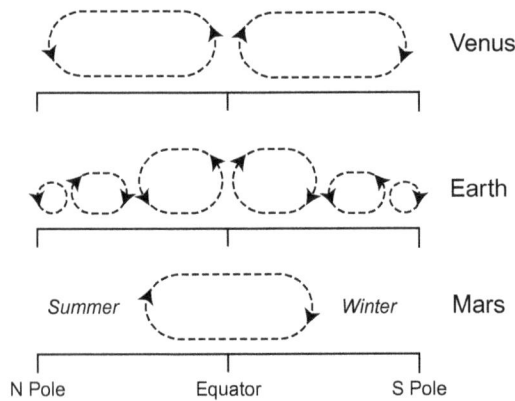

Figure 8.2   This very simplified picture of the circulation of the terrestrial planet atmospheres illustrates the difference between a 'Venuslike' circulation, based on an equator-to-pole Hadley cell in each hemisphere; an 'Earthlike' circulation with multiple cells, still essentially symmetrical about the equator; and a 'Marslike' circulation, with a single main cell spanning the equator. The Martian circulation is particularly variable with season, tending to a more equatorially symmetric circulation near the equinoxes.

The Martian general circulation is remarkable for its dependence on the surface topography, which is very pronounced, especially for a small planet. As on the Earth, airflow over mountains and other structures forces wave motions that propagate and grow, and these are an important part of the mean circulation. Other major factors are the very large day-night temperature contrasts — which can be as much as 100 K, because of the thin air, the variability in dust loading — and the seasonal pressure fluctuation due to the condensation of CO_2 at the winter pole. How these and other processes combine, and how the resulting Martian circulation responds, has been modelled using terrestrial general circulation models adapted for Mars, and is being refined using temperature soundings and wind data from recent and planned satellite and lander missions to Mars.

The radius of Mars is quite close to that of Titan (about 30% larger, in fact), but Mars' atmosphere is much thinner, by a factor of 200 or so in surface pressure. The strongest analogy between the two may be the effect of hills and valleys on the circulation of the atmosphere, although of course this depends on whether Titan's surface has any significantly elevated (or depressed) features, and how high and wide they are. We have yet to find any Titanic mountain that can match the great height of Olympus Mons (25 km above the surrounding plains) or the profundity of Valles Marineris (over 5 km deep), and it would be surprising if we do. So far, Cassini radar data shows Titan to be very smooth, with the highest features rising typically only about 100 m above their surroundings. In the very extensive continental surface feature known as "Xanadu", recently investigated, the surface is found to rise by only about 1000 m with respect to the surrounding terrain, although these results need to be refined. If this is a mountain, it remains unimpressive by comparison with

Mars or Io. However, Titan has surprised us more than once before, and most of the surface remains to be covered, including the higher northern latitudes where taller features may reside.

The gas giants of the outer Solar System have very deep atmospheres with no solid surface, and of course are much larger than Titan (Uranus is ten times, and Saturn twenty times, greater in diameter than Titan, for example). They all have internal heat sources, due most likely to slow contraction of the bodies, either in an overall sense or as a result of fractionation (separation of different constituents, with the heavier ones moving towards the centre) and phase changes in the interiors, with the consequent release of gravitational potential energy. These warm the atmosphere from below, with power comparable to that arriving from the Sun, and give rise to convective overturning of the atmospheres on a planetary scale. This, and the rapid rotation rates of these large objects, produces a general circulation in the form of elongated convection cells stretched around the planet in the equatorial direction. The rising branches of the cell have denser, higher clouds and this gives the planets their banded appearance, most prominent on Jupiter where the clouds form highest in the atmosphere.

After the bands, the most striking characteristics of outer planet cloud markings are the various spots associated with giant eddies. These come in a variety of colours, sizes and patterns of behaviour and in general are fairly mysterious. One, the Jovian Great Red Spot, is known to have persisted for hundreds of years at least. On Neptune, a small number of large spots dominates the appearance of the planet with the Great Dark Spot resembling its Jovian counterpart superficially, but not in detailed behaviour. Both are cousins to terrestrial hurricanes but a theoretical understanding of the relationship remains elusive until we have more data.

The area where the giant planets present a strong analogy for the study of Titan is in the atmospheric radiative transfer processes. On all five bodies, the dominant gas absorbing and emitting infrared radiation in the upper atmosphere is methane, while the contribution of haze and aerosol is also fundamental. Although Titan's atmosphere does not have anything corresponding to the enormous pressures that are experienced deep below the clouds on Jupiter or Saturn, the tropospheric pressures are sufficient that pressure-induced absorption bands are important, as they are on the larger bodies. These are due to energy level transitions within atmospheric molecules which are normally forbidden, but which are activated when the molecule is distorted by frequent collisions with others. The main difference between Titan and the better-studied example of Jupiter is that we are predominantly concerned with collisions involving nitrogen (e.g. N_2–N_2, or N_2–CH_4) rather than hydrogen (H_2–H_2 and H_2–He).

Atmospheric flows are subject to basic instabilities, generally classified as either *vertical, barotropic* or *baroclinic* instabilities. An example of the first of these would be the convective instability, which, as discussed above, gives rise to regular overturning motions in the troposphere. Barotropic instabilities are those that grow by receiving energy from the mean flow, as a result of the abrupt horizontal velocity gradients known as shear. Baroclinically unstable disturbances extract energy from the horizontal temperature gradient. Both barotropic and baroclinic disturbances can give rise to oscillations in the basic atmospheric fields (temperature, pressure, density and wind), and these are what we know as waves. They range in scale from the *planetary* waves, which span the globe in a few wavelengths and are a major factor in the general circulation, to small-scale non-linear motions whose behaviour is essentially chaotic, which mix the atmosphere, and which act as a frictional term in the equations of motion. Non-linear effects in waves can, under certain conditions, roll waves into vortices, some of which become the familiar cyclones seen in terrestrial weather charts.

Atmospheres become turbulent when the ratio of the wind velocity to the kinematic viscosity is greater than a certain value, called the Reynolds criterion. Turbulent flow is common on the Earth, and this is likely to be the case on Titan, also. Turbulence transfers heat, minor constituents and momentum in important amounts at every altitude in the Earth's atmosphere; in the lowest 1 km (the *boundary layer*) it is the dominant process responsible for exchange of these quantities between the land and ocean surfaces and the atmosphere. In the stratosphere, it is found that calculations of the zonal flow and mean meridional motions, which are important for the transport of minor constituents like ozone, resemble observations only if the heat and momentum fluxes due to eddies are incorporated. This is generally accomplished using mixing-length theory, or an empirically-determined eddy diffusion coefficient. Again, there must be quite close analogues of these processes at work on Titan.

The gravitational tide exerted by an adjacent celestial body is another relevant atmospheric forcing mechanism that is important and quantifiable. On Earth the influence of the lunar tide in the atmosphere is not as significant as in the ocean because other processes prevail, but on satellites of the giant planets the tide exerted by the planets can be huge. It is this effect that melts the interior of Jupiter's moon Io, producing volcanoes, and which is believed to maintain a liquid water ocean below the icy crust of Europa. On Titan, a variable tidal force is maintained as a result of the eccentric orbit of the moon around Saturn, causing a variation in the Titan-Saturn distance as well as exact position of the sub-Saturn point on Titan's surface. This causes the tidal wave to migrate eastward with a constant phase speed that is exactly half of Titan's orbital velocity.

A 3-dimensional dynamical model of the atmosphere predicts that the global surface pressure and horizontal wind patterns vary with the period of a Titan day, and that the wind in the north-south direction reverses twice per Titan day, much more frequently than the overturning in the Hadley cell. The tide may therefore be a major driver of the atmospheric dynamics, especially in the troposphere where the solar heating is weak. The tidal wave is also expected to propagate vertically and to affect the wind and temperature in the upper atmosphere. On the other end, tidal waves have been invoked to account for the formation of the dunes found on Titan's surface covering large areas (Chapter 9).

8.3 Titan's General Circulation

Comparisons with experience on Earth and other planets where atmospheric measurements have been made have led to various expectations for Titan, as have numerical experiments with models that apply the universal laws of atmospheric physics
and fluid dynamics to the known boundary conditions of solar heating, surface pressure and so on. Now, with Cassini and Huygens data, these expectations can be
tested and more detail added. It helps, especially since we are still very short of a lot
of fundamental information, to break complex questions of the global mean circulation into different regimes: the zonal circulation (motions parallel to the equator),
the meridional circulation (equator-to-pole), and differences between both of these
in the upper and lower atmospheres, where the seasonal forcing is expected to be
different.

Ultimately, of course, these must be combined to form a view of the full 3-D
picture. Some key phenomena stand out, such as the global super-rotation, a feature
of the zonal flow in which the mean wind accelerates with height to what seem at first
to be improbably high values. Terrestrial experience has taught us to examine the role
of waves and turbulence in the global climate, knowing that the long-term average
circulation is only half the story in determining the transport of heat, momentum, and
minor constituents around the globe. Finally, we have a special interest in seasonal
and day-to-day changes, and the episodic behaviour we know as weather, especially
near the surface, because these are a major feature of Titan's environment. Storms,
precipitation, and lightning are key examples that are coming under scrutiny with
the newly acquired close-range and *in situ* data.

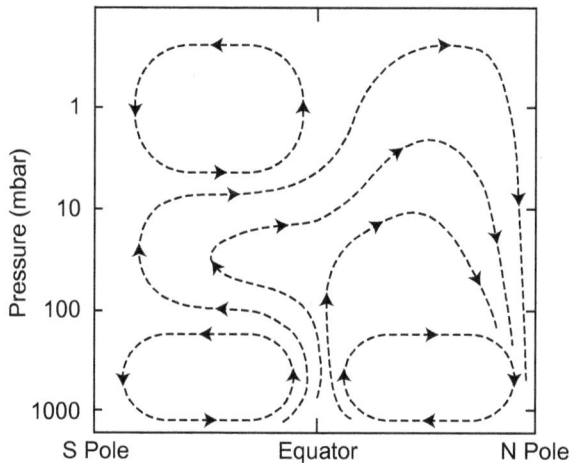

Figure 8.3 A schematic of Titan's general circulation, as postulated by Flasar *et al.* The flow is
seasonally dependent, and shown here for the northern spring. For about 80% of Titan's year, a summer
pole to winter pole circulation dominates, as in Figure 8.6.

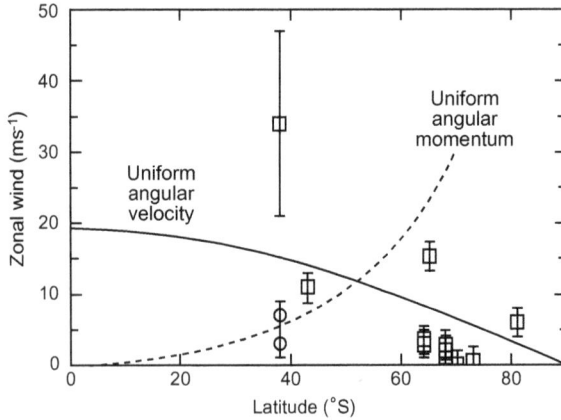

Figure 8.4   Wind measurements at middle- and lower-troposphere altitudes obtained by tracking discrete (squares) and streaky (diamonds) cloud features in Cassini images. The vertical lines are error bars and the scatter is due to real temporal variability, clouds at different altitudes with different mean winds, and tracking uncertainty (Porco *et al.*, 2005).

8.4 Zonal Motions

The faintly banded appearance of Titan's haze suggests rapid zonal motions, i.e. winds parallel to the equator. This impression is reinforced by the infrared temperature maps, which show very small contrasts in the longitudinal direction and rather large ones (of around 20 K) between the equator and the winter pole. Fast zonal circulation is also what we would expect by analogy with Venus, which is also a slow rotator with a thick, cloudy atmosphere and rapid zonal winds in the same sense as the rotation of the solid body. On Titan and Venus, thermally-induced pressure gradients are opposed by the centrifugal forces arising from the rapid rotation of the atmosphere, rather than the Coriolis force due to the surface rotation as on Earth, resulting in the condition known as *cyclostrophic* balance.

In cyclostrophic balance, the zonal winds can be obtained from knowledge of the latitudinal temperature gradients and the associated pressure differences to be balanced by the zonal wind flow. Since the vertical spacing of isobaric surfaces is compressed in cold regions and expanded in warmer regions, the pressure gradient that must be balanced by the zonal wind increases with height. The relation is quantitatively expressed in the thermal wind equation, which provides a means to compute the vertical shear of the zonal wind from the latitudinal temperature gradients along isobaric surfaces. Since the cyclostrophic form of the thermal wind equation is quadratic in the zonal wind, it can be integrated to yield the wind magnitude, but not its zonal direction.

The latitudinal temperature contrasts first seen by Voyager indicated zonal winds of the order of 80 m s^{-1} at 0.4 mbar. From Cassini flybys in July and December 2004,

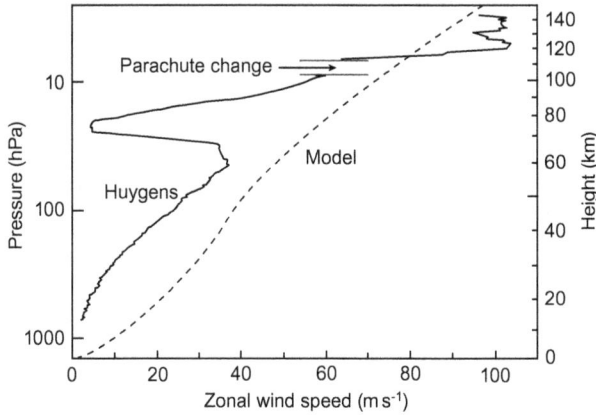

Figure 8.5 The zonal wind profile derived from tracking the descent of Huygens is compared with the model predictions made using Voyager temperature data by Flasar *et al.* (1997). The estimated uncertainties are $80\,\mathrm{cm\,s^{-1}}$ at high altitude down to $15\,\mathrm{cm\,s^{-1}}$ just above the surface (Bird *et al.*, 2005).

the CIRS temperatures at the 1 mbar level are lower at 60°N than at the equator by around 20 K, whereas the equator-to-pole difference is only 4–5 K in the southern hemisphere. The mean zonal winds inferred from this temperature field are weakest at high southern latitudes and increase towards the north, with maximum values at and mid-northern latitudes (20–40°N) of about $160\,\mathrm{m\,s^{-1}}$.

Stellar occultations are another indirect means to obtain the zonal winds. The atmospheric oblateness due to the zonal winds can be constrained from the analysis of the central flash, the increase of the signal at the center of the shadow (when the star is behind Titan) due to the focusing of the atmospheric rays at the limb. On July 3, 1989, Titan occulted the bright K-type star 28 Sgr, revealing fast zonal winds, close to $180\,\mathrm{m\,s^{-1}}$ at high southern latitudes, and close to $100\,\mathrm{m\,s^{-1}}$ at low latitudes. Other occultations occurred on December 20, 2001 and November 14, 2003. They seem to suggest a seasonal variation with respect to 1989. In 2001 a strong $220\,\mathrm{m\,s^{-1}}$ jet was located at 60°N, with lower winds extending between 20°S and 60°S, and a much slower motion at mid-latitudes. The CIRS data suggest that the strongest northern winds have migrated closer to the equator with respect to previous measurements, while the southern winds have weakened.

Since Voyager and occultation wind measurements could not provide the wind direction, determining this became a crucial goal for the Huygens mission. Before the probe arrived, different teams of ground-based observers had tried to measure the zonal winds directly by detecting the Doppler shifting of spectral lines emitted from the atmosphere. The first indication that the winds are prograde (in the same sense as the rotation of the surface) was obtained in 2001 by a group from NASA/GSFC using infrared heterodyne spectroscopy of Doppler-shifted ethane emission lines. Other Doppler studies probing different levels also found prograde winds, using

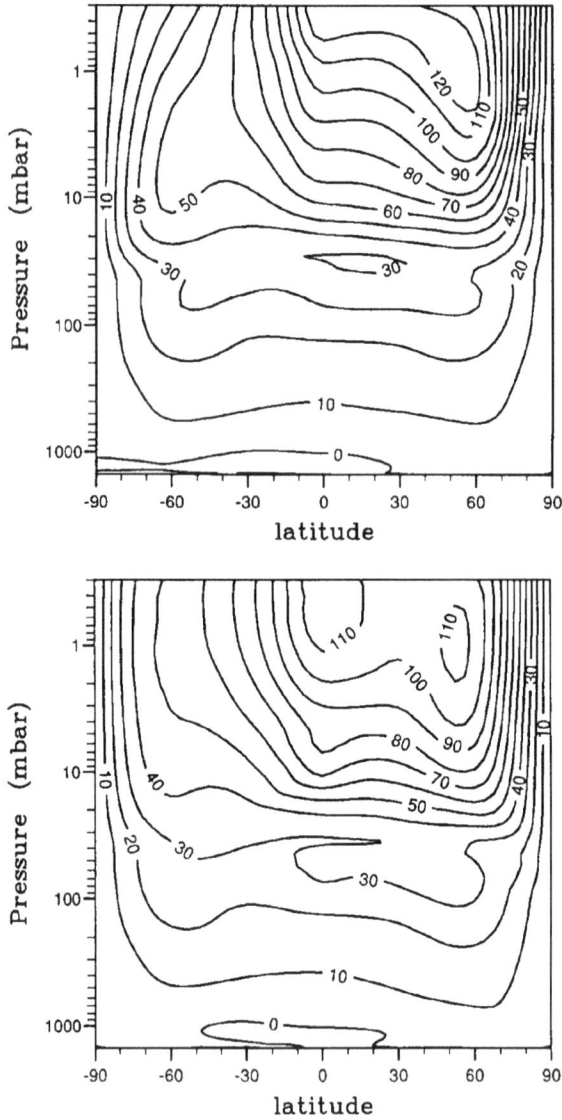

Figure 8.6 Zonal winds (ms^{-1}) computed by a Titan general circulation model, showing super-rotation and a mid-latitude jet. The top frame shows northern winter solstice and the lower frame the spring equinox (Hourdin *et al.* 1995).

millimeter-wavelength interferometry of nitrile lines, or from high-resolution spectroscopy of Fraunhofer solar absorption lines in the visible. The recent advances in adaptive optics also allowed for the first detections of tropospheric clouds from the ground, mainly at circumpolar southern latitudes, but obtaining wind speed and direction from these is poorly constrained since the clouds are few and far between.

The high spatial resolution Cassini ISS observations of the clouds confirm that they occur at around 25 km altitude and show slow, eastward motions. It is difficult to get good numerical values for the winds by tracking clouds even in these data, partly because they cannot be followed for long periods since each of Cassini's encounters with Titan lasts only a few days. The first results still have large uncertainties: one estimate by the Cassini imaging team for the zonal wind velocity at the equator, extrapolated under the assumption of solid-body rotation observations of clouds near the south pole, yielded $19 \pm 15 \, \mathrm{m \, s^{-1}}$. This is consistent with the result from tracking the descent of Huygens, which found near-equatorial winds (the probe entered at about $10°$S latitude) of about $5 \, \mathrm{m \, s^{-1}}$ at 25 km altitude. The profile measured by Huygens showed the winds decreasing with altitude, from $100 \, \mathrm{m \, s^{-1}}$ at 140 km down to almost zero at 80 km, then an increase up to $40 \, \mathrm{m \, s^{-1}}$ at 60 km before decreasing again to negligible zonal velocity at the surface. The winds are generally smaller than expectations from models, and the sharp dip in the wind velocity in the height range between 60 and beyond 100 km was a complete surprise (although a possible explanation in terms of atmospheric tidal effects has recently been advanced, see Section 8.7).

The few experimental determinations of winds on Titan, although quite uncertain, are consistent with high zonal winds at pressure levels in the 1–100 mbar region. An attempt to track faint features in the clouds led to inferred wind speeds in the 28 to $99 \, \mathrm{m \, s^{-1}}$ range, which if real would correspond to air circulating around the equator in about 2 to 5 days. This may be contrasted with Titan's solid-body rotation rate of 16 days, for a superrotation factor of 3 to 8 (compared to between 50 and 60 on Venus).

Titan's disc is slightly oblate, an effect which may be attributed to the super-rotation of the atmosphere. Accurate measurements along one chord can be made during a stellar occultation by Titan, and if the corresponding non- circularity is entirely due to zonal winds their velocity would be about $180 \, \mathrm{m \, s^{-1}}$ at 0.25 mbar. This is again much faster than the surface rotation speed of $12 \, \mathrm{m \, s^{-1}}$, and presumably, although the measurement is ambiguous with regard to direction, in the same sense. At this level the superrotation factor is about 15.

The preliminary cloud-tracked winds were prograde, like Titan, and the surface probably has to be the source of the atmospheric angular momentum, although the processes by which it is transferred and a balance maintained are quite complicated. The usual way to study this sort of situation is to attempt to construct a numerical model on a computer, whence one can see whether any reasonable combination of realistic physical behaviour can reproduce the observations. Models can also predict observable quantities in advance, and so be used to design experiments that test the theoretical aspects of the model.

The most sophisticated dynamical models of Titan's atmosphere are based on a general circulation model of the Earth, modified for Titan's radius of 2,575 km, obliquity of $26°$, and rotation rate of 16 days, as well as other factors including of course the composition, cloud and temperature structures dealt with in earlier chapters of this book. The dynamical code of the GCM is the same as for the Earth, based on the classical (so-called 'primitive') equations of meteorology, with the same parameterisations for processes like vertical mixing, horizontal dissipation, and convection, which cannot be calculated from first principles. The version developed at Laboratoire de Météorologie Dynamique in Paris starts from an atmosphere at rest and 'spins up' to produce a wind field featuring a strong zonal super-rotation, with winds of the same order as the observed ones. A compact region of high wind, known as a *jet*, appears at mid-latitudes; a similar feature (and with about the same wind speed, around $120 \, \mathrm{m \, s^{-1}}$) is known to occur on Venus.

The superrotation observed in the stratosphere, a dynamical state in which the averaged angular momentum is much greater than that corresponding to co-rotation with the surface, is difficult to explain and has defied our understanding in the much better documented Venus case, the paradigm of a slowly rotating body with an atmosphere in rapid rotation. General circulation models, however, have provided much insight into the mechanisms controlling Titan's atmospheric dynamics. Two independent studies using modified terrestrial models for idealized conditions of Titan radius and rotation rate both succeeded in obtaining superrotation. The Gierasch-Rossow mechanism, originally proposed in the context of Venus, offers an explanation. In an atmosphere at rest relative to the surface, the angular momentum is larger at lower latitudes due to the longer distance to the spin axis. When a symmetric Hadley cell sets in, with ascending motions at the equator and subsidence at the poles, the upward transport creates an excess of angular momentum at the upper layers. If momentum is conserved in the poleward branch of the cell, the zonal wind in a fluid ring is bound to increase as it approaches the spin axis, generating a prograde high speed jet. Momentum diffusion, however, depletes the total momentum before the cycle is completed in the lower, returning branch of the Hadley cell, and the momentum deficit would lead to retrograde winds at low latitudes unless some other, non-axisymmetric process, closes the angular momentum cycle. In more realistic Titan simulations by Hourdin and colleagues in 1995 such a process has been identified under the form of planetary waves, forced by instabilities in the equatorward flank of the high-latitude jet. Two factors play a key role in facilitating the acceleration process. On the one hand high altitude absorption processes decouple upper atmosphere dynamics from dissipation occurring at the surface layer, while on the other hand the slow rotation allows the Hadley cell to reach high latitudes, by reducing centrifugal forces in the poleward branch. A strong seasonal cycle due to Titan's obliquity of 26.7° was also established: during most of the Titan year, the meridional motion is dominated by a large Hadley cell extending from the winter to the summer pole, with the symmetric two-cell configuration typical of equinoxes occurring only in a limited transition period. In the model the jet is located close to 60° in the winter hemisphere, while the summer zonal circulation is close to solid body rotation.

8.5 The Meridional Circulation

The gradients and contrasts that are apparent in Titan's atmospheric distributions of composition, haze and temperature (Chapters 5–7) provide indirect evidence for the existence of an underlying meridional (equator-to-pole) circulation. Early theoretical studies of this by C. Leovy and J. Pollack, as long ago as 1973, used a radiative equilibrium model combined with fluid dynamical similarity methods to predict a weak axially symmetric circulation and weak horizontal temperature variations. This takes the form of a Hadley-cell regime extending from equator to pole in each hemisphere, with upwelling at the equator and downwelling at the poles. This, and the slow rotation of the solid body combined with zonal superrotation in the upper atmosphere, is the basis for the close analogy that is frequently drawn between the atmospheric circulations of Titan and Venus.

The deep atmospheres of both bodies are characterised by quite small thermal gradients of 2–3 K between equator and pole. These are presumably a result of the efficient heat redistribution in the Hadley circulation, since there would have to be much bigger temperature differences if heat were not being transported polewards by some efficient mechanism. Since Titan's slow rotation and small radius inhibit non-axisymmetric processes, such as baroclinic eddies, as a preferred mechanism for heat transport, considerable meridional motions are indicated.

It was further expected, from simple arguments based on a calculation of the radiative time constant as a function of atmospheric density, that the temperatures in the lower atmosphere of Titan should remain nearly constant throughout the long Saturnian year of approximately 30 Earth years. The calculated radiative time constant in the troposphere is large enough so that the heating effect of the Sun is averaged over the solar day, and the circulation should be symmetric about the equator if this were the only factor. In fact, due to the finite thermal inertia of the surface, the surface temperature may vary by a few K with season, and even slightly diurnally, as long as the thermal inertia of the surface is not too large. If there is a pole-to-pole temperature gradient at the surface, as predicted by some models, this would give rise to a pole-to-pole Hadley cell reversing after the equinox when the Sun moves to the opposite hemisphere. The north-south flow associated with this circulation is opposite to that in the stratosphere since at the lower boundary the flow has to compensate for the reversed flow further above. However, it was also shown that the latitudinal and seasonal variation in the stratospheric haze distribution could considerably affect the tropospheric Hadley circulation by modifying the radiative heating rate near the surface. In some scenarios, the single pole-to-pole circulation is predicted to disappear at the expense of multiple cells. This would weaken the seasonal variation in both east-west and north-south circulation.

The seasonal variation in the surface temperature may also affect the weather in the troposphere in a more direct way, through a role in the methane "hydrological" cycle in which the vertical and horizontal transport of methane will vary seasonally. As a result of seasonal surface heating or cooling the lapse rate — and hence the condition for dry and moist convection in the lower troposphere — slightly changes, affecting the development of convective methane clouds. Generally the lapse rate is predicted to increase in warm seasons and to decrease in cold seasons. The convective clouds observed by Cassini near the south pole have been interpreted as evidence of surface heating in summer and the summer pole is predicted to be the warmest place unless thick haze layers weaken the sunlight arriving at the surface. Conversely the air near the surface may become stably stratified in winter and convection may completely vanish.

Angular momentum transport by the Hadley circulation can also affect the direction of the zonal wind, in the lowest part of the troposphere where the centrifugal force becomes less important than the Coriolis force because of the weakness of the wind. In the presence of a pole-to-pole Hadley circulation the east-west wind in balance with the actual Coriolis force is dependent on the hemisphere because the Coriolis force changes sign across the equator, causing weak eastward wind in winter and weak westward wind in summer. This north-south difference disappears in the stratosphere, changing over to super-rotating winds in both hemispheres.

8.5.1 *The Hemispherical Asymmetry*

Periodic changes in Titan's disk-integrated brightness have been monitored by Earth-based observations since the 1970s (Chapter 3). Spatially resolved observations, starting with Voyager, provided an interpretation of the periodic changes of the disk-integrated brightness as the combined action of the high inclination of the rotation axis and the seasonally varying north-south asymmetry. The asymmetry that Voyager 1 observed in 1980, with a darker northern hemisphere in visible light, has since been observed to reverse, as Titan's season shifted from northern spring to present-day northern winter. When the Hubble Space Telescope first observed Titan early in the 1990s, a little over a quarter of a Titan year after the Voyager encounters, the northern hemisphere was found to be brighter than the south. The turnover was later also found to occur gradually, starting at higher altitudes in the atmosphere.

Whereas Voyager only observed up to red wavelengths, HST and adaptive optics systems from the ground also imaged Titan in the near infrared. At these wavelengths, the asymmetry is reversed, and indeed is somewhat stronger than in the visible. This is due to different variations of the atmospheric and haze brightness with wavelength. The atmosphere is bright at short wavelengths due to Rayleigh scattering and dark in the near infrared due to methane absorption, whereas the haze appears to be dark in blue and bright at red and longer wavelengths. Limb darkening is also strongly wavelength-dependent. The disk at UV and violet wavelengths is fairly flat, while it is near-Lambertian at green and red wavelengths. In the near-infrared, there is limb-brightening.

The hemispherical asymmetry in Titan's brightness at visible wavelengths is a clue to the nature of the general circulation. The two-dimensional dynamical model, coupled to a microphysical haze production model by P. Rannou and colleagues, explains both the seasonal reversal of the asymmetry and the detached haze layer in terms of a seasonally varying circulation. 80% of the time the mean transport in the upper atmosphere is from the summer pole to the winter pole, with a return branch at lower levels. During the changeover from one summer to the other, equator-to-pole Hadley cells occur in each hemisphere.

In the model, the photochemical production of aerosols peaks at an altitude around and above 400 km. The meridional wind in the upper branch of the Hadley cell is stronger in the production zone than below, and particles there are rapidly transported towards the pole, at a speed of around $2.5\,\mathrm{m\,s^{-1}}$. This is more than 100 times faster than the settling velocity of the aerosols, so they travel essentially horizontally to the polar region. During this process, coagulation is also taking place, producing larger and heavier particles. These sink as the air descends forming the polar hood, which Voyager and Cassini images covering the winter pole showed. The sinking action is augmented by enhanced radiative cooling at that pole, due to the emissivity of the aerosols themselves. The top of the polar hood cloud occurs about 100 km below the top of the detached layer in both the model and the images. As the season changes, shortly after equinox, the circulation reverses and an ascending motion sets in where the particles were previously descending. At the time of the transition the polar haze, which was previously descending, is then redistributed about a scale height below the production zone, becoming physically separated from the freshly created particles aloft.

In this 2-D model, the main haze layer on Titan is supplied by the transport of aerosol particles, originally produced in the upper detached layer, from the winter polar hood region towards the equator. The lower branch of the main circulation cell starts rising when it reaches the summer hemisphere and, since there is no source of particles in the lower atmosphere, and coagulation and precipitation of particles is continuously occurring, the rising air is relatively depleted in aerosol. This accounts for the relative brightness of the summer hemisphere.

An alternative explanation for the detached haze structure, in terms of the gravitational tides in Titan's atmosphere induced by its eccentric orbit, was advanced in 2006 by Richard Walterscheid and Gerald Schubert. In their model, the tides affect both the zonal winds and the vertical propagation of aerosols in a strongly height-dependent (and seasonally-dependent) fashion, which could produce the observed layering without, or in addition to, the global transportation mechanism discussed above. The modulation of the vertical wind profile by tidal dissipation also provides a possible explanation for the sharp drop in the zonal winds, from about $30\,\mathrm{m\,s^{-1}}$ to almost zero, measured by Doppler tracking of the Huygens probe during its descent.

8.5.2 *The Polar Vortex*

The heat transport in the Hadley circulation is responsible for the relatively small temperature contrast between the equator and poles. The Hadley cell must also transport chemical species and aerosols horizontally and vertically, and play a role

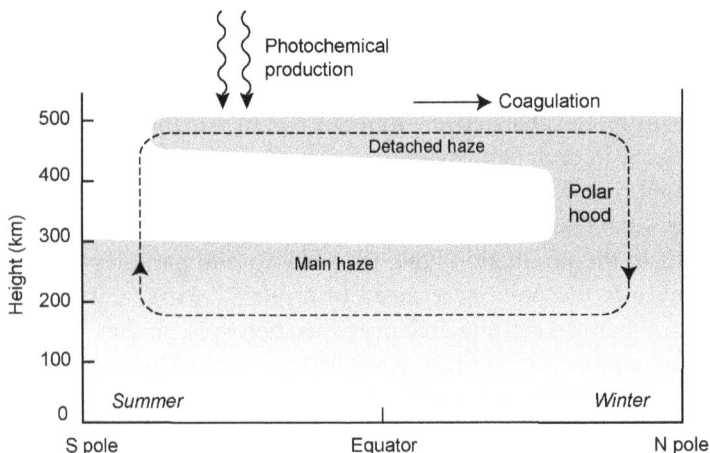

Figure 8.7   A model of the two-dimensional distribution of haze. The detached haze is a secondary layer found at a height of around 400 km overlying the main haze, which lies below 300 km. The haze is preferentially accumulated near the pole. The winds are from the south polar region (summer) toward the north polar region (winter).

in the methane condensation cycle. The strength of the circulation is sensitive to the heating rates, and hence to the abundance and distribution of major greenhouse gases in the troposphere, and this in turn affects the angular momentum transport, location and strength of local jets (east-west wind). Thus observations of meridional variations in the gases in Titan's stratosphere are crucial since they are tightly coupled with the circulation.

Ethane, propane and acetylene are the three most abundant hydrocarbons after methane and were found by Cassini/CIRS to exhibit no noticeable compositional variations in latitude, but to be homogeneously mixed from pole to pole with only a slight increase in the north. In contrast, Cassini spectra from the first group of Titan flybys suggest that diacetylene, methylacetylene and the nitriles tend to increase towards the winter north pole, while the mixing ratios of the hydrocarbons remain fairly constant from mid-latitudes to the south pole.

The molecular abundances found by Cassini at this epoch indicate an enhancement for some species in the northern stratosphere at high latitudes, albeit not as dramatic as at the time of the Voyager encounter. Such latitudinal contrasts observed in the chemical trace species may be explained by invoking photochemical and dynamical reasons. The UV radiation from the Sun acts on methane and nitrogen to form radicals that combine into nitriles and the higher hydrocarbons. This production occurs in the mesosphere at high altitudes (above 300 km or 0.1 mbar). Eddy mixing transports these molecules into the lower stratosphere and troposphere where most of them condense. Photodissociation by UV radiation occurs on timescales ranging from days to thousands of years. The combination of these processes leads to a vertical variation in the mixing ratio, which usually increases with height towards the production zone. In a purely photochemical explanation by Yuk Yung in 1987, the nitrile enrichment is due to reduced photodissociation in the lower stratosphere, which remains in the shade during the long polar winter night, while production still occurs in the upper atmosphere, continuously exposed to the solar flux. However, three-dimensional computation of actinic fluxes by S. Lebonnois and D. Toublanc suggested that this mechanism alone cannot explain the latitudinal contrasts and that circulation must intervene. Simulations by F. Hourdin and colleagues that coupled photochemistry and atmospheric dynamics provide a consistent view: competition between rapid sinking of air from the upper stratosphere in the winter polar vortex and latitudinal mixing controls the vertical distribution profiles of most species. The magnitude of the polar enrichment is controlled by downwelling over the winter pole, which brings enriched air from the production zone to the stratosphere, and by the level of condensation. Short-lived species are more sensitive to the downwelling due to steeper vertical composition gradients and exhibit higher contrasts.

As F.M. Flasar and colleagues pointed out, for the enhancements to persist at high northern latitudes, there must be no lateral mixing with the atmosphere at other latitudes. Titan's strong circumpolar winds could be responsible for this isolation, which in the case of the Earth leads to the formation of the Antarctic ozone hole. The cold polar temperatures of the Earth's Antarctic cause the formation of strato-spheric clouds, which denitrify the polar atmosphere by heterogeneous chemistry, liberating chlorine which irreversibly destroys O_3. This occurs because the nitrogen compounds contained in the warm air coming from lower latitudes to be mixed into the high southern atmosphere. Although the mean meridional circulation itself is slow, mixing by planetary waves can be fast. The circumpolar vortex associated with the cold polar temperatures inhibits this wave-induced mixing. The ozone hole is finally destroyed when spring sets in and the planetary waves, forced by topographi-cal and thermal contrasts, disrupt the vortex and mix high- and low-latitude air. This type of disruption may also occur in Titan's stratosphere. It may be possible in the future to find confirmation of all this by detecting a weak thermal signature of the planetary-scale waves in the zonal structure of the temperature field.

Thus, Titan's winter polar atmosphere may be analogous to the terrestrial Antarc-tic ozone hole, albeit with different chemistry. Already at the time of the Voyager encounter, in 1980–1981, infrared observations showed a north pole coming out of winter with cold stratospheric temperatures, strong circumpolar winds, and an enhanced concentration of several organic compounds. This suggests the isolation of the winter polar atmosphere from the other latitudes.

8.6 Vertical Motions

The troposphere on Titan, when defined as the region above the surface where temperature decreases with height, is about 40 km deep. Here, the atmosphere tends to be 'optically thick', favouring vertical heat transfer by convection, and suppressing loss to space directly by radiation. Under such conditions, simple theory predicts

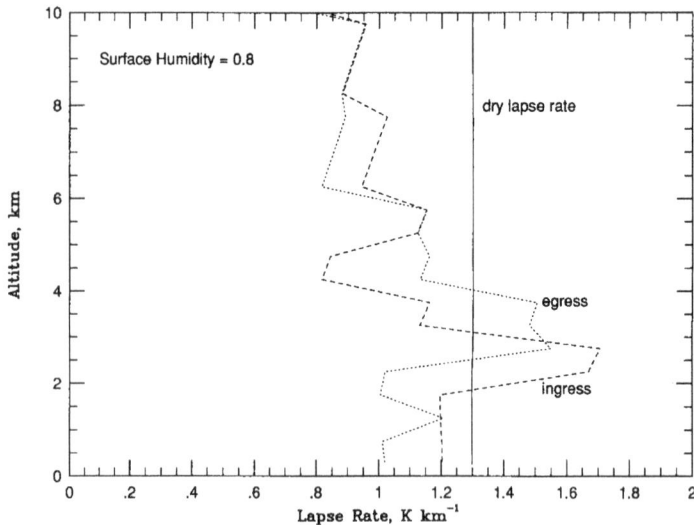

Figure 8.8   The measured lapse rate, i.e. the rate of change of temperature with height, in the lower atmosphere of Titan, from Voyager radio occultation profiles. Values greater than the appropriate moist adiabatic lapse rate are unstable against convection (McKay *et al.*, 1997).

that the temperature falls with height at a steady rate Γ, the adiabatic lapse rate, given by the need for the buoyancy of an air parcel to just balance the vertical pressure gradient. On Titan Γ is about 1.4 K km^{-1} for dry air, compared to $\sim$10 K km^{-1} for the Earth. The difference is due mainly to the smaller acceleration due to gravity on Titan, since the bulk atmospheric composition (which is the other main parameter upon which Γ depends) is nearly the same.

The lapse rate in Titan's lower atmosphere was obtained as a function of height by Voyager radio occultation measurements. Because this was a single flyby, only one time and two (near equatorial) locations were probed. It is not easy to say what the full error bars are on the points in the profiles, but, examining the fluctuations and bearing in mind that the lapse rate is the result of taking the difference of two numbers, with their own errors and uncertainties, it is probably at least a few tenths of a degree per kilometre. With this in mind, we might cautiously conclude that Titan's atmosphere appears to be unstable against dry convection below about 4 km, and unstable against moist convection to some considerably higher level, between 20 and 30 km above the surface. Moisture in this case means methane, of course, and we have already seen that a case exists for super-saturation of that condensable species in Titan's lower atmosphere.

The lapse rate profile measured directly inside Titan's atmosphere by Huygens is generally similar to the Voyager profile, both being consistent with moist convection on Titan, with wet air rising to cooler levels, which would be expected to lead to cloud formation with occasional episodes of rainout. The vertical heat transfer implied by the apparent existence of a dry adiabatic lapse rate region, which was

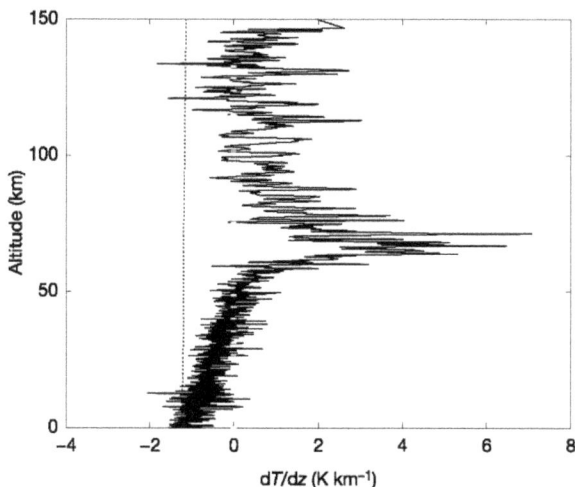

Figure 8.9   The lapse rate derived from the temperature profile measured by HASI during the Huygens descent. The large peak between 80 and 60 km is probably associated with the large vertical shear of the wind speed measured by the Doppler Wind Experiment (Fulchignoni *et al.*, 2005).

suggested by the Voyager radio occultation experiment, implies in turn that there is a significant source of atmospheric heating at the surface of Titan. The only really plausible source of this is sunlight, which must penetrate the haze layers and reach the surface in significant amounts, probably about 10% of the incoming sunlight at the top of the atmosphere. Knowing this places important constraints on the optical thickness of the aerosol in the atmosphere of Titan and hence on the number, size and composition of droplets.

The aerosol layers could still be quite thick; on Venus, enough sunlight penetrates through the haze and cloud layers to the surface to read a newspaper (the Russian scientists who first measured this described it as 'like noon in Moscow just before a thunderstorm'). The solar illumination at the surface of Venus is about the same as it is *outside* the atmosphere on Titan, so we expect the natural lighting on the surface of Titan, at one-tenth of that, to be pretty dim — explorers may have to carry a torch. The key factor determining how many solar photons get through the haze is whether the cloud droplets scatter sunlight conservatively, in which case there could be a lot of them and the light would still diffuse through, or whether they are strongly absorbing. On Venus, the droplets are good conservative scatterers at short wavelengths and good absorbers at longer wavelengths, ideal for producing a strong greenhouse effect. Something similar (but not identical: the droplets on Venus are made of sulphuric acid solution, while those on Titan are condensed hydrocarbons) may well be happening on Titan. Any model of the cloud and haze structure on Titan has to have the right radiative transfer properties to produce the observed greenhouse effect, quite a rigid constraint. It seems clear that Titan has a substantial greenhouse, although smaller than that of the Earth, and very small compared to

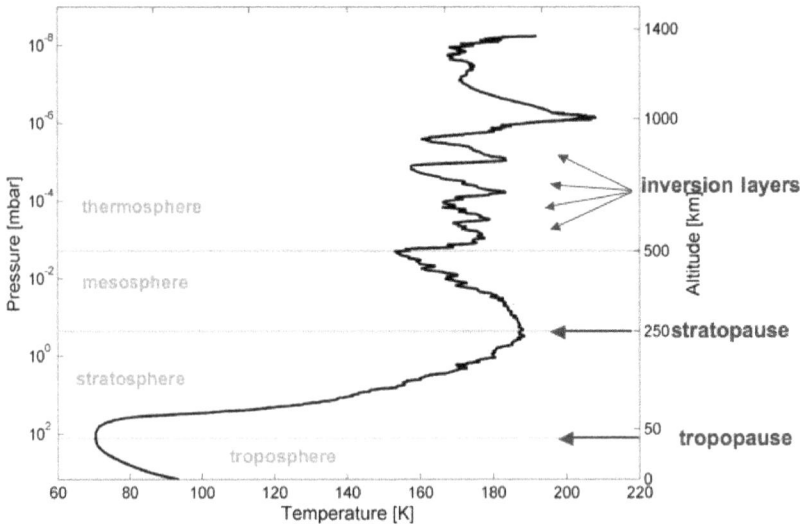

Figure 8.10   The vertical temperature profile measured by the Huygens probe during its descent through Titan's atmosphere (Fulchignoni *et al.*, 2005).

Venus, since the surface temperature of 94 K is significantly higher than its overall external radiative balance temperature of about 82 K, the error in the latter being estimated as a few degrees, mostly due to the uncertainty in Titan's wavelength- and spatially-integrated albedo.

8.7 Waves, Tides and Turbulence

Waves in the atmosphere are classified into at least twenty different types; we will consider only two prominent examples. *Gravity* waves arise when parcels of air of different density to their surroundings oscillate under the control of buoyancy forces. The motion is therefore at right angles to the direction of propagation. A particular example would be the lee waves which form behind a mountain, as a result of relatively dense air being forced to ascend to cross the obstruction, and then falling freely on the other side. There is ample evidence for gravity waves on the terrestrial planets and it would be surprising if they do not occur on Titan also, produced basically by flow over surface features and/or flow around weather systems in the atmosphere.

   Many of the large-scale features that are seen, for example in satellite maps of atmospheric temperature fields on the Earth, are *Rossby* waves. These arise as a result of the Coriolis effect produced by the rotation of the planet, which generates a restoring force on an eastward flow that is perturbed, for any reason, away from the purely east-west direction. The force on a westward flow is away from the original direction, so the air mass cannot oscillate and westward-propagating Rossby waves do not occur. This explains the absence of planetary wave activity in Earth's stratosphere in the summer, when the mean flow is predominantly to the west.

Rossby waves are the most frequently cited reason for the large horizontal Y-shaped feature that dominates the appearance of Venus when it is imaged at ultra-violet wavelengths. In that case, a similar dynamical feature might be expected to occur on Titan, although because of the different composition of the clouds it might be more difficult to observe and has yet to be found. The 'Y' on Venus is seen in ultraviolet images that exhibit contrast produced by variable amounts of unidentified sulphur compounds in the upper cloud and haze layers. These absorb strongly at wavelengths shorter than about $0.320\,\mu m$, and act as tracers for the motions. There is no corresponding absorber in the haze layers on Titan, but some of the trace atmospheric gases are variable across the globe, and methane condensate clouds have been seen moving with the wind near the surface.

Direct evidence for wave processes in Titan's atmosphere remains scarce, despite their importance in the maintenance of super-rotation. Since baroclinic processes are excluded, essentially barotropic waves should be expected as the principal carrier of momentum from high to low latitudes. Modeling predicts wavenumber-2 waves with an amplitude for the zonal component of about 10% of the mean wind speed, and in principle they can be inferred from horizontal maps of temperature and trace species exhibiting strong latitudinal contrasts. The first Cassini/CIRS temperature maps at 1.8 mbar do show spatial inhomogeneity, but long time series and better spatial coverage are needed to constrain spatial and temporal variations.

Another relevant non-axisymmetric phenomenon in Titan's troposphere is the gravitational tide exerted by Saturn. The eccentric orbit of Titan around Saturn gives rise to a tidal force, resulting in periodical oscillation in the atmospheric pressure and wind with a period of a Titan day (16 days), among which the most notable effect is the periodical reversal of the north-south component of the wind. In the lower atmosphere the effect of this tide is modest, with a maximum temperature amplitude about 0.3 K and winds of $2\,ms^{-1}$. Computations by Darrell Strobel indicate, however, that tidal waves attain high amplitudes in the upper atmosphere and could trigger haze formation through temperature perturbations.

Temperature inversions have been detected in both the Huygens HASI measurements and in stellar occultation data. Inversion layers were present close to 510 km altitude in HASI and 2003 occultation data, and at 425 and 455 km in 1989 occultation light curves. Vertical wavelengths were on the order of 100 km.

8.8 The Weather Near the Surface

Weather phenomena, in the form of precipitation and storm activity, appear to be plentiful on Titan, further reinforcing the terrestrial analogy. While direct observations of weather require the kind of long-term, close-up observations that are not

possible with any mission so far flown to Titan, the indirect inferences are convincing and strongly suggest the sort of dynamic environment that would amply repay
the deployment of long-lived surface or floating stations on Titan in the near future
(see Chapter 11).

There seem to be three kinds of precipitation. The first is the slow drizzle of
hydrocarbons, possibly including heavy, relatively complex molecules, from the
planet-wide haze layers. It is also possible that there is a near-global layer of methane
ice cirrus, below the hydrocarbon haze layers. The profiles measured by Huygens
show that the methane relative humidity, with respect to pure methane condensation,
is around 80% at altitudes from 8–16 km, but 100% between 20 and 30 km. These
conditions correspond to the condensation of methane ice over the higher range,
a condition that may apply over much of Titan. The growth of the crystals would
eventually lead to some of them precipitating out, in which case (as on Earth)
heavy ice particles falling through the relatively warm lower atmosphere would melt
before reaching the surface. The amounts are likely to be such that this amounts to
a moderate drizzle rather than serious rainfall.

The 80% relative humidity measurement over the range from 16 to 20 km has
led to the suggestion by T. Tokano and colleagues that a liquid solution of nitrogen
dissolved in methane may condense out in this vertical range, saturating the atmosphere through which Huygens passed with methane–nitrogen liquid condensate.
The resulting tenuous clouds may also produce a planet-wide drizzle.

In this lower condensation region, the atmosphere is much more likely to be
unstable with regard to convection, especially under conditions of strong heating.
The rising air can then bring large amounts of moisture up to form dense clouds.
Cassini observations have confirmed inferences from Earth-based observations that
these clouds occur, and that they are locally thick but patchy in coverage, occurring
mainly in middle and high southern latitudes where it was summer at the time. It
seems that locally high surface temperatures, combined with a source of methane
on or in the surface, leads to the rapid growth of as convective storm analogous to
thunderstorms on the Earth, and accompanied by the same kind of torrential rain.
The cloudy, stormy region will migrate from one hemisphere to another with the
seasons, producing local flooding for a short time each year. This inferred behaviour
has led to references to 'methane monsoons' on Titan, with the fact well noted
that this sort of behaviour explains why the landscape viewed by Huygens features
river valleys and other features apparently produced by copious amounts of running
liquid, although none could be seen at the time.

The anatomy of methane storms on Titan was explored by R. Hueso and
A. Sanchez-Lavega in 2006 using a mesoscale model with a cloud lifetime of hours
and spatial scales of 10 m, and a detailed microphysical treatment of methane droplet
growth. As on Earth, the presence of particles to act as cloud condensation nuclei is
essential for cloud formation and growth, via the processes of condensation, evaporation and precipitation as well as coalescence of particles of different sizes falling
at different terminal speeds. The nuclei on Titan are probably the drizzle from the

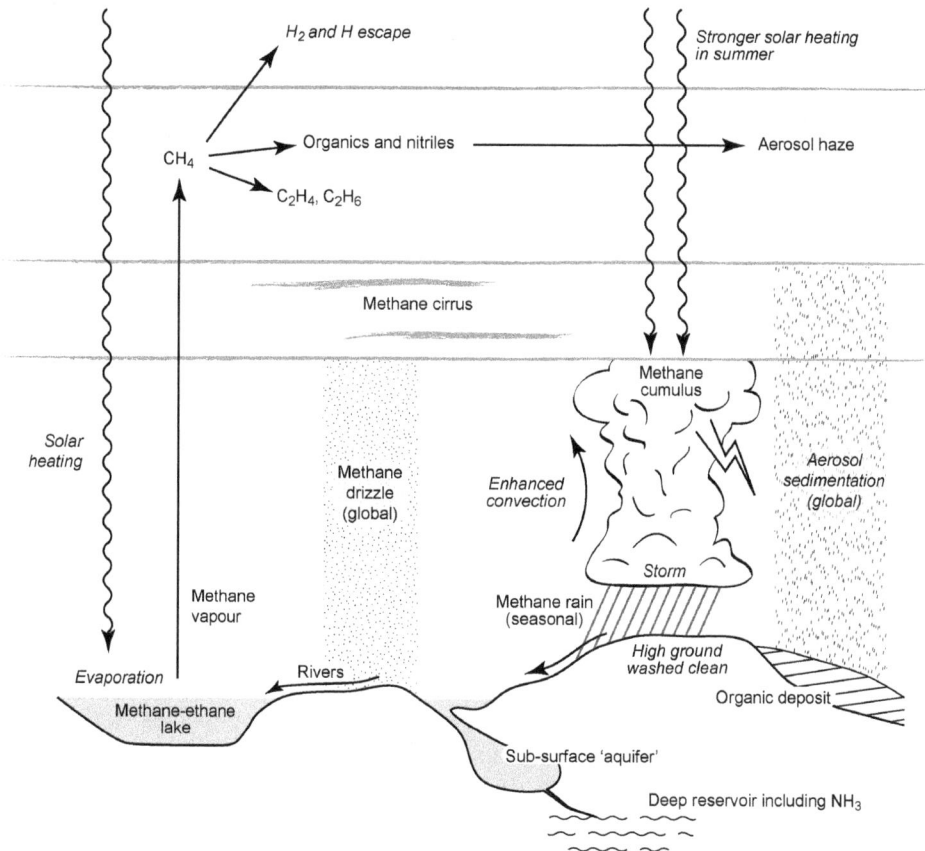

Figure 8.11   Some of the key features of Titan's meteorology that are discussed in the text are illustrated in this cartoon.

overlying organic haze. Problems are encountered in formulating a model treatment of the antifreeze effect of nitrogen dissolved into methane, which as discussed above probably results in liquid methane droplet formation below an altitude of 15 km, and of the ice microphysics of methane that dominates at higher altitudes, both processes that remain poorly known.

Despite these and other uncertainties, the results of the model show that severe convective storms accompanied by heavy precipitation of methane can occur in Titan under the right conditions. For a relative humidity of 90%, a small temperature perturbation of 0.5 K was found to trigger vertical motions with velocities of up to 18 ms^{-1} and rapid particle growth. The cloud tops reached altitudes of 30 km before dissipating in 5–8 hours as rain. The fall of drops 1–5 mm in radius produced rainfalls on the surface as high as 110 kgm^{-2}, comparable to flash flood events on Earth.

During the seasons when methane clouds and rain are absent, the estimated production rates for droplets in the ubiquitous haze layers suggest that the precipitation

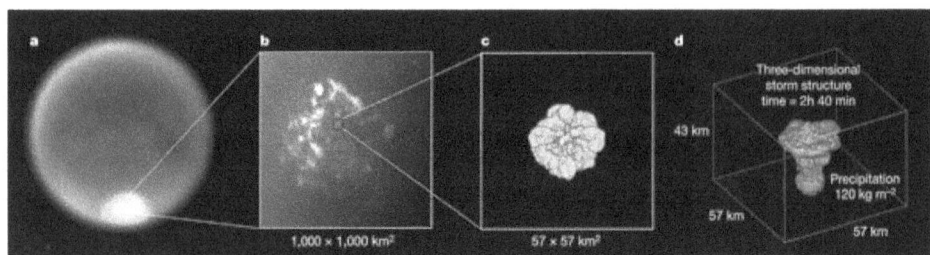

Figure 8.12   Storms on Titan, as seen by Cassini on the left and as modelled by by R. Hueso and A. Sanchez–Lavega (2006) on the right.

of methane and higher hydrocarbons is quasi-continuous on a global scale. If particles are created continuously by photochemistry near 400 km and by the freezing of atmospheric methane near 25 km, then they should coalesce into droplets of various products, ethane and higher hydrocarbons and nitriles, and rain out. This then accumulates on the surface to form lakes, which apparently either re-evaporate or drain into underground 'aquifers'. These subsequently provide a source of fresh methane, to buffer the supply in the atmosphere. The more complex organic molecules remain condensed and are frozen or bound into the crust. These accumulations, probably oily or tarry in nature, are darker in colour than water ice, particularly as exposure to solar photons continues to convert them to more complex species in the solid or liquid state. Thus, the darkest regions on Titan's surface may be those with the largest accumulations. The light regions, on the other hand, might be the higher ground from which the dark organics have been washed down into the valleys, lakes and seas by the seasonal methane downpour.

This cannot be a closed system, obviously, as methane is continuously used up, to produce first the hydrocarbon haze and then the surface deposits. Some of the latter, including methane itself, can evaporate and return to the atmosphere, but the denser materials are removed permanently. Calculations show that the amount of methane in the atmosphere at present should be removed after a lifetime of the order of 100,000 years, very short compared to the age of the Solar System. This is why, during the years before anyone knew what lay beneath Titan's shroud of haze, planetary scientists leaned towards the concept of a liquid methane ocean, with an admixture of higher hydrocarbons such as C_2H_6, which continually replenished the atmosphere. This, too, would be depleted eventually, of course, but it appears that has not happened yet. Unfortunately for that convenient theory, ground-based and Cassini radar measurements and images have ruled this out by showing most of Titan's surface to be solid (see Chapter 9). We are left to assume that the methane supply must be mostly below ground, soaked into a porous regolith, which may have large 'aquifers' in contact with the atmosphere through cracks or channels in the icy crust.

In 2006, R. Lorenz summed up the weather on Titan, thus: long, dry winters, with short, stormy summers featuring torrential downpours. Boating is possible for a time after these, but the surface accumulations of dirty liquid methane soon drain below ground or evaporate. During the storms, winds are likely to be fierce, but most of the time they are quite gentle, at least at the surface where the atmosphere is dense and the horizontal temperature contrasts are small, if not in the upper atmosphere. A global surface wind cannot be absent, however, because drag between the air movement and the surface is an important part of the momentum balance equation for the circulation, and required by all of the atmospheric models. Values of the order of a few metres per second seem most likely, with the variability unknown. Indeed, we witness the manifestation of this effect in the dunes observed on Titan's surface, which may be due to the winds caused by Saturn's tides.

8.9 Does Lightning Occur on Titan?

Does lightning occur in all substantial, cloudy planetary atmospheres, and in particular, is the phenomenon present on Titan? This is an interesting question, not just in terms of understanding the processes involved, but because lightning can be involved in the production of certain chemical species, possibly including some of those in the chain which led to the genesis of life on Earth.

Of the seven main candidates, Venus, Jupiter, Saturn, Uranus, Neptune, Earth, Mars, and Titan, lightning has probably been detected in six, although, besides Earth, optical evidence has been found only on Jupiter. Most of the evidence comes from radio or plasma wave evidence recorded by Voyager 1 and 2 and the Galileo flyby of Venus. The high-pitched radio waves known as 'whistlers', which (on Earth at least) are produced by lightning flashes, have been reported on Jupiter, Neptune and — perhaps — on Venus.

Voyager did not detect any lightning on Titan. The PRA instrument set an upper limit on the total energy per flash in Titan lightning of about 10^5 J (or 10^4 times weaker than terrestrial), and a flash rate of <266 flashes per second (3 times that of Earth). This left 3 possibilities: lightning is not present on Titan at all; lightning exists on Titan but could not be detected due to its low intensity; or lightning occurs only very infrequently and was not observed during the brief (1 hour) flyby. To see which of these is the most likely, we need to examine the lightning production scenario and its prerequisites.

Again we rely mainly on experience gained on Earth to try to picture what may be happening on Titan. Lightning is a pulse of high current going through an electrified channel (called a "stepped-leader"), connecting two regions already electrically charged, cloud to cloud or ground to cloud. The charging mechanisms (first microscopic, then followed by a macroscopic separation of the charges) are not perfectly understood, even on Earth.

The positive charges rise, while the negative ones descend. The mobility of positive and negative charges is, in general, different and therefore, due to convection and gravitation, this large-scale movement results in a charge separation by a distance that can be up to the order of one atmospheric scale height (16 km on Titan). There results an electric field and a discharge when the electric field induced by the separate charge exceeds the breakdown level of the atmosphere.

The discharge produces heating of the environment of the channel up to 30,000 K and hence production of visible light, expansion of the heated region at supersonic speed, producing a shock wave (thunder), and electromagnetic emission whose spectrum shows a peak between a few or a few hundreds of kHz followed by a decrease in the flux proportional to 1/frequency (or sometimes 1/frequency^2). The electromagnetic phenomenon lasts typically (within a factor of 2) about 100 ms and the associated mechanical phenomena decrease very rapidly (roughly exponentially) with time and distance. The peak frequency and the intensity of the radio emission depend on various atmospheric parameters; on Earth, the peak is situated at about 100 kHz.

The production of lightning requires the presence in the atmosphere of at least one species suitable for electrification, i.e. with high dielectric constant and easily polarisable. Such materials are to be found among the asymmetric molecules such as H_2O, H_2SO_4, and NH_3. The necessary electrification of the atmosphere may be due to the presence of (in decreasing order of efficiency): ionised molecules or aggregates, charged aerosols, ice particles, charged raindrops, crystals and snowflakes, or rain and hail. The energy for the large-scale movement leading to the charge separation can only exist if convection is efficient enough, that is if the solar input is high, or if there is an internal source of energy as on Jupiter and Saturn. Thus, lightning on Earth is more frequent when and where high temperatures occur and where therefore convection is maximum: equatorial and temperate zones, continents and mid-day hours.

The major components on Titan, molecular nitrogen, methane, molecular hydrogen, carbon monoxide and hydrogen cyanide all tend to have low dielectric constants, while species like NH_3 and H_2O that are more liable to electrification have very low abundances or do not exist. Electrification on the microscopic scale is possible, however. The existence of an ionosphere with a maximum electron concentration of about 3000 cm^{-3} ensures the presence of ionised molecules and aggregates, as well as charged aerosols. Many of the species present in Titan's atmosphere condense in its lower part and give rise to ice particles, crystals, droplets, etc. Methane drops (supposing that condensed CH_4 is the main cloud material on Titan), can act as charge separators as they fall down in the atmosphere, although calculations suggest that the maximum charge on a methane drop is around ten times less than that on a typical water drop on Earth. Convective clouds are also more common on Earth than on Titan, where they have so far only been observed near the south pole during the summer season, that is, when the Sun is overhead and heating is a maximum, suggesting any they may be confined to the most optimum conditions.

The chances of detecting the electrical activity that is characteristic of lightning were always higher for Huygens, during the probe descent, than for Cassini, looking down from orbit. Even if the flashes occurred only in the convective region near the south pole, the low-frequency electromagnetic waves it produces could easily propagate from the south pole to the Huygens entry location and be detected there by the Permittivity, Wave and Altimetry package. PWA measured the electrical state of the atmosphere below 140 km, and found that the conductivity peaks at ~60 km. The electric field due to natural wave emissions was measured during the descent in two frequency ranges, 0–11.5 kHz and 0–100 Hz, specifically to investigate in situ lightning and related phenomena like coronal discharges that would produce electromagnetic waves, excite global and local resonance phenomena in the surface–ionospheric cavity and could drive a global electric circuit. Several impulsive electrical events similar to terrestrial 'sferics' were observed during the descent, and narrow-band wave emission, reminiscent of a possible resonance generated by lightning, was seen near 36 Hz.

Huygens had a relatively brief observing opportunity compared to Cassini, and determining the frequency and global distribution of lightning remains important once it has been detected. The RPWS on the orbiter has a special millisecond mode for the detection of the short radio impulses from lightning flashes, although the detection of the lower frequency emissions is inhibited by the shielding effect of Titan's ionosphere. Optical detection of lightning on Titan by Cassini's camera is still a possibility, although the overlying atmospheric opacity remains a problem.

...A reddish colour dominated everything, although swathes of darker, older material streaked the landscape. Towards the horizon, beyond the slushy plain below, there were rolling hills with peaks stained red and yellow, with slashes of ochre on their flanks. But they were mountains of ice, not rock....

Stephen Baxter, *Titan*

9.1 Introduction

The information we have about the composition of the planetary and satellite surfaces in the outer Solar System has largely come from observations in the visible and near infrared of the reflectance spectra of these bodies, by comparing them with laboratory spectra of possible analogue material. The characteristic water ice absorptions at 2.0, 1.5 and 1.25 μm are observed in the spectra of most of these satellites, with significant band depths. The three largest Galilean satellites of Jupiter were found in this way to have abundant H_2O ice on their surfaces, with Europa having the purest icy surface and Callisto the 'dirtiest' one. The satellites of Saturn (Enceladus, Dione, Rhea and Tethys) have also H_2O icy surfaces, like those of Europa or Ganymede but with finer ice grains. Iapetus and Hyperion are different, the former exhibiting two different hemispheres, one bright, one dark, while the latter shows an H_2O ice surface contaminated with a reddish-coloured material. The water ice bands are stronger on the trailing sides of Dione, Rhea and Tethys than on their leading sides, and something similar is observed on the Galilean satellites.

Titan's surface however remains much of a mystery still to this day. The cloud cover meant it was not directly observable by remote sensing in the visible, and astronomers have only recently (since 1990 or so) become aware of techniques that allow them to recover information on this hidden ground. As this information slowly came in and was analysed (see Chapter 3), it tended to produce more bold speculation than scientific models. The Cassini–Huygens mission has successfully penetrated Titan's atmosphere from Saturn orbit since 2004, and all the way down to the surface with the Huygens probe in 2005. Since then, Titan's surface has been exposed to the "eyes" of radar, imagers and spectrometers. Now we have a better picture of the surface morphology, of the craters, dunes, lakes and pebbles that shape the ground, and we know something of the material that lies on the surface. Yet we still have only a very incomplete picture of the ground composition and of the interior structure. Cassini is likely to operate for several more years and cover an increasingly large fraction of Titan's surface; as these data are analysed, they will gradually reign in the 'runaway imagination' effect.

9.2 Remote Sensing of the Surface

Most pre-Cassini models suggested that the surface of Titan should be coated with atmospheric debris and condensates accumulated over the ages. The idea of an extensive ocean covering the ground, like that suggested by Jonathan Lunine in 1983, was an attractive and elegant concept that remained in vogue for a long time. This ethane-methane ocean could serve as both the source of methane and the sink for photolysis. At the same time, much of the outer part of the solid body of the satellite should, to be consistent with the observed mean density, consist of a thick layer of water ice.

Ground-based and Earth-orbital spectroscopy and imaging have indicated since the beginning of the 1990s that the hydrocarbon ocean is well hidden, if it is present at all, and that the ice could be at least partially exposed on Titan's surface. Even a shallow global ocean was shown to be inconsistent with the constraints imposed by Titan's orbital characteristics: the tidal action on an ocean less than 100 meters deep would have dissipated Titan's eccentricity of 0.03 (where 0 is circular and 1 is parabolic) long ago. Of course, this assumes that the eccentricity originated early in Titan's history, and was not introduced recently, for example by a large impact. It is also dependent to some extent on assumptions about Titan's interior and on whether or not the ocean is confined in basins; given the right conditions, the tidal argument could be overcome, but the sum of information pointed towards the absence of a global ocean.

Also, the first remote sensing technique to be used for sounding Titan's surface, Earth-based radar, indicated that the surface could be non-uniform and mostly solid with small lakes, if any. Indeed, the radar echos obtained by Muhleman and colleagues in 1990 using the National Radio Astronomy Observatory's Very Large Array in New Mexico as a receiver of the signal transmitted to Titan by the NASA Goldstone radio telescope in California, were among the first evidence against the global ocean model of the surface. Then, in 2003, radar measurements from the Arecibo Observatory in Puerto Rico revealed a specular component from 75% of the regions observed, globally distributed in longitude at about 26°S. These were interpreted as indicative of extensive regions of dark, liquid hydrocarbon on Titan's surface. However, the idea of a widespread surface liquid was challenged again in even more recent observations from the ground that failed to find any such signatures and proposed instead that very flat solid surfaces could be causing the radar evidence.

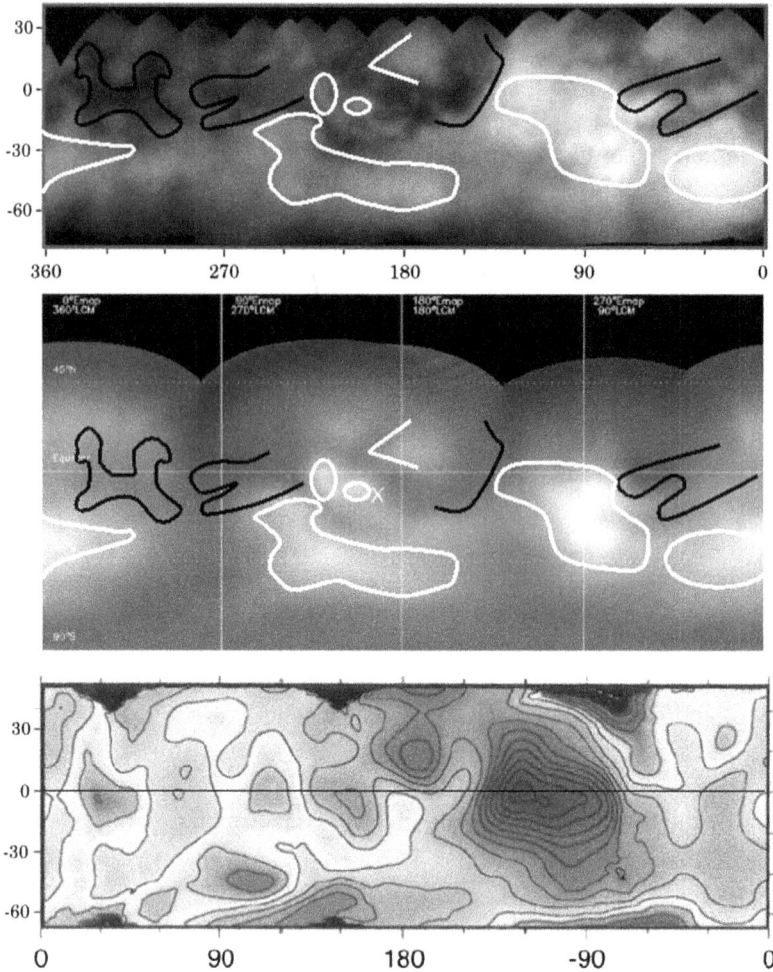

Figure 9.1   Three maps of Titan's surface taken with the Cassini/ISS at $0.94\,\mu$m (top); the adaptive optics system NAOS at the VLT at $1.28\,\mu$m (middle) and the HST NICMOS at $1.6\,\mu$m (bottom), showing coherent surface features. The bright areas Xanadu Regio is observed near $110°$ LCM; the Huygens landing site is near $192°$LCM and $10°$S (Porco *et al.*, 2005; Coustenis *et al.*, 2005; Meier *et al.*, 2000).

Only relatively recently has it become certain that Titan's surface is not homogeneous. The first spectra taken inside the methane windows showed a strong variation over Titan's surface as a function of longitude, with one side brighter than the other. Follow-up exploration of Titan's surface with ground-based and HST images show it to be covered with bright regions separated by darker areas, probably due more to the presence of surface materials with different albedos rather than to topography, since Titan's icy bulk cannot support very high structures. A possible elevation of about 3 km has been deduced from models as the theoretical maximum, while only about 1-km-high mountains have been found so far in Cassini

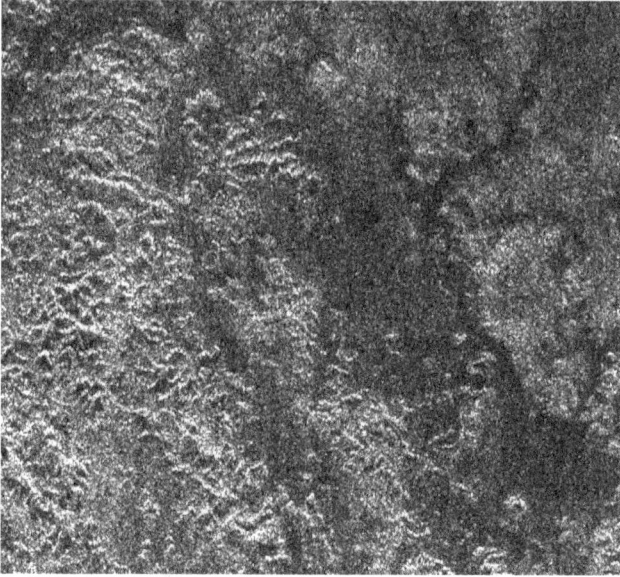

Figure 9.2   The mountains of Xanadu, seen in detail by the Cassini radar in April 2006. They are characterized by a rough and chaotic appearance, lacking any clear tectonic pattern, and are dissected by narrow channels that seem to have been caused by liquid erosion (NASA/JPL/SSI).

observations. The idea that the albedo differences are connected with these modest rises and falls remains persuasive, however: the icy hills might be washed clean by methane rain, for instance, while the dark sediments collect in the shallow valleys below.

The ISS and VIMS cameras on Cassini generally confirm the ground-based adaptive optics and the HST first results, finding dramatic changes in surface albedo within the same regions. The resolution achieved by ISS and VIMS is around a kilometre on Titan's surface, while from the Earth, the best one could hope for was a spatial resolution of a few hundred kilometres, so the spacecraft images give additional detail of the borders between different regions. Among other features, the large bright area around the equator first observed by the HST and using adaptive optics in 1994 is resolved and observed in detail by Cassini instruments, including the radar.

9.3 Huygens Takes a Plunge

While the Cassini orbiter has provided detailed views of Titan's surface in the visible and in the near-infrared with its camera, mapping spectrometer and radar, the Huygens probe, descending through the atmosphere on January 14, 2005, returned the most extraordinarily detailed images and can claim to have landed upon the farthest location a human-made vessel has ever achieved and to have seen the local terrain on

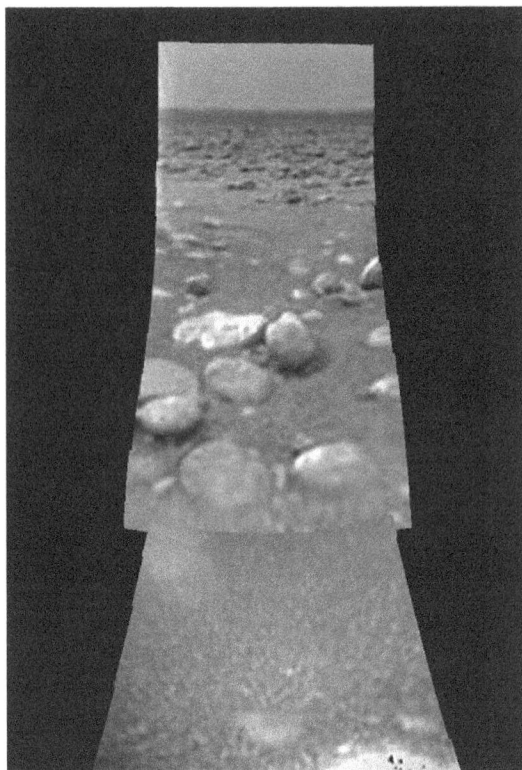

Figure 9.3   Titan's surface after the landing of the Huygens probe. The 'rocks' or pebbles are probably made of dirty water ice, and the darker riverbed is thought to be methane-wet icy 'sand' (Tomasko *et al.*, 2005).

this most distant world for the first time. The features it saw on Titan's surface were more complex than anyone expected, offering an exotic view of a land that really does seem to resemble the landscapes on Earth, with mountains, lakes, shorelines and outflow channels, but where methane plays the role water does on Earth. The detection of Argon 40, and observations of icy 'lava' flows from cryo-volcanoes, reveals that the interior of Titan is geologically active, and in fact as complex as any of the terrestrial planets, and (geographically, at least) more similar to our own planet than either Mars or Venus.

While the first Cassini observations of the surface were still being analysed, the breath-taking plunge of the Huygens probe into Titan's atmosphere took place on January 15, 2005. The probe flew over an icy surface, floated down and drifted eastwards for about 160 km, its highly advanced suite of six instruments gathering information about Titan's atmosphere and winds. After a two-and-a-half-hour descent, the ESA spacecraft landed at 10.3°S and 192.3°W among Titanian rocks and mud. The HASI instrument measured the surface temperature and pressure at the landing site: 93.65 ± 0.25 K and $1,467 \pm 1$ atmospheres.

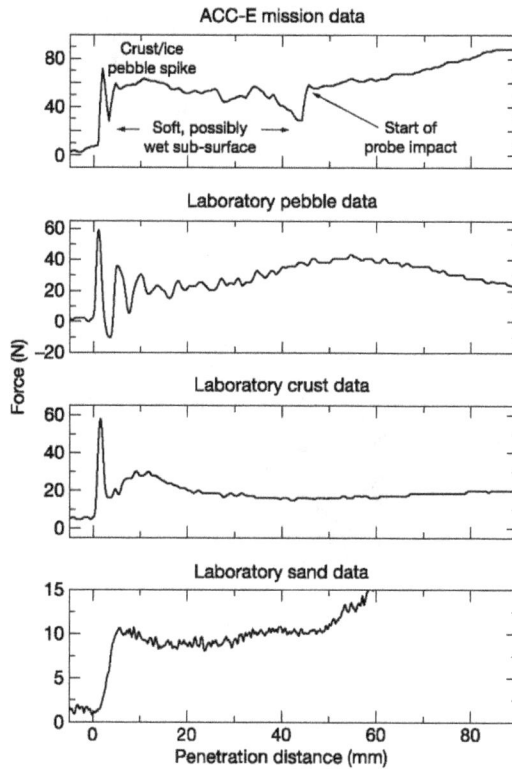

Figure 9.4   A comparison of SSP penetrometer force profiles for Titan and for laboratory analogues. Top to bottom: Titan data, laboratory data for impact onto a pebble, impact onto a surface crust layer, and impact onto sand. The probe protruded to give 55 mm of undisturbed penetration before the main structure contacted the surface (Zarnecki *et al.*, 2005).

The acoustic sounder in the Surface Science package first detected the ground from 88 metres in altitude, its signal revealing a relatively smooth, but not flat surface below. The first part of the probe to touch the surface was the penetrometer, also part of the SSP. With a landing speed of about $5\,\mathrm{m\,s^{-1}}$ the front of the probe followed and hit the surface, then slid slightly before settling. Amazingly, it continued to take measurements for more than 3 hours, providing the "ground truth" for the orbital measurements in terms of composition, structure and geomorphology. The DISR camera took several pictures of its surroundings: a rather Mars-like landscape, complete with a dark, sandy surface strewn with brighter rounded rocks. The largest of these are about 15 cm in diameter, and most likely they consist primarily of water ice, with dissolved impurities and a coating of tholin-type material drizzled onto them from the ubiquitous haze. The underlying material, for which the best current earthly analogue would be gravel, wet sand, wet clay, or lightly packed snow according to the impact data, is probably also mostly composed of dirty water ice, weathered from the rocks and cliffs by the Titanian analogue of such processes on

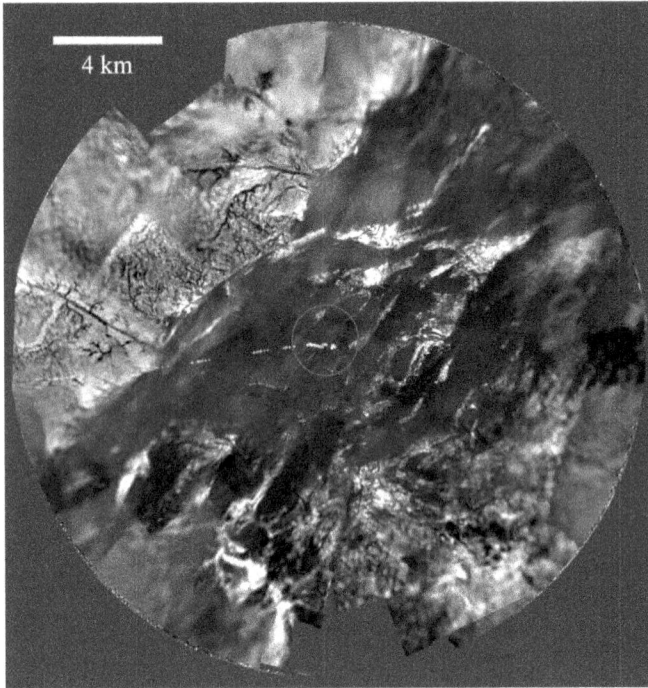

Figure 9.5   Titan's surface as viewed by the Huygens/DISR cameras from a distance of 8 km in altitude (Tomasko *et al.*, 2005).

Earth. The penetrometer data show an initial spike, that suggested at first the soil had a hard crust on top of softer material, but the spike was later interpreted as more likely to indicate that the probe hit one of the icy pebbles littering the landing area before sinking into the softer, darker ground material.

While there were no open bodies of liquid where Huygens landed on Titan's surface, the ground did show evidence of methane 'humidity' since a 40% stepwise increase in methane abundance was measured by the GCMS right after landing. The surface is organic-rich, with trace species such as cyanogens and ethane detected on the ground. A near-by dark area closely resembles a dry lake, possibly like those shown in recent Cassini/ISS images near Titan's north pole, some of which may still contain hydrocarbon liquid. The nature and extent of the exchange of condensable species between the atmosphere and the surface and the dynamic equilibrium that exists between the two is linked to the mystery of the methane liquid reservoir.

In spite of a few misadventures (the loss of the Sun sensor measurements and of about half the images from Channel B, and the unexpectedly erratic motion of the probe), the DISR imager and spectrometer gathered a precious set of data, starting from the first image of the surface from 49 km above, down to the unprecedented overview recorded a few km over the Huygens landing site, and through the lamp-on data recorded below 700 km in altitude to the final picture on the surface. Panoramic mosaics constructed from a set of images taken at different altitudes show brighter regions separated by lanes or lineaments of darker material, interpreted as channels, which range from short stubby features to more complex ones with many branches. This dendritic network apparently was caused by rainfall creating drainage channels, implying a liquid does flow at some times and places on Titan's surface, although none seemed to be present at the time of the landing. The stubby channels are wider and rectilinear, often starting or ending in dark circular areas suggesting dried lakes or pits. No obvious crater features were observed in the landing site district.

Stereoscopic analyses of the DISR images indicate that the bright area cut through with dendritic systems is 50–200 m higher than the large darker plane to the south. If the latter feature is a dried lakebed, it seems too large by Earth standards to have been created by the creeks and channels seen on the images and could be due to larger, unseen rivers or possibly to a catastrophic event in the past. The dark channels could be due to liquid methane irrigating the bright elevated terrains before being carried through the channels to the region offshore in south-easterly flows. This migration towards the lower regions leads to ice being exposed along the upstream faces of the ridges. The slopes are generally on the order of 30°. Some of the bright linear streaks seen on the images could be due to icy flows from the interior of Titan emerging through fissures.

The spectra acquired during the descent gave information on the surface, as well as on the atmospheric properties. Indeed, it was shown from spectral reflectance data of the region seen from the probe that the differences in albedo were related to differences in topography which in turn can be connected to the spectral behaviour of the ground constituents. The higher brighter regions were found to be redder than the lowland lakebeds, as are the regions near the mouths of the rivers. The water ice absorption band at 1.5 μm was observed in spectra taken by DISR, consistent with the earlier findings from ground-based observations. A good match with the DISR spectra also requires, in addition to water ice, the presence of an absorber resembling laboratory tholin, plus an unknown blue material on Titan's surface. No combination of any known ice and organic material has been found that reproduces the characteristics of the blue material, but attempts are being made to identify or synthesize it in the laboratory.

Although many questions still remain about the sequence of flooding and the formation of all the complex structures observed by Huygens, the new data clarify the picture we have of Titan today and at the same time enhance the impression that by studying Saturn's satellite we are looking at an environment with many facets that resemble the Earth. The question that arose in almost all press conferences that the scientists gave was: "what about possible past, present or future life on Titan?" One of the elements in the negative response (at least so far as present or past life is concerned) was provided by the GCMS in the ^{13}C/^{14}C isotopic ratio which showed that no active biota exist on Titan and that the methane has a non-biologic source. No "little orange men" were photographed on Titan, either.

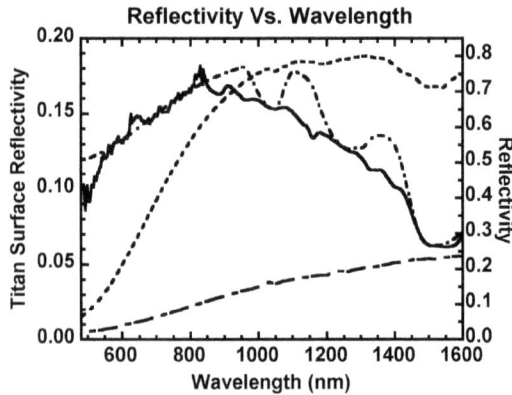

Figure 9.6   Surface reflectivity as measured after landing (solid line). It is compared with a simulation (dash-dotted line) of a mixture of large grained (750 μm) low temperature water ice, yellow tholins, and an unknown component with featureless blue slope between 850 and 1,500 nm. Spectra of two different organic tholins (yellow tholin — dashed line, and dark tholin — dot-long dash line, from Bernard *et al.* (2005)) are also shown for comparison (reflectance scale reduced by a factor of 4). The low reflectivity near 1.5 micron is compatible with the water ice absorption (Tomasko *et al.*, 2005).

9.4 Naming Distant New Places

The combined observations from the orbiting spacecraft and from the probe/lander have revealed many new surface features prominent enough to require, following well-established tradition, names of their own. The most prominent is the bright equatorial region, centred at 10°S and 100°W and now officially named "Xanadu Regio". The mid-latitude regions around the equator on Titan are rather uniformly dark, while the poles are relatively bright. Both exhibit many complex geomorphological features, from dunes to craters, lakes to rings, channels to volcanoes, and simple albedo features — the "bright" and "dark" spots — that have found identities.

The different morphological features on Titan are named after sacred or enchanted places in world mythologies and literature. Tables 9.1 to 9.6 list the names currently attributed to Titan surface features, as decided by the International Astronomical Union.

There is also an arc-shaped feature which has been named "Arcus Hotei" after a Japanese God, and a small plain (lacus) named after Lake Ontario in Canada. Nicknames have been informally used for some of the other features:

- 'The Sickle': a large, dark, sickle-shaped region identified by the Hubble Space Telescope.
- 'Dog and Ball', 'Dragon's Head': large, dark, roughly equatorial regions identified by the European Southern Observatory's Very Large Telescope, named for their distinctive shapes.
- 'Si-Si the Cat': a region that appears dark in radar images, named after a researcher's daughter who said it looked like a cat.

Table 9.1   Albedo features on Titan.

Bright albedo feature	Named after mythological sites
Adiri	Adiri, Melanesian paradise
Dilmun	Dilmun, Sumerian heaven
Quivira	Legendary city in southwestern America
Tsegihi	Navajo sacred place
Xanadu	An imaginary country in Coleridge's Kubla Khan

Dark albedo feature	Named after mythological sites
Aaru	Aaru, Egyptian paradise
Aztlan	Aztlán, mythical Aztec homeland
Belet	Malayan paradise
Ching-tu	Chinese Buddhist paradise
Fensal	Norse heavenly mansion
Mezzoramia	Italian legend of oasis of happiness in Africa
Senkyo	Japanese paradise
Shangri-la	Tibetan paradise

Table 9.2   Bright spots on Titan.

Facula (bright spots)	Named after archipelagi on Earth
Antilia Faculae	Antillia, mythical Atlantic archipelago
Bazaruto Facula	Bazaruto, Mozambique island
Coats Facula	Coats Island, Canada
Crete Facula	Crete, Greek island
Elba Facula	Elba, Italian island
Kerguelen Facula	Kerguelen Islands, French sub-Antarctic island
Mindanao Facula	Mindanao, Philippine island
Nicobar Faculae	Nicobar Islands, Indian archipelago
Oahu Facula	Oahu, Hawaiian island
Santorini Facula	Santorini, Greek island
Shikoku Facula	Shikoku, Japanese island
Texel Facula	Texel, Dutch island
Tortola Facula or "The snail"	Tortola, British Virgin Islands
Vis Facula	Vis, Croatian island

Not everything on Titan's complex surface has found already a name, but the idea is for yet other features to be named after deities of happiness, peace, and harmony in world cultures. The harmonious collaboration among Cassini–Huygens investigators makes spreading a message of such virtues to the world very appropriate.

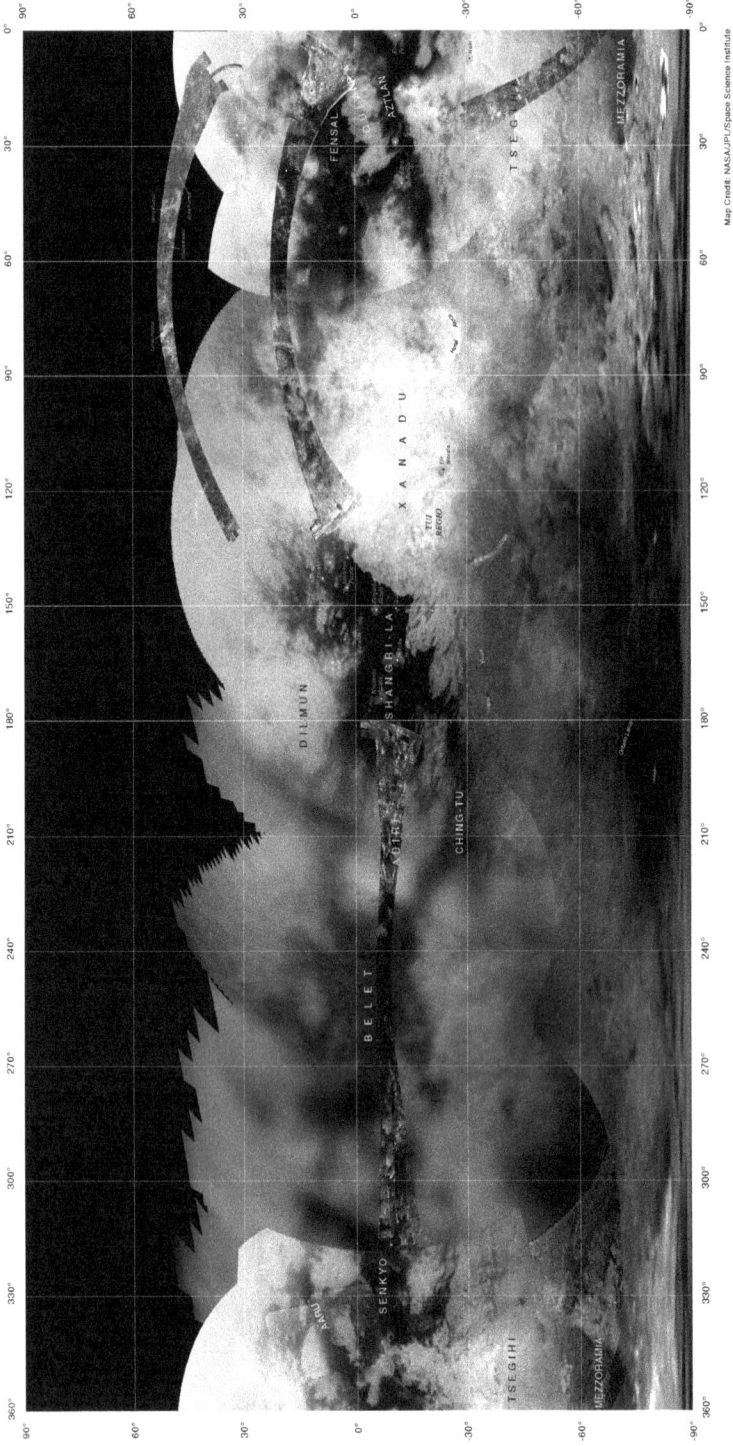

Figure 9.7   Names of different locations on Titan on a combined ISS-RADAR map (NASA/JPL).

Table 9.3   Dark spots on Titan.

Macula (dark spots)	Named after deities of happiness, peace and harmony
Eir Macula	Norse goddess
Elpis Macula	Greek god
Ganesa Macula	Hindu god
Omacatl Macula	Aztec god

Table 9.4   Ring features on Titan.

Ring feature	Named after deities of wisdom
Guabonito	Taíno sea goddess
Nath	Irish goddess of wisdom
Veles	Slavic god

Table 9.5   Streaks of colour regions on Titan.

Virga (streaks of colour)	Named after rain gods
Hobal Virga	Hobal, Arabian rain god
Kalseru Virga	Kalseru, Australian Aborigine rain god
Shiwanni Virgae	Shiwanni, Zuni rain god

Table 9.6   Craters on Titan.

Crater	Named after deities of wisdom
Mernva or 'Circus Maximus'	Mernva, Etruscan goddess
Sinlap	Sinlap, Kachin spirit

9.5 Evidence for Geological Activity

9.5.1 *Albedo Variations*

The most accepted cause for the darker regions remains the accumulation of hydro-carbons (in liquid or solid form) that have precipitated down from the atmosphere. For the brighter regions, the task of interpreting the data is more difficult. The idea that they could be associated with topography and more exposed ice tends to be in agreement with Huygens/DISR stereoscopic imaging reports that the brighter terrain is more elevated than the darker, smoother and lower-ice regions. The exact type of ice that can satisfy the constraints imposed by all the observations is not easy to determine. Besides water ice, hydrocarbon ice has been invoked on the basis of Xanadu appearing bright at all the near-infrared wavelengths observed to date.

9.5.2 *Craters*

The Cassini instruments have found little evidence for heavy cratering on either the bright or the dark areas of Titan so far, indicating that the surface of Titan is either young (which in planetary geology means completely resurfaced in less than a billion years) or highly eroded and modified.

A few features interpreted as impact craters have been spotted (Table 9.6), most notably the 440-km diameter crater named Circus Maximus seen by Cassini's Radar and VIMS during two separate flybys in early 2005. The incoming comet or asteroid that produced such a large crater must have been at least 10 km or more across, and have arrived fairly recently, since there must have been many other impacts whose record has been obliterated by Titan's active environment. The albedo patterns indicate that the terrain inside the crater is rough, with the brightest parts

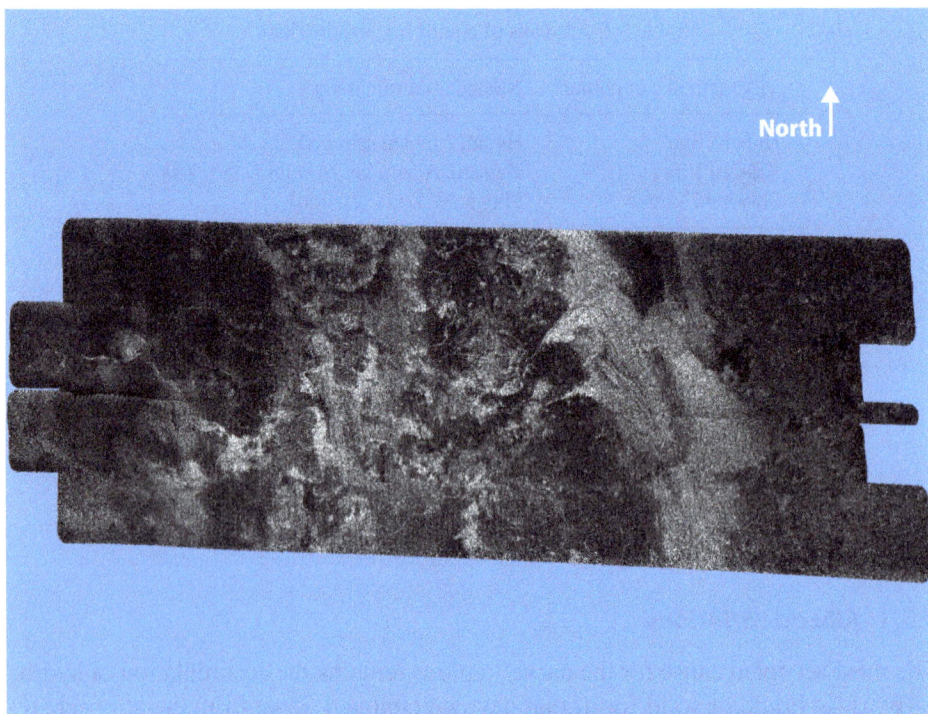

Figure 9.8   Cassini RADAR image of an area in the northern hemisphere of Titan, showing highly contrasted terrain with a variety of geological features, including the large crater-like 'Circus Maximus' to the upper left. The image is about 150 km wide and 250 km long, and is centred at 50 N, 82 W, a region that has not yet been imaged optically. The smallest details seen on the image are about 300 m across (NASA/JPL).

tilted towards the radar during the observations. There is also evidence for different material on the crater floor from that in the ejecta, the latter presumably excavated from deeper strata. A smaller crater of about 40 km has also been observed, this time exhibiting a parabolic-shaped ejecta blanket suggesting an oblique collision.

Nevertheless, craters identified by the RADAR, VIMS or the ISS are rare. This may mean that the surface of Titan is young (less than a billion years) or highly eroded/modified by a combination of slow, plastic flow of the surface material, presumably mainly ice; erosion by wind and rain, and resurfacing by fresh material from precipitation or from volcanic venting. Titan may be uniquely efficient at healing scars caused by impactors, plus of course the atmosphere acts as a partial shield against all but the largest incoming objects, an advantage not enjoyed by any other icy satellite. The main benefit, so far as protecting the surface from massive impact features is concerned, may be the effect of the atmosphere in breaking up large meteors into smaller pieces, especially if they are loosely aggregated to start with, or held together by ice, like many comet nuclei. Saturn's gravity can also disrupt comets by tidal forces, as Shoemaker-Levy-9 was by Jupiter, to form clusters or chains of smaller pieces. Strings of craters on some of the airless icy satellites were probably produced by collisions with chains of bodies produced in this way.

The readiness with which craters on Titan are erased or filled in after they form depends to some extent on how large they are, of course; small, shallow craters are more easily obliterated. The crater density at small sizes is also sensitive to Titan's climate history — if Titan had a thinner atmosphere in the past, more small impactors may have been able to reach the surface. Impact basins, even if not formed on liquid-covered surfaces, may be subdued due to the effects of viscous relaxation. Although Titan's present surface temperature is low enough that water ice is very hard, the possible presence of ammonia hydrates and other impurities in the crust may make the ice more mobile. Also, it could be considerably warmer at a depth of a few kilometres, due to geothermal gradients. The solubility of ice in hydrocarbons is smaller than that of most rocks in water. Thus, except where the surface is more susceptible to erosion, due to organic deposits or perhaps water-ammonia ice, Titan's topography should not be significantly modified by erosion, or rain.

The effect of a warmer interior, as observed on other icy satellites, is to cause the floors of large craters to dome upwards. While small, bowl-shaped craters may be filled with rainfall to form simple lakes, craters larger than a few tens of kilometres in diameter may have an island in a ring-shaped lake where the centre of the dome surfaces. In time, tectonic activity could distort these and change their shape, also they could breach or overfill, bursting their banks and becoming the sources of rivers and other fluvial features. Other lakes could be on the receiving end of these flows, perhaps forming cascades in which the liquid eventually flows underground.

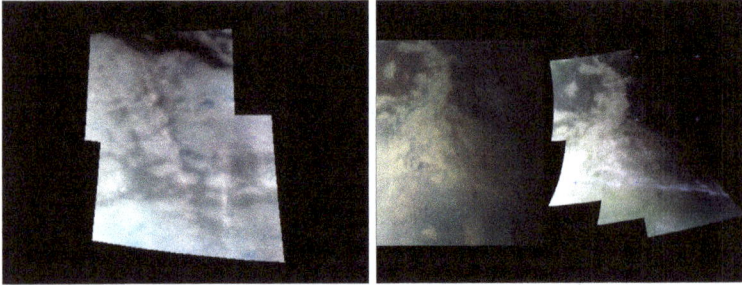

Figure 9.9   Left: A massive mountain range, observed by Cassini/VIMS on December 12, 2006, nearly 100 miles long and located just south of Titan's equator. Right: This VIMS image set was taken at a distance of 15,000 km from Titan and shows two views of a mountainous area with, near the bottom of the right image, a band of bright methane clouds (NASA/JPL).

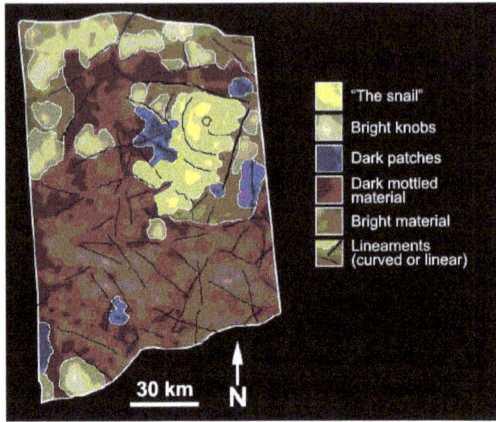

Figure 9.10   A circular feature from which bright flow patterns emanate was interpreted as a cryovolcano, in which icy 'lava' has been forced to the surface by expanding gases, by Cassini/VIMS team members (C. Sotin *et al.*, 2005). From infrared images a geological map was obtained, shown here, and indicating that the circular feature shows several episodes of activity on the volcano (NASA/JPL/University of Arizona).

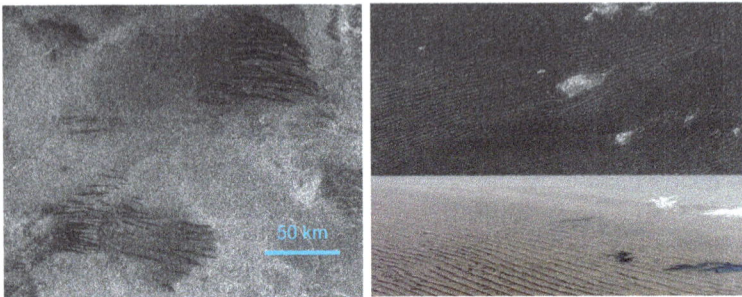

Figure 9.11   Left: Longitudinal aeolian (wind-formed) features seen in a Cassini radar image ~250 km wide, centred near 20°N and 95°W. Such 'dunes' occur as vast sand seas in the equatorial region (Lorenz *et al.*, 2006). Right: sand dunes on Titan compared to Namibia on Earth. The bright features in the upper radar photo are not clouds but topographic features among the dunes (NASA/JPL).

9.5.3 *Mountains and Volcanoes*

During several flybys designed to obtain the highest resolution infrared views of Titan yet, Cassini resolved surface features as small as 400 m (1,300 feet). The images revealed a large mountain range, about 150 km long (93 miles), 30 km (19 miles) wide and about 1.5 km (nearly a mile) high. Deposits of bright, white substance, which may be methane "snow", or alternatively exposures of some underlying material, lie at the top of the mountain ridges. Water ice, as hard as rock at Titan temperatures, is probably the main mountain-building material, coated with different layers of organics falling out of the atmosphere as rain, dust, or smog onto the valley floors and mountain tops, which are coated with dark spots that appear to be brushed, washed, scoured and moved around the surface. The mountains probably formed when material welled up from below to fill the gaps opened when tectonic plates pull apart, similar to the way mid-ocean ridges are formed on Earth.

Separately, the radar and infrared data are difficult to interpret, but together they are a powerful combination. In the infrared images, one can see the shadows of the mountains, and in radar, one can see their shape. But when combined, scientists begin to see variations on the mountains, which is essential to unravelling the mysteries of the geologic processes on Titan. Near the wrinkled, mountainous terrain are clouds in Titan's southern mid latitudes whose source continues to elude scientists. These clouds are probably methane droplets that may form when the atmosphere on Titan cools as it is pushed over the mountains by winds.

If there is liquid below the surface and an internal source of heat, Titan may have or have had cryovolcanoes at some time in the satellite's thermal history. If a liquid — methane for example — is close to its boiling point, perhaps as a result of the higher pressure at some depth in the crust, then a relatively small amount of heat may vaporise it and increase the pressure dramatically. If the vapour can escape through cracks to the surface, then it may expel a mixture of other species (including water, perhaps) in liquid and gaseous form. Thus, methane and water might have similar roles on Titan to those of water and lava in terrestrial volcanoes.

And indeed, Titan may well exhibit cryovolcanic activity, in which gases and solid material are expelled, possibly explosively at times, through vents on the surface. The driving force in the 'cryo' form of volcanism is likely to be the vaporisation of methane or ammonia ices, and the 'lava' ammonia-rich water that soon freezes on the surface. A bright circular structure (about 30 km in diameter) found in the VIMS spectral images has been interpreted by C. Sotin and colleagues as the first discovery of a cryovolcanic dome in an area of Titan dominated by extension (stretching of the surface due to some underground upheaval, possibly a convection plume). To explain the cryovolcano, the VIMS team further hypothesized that the dry channels observed on Titan are related to upwelling "hot ice" and contaminated by hydrocarbons that vaporize as they get close to the surface (to account for the methane gas in the atmosphere), mechanisms similar to those operating for silicate volcanism on Earth (using tidal heating as an energy source) may lead to flows of non-H_2O ices on

Titan's surface. Following such eruptions, methane rain could produce the dendritic dark structures seen by Cassini–Huygens. If these structures are indeed channels, they could have dried out due to the short timescale for methane dissociation in the atmosphere.

However, despite the image which this conjures of spectacular eruptions on Titan, it is more likely that the water-ammonia liquid seeps gently to the surface and freezes to form a lake or a flat dome, rather like the corresponding features found on Venus. The solubility of the most likely volatile, methane, is small in water or water-ammonia solutions, so the gas supply released from the 'magma' column as it rises to smaller pressures will be too small to accelerate and disrupt the flow. Also, the surface pressure on Titan is high, compared to Triton, for example, where cryovolcanism fuelled by boiling liquid nitrogen expels the vapour into a near-vacuum. Finally, the analogy with Europa suggests that the cracks through which the liquid magma reaches the surface are more likely to be large than small, possibly because small cracks would freeze up and seal before much liquid could pass through. Deep cracks tens of kilometres long and hundreds of metres wide, on the other hand, could transfer hot water to the surface with relatively little loss at a plausible rate of flow. Thus we expected even before Cassini to see frozen flood plains and shallow shield volcanoes on Titan, rather than tall, conical ash piles like the Earth.

Studying volcanism on Titan is important, not only to understand the thermal history of Titan (which since it differs in its incorporation of volatiles from the Galilean satellites, must have evolved differently) but also how volatiles — in particular, methane — are delivered to the surface.

9.5.4 Dunes

Cassini scientists soon noticed that many of the dark patches on Titan's surface had a particular appearance that initially led to calling them "tiger scratches" or "cat scratches". These consist of linear dark features like those visible across a large part of the desert to the west of the large crater Circus Maximus, which were the first examples detected. Their similarity to dunes as observed on Earth led some scientists to hypothesize that they could be due to fine-grained material, perhaps tiny grains of ice or hydrocarbons, deposited by fluid flows across the surface. Soon after, the Cassini radar found vast swaths of longitudinal dunes on Titan sculpted like Namibian sand dunes on Earth. These phenomena are one of the few features that are discernible in images returned by all three instruments that scrutinise Titan's surface: VIMS and RADAR on the orbiter, and DISR on the probe.

The formation of dunes on Earth and Mars is well understood today, but the process on Titan must be very different. On Earth or Mars, heat from sunlight drives the winds that form sand dunes. On Titan, sunlight is too weak to do the job. Instead, the tides due to Saturn's gravity push and pull on Titan's dense atmosphere as the moon traces an elliptical orbit around the planet. The tidal forces on the atmosphere are 400 times stronger on Titan than on Earth, where lunar gravity does have an effect on the winds, but one that is far less than the thermally driven tide from the Sun.

By Earth standards, wind speeds on Titan — estimated to be typically about 1 mile an hour, and no stronger than that at the Huygens landing site — would barely qualify as a breeze. But Titan's atmosphere is so dense and its gravity relatively weak that even such a feeble wind could be enough to sweep small particles for long distances across the surface. The typical size of sand grains is about the same everywhere; whether on Titan, Mars, or Earth, they are roughly the size of granulated sugar. Provided it is dry and not sticky, Titan's winds might be strong enough to move such material around the surface, and the new images confirm that this is the case. It is somehow reassuring to think that familiar, fundamental processes operate in such drastically different environments.

This raises the question again of where the sand comes from and what it is made of. Current best guesses are of two types, the first that the grains are somewhat like hailstones, originating in the haze through sunlight-driven chemical reactions in Titan's stratosphere and rained to the surface over time. Alternatively, liquid methane flowing on the surface could erode exposed formations of icy bedrock. The explanation may be that the atmospheric circulation patterns dry out regions around the equator but promote torrential storm activity near the poles. These cloudbursts, though relatively rare, could provide fluids that flow with erosive force sufficient to grind loose solids on the surface into grains.

9.5.5 *Lakes*

Finding the source of the hydrocarbons that feed the complex organic chemistry in Titan's atmosphere has been a major goal for the Cassini mission and related observations for the longest time. Scientists had predicted, but had no confirmation until now, that pools of liquid were contributing to the high concentration of methane and other hydrocarbons in Titan's atmosphere. It was believed that Titan's methane had to be maintained by liquid lakes or extensive underground 'methanofers', the methane equivalent of aquifers. The first searches were disappointing since no specular reflection could be found nor any indication of seas at the lower latitudes observed by the spacecraft. After a brief indication of one possible lake-like feature in the southern hemisphere of Titan found by the Cassini/ISS, numerous well-defined dark patches resembling lakes were discovered during a July 22, 2006, flyby in radar images of Titan's high latitudes. Dark regions in radar images generally mean smoother terrain, while bright regions mean a rougher surface. Some of the new radar images show channels leading in or out of a variety of dark patches. Thus, the liquid nature of the dark areas is indirectly inferred from the presence of these rivers seemingly supplying the "seas" with at least part of the liquid. The shape of the channels also strongly implies they were carved by liquid.

This area of Titan had been in winter's shadow since before Cassini arrived, and the spacecraft had not flown over it before. During the flyby, Cassini's radar spotted

Figure 9.12   Lakes at Titan's north pole as seen by the Cassini radar in July 2006 (NASA/JPL).

Figure 9.13   VIMS (top), DISR and RADAR (bottom) coverage of the region of the Huygens landing site (marked +). Arrows show correlated features visible in pairs of images (Soderblom *et al.*, 2007).

several dozen lakes as small as 0.6 miles wide, with others nearly 20 miles wide. The biggest lake was about 62 miles long and may be only partly wet. Some of the dark patches and connecting channels are completely black — they reflect back essentially no radar signal, which means they must be extremely smooth and might contain liquid. In some cases rims can be seen around the dark patches, suggesting deposits that might form as liquid evaporates. On February 27, 2007, a new view of Titan's north polar region by the Cassini Camera, revealed a giant lake-like feature approximately 1,100 km across.

The lakes are the best prospect yet as the source of the hydrocarbon smog in the moon's atmosphere. At Titan's frigid temperatures, about minus 180° Celsius, the liquids in the lakes are most likely methane or a combination of methane and ethane. Thus, these findings corroborate the suggestions made after the Huygens landing on Titan's surface, that methane plays the role of water on Titan in a similar cyclic pattern: the liquid methane evaporates, rises into the atmosphere where it contributes to the photochemistry before it condenses and rains down to replenish the ground liquid. An alternative solution would be for a liquid-methane sub-surface table to exist, which would then fill surface cracks, such as craters or volcanic calderas. Whatever the forming procedures, if the liquid is confirmed, it means that we've now seen a place other than Earth where lakes are present.

Why do the lakes appear only at Titan's northern hemisphere at the current epoch? It may be due to a seasonal phenomenon where during Saturn's 29.5-year revolution around the Sun liquid can persist on the surface only in the winter (which prevails in the northern hemisphere currently), and that it shrinks in the spring as a result of increased evaporation, to vanish altogether in the summer. If the lakes come and go with the seasons, they probably also wax and wane over shorter periods of time with the weather. Winds might alter the roughness of their surfaces, a phenomenon that Cassini may see by repeated coverage that involves passing over a lake in a different direction, with the effect of prevailing winds changing the apparent brightness of the lake. On later passes toward the end of its nominal mission in 2008, Cassini might also see changes in the shape or size of lakes as winter yields to spring in the northern hemisphere. However, it will still only have mapped about 15% of the total surface: the extended mission will be able to gain larger and longer coverage and then we may better comprehend the evolution of these lakes.

9.6 The Nature and Composition of the Surface

The diversity of the terrain on Titan pictured by the Cassini–Huygens instruments sets a challenge that goes beyond anything that had been speculated about. With impact craters, dark plains with some brighter flows, mysterious linear black features possibly related to winds, dunes, and a host of possible agents: solids, liquids, ices, precipitation, evaporation, flow, winds, volcanism, etc. to be included, Titan has proven to be a much more complex world than originally thought and much tougher to interpret.

As we have seen, attempts have been made to explain Titan's surface region images and spectra variations by compositional mixtures based essentially on water ice and aerosol dust. The purpose of some more recent studies has been to explore the spectral properties of the bright and dark, dune and non-dune, liquid and solid areas, and to hypothesize surface composition based on differences in appearance. This approach is most fruitful if it focuses on the Huygens landing site, where 'ground truth' data is also available. Cassini-Huygens IDS Larry Soderblom and colleagues performed such a study using a colour-coded combination of data from VIMS, DISR and RADAR. The authors argue that the brighter areas found on the selected images are water-free and covered with aerosol dust; the dark brown areas (associated with the dunes) contain less water ice than the dark blue ones where a mixture of water ice and other material is expected. If the rate of deposition of bright aerosol is ~0.1 μm/yr, the surface would appear coated to optical instruments in a few years unless some process is cleansing them, hence the dark dunes must be mobile on this timescale to prevent the accumulation of bright coatings. Likewise, regular rain or run-off must be cleaning the dark floors of the incised channels and dark scoured plains that DISR imaged at the Huygens landing site. Huygens landed in a region of the VIMS bright and dark blue materials a few kilometres south of bright highlands and about 30 km south of the nearest occurrence of the VIMS-dark brown dunes that were discovered in the RADAR images.

A strong correlation between the dark brown units and dune fields was discovered in the RADAR SAR images. On the contrary, the bright and dark blue units observed by VIMS show no obvious correlation with SAR images, at least in these areas. The Huygens landing site is situated in a combination of VIMS bright and VIMS dark blue units, in a dark scoured plain that Soderblom and co-workers suggest to be richer in water ice than either the brighter regions or the dark dunes. The identification of water ice absorption at 1.54 μm in the DISR near-infrared spectrum of the surface acquired after landing is consistent with the dark blue unit containing water ice, with the absence of the weak ice bands at 1.04 μm and 1.25 μm possibly due to intimate mixing with dark hydrocarbon materials.

Soderblom and colleagues thus proposed a simple model for the composition of Titan's surface in which (1) the bright regions are covered with deposits of bright atmospheric aerosol tholin dust, (2) the dark regions like the dunes consist of large 100–300 μm grains; and (3) the regions that appear dark blue to VIMS are a dirty water ice substrate that is variably coated or mixed with the bright and dark material. The combination of bright material and dark floors of the channels in the DISR images of the highlands north of the landing is consistent with a blanket of fine grained water-free atmospheric aerosol dust deposits that are washed out of channel floors and off the dark plain where Huygens landed by methane rivers and floods, exposing dark water-ice rich substrate. Further, they speculate that some set of physical and chemical processes is converting the bright fine-grained aerosol into the coarser, dark hydrocarbon/nitrile saltating particles that make up the dunes.

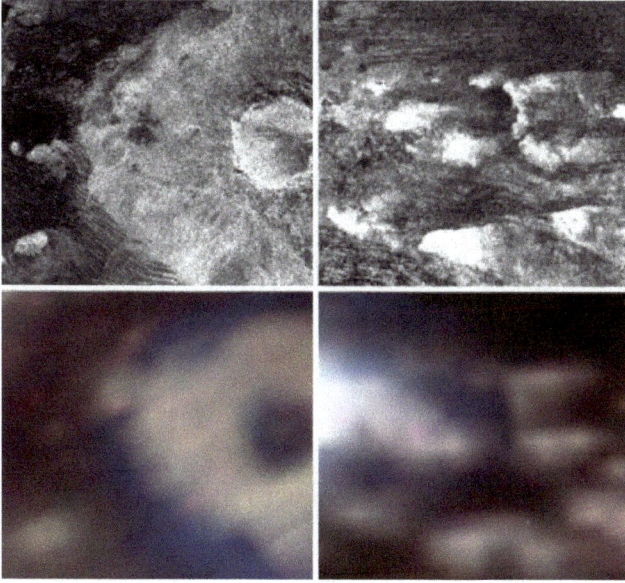

Figure 9.14 VIMS-RADAR correlations for (left) the Sinlap crater at 10°N and 15°W; (right) the region named Guabonito at 10°S and 150°W, approximately 150 km east of the Huygens landing site (Soderblom *et al.*, 2007).

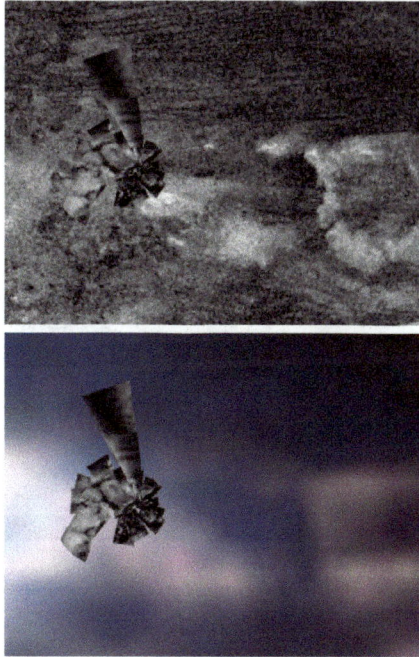

Figure 9.15 Correlation of the DISR images with RADAR SAR (upper) and association with VIMS bright and dark blue surface units. The Huygens landing site is ~30 km south of the nearest dunes (Soderblom *et al.*, 2007).

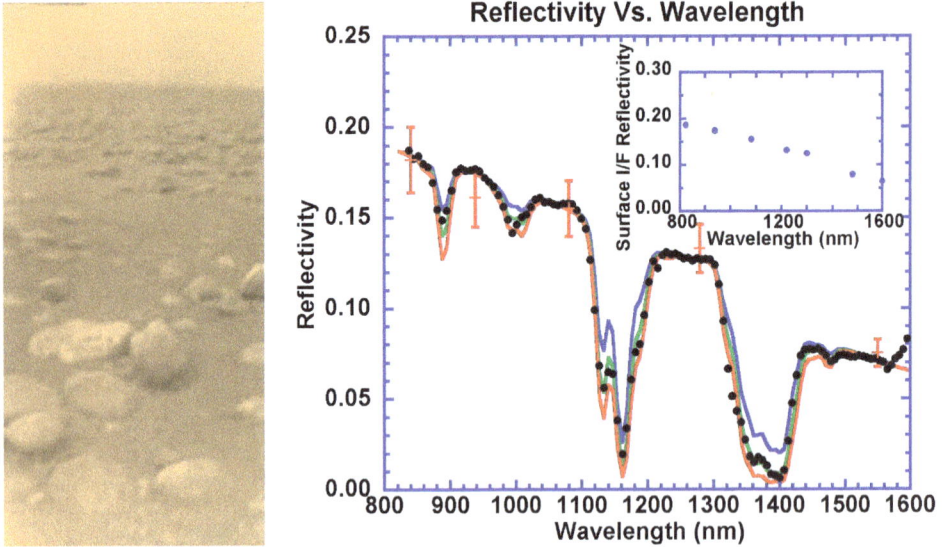

Figure 9.16   Left: Titan's surface after the landing of the Huygens probe. The 'rocks' are probably made of dirty water ice, and the darker riverbed is thought to be methane-wet icy 'sand'. Right: Downward-looking spectrum by DILR from an altitude of 21 metres compared to three models, having 3%, 5% (best fit), and 7% methane mole fraction (Tomasko *et al.*, 2005).

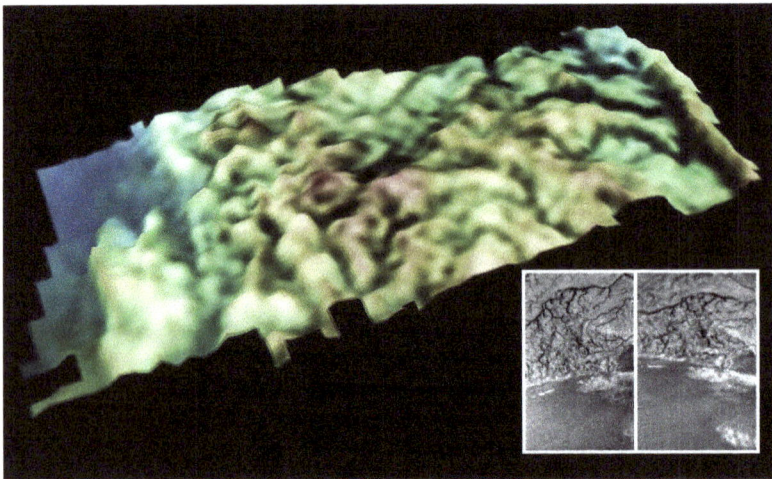

Figure 9.17   Perspective view of Titan's surface using a topographic model of the highland region to the north of the Huygens landing site seen with DISR. A lowland plane or lakebed is to the left side of the display, the northern highlands (with the dendritic channels) to the right (Tomasko *et al.*, 2005).

9.7 The Interior of Titan

Obviously, we have no direct knowledge of Titan's interior structure, and only one really strong constraint — the mean density, which comes from accurate measurements of the radius and the mass of Titan by spacecraft — to help us when we try to build models. Table 9.7 summarizes some of the existing observational constraints.

As noted at the beginning of this chapter, a mean density of $1.88\,\mathrm{gm\,cm^{-3}}$ for Titan means that the solid body is roughly 0.5–0.7 silicate by mass if the remainder is mostly water ice, the most abundant lightweight solid material in the Solar System. Thus we can picture Titan as a ball of rock about the size of our Moon (1,700–1,800 km), overlaid by a thick mantle (800–900 km) of impure water ice, possibly liquid at some depth. This assumes that the heavy silicate compounds migrated to the centre of an early, molten body, and formed a discrete rock core. To look more closely at the possible interior structure, we need to consider all of the possible formation scenarios for Titan. The principal options are: warm or cold accretion, mainly solid phase or in the presence of a thick primordial atmosphere, and homogeneous or heterogeneous (i.e. with the various components, e.g. rocks and ice, well mixed, or not).

Although it seems likely that Titan was hot at some stage during its formation, we cannot be sure that it did not form by cold accretion, where material came together so slowly that the net temperature rise was small and icy material stayed frozen. There are several arguments against this, however. Firstly, most of the dynamical models of satellite formation in the proto-Saturn nebula require an object as large as Titan to have been assembled quite quickly, certainly fast enough to produce a lot of heating. The pieces of rock and ice, in varying proportions, coming together to form Titan or any of the planets (called 'planetesimals') would, according to current thinking, have been quite large, like present-day asteroids or comet nuclei. They would have trapped the heat produced by the gravitational energy released because their size was much bigger than the thermal skin depth (the characteristic distance separating low and high temperatures in a solid body of a given material) of the newly formed moon. It is quite easy to calculate how much gravitational energy is released when an appropriate number of these objects is brought together to form something as large as Titan; most of this is converted to heat. One estimate puts the resulting temperature rise as high as 1,600 K, if there were no losses, and still several hundred degrees of K when losses are estimated. Finally, it is harder to understand the formation of Titan's present-day atmosphere in a cold accretion model, where all volatiles remained frozen. The next simplest concept is one in which Titan warms up during formation and melts. Models of this process invariably lead to differentiation inside Titan, i.e. separation of the rocky material from the ice. This means that we do not need to worry much about heterogeneous versus homogeneous accretion, since the end result would be much the same. It is not too hard to imagine a molten Titan going through differentiation beneath a thin, frozen crust at the interface with cold space.

Table 9.7   Summary of the observational constraints available for the evolutionary model of Titan by Tobie *et al.* (2005).

Instruments	Observational constraints	Interpretation in the framework of the evolutionary model
Huygens: DISR (Descent Imaging-Spectral Radiometer) [Tomasko *et al.*, 2005]	Dried, ancient shoreline, Fluvial networks, Eroded blocks.	Presence of liquids on the surface after long periods of outgassing. Saturation of the atmosphere by large eruptions of methane. Fluvial erosion owing to higher precipitation rate in the past.
Huygens: GCMS (Gas Chromatograph Mass Spectrometer) [Niemann *et al.*, 2005]	$^{14}/N/^{15}N \sim 183$, $^{12}C/^{13}C \sim 82.3$, $CH_4/N_2 \sim 1.5\%$–5%, $^{40}Ar/N_2 \sim 4.3 \times 10^{-7}$.	Formation of a NH_3-rich primitive atmosphere, followed by conversion into N_2. Late methane outgassing, not a relic of the primitive atmosphere. Outgassing still sustains a few % of methane in the atmosphere. Extraction of radiogenic argon from the ocean during methane outgassing events.
Huygens: SSP [Zarnecki *et al.*, 2005] Cassini: VIMS/RADAR/ISS (VIMS: Visual and Infrared Mapping Spectrometer) (ISS: Imaging Science Subsystem) [Sotin *et al.*, 2005; Griffith *et al.*, 2005; Elachi *et al.*, 2005; Porco *et al.*, 2005]	Soft (granular) solid soil. Albedo variations within both bright and dark terrains: no extensive bodies of liquids on the surface. Cryovolcanic edifices and cryovolcanic flow. Few impact craters. Variation of the surface composition and texture. Cloud structures from the south pole to mid-latitudes.	Floods of liquid methane from intense precipitation. Current outgassing not sufficient to lead to liquid methane accumulation on the surface. Cryovolcanism induced par thermal ice plumes, possibly favoured by the presence of ammonia. Relatively young surface, widespread resurfacing roughly 2 billion years ago. Fluvial erosion, cryovolcanic activity, tectonism etc. Injection of methane and saturation of the atmosphere.

(Continued)

Table 9.7   *(Continued)*

Instruments	Observational constraints	Interpretation in the framework of the evolutionary model
Ground-based telescopes [West *et al.*, 2005; Roe *et al.*, 2005; Schaller *et al.*, 2006]	No IR specular reflection: no extensive bodies of liquids on the surface. Geographic control of mid-latitudinal clouds. Cloud outburst above the south pole.	Current outgassing insufficient for liquid methane accumulation on the surface. Cryovolcanic eruptions at (350°W, 40°S). Large amount of methane in the atmosphere.

The models that look more promising, however, are those in which accretion occurred within a gaseous environment, one thick enough that the heat being released from the condensed body below could not escape directly to space but only by heating the atmosphere. The accreting planetesimals nearly all come from Saturn orbit; other trajectories would deliver so much energy that they would tend to blow off the atmosphere. The surface temperature and pressure would have been much higher than at present, probably hundreds of degrees K and hundreds of bars, and multiple layers of cloud (including water clouds) would have been present. Convection carried the heat upwards to high altitudes where the atmosphere and cloud tops were colder (Titan sounds like Venus in this scenario), there to radiate away to space. Down at the surface, the temperature is high enough to maintain a liquid water-ammonia ocean for tens of millions of years, during which time Titan cools and the surface finally freezes. So we arrive at a family of evolutionary models in which Titan formed inside the proto-Saturn nebula, a flattened disk of gas, rich in ammonia and methane, and dust, surrounding the forming planet during its early stages of contraction. Titan would have been a hot object in the early period after its formation, and more homogeneous in composition. As it cooled, the heavy, rocky material would tend to sink to the centre leaving the lighter materials, mostly water with an estimated few percents by weight of ammonia in solution, to form the mantle. Irradiation of the atmosphere and surface by solar ultraviolet radiation, cosmic rays and charged particles from Saturn's magnetosphere could have dissociated enough ammonia to form a thick nitrogen atmosphere, the precursor of that which we now observe, helped by the energy and vapour released from an intense meteoritic bombardment.

A model of Titan's evolution, focusing on the source of the methane that plays a role similar to water on Earth, was created by Gabriel Tobie and colleagues. According to this, methane would have been released during three episodes: the first occurring just after the formation of the rocky core and the accumulation of methane clathrate (methane trapped in water ice) above the ammonia-water ocean;

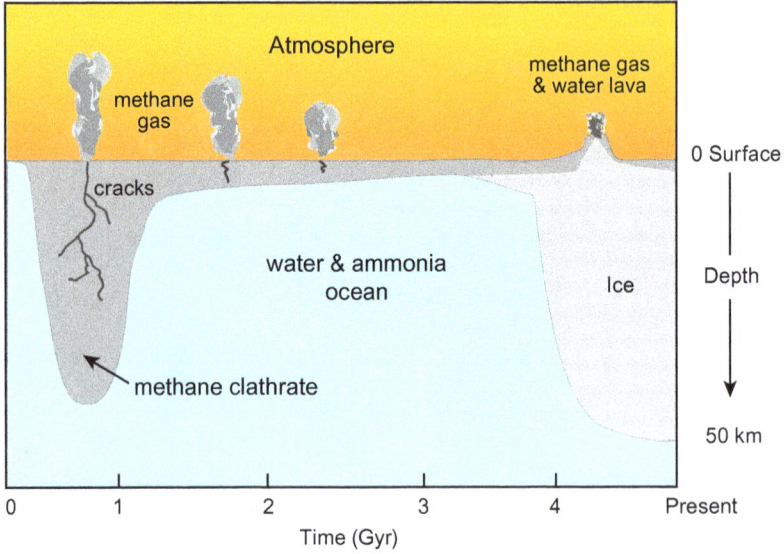

Figure 9.18   Evolution of the interior and of the methane outgassing rate during Titan's history, based on the model by Tobie *et al.*, 2006.

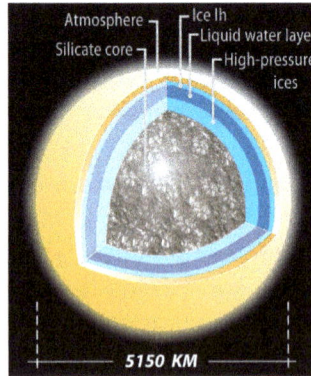

Figure 9.19   A possible model of Titan's interior, with a rocky core and a sub-surface ocean (Tobie *et al.*, 2005a).

the second when thermal convection initiates within the rocky core, the third, still active, induced by thermal instabilities within the icy crust. These three outgassing episodes provide enough methane to counterbalance the CH_4 loss rate in the upper atmosphere and to sustain a few percent of methane in the atmosphere, and should result in the formation of cryovolcanic edifices similar to the one observed by VIMS during the October 26, 2004 flyby.

In more detailed interior models, the ice layers outside the rocky core have locations, crystal structures and extents that depend on the distribution of pressure and temperature. They also contain ammonia and methane clathrates, in which the NH_3 and CH_4 molecules are trapped inside the water ice lattice, and perhaps other ices altogether. The heavier core may itself be differentiated, with, for example, iron and sulphur compounds and other metals at the centre, overlaid by the silicates and other rocky materials (the mantle).

A particularly interesting question, to which models provide a tentative answer, is whether some of the layers of ice below the surface of Titan are liquid. In other words, Titan may have a global ocean after all, but at some depth below the surface rather than on top of it. The possibility is quite a strong one; the amount of heat released inside Titan need not be enormous, since 100 km or so of ice above the top of the ocean provides good insulation from cold space. The presence of impurities, such as ammonia, makes easier for water to remain liquid. The Galileo orbiter has found persuasive evidence for a sub-surface ocean of this kind on Jupiter's icy satellites, Europa, Callisto and Ganymede.

A long search has been performed for this sub-surface ocean, first through Cassini Radio Science observations of the gravity field and determination of Titan's tidal distortion. Recently this search has paid off, as Cassini radar observations indicate that Titan's rotation speed is variable by several tenths of a degree per year. The radar can see through Titan's dense haze, detailing hidden surface features and establishing their locations on the moon's surface. Now it has discovered evidence of an underground ocean in imaging data from 19 Titan flybys between October 2005 and May 2007. The early observations established the locations of 50 landmarks on Titan's surface (lakes, canyons and mountains) and in later flybys these had shifted from their expected positions by up to 30 kilometres. This must mean that Titan's surface is decoupled from the interior by the presence of a sub-surface fluid layer, tentatively identified as an ocean of liquid wafer mixed with ammonia lying about 100 km below the surface.

In one current model, a liquid water-ammonia ocean lies some 100 km beneath the surface of Titan, occupying a depth of around 200 km above a high-pressure-induced phase water ice mantle, itself situated right above the rocky outer core (at about 1,700–1,800 km from the centre). The depth of the ocean depends, of course, on what is assumed about the source of interior heat. Possible sources include radioactive decay in heavy elements in the core, tidal heating due to perturbations of Titan's orbit around Saturn, and primordial heat. The heat loss is in fact mainly by slow thermal convection and thermal diffusion through the outer ice layer. Model studies do suggest that, if Titan ever had a liquid ocean in its evolutionary history, which is likely, as we have seen, then it should still be present, albeit thickly roofed-over with ice. The even more provocative question of whether such an ocean might contain primitive life is one to which we will return later.

CHAPTER 10

Titan's Origin and Evolution in the Solar System

O, aching time! O moments big as years!
All as ye pass swell out the monstrous truth,
And press it so upon our weary griefs
That unbelief has not a space to breathe,
Saturn, sleep on!

John Keats, *Hyperion*

10.1 Introduction

The study of planets and their environments as a family, the science of Comparative Planetology, seeks to explain the similarities and differences between Solar System bodies, and relationships to things like planetary radius and distance from the Sun, by a unified theory of planetary physics and an in-depth comprehension of how the Solar System formed and evolved. For the present, when our knowledge is too limited for that, we can still use comparisons to focus our attention on the key questions.

Titan seems to fall into a category that is intermediate between terrestrial planets, where the atmosphere consists of very minor constituents of the planet, and the giant planets, where the atmosphere is intimately related to the deep interior. With respect to the other satellites, the mean density of $1.86\,\mathrm{g\,cm^{-3}}$ (twice that of water) and radius of 2,575 km places Titan somewhere between Ganymede and Callisto. These two satellites of Jupiter are believed to consist of rock (silicates and iron) and water ice (25–50% by mass). The similarities between the three satellites may suggest that they have similar interior properties, but, if Titan formed further from the Sun and therefore at a lower temperature than Ganymede and Callisto, it may have incorporated other ices having similar densities to water ice (ammonia–water and methane–water mixtures). All three satellites are silicate-enriched relative to cosmic abundances, which predict 52% water ice versus 40% rock by mass.

Titan is unique in being the only satellite with an atmosphere that is thick compared to that of the Earth. Furthermore, Titan's atmosphere is particularly interesting because it consists primarily of molecular nitrogen, N_2, making Titan and Earth the only two known planetary-scale bodies with substantial atmospheres for which this is true. Neptune's satellite Triton, and perhaps Pluto also, have nitrogen-dominated atmospheres, but these are very thin. Recently, molecular nitrogen has also been detected in the extremely tenuous atmosphere of Enceladus, which consists mainly of water vapour.

Titan is sometimes compared to Venus, not because it has the same composition, but because both have thick, hazy atmospheres on slowly-rotating solid bodies. Many aspects of the boundary conditions for the atmospheric circulation are therefore similar, and, as we have seen in Chapter 8, the dynamics of Titan's atmosphere resemble Venus in some ways.

In order to better understand the mechanisms that have led to the formation and evolution of Titan to its present state, in this chapter we collect together what is known about Titan's origin and its relationship to its immediate neighbourhood (the Saturnian System), and to other comparably-sized bodies (Venus, Mars, Ganymede, Triton, and Pluto), with particular emphasis on its similarities and differences with the Earth, as well as with its parent planet, Saturn. We may then begin to see how the properties of the Solar System as a whole, and Titan's place within it, makes sense — or not, as the case may be.

10.2 Relations Among Solar System Bodies

The atmospheres of the objects making up the Solar System separate into two main classes: those of the terrestrial or Earthlike planets Venus, Earth and Mars, and those of the outer gas giants Jupiter, Saturn, Uranus and Neptune. Mercury, Moon, and Pluto, along with most of the satellites of the outer planets, could be considered another class of smaller, virtually airless planetary bodies. Titan, with its thick

Figure 10.1   Titan and Earth's Moon: their sizes compared (NASA/JPL).

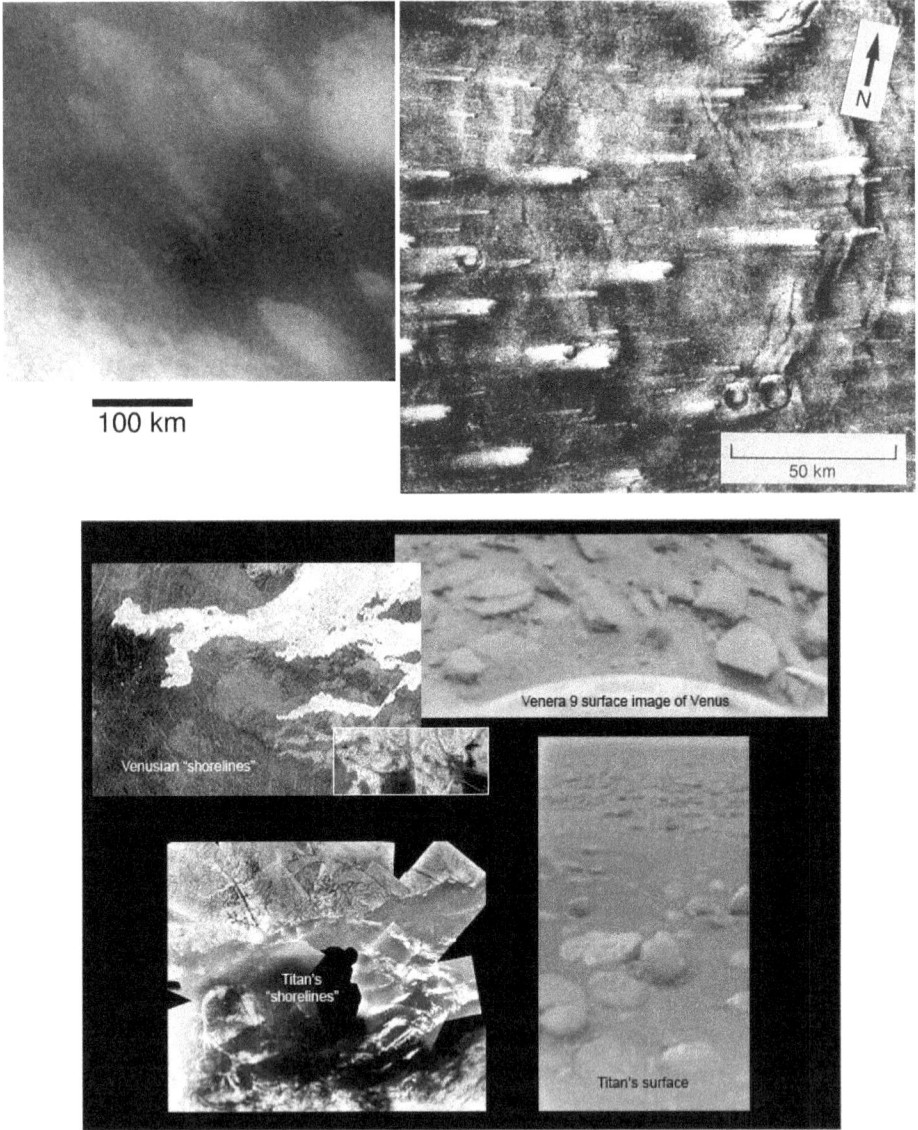

Figure 10.2   The upper image pair compares a Cassini image of streaked terrain on Titan in the region where the Huygens probe landed, to a picture from the Viking 1 orbiter showing similar streaks on Mars. In both cases the fluid flow, probably wind in both cases but on Titan possibly liquid flow as well, is from west to east. In the lower set, Titan's surface features (volcanoes, rivers, etc.) are compared to those found on Venus (NASA/JPL/ESA/ASI/University of Arizona).

Earthlike atmosphere, resembles the terrestrial planets, although as a solid body it might be considered to belong with the other large satellites, Io, Europa and Ganymede of Jupiter, its neighbour Enceladus, and Neptune's Triton. A family might also be established on chemical grounds by assembling Titan with Earth, Triton

Figure 10.3   Saturn: Titan is about 20 times smaller in diameter, and lies about 20 Saturn radii away from the centre of the planet, in the plane of the rings but well outside them. This image taken by the Cassini cameras on February 18, 2005, shows the shadows cast by the rings onto the northern kronian hemisphere and details of the banded atmosphere (NASA/JPL/SSI).

and Pluto, the other objects in the Solar System possessing nitrogen-dominated atmospheres.

10.2.1 *The Formation of the Solar System*

After centuries of controversy, involving most of the leading scientists of each era, it is now generally believed that the Sun and the planets evolved from a cloud of interstellar gas and dust known as the solar nebula. The solar nebula itself formed when part of the interstellar medium was compressed due to a disturbance, perhaps the shock wave from a nearby supernova. It then went on to collapse under its own gravity, forming a turbulent disk of gas and dust orbiting a central conglomeration. As the cloud collapsed, the centre became more compressed, releasing heat that was increasingly unable to escape through the opacity of the surrounding cloud, until eventually the dust near the centre vaporized, and a protostar formed. The initial collapse is thought to have occurred in less than 100,000 years.

Centrifugal force kept some of the gas and dust in the rotating disk from falling in to the forming star. The material in the outer parts of the accretion disk radiated heat to space, leading to a radial temperature gradient. The metal, rock and ice present as gas condensed out into tiny particles, at different rates depending on the distance from the hot young Sun. These particles probably included grains of high melting point material like diamond and silicon carbide, which are found today in meteorites. The metals condensed almost as soon as the accretion disk formed, according to interpretations of the isotopic ratios in meteorites; the rock condensed a few hundred million years later, and the ice later still, and at greater orbital distances.

The condensed dust and ice particles initially had a porous structure that helped them to stick together when they collided randomly with each other, forming larger particles. Eventually, objects the size of boulders or small asteroids were produced that were massive enough to attract each other through gravity, accelerating the rate of growth. The larger objects tended to sweep up all of the solid matter in or close to their own orbit, their ultimate growth being limited by the material available at their distance from the star. The sizes that were reached in this way were perhaps comparable to the Moon in the inner Solar System and one to fifteen times the size of the Earth in the outer Solar System. The main reason for the difference is that, beyond the 'snow line' that lay between the current orbits of Mars and Jupiter, it was cool enough for water ice to condense, making much more accretable matter available. As a result, larger 'planetesimals' formed at large distances from the Sun, trapping not only the solid matter in their vicinity but also the gaseous material, even the lightest elements hydrogen and helium. These eventually became the gas giant planets.

Models of the formation process show that the protosolar cloud must have been many times more massive than the contents of the present day Solar System or the latter could not have formed. Most of the mass must have been removed somehow, at some stage. Stars similar to the Sun elsewhere in the Universe, including one in T Tauri after which the effect is named, are observed to develop a very strong outflow of charged particles, rather like the present-day 'solar wind', only more vigorous, when they are about 1 million years old. This tends to sweep away gas from the protoplanetary nebula. The gas giant planets formed before this process was complete by becoming massive quickly enough to attract the nebular gas as well as the icy planetesimals in their neighbourhood. The rapid growth of the gas giants was not matched by the solid protoplanetary bodies of the inner Solar System, which took longer, perhaps another hundred million years, to collide with each other and form the four inner planets in stable orbits. These planets melted and their interiors differentiated as a result of heating by the collisions forming them, and by the decay of radioactive elements in their interiors. As they cooled and their surfaces solidified, they swept up the remaining smaller objects leaving the cratering record which we see today.

10.2.2 The Terrestrial Planets and Titan

Like Earth, Venus and Mars are rocky planets with secondary atmospheres; that is to say, any atmosphere that formed with the planet has been lost, and the gaseous envelopes which now exist were exhaled from the interior at a later date, augmented to an unknown degree by the infall of icy material as comets. The composition of Earth's atmosphere is, of course, much modified by the presence of life, and to a lesser extent by the existence of free water on the surface. Neither Mars nor Venus has either feature, and their atmospheric compositions are similar, consisting of carbon dioxide and nitrogen with traces of the rare gases and some photochemically-produced species such as carbon monoxide and ozone. Carbon dioxide is a heavy gas and therefore easily retained; it is a major product of planetary outgassing and, in the absence of liquid water, relatively stable. If water is present, CO_2 tends to dissolve, forming carbonic acid, H_2CO_3, and eventually solid metallic carbonates like calcium carbonate, $CaCO_3$. This type of process has trapped most of the CO_2 from the atmosphere of the early Earth, but it is highly debatable to what extent this has happened on the other terrestrial planets. Venus still has huge amounts of atmospheric carbon dioxide, forming a thick envelope that raises the surface temperature, through the greenhouse effect, to around 730 K. This high surface temperature encourages further outgassing, and active volcanism on present day Venus is also a real possibility. Venus ends up with a surface pressure nearly 100 times that on the Earth, where most of the originally outgassed CO_2 is now in mineral form.

The Earth's primitive atmosphere probably consisted primarily of carbon dioxide, methane, ammonia and water vapour, with no free nitrogen or oxygen. The large proportion of nitrogen now present arose as a consequence of the photodissociation of ammonia into nitrogen and hydrogen. The latter gas is so light that it can escape from the Earth even at the present time, whereas most heavier gases are gravitationally trapped now that the Earth is cooler and the Sun less active. Most of the methane was oxidised to form carbon dioxide and water vapour, although enough is produced (for example, by the decay of vegetable matter) to leave a substantial admixture in the present day atmosphere. The existence of free oxygen is a consequence of the presence of life, in the form of planets which convert carbon dioxide into oxygen by photosynthesis. Thus, we owe the present state of the terrestrial atmosphere to a series of complex and dynamic processes. There is evidence that the atmosphere is not entirely stable, but continuing to evolve, perhaps towards a state less well suited to human comfort and survival.

Because Titan has a substantial nitrogen-dominated atmosphere, the satellite is sometimes thought of as a large-scale laboratory for scientists to study the chemical evolution of an atmosphere similar to that of the primitive Earth. Of course, there are differences as well as similarities (Table 10.1); the satellite is half the Earth's size, while its atmosphere is denser at the surface, and more extensive than the Earth's, reaching up towards space more than 1500 km from the surface, compared to about 100 km on the more massive body. But the biggest difference is the temperature — Titan is a water world, like Earth, but Titan's water is frozen as hard as steel.

The general differences between the solid bodies of Titan and the Earth can be reasonably well understood in terms of the models of Solar System formation we have already discussed.

Table 10.1   Orbital and physical properties of Titan, Earth and Venus.

	Titan	Earth	Venus
Semi-major axis of orbit (AU)	9.54	1	0.7233
Sidereal period of revolution (yr)	29.46	1	0.6152
Rotation period (day)	15.945	1	243.05
			(retrograde)
Equatorial radius			
Surface	2575	6378	6051
Clouds (km)	2775	6393	6118
Orbital eccentricity	0.029	0.017	0.007
Inclination of the orbit to the ecliptic plane (°)	19′8	0	3°23′
Oblateness	< 0.014	0.0034	0.0000
Obliquity (°)	26.7	23.45	177.34
Volume (Earth = 1)	0.065	1	0.869
Mass (Earth = 1)	0.022	1	0.815
Mean density (10^3 kg m^{-3})	1.881	5.52	5.26

Firstly, being more than nine times farther from the Sun, Titan receives only 1.1% of the solar flux that reaches the Earth. Thus the effective temperature at which Titan radiates heat to space, to remain in equilibrium with the sunfall, is only about 83 K, less than a third of the figure for Earth. Within the atmosphere, Titan's thermal structure is like a colder version of the Earth's. Just as ozone absorbs solar UV on Earth, and leads to a warm stratosphere, the haze on Titan fulfills the same role. The temperature decreases with altitude in the stratosphere from 600 km down to the tropopause near 40 km, where the minimum value of 71 K is found. Below this level, the greenhouse effect causes the temperature to rise again until it reaches 94 K at the surface for a pressure of 1.5 bar (Table 10.2). Water vapour, the principal greenhouse gas on Earth, has an analogue in methane on Titan, also a major greenhouse gas, and also condensable, apparently to the extent of forming lakes on the surface. Carbon dioxide, a non-condensable greenhouse gas on Earth, has an abundance controlled by the balance between removal (by weathering of silicate rocks) and production (by combustion and volcanic venting). On Titan, non-condensable hydrogen contributes to the greenhouse effect, its abundance a balance between production by photochemistry and escape into space.

The meridional temperature distribution at the time of the Voyager encounter near the spring equinox on Titan was not symmetric about the equator. Latitudinal variations were measured in the stratosphere that showed the north polar region to be 15–20 K colder than the south, evidence for seasonal behaviour and dynamical inertia (the lag between the longest day and the warmest) similar to the properties of the seasons on Earth. The minor species in Titan's atmosphere (Table 10.3) also show latitudinal variations as some do on Earth, again due to seasonal effects. The orange-yellow haze that covers the satellite indicates the presence of aerosols of photochemical origin in its upper atmosphere, related to the volcanic aerosols and

Table 10.2   Atmospheric and surface data for Titan, Earth and Venus.

	Titan		Earth		Venus	
Bond albedo	0.29		0.36		0.77	
Cloud top (km)	20–40		—		~160	
Surface (equator)						
Pressure (bar)	1.496		1.103		92	
Temperature (K)	94		288		733	
Atmospheric temperature	$T(K)$	Alt (km)	$T(K)$	Alt (km)	$T(K)$	Alt (km)
Tropopause	71	40	217	15	250	90
Stratopause	180	200	247	50	—	
Mesopause	150	490	190	80	—	135
Thermopause	152	1500	1000	700	300	
Surface gravity (m s^{-2})	1.35		9.81		8.87	
R(exobase)/R(planet)	1.6		1.08		1.03	

Table 10.3   Chemical composition of the atmospheres of Titan, Earth and Venus.

		Titan		Earth		Venus
Major constituents (%)	N_2	90–97		78		3
	CH_4	2–7	O_2	21	CO_2	96
Minor species (%)	Ar	? (< 10)		0.93		0.007
	H_2O	~ 10^{-9}		0.1–3		0.003
	SO_2	—		0.2×10^{-7}		0.015
	H_2	0.2		0.5×10^{-4}		1.6×10^{-3} (?)
	CO	10^{-6}–10^{-4}		10^{-5}		0.002
	OCS	—		0.5×10^{-7}		0.002 (?)
	CO_2	10^{-8}		0.034		
			CH_4	1.7×10^{-4}		

anthropogenic smog found on Earth. The clouds and rain, in otherwise clear air, observed by Cassini near the surface on Titan are the result of methane, rather than water, condensation, but still are analogous to the terrestrial phenomena.

The low temperatures on Titan slow down most chemical reactions considerably, and this is likely to have prevented the formation of complex molecules required in any life-chain. Molecular oxygen is almost completely absent, and common oxygen compounds like water and carbon dioxide have such low vapour pressures that they condense on Titan and are at most trace constituents of the atmosphere. If there is liquid water on Titan, it must be at considerable depths below the surface where a combination of high pressures, interior heat release and dissolved impurities prevents it from freezing.

The low temperature and gravity produce a large atmospheric scale height on Titan. Unlike the Earth, where the atmosphere is sometimes likened to the skin on an apple, Titan's atmosphere extends to a substantial fraction of the solid body's radius. The total mass of Titan is about 1/50 that of the Earth, the radius is 40%, and the surface gravity is about 7 times less. These factors, together with temperatures that are typically about 3 times lower, on the absolute scale, lead to an atmospheric scale height that is around four times as large as that on Earth or Venus. The extensive, cloudy nature of Titan's atmosphere led early Earth-based telescopic observers to overestimate the diameter of the satellite, making them think Titan was the largest satellite in the Solar System, whereas in fact this distinction belongs, by a small margin, to Jupiter's Ganymede.

There are a number of additional reasons that conditions on present day Titan do not replicate those on the primordial Earth. The nitrogen-methane composition of Titan's atmosphere lacks the substantial amounts of carbon monoxide, carbon dioxide and water vapour thought to exist in the primitive Earth's atmosphere, not only because of condensation but because the state of oxidation between the inner and the outer solar system atmospheres is very different. Carbon is present in its fully oxidized state, carbon dioxide, in all the terrestrial planets and in its fully reduced state in most of the outer Solar System objects. This difference could reflect the ability of water to buffer the oxidation state of the carbon species in the atmospheres of the terrestrial planets, while the oxidation state of the carbon in the outer Solar System atmospheres may be primordial. The present-day Earth is unique in having a high abundance of molecular oxygen, reflecting the Earth's exclusive position in the solar system as an abode of advanced forms of life.

On all planets, the flux of carbonaceous material arriving in meteorites, comets, etc. is smaller today than in the past and has probably always been different at Titan as compared to Earth. The material exuded into the atmosphere from within the two solid bodies due to volcanic and tectonic activity will also have evolved differently over time. Titan may have undergone early dramatic changes followed by tectonic quiescence, whereas the Earth is still active. The question of present-day volcanic activity on Titan remains open, but there is indirect evidence (for instance the high abundance of methane in the atmosphere despite the rapid rate of removal by photolysis) for some outgassing from the interior that continues to this day.

A final, possible resemblance between the two bodies, which is unique in the Solar System, is the co-existence on the surface in both cases of dry land and liquid seas. As we have seen, the most recent observations of Titan rule out early theoretical models which suggested the presence of large liquid hydrocarbon reservoirs covering the surface, although the existence of clouds, aerosols and large amounts of methane in the atmosphere still seem to imply that reservoirs of liquids should exist on, or at least in contact with, the surface. As evidence accumulates from radar and near-infrared observations, a mostly solid surface is revealed, probably composed of water ice, mixed with other compounds, covered with condensed solid or viscous hydrocarbons, possibly with rocky material exposed in places. Lakes, some of them quite large, and rivers, do occur but are short-lived and dry out regularly on

Table 10.4   Surface composition of Titan, Earth and Venus.

	Titan	Earth	Venus
Seas	Hydrocarbons (?) $(C_2H_6$-CH_4-$N_2)$	H_2O	none
Ices	H_2O, C_2H_2, CO_2, ...?	H_2O	none
Crust	dirty ice, minerals?	minerals	basaltic rocks

seasonal timescales. A porous icy or rocky regolith may store a large liquid hydro-carbon reservoir, continuously supplying methane and other volatile species to the atmosphere. Putting this together with the rich display of atmospheric phenomena to be expected, whatever the details turn out to be, an observer on Titan is likely to see as Earthlike a terrain as any in the Solar System (Table 10.4).

If Titan's environment is in some ways closer to the Earth's than any other known planetary object, it should provide some coded insights into the conditions which prevailed on Earth during its early history and into the chemistry which then gave rise to life. Venus is another planet that is often compared to the Earth, being a rocky body of the same size and similar composition. In terms of atmospheric dynamics, however, Venus is more like Titan: a slowly-rotating solid body with a deep cloudy atmosphere, super-rotating at high levels. These dynamical similarities between Titan and Venus were discussed in Chapter 8.

On Venus, three main cloud layers cover the entire planet, occupying the height range from about 30 to 70 km above the surface. The main constituent appears to be concentrated sulphuric acid solution, about 75% H_2SO_4 and 25% H_2O, at least in the uppermost layer where spectroscopy and polarimetry measurements have been made. The droplets there are fairly constant in size, at about 1.1 μm radius; lower down, larger particles up to about 30 μm in radius have been detected by instruments on descending probes. A rather complex, altitude-dependent cloud chemistry, involving photochemical processes, gaseous SO_2, and possibly elemental sulphur, is implied in order to produce and maintain clouds with this composition and structure.

Of course, the cloud composition and microphysics is not the same on Venus and Titan. The latter is too cold for aqueous, sulphurous clouds like those found on Venus. The lower stratosphere of Titan is expected to be at least as stable against turbulence as that of Venus, and aerosol particles are just as abundant, yet cloud particles are probably larger. Particle radii of about 5 μm and about 10 μm are found, respectively, for condensed dicyanogen and ethane in the north polar hood of Titan. On the other hand, about 1 μm radius particles are the rule for the upper clouds on Venus, although some larger particles (typically with radii of about 3.5–4 μm, with a few up to 35 μm) occur in the lower cloud deck. Ice cloud particles are typically larger than liquid cloud droplets — cirrus clouds on Earth have non-spherical particles with characteristic dimensions of hundreds of μm, and clouds which appear to contain particles of ethane ice with radii about 75 μm or so have been inferred to exist on Titan.

Another important difference is that Venus' cloud particles are bright, scattering most of the solar spectrum, while Titan's are dark (especially at blue wavelengths) and absorb it. While Venus' atmosphere falls off rapidly with altitude, the aerosol on Titan warms the stratosphere and gives a large scale height at high altitudes, so the atmosphere tails off only slowly into space.

Mars is something of an enigma in having such a low surface pressure, another factor of 100 down on the Earth. In part, this might be explained by the relatively small size of the planet, causing volcanism to end early in its history and rendering the escape of gases to space easier. Titan is even smaller, of course, and has upper-atmospheric temperatures that are comparable with those on Mars, yet has retained a much denser atmosphere. Another factor is that Mars is cold enough so that its principal constituent, CO_2, condenses on the winter pole, a phenomenon which leads to a huge annual swing in the surface pressure all over the planet, and means that some fraction of the atmosphere can be trapped at all times as ice. An additional and, in some models of Mars' evolution, an essential feature, is that Mars too may have formed carbonate rocks during an early epoch when the surface was partially flooded. Considerable evidence can be seen in the geological patterns of the surface, which in places on Mars are strongly suggestive of rivers, lakes and seas. How these were formed and how long they persisted is a matter of much active enquiry and debate.

Mars' atmosphere exhibits a variety of thin clouds at various locations, most spectacularly downwind of the giant volcanoes that tower up to 25 km above the mean surface height. Clouds of both water and CO_2 ice occur. More important than either, in terms of its effect on conditions on the surface, is the load of dust which the atmosphere carries at all times. During storms, this obscures the surface from above, and extends over huge areas, sometimes the entire planet, and to heights of 30 km. On Mars, the dust alone exerts a considerable 'greenhouse' warming of the surface, as much as 20 K under high-dust conditions, making it more important than CO_2. The combination of gaseous and particulate absorbers has some parallels with the situation on Titan, although obviously the details are very different.

There are also parallels when we consider why Titan and Mars now appear to be so dry compared to Earth. For Titan, it may be a seasonal effect. It is thought that on Mars, most of the water that probably once flowed on the surface still exists on the planet, but now as polar cap ice and lower latitude sub-surface deposits. This is in contrast to Venus, which appears to have lost most of the water it ought to have outgassed along with its carbon dioxide, although a little is still present in the cloud droplets and as vapour. In part, the loss is due to Venus' closeness to the Sun. The high atmospheric temperature and intense ultraviolet fluxes allow greater upward transfer of water vapour, a higher rate of dissociation in the upper atmosphere, and a higher rate of loss to space of the hydrogen atoms produced as a result, compared to the Earth. Such loss probably did not occur on Mars or Titan, to anything like the same degree, so both bodies probably have a very large inventory of frozen water.

10.2.3 *Titan and the Outer Planets*

When they formed, the gas giant planets Jupiter, Saturn, Uranus and Neptune were large enough and cold enough to retain the lighter elements, predominantly hydrogen and helium, which were present in the primordial cloud from which the Solar System formed; those elements were lost from the inner planets. The outer planets are, in consequence, mostly atmosphere, increasing in density and temperature with depth to very high values, and with the heavy elements concentrated near the core.

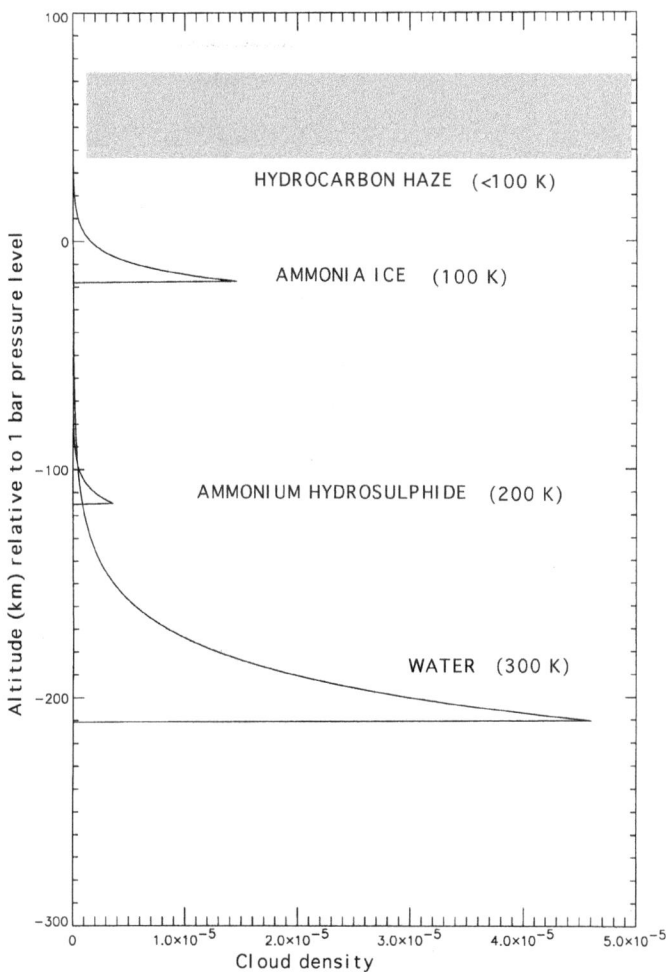

Figure 10.4   A model of the vertical cloud structure on Saturn, showing four layers of cloud with different compositions, and the approximate temperatures at which they form.

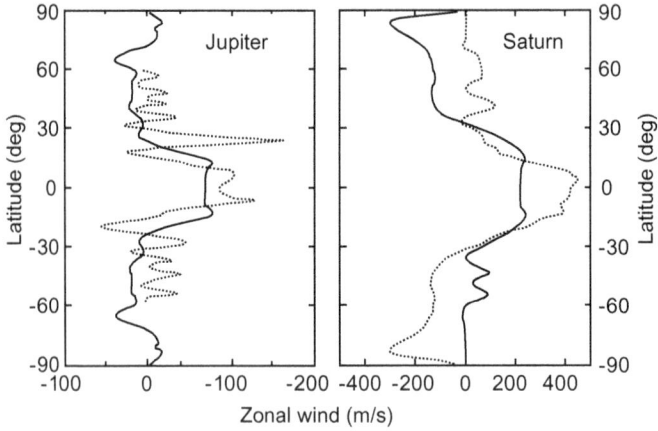

Figure 10.5   Cloud-top wind velocities on Jupiter and Saturn.

Figure 10.6   The swirling hurricane-like vortex observed by Cassini at Saturn's south pole. The vertical structure of the clouds is highlighted by shadows (NASA/JPL).

The elemental abundances in Jupiter were long thought to be about the same as in the Sun, as would be expected if star and planets formed from the same mass and Jupiter retained all of its original material by virtue of its strong gravity field. Most of the elements are combined with hydrogen to form the most stable compound, usually the fully hydrogenated form. For example, most of the nitrogen is found as ammonia, NH_3 and most of the carbon as methane, CH_4. This makes it possible to estimate the N/H, C/H and other elemental ratios, by measurements of these common compounds in Jupiter's atmosphere, using spectroscopy and direct measurements by probes, like that deployed by the Galileo mission in 1995. In fact, the common elements were found to be several times more abundant in Jupiter than in the Sun, a finding that can be explained by some versions of the planetesimal hypothesis of Solar System formation outlined earlier in this chapter. Composition measurements, including the isotopic ratios in each element (including the D/H ratio in hydrogen), are thus found to be powerful tests of theories of planetary evolution. This has great relevance to Titan, of course, as we have already discussed in detail in Chapter 6.

Many other species, in particular more complex hydrocarbons, are present on Jupiter and the other giant planets, in the thinner upper atmospheres where enough solar radiation penetrates to drive photochemical reactions. This process, too, is directly relevant to haze production on Titan, as discussed in Chapter 7, although the Jovian and Saturnian upper atmosphere hazes, since they form in a predominantly hydrogen atmosphere, will not have the same detailed composition as those on Titan. Fewer nitrogen compounds are likely to be involved, for example, although it will be a long time before the compositions of any of the photochemical hazes in the outer Solar System are known well enough to make detailed comparisons.

All of the gas giant planets are cloud-covered, as rising motions from the interior bring air rich in condensables such as water and ammonia up to cooler levels where they condense. Jupiter is thought to have at least three layers of cloud in the region visible to the outside observer, and many more below that. The uppermost layer consists of ammonia ice crystals, the one below that of hydrogen sulphide in combination with ammonia as ammonium hydrosulphide, NH_4SH, and the lowest of the three of water. All are bound to have impurities, and the bright coloration of many of the Jovian clouds is testimony to that fact. The so-called chromophores are not yet definitely identified, but various forms of elemental sulphur may contribute to the yellows and browns, while the red spot may be coloured by phosphorus produced photochemically from phosphine. Again, we lack the detailed knowledge of composition to do more than speculate about any relationship to the colour of the hazes on Titan. Current models suggest that the orange colour of Titan has its origin in its hydrocarbon and nitrile-derived 'tholins', while the yellows and browns seen in the clouds of Saturn, after allowing for the overlying thin veil of organic haze mentioned above, are more likely to be derived from sulphur compounds.

As Titan's parent body, Saturn deserves a detailed description, although in fact much of our understanding of its atmosphere, and interior is derived by extrapolation from Jupiter, which it closely resembles. Jupiter is about 20% greater in radius, but cannot match Saturn's beautiful and extensive ring system. Jupiter is only half the distance from the Sun as Saturn and more extensively explored, because it is more accessible to Earth-based telescopes, and because of several years of detailed study by the Galileo Jupiter orbiter, plus the one of only two entry probes to date into the outer planet atmospheres (the other is Huygens at Titan in 2005). Saturn is, of course, being intensively studied by the Cassini orbiter, which spends only a small fraction of each orbit in the vicinity of any of the satellites, including Titan, but is always within observing range of the planet. The detailed picture of Saturn which is emerging allows the key differences between Saturn and Jupiter to be appreciated and, eventually, understood.

Known species in Saturn's atmosphere, detected by Earth-based and space spectroscopy, include methane (and its isotope, monodeuterated methane), ethane, acetylene, ammonia, phosphine, arsine, and germanium. More recent observations showed that water, carbon dioxide, methylacetylene and diacetylene also exist on Saturn, while it lacks ethylene, hydrogen cyanide and other nitriles. Cassini observations are continuing to add to this inventory, searching for higher hydrocarbons and nitrogen-containing molecules such as HCN, and possibly SiH_4 as well as GeH_4. In addition, better values of the isotopic ratios D/H and $^{13}C/^{12}C$ are being obtained. However, the key abundances for understanding the deep structure and the formation history of the planet (and hence making a key contribution to studies of the formation of the Solar System) are not likely to be accessible. The noble gases and their isotopic ratios are not readily detectable spectroscopically, and the principle compounds of oxygen, nitrogen and sulphur are likely to be hidden from us by the upper cloud layers of frozen ammonia and, in the stratosphere, hydrocarbon hazes. Cassini was originally designed to carry a Saturn entry probe, as well as the Huygens Titan probe, but this was deferred in order to reduce cost and complexity and will now be part of some future mission to the Saturn system.

Clouds are a complication for interpreting remote sensing observations of temperature and composition, but are also an important subject of study in their own right. The vertical distributions of cloud material can be predicted by chemical equilibrium calculations, usually by assuming that parcels of air rising through the

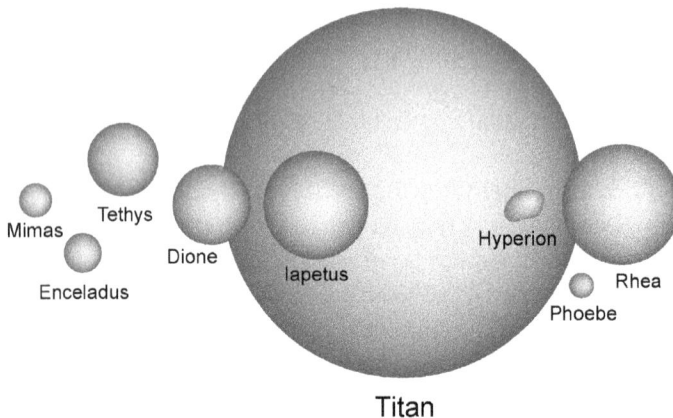

Figure 10.7   The relative sizes of the larger satellites of Saturn, showing how Titan dominates.

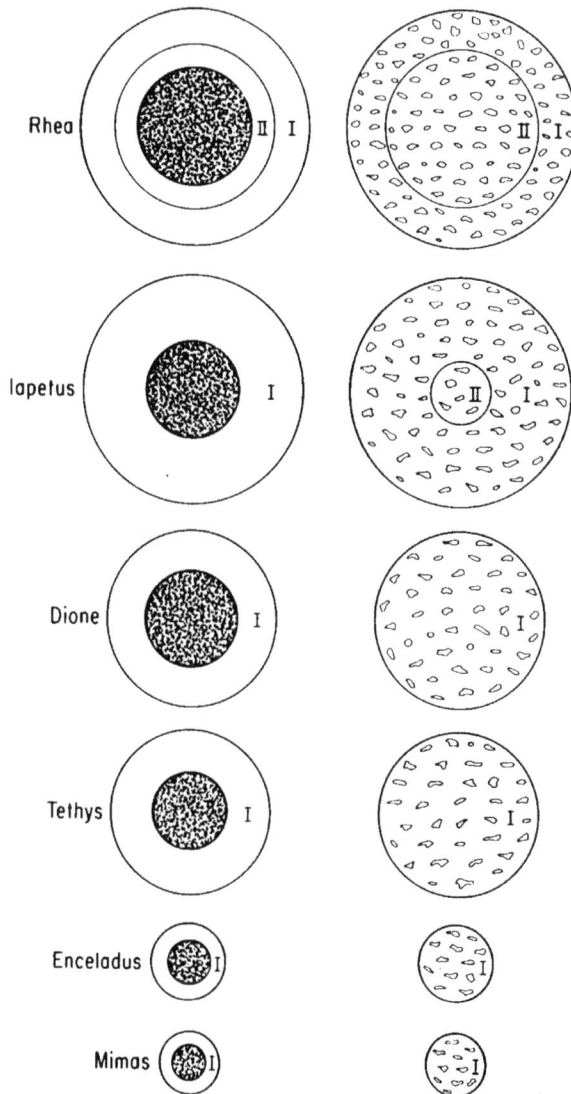

Figure 10.8 Models of the expected interior structure of Saturn's satellites, differentiated on the left and homogeneous on the right (Schubert *et al.*, 1986).

atmosphere reach chemical and thermodynamic equilibrium at each level. In particular, any condensable component of the atmosphere which is raised to a temperature level where it becomes super-saturated is restored to the stable point on its phase diagram by forming cloud particles until the mixing ratio is again in equilibrium. The results of this type of calculation provide an indication of the relative amounts of condensation at each level, but can not tell us very much about absolute abundances of cloud material since this must depend on unknown details of the vertical motions and cloud microphysics, which are not included in the model. Three cloud layers are predicted to fall within the range which is accessible to remote sensing observations on Saturn, a thick water cloud at the bottom, ammonia ice clouds at the top, and ammonium hydrosulphide (NH_4SH, a compound of ammonia and hydrogen sulphide) in between. The main difference from Jupiter is the vertical scale; on Jupiter the cloud layering is much more compact. For example, on Saturn the water cloud base is predicted to be about 200 km below the one bar level, some 150 km deeper in the atmosphere, relative to this pressure level, than on Jupiter.

All of the giant planets, and Titan, have a stratospheric haze, thought to be composed of photochemical products of methane in each case although, as already noted when comparing Jupiter and Titan above, the haze composition and other properties will vary on each body. For instance, due to the generally lower temperatures, the haze could be deeper on Saturn than on Jupiter. This may account in part for the relatively bland appearance of Saturn compared to Jupiter, since the haze partially obscures the belts, spots and other features in the thicker clouds below. The stratospheric haze may therefore account for the fairly uniform 'butterscotch' colour of Saturn. The Cassini instruments are seeking to characterise the vertical distribution and optical properties of the stratospheric haze on both Saturn and Titan, and thereby gain clues about its probable composition. The spectral properties of the material in the liquid droplets may not be completely diagnostic, but will be augmented by determinations of the gaseous hydrocarbons present and their relative abundances. From these, and the appropriate theory, a picture of the chemical sequences producing hazes on both bodies can be constructed.

The main cloud deck on Saturn, that is, the one in which the banded features and giant eddies that we observe from outside the atmosphere arise, is likely to consist mainly of ammonia, since this species is observed spectroscopically in the atmosphere, and is expected on theoretical grounds to condense near the 1 bar level. Ammonia ice is white, and cannot account on its own for the colour contrasts on Saturn, which can be seen in the Voyager images through the overall yellowish tint probably imparted by the overlying hydrocarbon haze. More so than for Jupiter, images of Saturn need to be very highly 'stretched' before they show much colour contrast. The large 'red spot' observed by Voyager in the 55 S region is not only smaller than its Jovian counterpart (but still on a scale comparable to the size of the Earth), but not really any redder than some of the nearby cloud bands. Similarly, the differently-coloured bands are probably separate convection cells, so may have different contaminants, or different microphysical properties, or both.

Several of the most fundamental questions concerning dynamics apply equally to all of the major planets, and can be addressed through a comparative study of each planet at a similar level of detail given the sustained campaign of observations of Jupiter from the Galileo orbiter, and now Cassini is well on the way to providing at least as much information for the dynamicist about Saturn's atmosphere. One key question is whether the cloud-level circulation resides in a shallow atmospheric layer, separate from the deep interior, or is the outermost part of a flow which extends right down to the molecular hydrogen fluid layers at great depths, temperatures and pressures. By investigating all of them, we should get a better idea of the cause of the organised cloud bands on the major planets, what determines their scale and stability, and of the energy sources which drive the cloud-level circulation. The indications from the Galileo probe that the cloud-level zonal wind on Jupiter penetrates below the visible cloud tops suggest the possibility of deep circulation patterns. Such patterns are expected to be coherent in the direction parallel to the rotation axis of the planet, and might be different (for example, because of difference in wave propagation characteristics and in stability criteria for barotropic disturbances) on either side of the tangent cylinder which intersects the outer boundary of the metallic hydrogen core. On Jupiter this boundary would appear near the surface at a latitude of around 40°, whereas on Saturn it should appear much closer to the poles around 70°, owing to the smaller core size.

The zonal winds on Saturn are stronger and more extensive in latitude than on Jupiter. This may reflect the larger vertical scale height on Saturn compared with Jupiter (a consequence of the lower gravity), but may also be related to the substantially deeper molecular hydrogen envelope on Saturn which gives the internal convection more opportunity to build up a substantial reservoir of momentum against reduced effects of friction and dissipation. Until we can probe deep enough into Saturn's atmosphere to discriminate directly between deep or shallow models for this structure, there may be some clues to be gleaned from observing how the circulation patterns change between low and high latitudes (suggesting a signature of the 'tangent cylinder' mentioned above). The equatorial jet on Saturn is of particular interest, since it is prograde and much stronger than the equivalent feature on Jupiter. Waves and eddies must almost certainly play a crucial role in its generation, and observations from Cassini over a sustained period of time will reveal some new information on the interaction between waves and the equatorial jet, at least in the upper troposphere and lower stratosphere.

Saturn's atmosphere shows a number of large spots, which are cloud features associated with compact oval eddies, similar to the Great Red Spot (GRS) and White Ovals on Jupiter. Saturnian red spots are smaller and not as red as the GRS, and maybe not red at all in the sense of requiring a unique chromophore — the colours in some of the neighbouring regions are nearly the same. Cassini is assembling the detailed observations of these features — visible morphology, cloud motions and structure, thermal structure and compositional variations — which will allow them to be classified, characterised and hopefully understood. Models of these kinds of feature are becoming increasingly sophisticated and refined, and the observations from Cassini will enable a detailed comparison to be made between the Saturnian and Jovian versions of these features via these kinds of model.

Finally, Saturn has a number of unique dynamical features which were discovered during the Voyager flybys and which have no obvious counterpart on Jupiter. At mid-to-high latitudes, Voyager found a quite different type of wave structure to those on Jupiter — the 'ribbon waves' at around 47°N, and the 'polar hexagon', a regular six-sided pattern around the pole at a latitude of about 70°N. This puzzling pattern has to be some kind of planetary wave, possibly one trapped in some kind of shallow channel which forms as part of the overall banded structure of Saturn. It is still quite hard to explain theoretically how the geometrical shape arises, and why there is no analogue on Jupiter or Earth.

A related feature is the vortex discovered by Cassini in Saturn's south polar region in November 2006. The vortex resembles a swirling hurricane with winds blowing at 550 km/h and displays two spiral arms of clouds and a well-developed eye ringed by towering clouds. The vortex is 8,000 km across, much larger than anything similar observed on Earth, although eye-wall clouds are a common feature of hurricanes on Earth. On our planet, such systems form where moist air flows inward across the ocean's surface, then rises vertically and causes rainfall around an interior walled region of descending air, which is the "eye" itself. On Saturn, the polar downwelling is likely to be part of the planet-wide general circulation, in common with the vortices seen over both poles on Venus.

The overarching question concerning the dynamics of Saturn's atmosphere is whether Saturn is just a smaller, colder version of Jupiter or whether there are major differences which cannot be explained by extrapolating our increasingly more detailed understanding of Jupiter. Even before Cassini arrived at Saturn, we were able to use the success of Galileo to predict how the new models of Jovian dynamics would account for features such as the morphology of the bands, the rapid zonal winds speeds, the nature of the compact eddies (giant oval spots) and so on. The meteorological activity seen in pictures of Saturn may, for example, have its origins in bubbles of moist gas rising from deep in the atmosphere, made buoyant by the release of internal heat. Similar phenomena can be seen on Jupiter, probably on Neptune and possibly also on Uranus. The composition of the gaseous atmosphere and clouds in the neighbourhood of these features may contain information about the composition at depths otherwise inaccessible to remote sensing. Furthermore, maps of the distribution of transient chemical species can also be used as tracers of horizontal mixing and transport e.g. within and around the bands and compact oval eddies. Such information can give important clues as to what processes may be sustaining the major eddies against dissipation. The Cassini data on temperature, cloud properties and compositional mapping should allow us to test such extrapolations.

Although the present-day atmospheres of Saturn and Titan are so different, with the cold, massive gas giant planet having retained the light elements hydrogen and helium in abundance, if we were to imagine Titan's atmosphere removed and replaced with an unlimited amount of gas from Saturn, the heavy components, primarily ammonia and methane, would be retained and photolysis would eventually produce nitrogen and higher hydrocarbons — including those which condense as aerosols — so the result might be the atmosphere we see today.

10.2.4 *Titan and the Other Saturnian Satellites*

When the Cassini–Huygens mission was launched on its 7.5-year-trek, there were 18 known satellites orbiting the planet. By the time the spacecraft had arrived in the Saturnian system the number had risen to 31, with 13 more moons found using ground-based observations, mostly by astronomers from the University of Hawaii.

Titan is not only Saturn's largest satellite, it has more mass than the rest put together. Titan orbits at a distance of about 1.2 million km from its primary, or some 20.6 Saturn radii. This could be compared with Ganymede, which is about 1.0 million km, or 15 Jupiter radii, from that planet. In comparing Titan to Saturn, the magic number is 20: Titan is approximately 20 times smaller in radius as well as being at a distance of roughly $20R_s$. The mass of Saturn is larger than that of Titan by a factor of about 4,000, or $\frac{20^3}{2}$.

Titan's orbit around Saturn is significantly eccentric, with the radius of the orbit varying during each orbit by as much as 2.8%, which is too large to be due to perturbations by existing bodies, at least if they have always been in their present locations. Titan may have been knocked sideways by the impact of a large body such as a comet. If so it must have been relatively recently since tidal forces act constantly to circularise the orbit.

No integrated theory exists which explains the relationship of Titan to the other satellites and rings of Saturn. Why did this particular pattern emerge, with one very large satellite and the relatively dense rings, so much thicker than those of any other planet that they were for a very long time believed to be unique in the Solar System? The answer to questions like these lies in first achieving a much more detailed understanding of the present-day composition and constitution of Titan and its siblings; its parent, Saturn, and the other planets and their systems of moons and rings; and working back to a model of the formation and evolution of the whole Solar System. In addition to Titan, Saturn has six mid-sized satellites with diameters of more than 500 km (Table 10.5). Into this group fall Rhea and Iapetus (each less than a third of Titan's size), Dione and Tethys (with diameters close to 1,000 km) and Enceladus, the smallest of the mid-size satellites. Janus, Epimetheus, Mimas, Prometheus, Hyperion, and Phoebe are smaller but still reasonable-sizable objects, with diameters of more than 100 km. The first four orbit close to the planet and are therefore difficult to observe from the Earth because of scattered light from Saturn, while Hyperion is ten times and Phoebe 100 times further away from Saturn than Prometheus, the innermost of this size group. Most among these, and the even smaller icy satellites Pan, Atlas, Pandora, Telesto, Calypso and Helene, were not discovered before the space missions of the 1980s.

Cassini has added 4 new satellites to the Saturnian family of 31 officially recognised and named moons known by 2005. The two smallest satellites yet detected around Saturn were found on Cassini/ISS images in May 2004. Named Methone and Pallene, the two moons span 3 and 4 km in diameter respectively and orbit around the primary at distances of 194,000 and 211,000 km, between the larger moons Mimas and Enceladus. In September of 2004, Polydeuces was found orbiting Saturn at 377,400 km, in the F-ring region. Finally — for now — 7-km across Daphnis appeared in May 2005 on images showing the Keeler gap in Saturn's outer A ring, at 136,505 km from the centre of Saturn. Table 10.5 summarises the radius, mass,

Table 10.5   Data on Saturn's moons as known in mid-2007 ("retro" stands for retrograde motion).

Name	Date of discovery	Radius (km)	Mass (kg)	Orbital distance (km)	Orbital period (days)
Albiorix	2000	13		16,394,000	
Atlas	1980	$18 \times 17 \times 13$	2×10^{15}	137,670	0.6
Calypso	1980	$15 \times 8 \times 8$	8×10^{17}	294,660	1.89
Daphnis	2005	3.5		136,530	
Dione	1684	559	1.1×10^{21}	377,400	2.74
Enceladus	1789	256	1.2×10^{20}	237,378	1.37
Epimetheus	1966	$72 \times 54 \times 49$	5.3×10^{17}	151,422	0.69
Erriapo	2000	4.3		17,604,000	871
Helene	1980	$18 \times 16 \times 15$	8×10^{17}	377,400	2.74
Hyperion	1848	$180 \times 140 \times 112$	8×10^{17}	1,481,000	21.28
Iapetus	1671	718	1.6×10^{21}	3,561,300	79.33
Ijiraq	2000	5		11,442,000	450
Janus	1966	$98 \times 96 \times 77$	1.9×10^{18}	151,472	0.69
Kiviuq	2000	7		11,365,000	449
Methone	2004	1.5–2	?	194,000	1
Mimas	1789	196	3.80×10^{19}	185,520	0.94
Mundilfari	2000	2.8		18,710,000	951 (retro)
Narvi	2003	3.3		18,722,000	956 (retro)
Paaliaq	2000	9.5		15,198,000	687
Pallene	2004	2		211,000	1,1
Pan	1990	10	5×10^{15}	133,583	0.57
Pandora	1980	$57 \times 42 \times 31$	1.5×10^{17}	141,700	0.63
Phoebe	1898	110	4.0×10^{17}	12,952,000	550.4 (retro)
Polydeuces	2004	6.5	?	377,400	2.74
Prometheus	1980	$72.5 \times 42.5 \times 32$	1.9×10^{17}	139,350	0.61
Rhea	1672	765	2.31×10^{21}	527,040	4.52
Siarnaq	2000	16		18,195,000	893
Skathi	2000	3.2		15,641,000	729 (retro)
Suttung	2000	2.8		19,465,000	1017 (retro)
Tarvos	2000	6.5		18,239,000	926
Telesto	1980	$15 \times 18 \times 12$	8×10^{17}	294,660	1.89
Tethys	1684	530	6.1×10^{20}	294,660	1.89
Thrym	2000	2.8		20,219,000	1089 (retro)
Titan	1655	2,575	1.35×10^{23}	1,221,850	15.94
Ymir	2000	8		23,130,000	1312 (retro)

distance from the centre of the planet, discoverer and the date of discovery of each of the 35 satellites of Saturn.

More recently discovered satellites for which very little is known to date include such exotic names as Aegir (S/2004 S10), Anthe (S/2007 S4), Bebhionn (S/2004 S11), Belgelmir (S/2004 S15), Bestiq (S/2004 S18), Farbauti (S/2004 S9), Fenrir

(S/2004 S16), Fornjot (S/2004 S8), Greip (S/2006 S4), Hati (S/2004 S14), Hyrokkin (S/2004 S19), Jarnsaxa (S/2006 S6), Kari (S/2006 S2), Loge (S/2006 S5), Skoll (S/2006 S8) and Surtur (S/2006 S7).

Most of Saturn's satellites, including Titan, exhibit synchronous rotation, which means that they spin once for each orbit of the planet, and so always keep the same face towards Saturn. The exceptions are Phoebe, and the chaotic Hyperion. All but Iapetus and Phoebe have nearly circular orbits and lie in the equatorial plane. They all have densities which are so low that they must be made mostly of ice, but not so low that they could be made of ice alone. The most probable structure for each of them (and Titan too, see Chapter 9) is a rocky centre surrounded by a thick shell of ice. Taking typical densities for rock and ice, and knowing the size of the satellite and its mass, a model for the relative sizes of the rocky core and icy mantle can be worked out.

Most of the satellites are bright, with albedos that reflect 60 to 90% of the light of the Sun. Prior to the arrival of Cassini, we already knew from high-resolution photometry and spectroscopy that the brightest among Saturn's satellites contain water ice. However, the four outer satellites are darker, especially Phoebe which reflects only 2% of the light that strikes it. These dark satellites probably did not form with Saturn, but may be captured asteroids instead. More recent measurements were devoted to the study of the smallest among them: Rhea, Hyperion, Tethys, Dione and especially Iapetus, a mysterious object with an extremely bright side and a much darker one. Hyperion's spectrum in the visible range and in the near infrared is compatible with that of Titan, showing a brighter leading and a darker trailing hemisphere. Hyperion is in resonance 4:3 with Titan and it has been suggested that Hyperion material was deposited in the past on Titan, as a reason for the hemispheric asymmetry exhibited in the latter's spectrum. Hyperion's spectrum is also similar to that of Iapetus in terms of cratering density, although its albedo is much higher (0.21 in the visible, lower to all satellite albedos but for the ones of Phoebe and Iapetus). Its red colour and intermediate albedo value suggest the presence of an additional component (besides water ice) on the surface of Hyperion. A partial coverage by Phoebe-originating material has been suggested in this case also, as on Iapetus.

Of course, the ice and the rock will not be pure water or silicate but may contain many impurities. The icy surfaces are obviously contaminated by materials of various colours with distinctive infrared spectral properties. Contrary to expectations, the Voyager and Galileo spacecraft revealed in 1980 and 1997 a wide range of geological processes on the moons of the outer planets and showed that several of the larger Kronian satellites may have undergone melting of their mantles and subsequent differentiation and resurfacing by liquid water or water-silicate mud. Flows of these liquid components resemble the lava flows on the Earth and the Moon, produced there by the melting of silicate rock mixtures. The ridged and grooved terrain found on Enceladus and Tethys may be the result of tectonic activities common to many parts of the Solar System. Explosive volcanic eruptions may be occurring on Enceladus, as on Earth, Io and Triton. Dark organic material and hydrated silicates are likely candidates for the composition of the external objects of our Solar System and could explain the albedos of both Hyperion and Iapetus. It is worth noting that Hyperion is the satellite closest to Saturn after Iapetus and, consequently, it is quite conceivable because of this proximity, that its surface may be covered by a similar material, which, however, was not deposited in the case of Hyperion preferentially on one of the hemispheres, due to its chaotic rotation.

Thus, among the smaller medium-sized icy satellites of Saturn, Hyperion and Iapetus present the mysterious hemispheric asymmetry that is also found on Titan. The observational data that we possess today have not provided yet a valid explanation as to the composition of the surface of these satellites, in particular the nature of the very dark component observed on all of them. It is quite possible that the surfaces of Hyperion, Iapetus, Titan and Phoebe contain common features that speak of their shared evolutionary history.

In the young Solar System, meteoritic bombardment was intense and frequent causing the surfaces of the solid bodies, including the icy satellites, to become heavily cratered. Nowadays the bombardment continues but at a significantly slower rate. Knowing something about the flux of impacting material and by enumerating the number of craters on a solid surface, geologists can recover information on the origin and epoch of formation of ground portions. Impacts have several consequences: they pulverize the rock and the icy components of a surface, forming a fine material that covers it, the regolith; they tend to excavate and expose fresh material on the surface; they also cause the vaporization and escape to space of volatiles, and they leave on the ground a deposit of opaque and dark material. As a consequence, the larger satellites of Saturn, like those of Jupiter, have a brighter 'leading' than 'trailing' hemisphere (the leading side is the hemisphere seen from the Earth when the satellite moves towards our planet; the trailing side is the opposite). This is the case for Hyperion and Titan, for instance, but not for Iapetus, as will be discussed later.

The hemispheric asymmetry is believed to be due to a higher rate of meteoritic encounters on the leading side. The fact that well-formed craters, similar to those observed in the inner solar system bodies, are also observed on the Galilean and Kronian satellites, suggests that the surface ice is strong enough to sustain immense impacts, perhaps because it is mixed with silicate contaminants or other impurities, in addition to being very cold. Surface gravity is apparently not enough in these cases for viscous relaxation to have eradicated the impact features. Mimas is covered with high-rimmed, bowl-shaped craters, one of which (Herschel) occupies a third of its surface. The number of craters found on Mimas suggests that the satellite has undergone numerous episodes of resurfacing.

All of the Saturnian satellites are being visited by Cassini, as the spacecraft follows its orbit around Saturn (see Table 4.4 for the flyby dates).

10.2.4.1 Iapetus

Iapetus has a number of curious properties: it is the only large satellite of Saturn in a highly inclined orbit; it is less dense than any other object of similar albedo; and its two faces are radically different, one bright and the other very dark. The surface

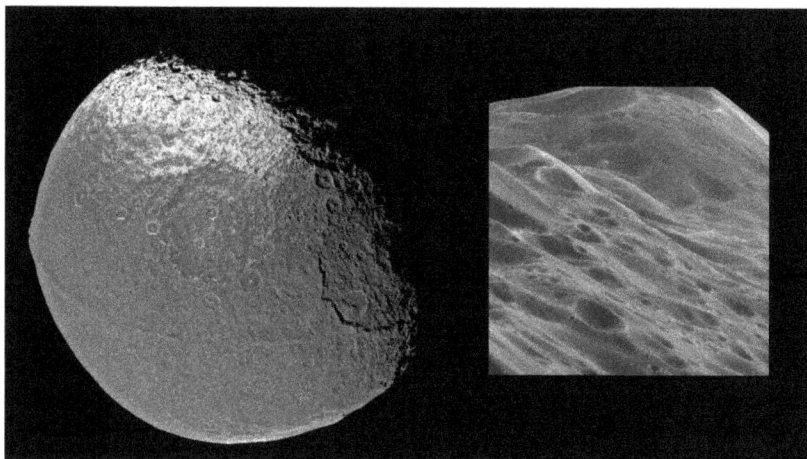

Figure 10.9   Cassini/ISS images of Iapetus. The view on the left taken on New Year's Eve 2004 is centred on the moon's equator and on roughly 90 degrees west longitude and shows both bright and dark terrain, the remarkable ridge about 20 km wide and 20 km high that coincides with the geographic equator, and the large crater called Cassini Regio. On the right is a more recent view taken on September 10, 2007, from a distance of 3,870 km with resolution down to 23 metres (NASA/JPL/SSI).

albedo on the trailing hemisphere varies from 0.4 to 0.5, values typical for ice-covered objects, to as low as 0.02 in the central parts of the leading hemisphere. As early as 1671, when Cassini discovered Iapetus, it was noticeable that at one point on its orbit around Saturn, the satellite seemed quite bright, whereas on the opposite side of the orbit, it almost disappeared! Cassini correctly deduced that this indicated a marked difference between the moon's two hemispheres, with the trailing hemisphere, the side viewed from the Earth when the satellite moves away from us, bearing highly reflective material, while the other (the one that faces forward in the direction of the motion of Iapetus around Saturn, and which might conceivably sweep up debris as the satellite travels in its orbit) was of a darker nature.

Images of Iapetus obtained by Voyager confirmed these observations and show that the bright part of the satellite, especially around the north pole, is covered with craters, reminiscent of Mercury or the Moon, or Jupiter's satellite Callisto. The other side is not only much darker (by about a factor of 5), but also covered with a thick layer of redder material, with a symmetrical distribution about the apex of the orbital motion. In the craters found near 200°E longitude, the shape and orientation of the dark deposits seem to cover bright terrain, which suggests that they are more recent. The mean density of the satellite is $1.16 \, \mathrm{g\,cm^{-3}}$, somewhat less than the other icy satellites of Saturn, and suggesting a composition of almost pure water ice.

Voyager was unable to map the surface on the whole of the leading hemisphere, but Cassini made one flyby on January 1, 2005 at a distance of 122,647 km from the surface. Another flyby in September 2007 had Cassini pass at just 1,227 km from Iapetus, providing astonishing close-up images of the satellite. Although it is no longer uncharted land, the equatorial ridge that bisects Iapetus' terrain, and how it fits into the story of the moon's strange brightness dichotomy remains a mystery. Also puzzling is the origin of the dark territory of Cassini Regio on Iapetus and the nature of the dark material. The observed symmetry with respect to the direction of the orbital motion of the satellite strongly argues in favour of an external origin for this material, or at least an external controlling mechanism. Studies of the light reflected from the dark face of Iapetus, using the large infrared telescopes in Hawaii, show that it is very red in the visible and near infrared. A fit to the spectrum was obtained at first by using a mixture of polymers (10%) and hydrated silicates (90%), simulated in the laboratory. A more recent study, however, achieves a better agreement with a mixture of HCN polymers, organic residuals and water ice.

A possible explanation for the bizarre appearance of Iapetus might lie with Phoebe, the outermost satellite, which has a retrograde orbit and a surface albedo similar to, but a little less red than, that of Iapetus. Phoebe looks like a captured asteroid, and as such probably consists of dark meteoritic material. If dust is eroded from Phoebe by some mechanism then it will tend to be pulled inwards by inter-actions with photons from the Sun and Saturn, the so-called 'Poynting–Robertson' effect, and could be swept up by Iapetus. It is also possible that the dark material is native to Iapetus, and is being exposed by the removal of ice due to the impact of dust from Phoebe and elsewhere. A recent study of some UV Voyager archived images indicates that both might be true, suggesting that the dark deposit is a mix-ture of Phoebe material and a pre-existent non-volatile constituent, uncovered by the impacts. A candidate for the latter could be methane erupting from the interior and subsequently being darkened by ultraviolet radiation. To explain the colour of Iape-tus, it is conceivable that dark particles ejected from Phoebe may have undergone chemical changes that rendered them redder.

10.2.4.2 Rhea

Rhea is the second most massive satellite of Saturn, but still possesses less than 2% of Titan's mass. The surface of Rhea bears distinct signs of multiple impacts and is heavily cratered. One hemisphere shows bright wispy streaks that may have been caused by ejecta from impacts or perhaps have been caused by the condensation of volatiles leaking from the interior as a sign of internal activity. Parts of the ground are covered with large craters while other regions lack them. The larger craters are probably the signatures of early bombardment, while the smaller ones may have been more prevalent during a later episode of collisions. In many respects Rhea looks very much like the Moon or Mercury, though the morphology of individual craters is somewhat different.

During a Cassini flyby on November 26, 2005 (at a distance of only 500 km), the observations of Rhea's surface indicated the presence of two geologically different areas, one with large craters and another one with less than 40 km in diameter craters, suggesting — for the latter — a major resurfacing event.

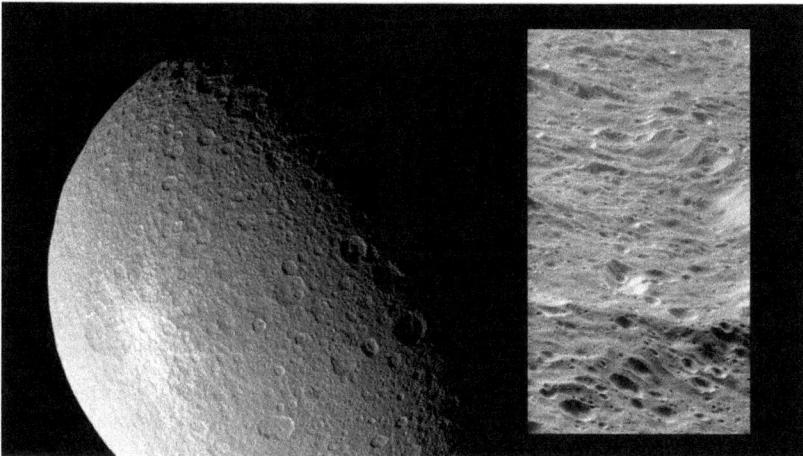

Figure 10.10   Left: The leading hemisphere of Rhea, the second largest satellite of Saturn, is heavily cratered and shows bright and dark streaks on its surface. The Cassini camera aquired this image showing a bright and rayed (hence young) crater. Right: A sea of craters observed on Rhea during a close Cassini flyby in November 2006. This high-resolution image shows the surface of the scarred face of Rhea to be filled with rolling mounds and many smaller craters (NASA/JPL/SSI).

10.2.4.3 Dione

Dione appears to be quite similar to Rhea, showing a wide diversity of morphology. It also shows parts in which craters seem to be too few in number, as though some have been obliterated or covered up, indicating that several periods of resurfacing have occurred during the first billion years of its existence. It therefore appears that, although small, these bodies nevertheless exhibited significant internal activity in some distant past. This could be the case if the relatively high densities have provided added radiogenic heat from siliceous material to spur this activity.

The leading hemisphere of Dione is about 25% brighter than the other and less heavily cratered than the trailing side, as demonstrated by Cassini views taken during a 1,000-km flyby in October 2005. This is contrary to what was expected prior to Cassini's arrival. The anomaly suggests that during the period of heavy meteors bombardment, Dione was tidally locked to Saturn in the opposite orientation and that due to impact(s), it rotated to its current position.

10.2.4.4 Tethys and Mimas

Icy Tethys exhibits a large impact feature, similar in size to that observed on Mimas. The latter is much the smaller of the two satellites, and the largest crater, Herschel, is so large relative to the radius of the body that it seems quite improbable that Mimas could receive such a blow and survive. Tethys, like Dione and Rhea, is covered with many impact craters, one of which, Odysseus, is the biggest such known formation in the Solar System. Also on Tethys a huge trench formation, the Ithaca Chasma (named for Odysseus' home country) is found, its origin perhaps something like

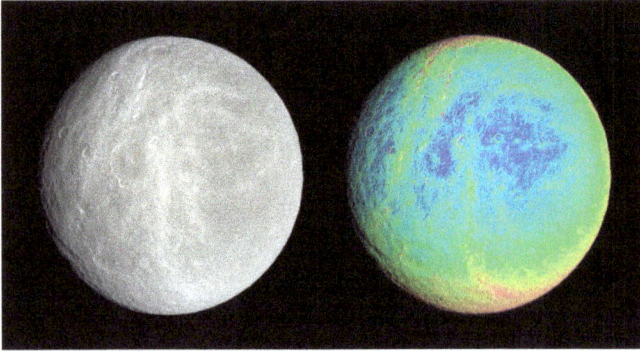

Figure 10.11   Side-by-side natural colour and false-colour views highlight the wispy terrain on Rhea's trailing hemisphere. The combination of colour map and brightness image shows that colours vary across the surface of Rhea, which may be due to subtle differences in the surface composition or the sizes of grains making up the icy surface material. The images were taken with the Cassini spacecraft narrow-angle camera on January 17, 2007 (NASA/JPL/SSI).

Figure 10.12   Dione is also heavily cratered, with bright streaks, visible here towards the limb. This extensive canyon system, photographed by Cassini on September 25, 2006, is located on a region of darker terrain than the rest of the moon, on the Saturn-facing hemisphere (NASA/JPL/SSI).

Figure 10.13   Natural colour and false colour images of Tethys, showing the gigantic impact structure called Odysseus (450 km across) on its leading side. Inside the basin, cliffs, mountains and smaller craters are visible. The images were taken on December 24, 2005 (left) and a year later in December 2006 (right) by the Cassini camera (NASA/JPL/SSI).

Figure 10.14   Left: Mimas, featuring the crater (130 km across and 10 km deep) named after Herschel, who discovered the satellite on September 17, 1789. Right: False-colour view of Mimas showing differences across the satellite's surface. These differences are not yet understood but are likely to be due to variations in surface composition (NASA/JPL/SSI).

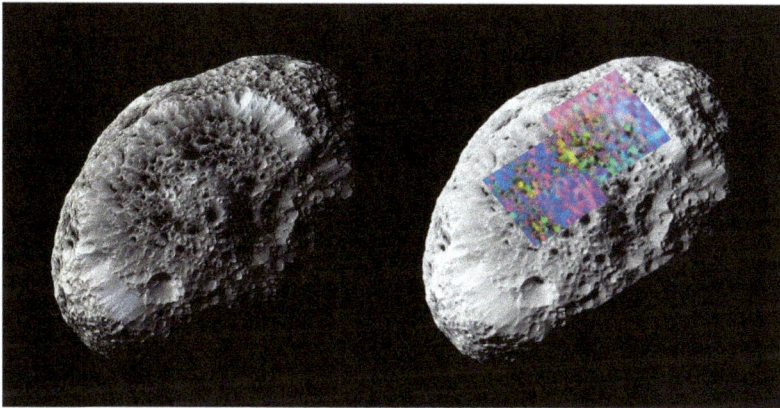

Figure 10.15   Left: Cassini/ISS viewed Hyperion's tumbling form on September 26, 2005. In natural colour, Hyperion appears reddish. Right: The colour map shows the composition of a portion of Hyperion's surface determined with Cassini/VIMS. Blue shows the maximum exposure of frozen water, red denotes carbon dioxide ice ("dry ice"), magenta indicates regions of water plus carbon dioxide, and yellow is a mix of carbon dioxide and an unidentified material (NASA/JPL/SSI/U. Arizona/ARC).

that of the grooves on Enceladus. The density of Tethys suggests that the moon is composed almost entirely of water ice. Viscous relaxation and flow in the craters of Tethys under the influence of its strong gravitational field give them an aspect flatter than for Mimas or the Moon. Regions where craters are fewer give evidence for several episodes of resurfacing on both Tethys and Mimas. On the latter, the surface gravity is not sufficient to allow the topography to relax, and the craters tend to have high rims.

Figure 10.16   Left: Cassini observed Enceladus from a distance of 4.1 million km as it passed in front of Titan, a further 1.2 million km behind, on February 5, 2006. Right: Enceladus is only 500 km across, and very bright, with evidence of resurfacing episodes in the past and southern fractures across the southern hemisphere which correspond to the location of geyser-like eruptions feeding a thin, water-dominated atmosphere (NASA/JPL/SSI).

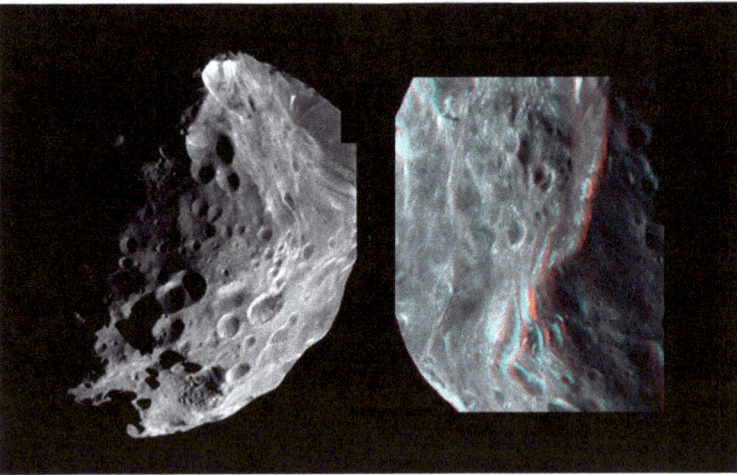

Figure 10.17   Left: Phoebe's dark and irregular surface as seen by the Cassini camera in 2005. Right: Phoebe's violent, cratered past is evident in this 3-D image of the tiny moon, showing bright material, likely to be ice, exposed atop a ridge-like feature, as well as around small craters and down the slopes of large craters. There are also bright streaks on steep slopes, perhaps where loose material slid downhill during the seismic shaking of impact events. This 3-D mosaic is made up of two images taken with the Cassini spacecraft narrow angle camera on June 11, 2004 (NASA/JPL/SSI).

Mimas is located at a distance of 185,520 km from Saturn and was discovered in 1789 by William Herschel. Its mean density is so low, at 1.17 g cm^{-3}, that it must be made primarily of water ice throughout, and its surface is so heavily cratered that it has most likely been collecting the impact features we see now since the time of

Figure 10.18   Titan and Jupiter's satellite Europa, approximately to scale (NASA/JPL).

Figure 10.19   Titan, Neptune's satellite Triton, and the minor planet Pluto, approximately to scale (NASA/JPL).

its creation, without any global melting or resurfacing episodes. One of the craters, named after Herschel, is surprisingly large — nearly one-third the moon's entire diameter and 10 km deep. It features a central mountain that rises 6 km above the crater floor. Traces of fracture marks can be seen on the opposite side of Mimas, presumably caused by the shock waves from the impact that produced the Herschel crater.

10.2.4.5 *Hyperion*

Hyperion is remarkable because of its chaotic rotation and its irregular shape — the largest non-spherical object in the Solar System. Although its global composition is uncertain, we know that water ice is present on the surface. Its spectrum in the visible range and in the near infrared is like that of Titan, showing a brighter leading and a darker trailing hemisphere (the opposite of Iapetus). Its irregular shape is evidence for heavy bombardment by meteors and indicates that the surface is very old. The eccentric orbit makes it subject to the gravitational forces from the primary planet, which causes Hyperion's rotational period to vary from one orbit to the other.

Hyperion's orbit is in resonance with that of Titan and the asymmetry in Titan's brightness could conceivably be due to the deposition of material from Hyperion. It has even been suggested that Titan's atmosphere originated in material which came in this way. Computer models in which Hyperion was larger initially, but suffered a catastrophic disruption during the formation period of the Saturnian system, show the fragments ejected falling on Titan over a time scale of about a thousand years. If Titan already had an atmosphere at that time, the shock effects of these impacts could have played a role in producing nitrogen in an ammonia-rich atmosphere. This sort of model has the advantage of helping to explain the remarkable features, not only of Titan, but also of Hyperion. If the latter has the composition of a 'native' Saturnian satellite (i.e., mostly ice) rather than an asteroid (like Phoebe) then its shape and motion do indeed suggest that it met with a major accident in its past.

Although it now seems that there are no very large oceans on Titan, small seas or lakes are not incompatible with the observations, and may already have been observed (Chapter 9). If such lakes do exist, they may actually be impact craters produced by pieces of Hyperion, filled in with liquid hydrocarbons from drizzle or rain. Hyperion is similar to Iapetus in terms of cratering density, and it has the same sort of red colour to its spectrum, although its albedo is much higher. Since Hyperion is the next satellite in, towards Saturn, after Iapetus, perhaps it is also partially covered by dark organic material and hydrated silicates from Phoebe. In the case of Hyperion, with its chaotic rotation, the deposition would not be expected to be preferentially on one of the hemispheres.

10.2.4.6 *Enceladus*

Although a small moon, Enceladus may well be one of the most interesting and beautiful examples among the primitive, airless worlds accompanying Saturn. Extremely close to its planet, at a distance of 237,378 km, Saturn's apparent diameter ($25°$) in Enceladus' sky is 50 times larger than that of the Moon seen from the Earth. Part of its surface was detected by Voyager and seemed entirely covered with vast plains of ice, an ice so pure and bright that it reflects almost all of the visible radiation incident on it (by comparison, our Moon reflects only 11% of the light it receives). The

Cassini spacecraft performed three close flybys in 2005, revealing not only a stunning, bright surface but also discovering a tenuous atmosphere, composed mainly of water, emerging from eruptions mainly in the south polar region. It was first detected by the Cassini magnetometer, which saw a bending of Saturn's magnetic field caused by the field lines coming in contact with a conducting object (the atmosphere). With a diameter of only 500 km, Enceladus does not have enough gravity to hold a permanent atmosphere, implying the presence of a continuous source of vapour.

The lack of impact craters on half of Enceladus' surface points to a relatively small age, less than a billion years, showing it has been subjected in the recent geological past to extensive resurfacing by some form of ice volcanism. The Cassini Cosmic Dust Analyzer detected a large increase in the number of particles near Enceladus, thus confirming that Enceladus is the source of particles for Saturn's E-ring. The Visual and Infrared Mapping Spectrometer determined that the large dark cracks, informally called "tiger stripes", at the south pole are very young and seem to have a continual supply of fresh ice, implying recent geological activity. The Composite Infrared Spectrometer then found that the south pole is much warmer than expected, suggesting an internal heat source. With all these facts together showing that Enceladus had a very interesting south polar region, soon afterwards in 2005 plumes containing water vapour and small ice particles were detected originating in the tiger stripes and forming geyser-like eruptions, indicating possible pockets of liquid underneath, but very close to, the surface.

10.2.4.7 *Smaller Satellites*

Smaller than Hyperion, Phoebe's eccentric (retrograde) orbit suggests it may be a captured object, coming from the outer reaches of the solar system, perhaps from the Kuiper Belt region. With a rotational period of about nine hours, Phoebe is the only medium-sized Saturnian satellite to exhibit a simple, asynchronous rotation. Phoebe is very dark and reflects only 6% of the sunlight that arrives in its vicinity.

The remaining Saturnian satellites are all less than 100 km in radius. Eight of these can be classified in three groups: the shepherding satellites, Atlas, Pandora and Prometheus, believed to play a key role in defining the edges of Saturn's A and F rings; the co-orbital satellites, Janus and Epimetheus, moving in almost identical orbits at about two and a half Saturn radii; and finally the Lagrangian satellites, Calypso, Helene and Telesto, are so called because they orbit in the Lagrangian points of the larger satellites Dione and Tethys. Lagrangian points lie about 60 degrees in front of and behind the larger body, and correspond to locations within an object's orbit in which a less massive body can move in an identical, stable orbit. The only other satellites in the Solar System with this property are the so-called Trojan asteroids orbiting in two of the Lagrangian points of Jupiter.

One shepherd, Atlas, lies several hundred kilometres from the outer edge of the A-ring, while the other two orbit on either side of the narrow F-ring, constraining its width and causing its kinky appearance. Janus and Prometheus, discovered in 1966 and 1978 respectively, may once have been part of a larger body that disintegrated following a major collision. Every four years, the inner satellite (which orbits slightly faster than the outer one) overtakes its companion. Instead of colliding, the satellites simply exchange orbits. The four-year cycle then starts all over again.

Tiny Pan was the last satellite of Saturn to be found in Voyager 2 images, detected in 1990 in pictures dating back to 1981. The 10 km radius object lies hidden within the A-ring, and contributes to keeping the Encke division clear of particles.

10.2.5 *Titan and Europa*

Jupiter's satellite family is even larger than Saturn's, with 61 moons discovered to date, although most of them are quite small. Of the four large Galilean satellites (named after Galileo, who discovered them in 1609) Callisto, Ganymede, Europa and Io, Europa is the smallest and the second closest to Jupiter. With a radius of 1560 km, it is slightly smaller than Earth's Moon. Objects of this size are expected to lose their internal heat in a relatively short period of time, making Europa particularly interesting because it apparently has a liquid water ocean at a relatively small distance below its icy surface. Tidal friction may have heated the interior of Europa enough to melt ice and produce a subsurface ocean, the same process that has rendered the rocky interior of its sibling Io completely molten. Since biologists indicate that the presence of some sort of liquid is required to serve as a medium for the chemical reactions needed to sustain life as we know it, Europa has been considered a rival to Titan (and now Enceladus) as favoured subjects of astrobiological study in the future.

10.2.6 *Nitrogen Atmospheres in the Outer Solar System*

In comparing Titan and Enceladus to other bodies in the outer Solar System, the small planet Pluto and Neptune's satellite Triton are sometimes singled out over the similarly-sized Galilean satellites of Jupiter, because of their common atmospheric composition. Indeed, with the Earth, these four objects are the only ones to possess an atmosphere essentially made of molecular nitrogen (Table 10.6). However, those of Triton and Pluto are very thin, because under the conditions at their surfaces, at their extra distances from the Sun compared to Titan (Table 10.7), the vapour pressure of nitrogen is very low. Thus, instead of the Earthlike values on Titan, surface pressures on Triton and Pluto are measured in microbars (Table 10.8).

Spectroscopic measurements from the Earth had demonstrated as early as 1983 the existence of nitrogen on Triton, possibly in the form of ice. It took however, almost ten more years of devoted observations of this faint world, to confirm the presence of solid N_2 and CH_4 on the surface. Similar observations of Pluto revealed that frozen molecular nitrogen and traces (of the order of one percent) of methane

Table 10.6 The bodies with a nitrogen atmosphere in the outer solar system are compared (and with Ganymede, the largest satellite, as a reference). Courtesy of S. Atreya.

Characteristic	Ganymede	Titan	Enceladus	Triton	Pluto
R_{planet}	$14.99R_s$	$20.25R_s$	$3.95R_s$	$14.33R_N$	[39.53 AU]
M [10^{22}kg]	14.82	13.5	0.011	2.14	1.31
R_o [km]	2631	2575	252	1352	1150
ρ [kg/m^3]	1936	1880	1608	2064	2030
g [m/s^2]	1.43	1.35	0.12	0.78	0.4
T_O [days]	—	—	—	—	[248.5 yr]
T_s [days]	7.16	15.95	1.37	5.877	[6.38]
i [deg]	0.18	0.33	0.02	157	17.14
e	0.001	0.029	0.005	0.000	0.25
A	0.4	0.2	1.4	0.4	0.52
v_O [km/s]	2.75	2.64	$0.235 (< v_T!)$	1.50	1.1
Surface T [K]	110	94	114–157	38	40
P	X	1.5 bar		16 μb	58 μb(var)
Atmosphere	O_3, (H_2O_2)	N_2, CH_4	H_2O, N_2, CH_4, CO_2, CO	N_2, CH_4	N_2, CH_4, (H_2O)

Table 10.7 Orbital data for Titan, Triton and Pluto.

	Titan	Triton	Pluto
Semi-major axis of orbit (AU)	9.54	30.06	39.44
Equatorial radius (km)	2575	1352	1137
Rotation period (day)	15.945	5.9 (retrograde)	6.4
Sidereal period (yr)	29.46	165	247.7
Mass (kg)	13.5×10^{22}	2.14×10^{22}	1.27×10^{22}
Density $(g\,cm^{-3})$	1.87	2.07	2.06
Orbital eccentricity	0.029	0.75	0.25
Surface gravity $(m\,s^{-2})$	1.35	0.78	0.60

Table 10.8 Atmospheric and surface structure and composition for Titan, Triton and Pluto.

	Titan	Triton	Pluto
Bond albedo	0.29	0.7	0.2–1.
Surface Pressure (bar)	1.5	14×10^{-6}	3×10^{-5} ?
Temperature at 1 mbar (K)	150–170	41	104 ± 21
Surface temperature (K)	94	38 ± 3	35–55
T_{max} (K)	185 ± 20	102 ± 3	?
Exobase height (km)	1600	930	?
$R_{exobase}/R_{planet}$	1.6	1.7	$\gg 3$?
Atmospheric composition	90–97% N_2	99% N_2	99% N_2
	1.5% CH_4	10^{-4}% CH_4	5×10^{-3}% CH_4
	0.1% H_2	10^{-4}% H_2	? H_2
	10^{-6}–10^{-4}% CO	? CO	? CO
Ocean	Non-global; hydrocarbons $(C_2H_6–CH_4–N_2)$	frozen	frozen
Ices	C_2H_2, CO_2, H_2O, + ?	N_2, CO, CH_4, CO_2	N_2, CO, CH_4
Crust	dirty ice, minerals?		

ice are also present on that surface, along with carbon monoxide ice. Compared to Triton, Pluto has three times more methane ice and five times more carbon monoxide ice respectively (Table 10.8). When the Hubble Space Telescope obtained images of the Pluto-Charon system, visually separating the two objects for the first time, Pluto's surface showed hints of surprisingly high-contrast morphology.

In 1992, following the encounter of Voyager 2 with Triton, this remarkable satellite of Neptune became a more familiar object in our Solar System. Its extremely tenuous atmosphere, mainly comprising molecular nitrogen and methane (the same as Titan, see Table 10.8), is nevertheless the site of occasional clouds, and winds blowing from the south pole towards the equator above an agitated surface with at least three active geysers. Surprisingly enough, in this inclement world, the atmosphere — with a pressure of about 16 ± 3 microbars of N_2 — is almost in vapour pressure equilibrium with the surface, heated to a modest 38 K.

As on Titan, on Triton molecular nitrogen and methane break up and recombine under the action of photolysis, producing other organics such as hydrocarbons and nitriles, which are responsible for the pink or pale yellow colours of the surface. But the bulk ratio of rock to ice varies, with Titan containing less rock, since it has a smaller density. The conditions prevailing during the formation of these objects must have been different in view of their characteristics today. In part because of its unusual retrograde orbit about Neptune, it is believed that Triton might have been captured after Neptune formed. In that case, Triton and Pluto may have very similar histories, and indeed they do have almost identical mean densities (Table 10.7), which might mean the same overall composition.

The surface appearance of Pluto remains mostly unknown, because it has not yet been visited by a space probe and is hard to view from Earth due to its faintness (magnitude 13.5). With its satellite Charon, it constitutes a most unusual planetary system about which we have recovered most of our present-day understanding through a series of eclipses and occultations that took place at the end of the 1980s. Thanks to these events, it was possible to measure the radii and bulk densities of Pluto and Charon, and to obtain some indications as to their global compositions, which appear to be quite similar for both these objects. Pluto's atmosphere, which is gradually frozen on the ground as the dwarf planet follows its eccentric orbit away from the Sun, is mostly molecular nitrogen, with a trace of gaseous methane. The surface temperature is currently approximately 40 K, which translates (under vapour pressure equilibrium conditions between the surface and the atmosphere) into a nitrogen atmospheric pressure on the surface from 10 to 30 microbars and a gaseous methane mixing ratio of about 0.005.

We will gain more information on these mysterious worlds through the data that will be returned by the New Horizons Pluto–Kuiper Belt mission, launched on 19 January, 2006. This aims to make a fast flyby in July 2015 of the dwarf planet and its moons, Charon, Nix, and Hydra, and then continue its trip towards the rest of the Kuiper belt. The resolution achieved in the Pluto–Charon system will be in the order of 40 km on the surfaces of the solid bodies, bringing us close to these borderline worlds of our Solar System and allowing us to uncover their surface properties, the geology, internal structure and atmospheres. We shall then have finally explored *in situ* the farthest of the "historical" planets of our Solar System.

10.3 Titan's Origin and Evolution

Any theory of the origin and evolution of Titan has, first and foremost, to explain how the satellite — located 10 times further away from the Sun — managed to acquire an atmosphere which is thicker than the Earth's, and why this satellite should be so endowed while the similarly sized Jovian moons Ganymede and Callisto are essentially airless. Recent discoveries indicate that tenuous envelopes are also found around Enceladus (see Section 10.2.4.6) and Europa (Section 10.2.5). Two other outer planet satellites, Io (Jupiter) and Triton (Neptune), have thin, transient atmospheres fuelled by volcanoes, expelling sulphur dioxide and hydrogen sulphide, and nitrogen, respectively. The volcanoes are driven by tidal heating of the interior of the body in both cases, rather than primordial and radioactive heat as with the Earth. Titan would not be expected to have a great deal of either.

The origin of Titan itself, i.e. where the material that formed the satellite came from, and how it accrued to become the large size that it is, remains a problem that is at best only partially solved. Titan formed in the outer Saturnian sub-nebula, the disk of rock, dust and ice that accompanied the formation of Saturn, in which the temperature conditions were cool enough (below 60 K) to have given birth to the icy bodies we find throughout the entire Saturn system, from the ring particles to Phoebe. But other volatiles, borne by the ices in these bodies, reflect the fact that the planet at the system's centre was hot when it formed, so that during some period of Titan's evolution its atmosphere must have been substantially (more than 50 K) warmer than today. Substances such as methane, ammonia, nitrogen, carbon monoxide, and argon are expected to have been prevalent in the materials that formed the more distant satellites. Of these, methane, nitrogen, carbon monoxide and argon have been detected (the latter in very small quantities) on Titan, while ammonia is essentially absent from the atmosphere, but may exist in large quantities in the interior.

In the early stages of Titan's accretion in the proto-Saturnian nebula when Titan was small and its gravity weak, the rock-ice planetesimals accumulated slowly without melting. In later stages, as the incoming planetesimals impacted with more energy, in response to the gravitational pull of the growing satellite, melting and mixing would have occurred. Titan became massive enough to have undergone some internal differentiation, which concentrated ices towards the surface, leading to substantial outgassing, but still had enough gravity to retain all but the lightest gases. A primordial atmosphere of water vapour plus ammonia would likely have been present. The ammonia was likely converted into nitrogen by photolysis and other processes early in Titan's history. Within the first few hundred million years, the ice crust atop the mantle would grow. In the interior, as radiogenic heating warmed the ice and silicates, the more mobile ice deformed and allowed the core to overturn — the dense rock layer falling to the centre. Many models suggest that if ammonia is present, a mantle of liquid water persists to this day — perhaps 75 to 300 km below the surface.

The temperature of the proto-Saturnian nebula was likely lower than that of the proto-Jovian one, such that not only water, but also ammonia and methane (but probably not N_2), may have been incorporated into the nascent Titan. The NH_3 ices condensed from the nebula and later vaporised, with the subsequent photolytic conversion of ammonia into molecular nitrogen in the early phases of accretion on primordial Titan. To some extent this process may still be going on, since there is evidence that atmospheric methane, which is destroyed by solar ultraviolet radiation in a few millions of years, is being replenished by episodic outgassing of this gas from the interior, possibly accompanied by traces of ammonia.

The Galilean satellites, although of similar size, were probably too warm to incorporate very volatile species like methane and ammonia. Titan is a relatively long way from its parent, and so formed in the cooler part of the nebula from which it was accreted, which made it easier to collect methane and ammonia ices (in addition to water ice), for the development of an atmosphere. Additionally, Titan is much less susceptible to atmospheric losses, by virtue of its distance from the Sun, with a smaller exposure to radiative heating of the upper atmosphere and to solar wind particles. This makes the atmosphere cooler, so thermal escape is much slower. Also, Titan is fortunate in that it lies mostly outside the Saturnian magnetosphere, since if it were inside, energetic particles would have stripped away much more of the atmosphere.

The history of impacts on Titan, after it achieved something close to its present size, must also be different from the large Jovian satellites. Because Jupiter is deeper in the solar gravity well than Saturn, impactors enter the Jovian system much faster than they do the Saturnian system. At the same time, Titan is less deep in the Saturnian well than are the Galileans in the Jovian gravity well. The net effect is that impactors like meteors and comets strike the Galileans with much higher velocities. They are therefore much better able to erode (or 'blow off') the atmosphere, whereas for Titan the impactors move slowly enough that they typically add mass to the atmosphere, rather than remove it.

The other satellites of Saturn, and those of the two outermost giant planets, share some of Titan's advantages over the Galileans in terms of retaining an atmosphere. However, the fact that Titan is Saturn's only real big satellite helps to explain why it is unique among the moons of that planet in having a substantial atmosphere. It is not too surprising that the other Saturnian moons are airless, because they are relatively small, with correspondingly modest gravitational fields to retain any gaseous envelope. And while the large moons of Jupiter may have been too warm, Neptune's large satellite Triton, about half the size of Titan, is too cold. Triton could have an atmosphere as thick as Titan's were it not frozen out on the ground.

None of this explains why Titan has more atmosphere than Mars or Earth. Perhaps the atmosphere of Titan was frozen during the epoch when Mars' atmosphere, and to a lesser extent Earth's, was depleted by heavy bombardment. Alternatively, there may be a supply of the more volatile ices, such as ammonia, methane and carbon monoxide, trapped inside Titan by layers of frozen water and able to escape only slowly through cracks or cryovolcanoes. Then the atmosphere could have been replaced after the worst of the bombardment was over, as it was on Mars and Earth, but more effectively on Titan because of its larger reservoir of volatile species. It is also conceivable that Titan's atmosphere has been more efficiently replenished by incoming cometary material in relatively recent epochs, compared to the inner planets.

10.3.1 *Evolutionary Models for Titan's Atmosphere*

Among the several model scenarios that have been proposed to explain the creation of Titan and for the acquisition of its atmosphere, one suggestion was that Titan captured its envelope from the gases contained in the nebula surrounding Saturn during the accretion period when both grew to their present sizes. A problem with this idea is that the Kronian subnebula at the time of Titan's formation must have had the same elemental composition as the solar nebula. Then the noble gases abundance ratios on Titan should be similar to those in the Sun. That does not seem to be the case: for instance, the solar abundance of neon is almost identical to that of nitrogen, whereas Titan has a predominantly nitrogen atmosphere in which no trace of neon has been detected. It follows that direct capture of the gaseous envelope around Titan must be excluded.

There are two other possible sources for the volatiles that produced Titan's atmosphere: accretion from planetesimals condensed within the Saturnian subnebula, and input of material which condensed outside the Saturnian subnebula and arrived at Titan later. It is logical to assume that the atmosphere we witness today around Titan should be a combination of both: outgassing from the ices composing the interior following the accretion of a proto-Titan from 'local' planetesimals, and the subsequent delivery of additional frozen gases by cometary impacts. But did one or the other dominate?

The cometary hypothesis is attractive because it explains the thick atmosphere around Titan and the lack of one around Ganymede and Callisto by virtue of the different impact energies in the Jovian and the Kronian systems, which would cause the creation of an atmosphere in the case of Titan and the erosion of any acquired initially by the Galilean moons. However, while comets contain nitrogen both in molecular form, as ammonia, and as part of organic molecules, they have very little methane. If Titan's atmosphere was entirely the result of cometary impacts, the N_2 is fairly easily accounted for by a combination of direct delivery, vaporisation and subsequent photolysis of ammonia, and the breakdown of complex organic material during the impact. Explaining the methane is more difficult, probably requiring more complex processes to manufacture it from carbon monoxide since the latter is abundant in most comets. The really serious problem for this hypothesis, attractive as it is in offering an explanation as to why other large moons are not favoured with an atmosphere, is the difference that is found in the D/H ratio in comets and Titan (see Chapter 6).

As concerns the contribution of planetesimal degassing into the atmosphere of Titan, we need to examine the nature of the gases that we expect to condense initially in this sort of material. This depends strongly on the temperature and pressure at which the ice formed: laboratory measurements have shown that very low temperatures (less than 75 K) are required to allow for a substantial trapping of highly volatile species such as CH_4, CO and N_2. There is no indication that the subnebula around Saturn could have become that cold prior to the accretion period on Titan.

Therefore, we are led to the conclusion that the dominant carrier for nitrogen on Titan was rather NH_3 and other volatiles that can be trapped efficiently at higher temperatures. Subsequent photolysis of NH_3 on early Titan could easily produce the present amount of N_2.

10.3.2 *Origin of the Atmospheric Components*

The main constituents of Titan's atmosphere today are molecular nitrogen, methane and molecular hydrogen, followed by 16 known trace species (hydrocarbons, nitriles and oxygen compounds), one of which is only in solid form (C_4N_2). Recent atmospheric detections include CH_3CN, H_2O vapour, benzene and C_2HD (Table 6.1). There are many unsolved issues regarding the origins of each of these, including the major constituents, as we have seen above. It helps that our knowledge of the exact proportions of these species has greatly improved with the arrival of the Cassini–Huygens mission, since many of the instruments on board address questions related to this subject. Pre-eminent among these is, of course, the gas chromatograph mass spectrometer on the Huygens probe, which measured with precision the abundances (or upper limits) of a wide range of species and their isotopes on Titan.

10.3.2.1 *Nitrogen*

It is intriguing that Titan's primary atmospheric component is the same as the Earth's. The nitrogen found in the outer Solar System is generally assumed to have been present in the solar nebula as ammonia. However, it is difficult to be sure that the nitrogen we see now in Titan's atmosphere was not originally incorporated in the form of N_2 particularly if it could have been trapped inside less-volatile ices, such as H_2O, as clathrate compounds. Until recently, there existed three possible scenarios for this that might have been viable. The first, and least probable, is direct capture of N_2 from the solar or proto-Saturnian nebula. For this to happen, Titan would have to have formed by cold accretion, in a part of the nebula which never got warmer than about 60 K. Large amounts of nitrogen would probably have had to be trapped underground, again presumably as clathrate hydrates, to replenish the atmosphere in the event that the early bombardment removed most of the primordial atmosphere.

Most of the evidence favours warm accretion, however. Ammonia is far less volatile than nitrogen and would have been much easier to accrete as early Titan cooled. During a condensed but still warm period early in Titan's evolution, ammonia could have outgassed from the interior and into the atmosphere where it photodissociated into molecular nitrogen, allowing the hydrogen to escape. Models of this process have defined what 'warm' would mean in this context: a surface temperature of at least 150 K would have been needed to account for the amount of N_2 observed today in the atmosphere of Titan.

A third possible scenario has Titan's atmosphere forming from the same mix of gases that is present on Saturn. The heavy components, primarily ammonia and methane, would be retained, while hydrogen and helium were lost (on Titan, but not on more massive Saturn). Photolysis would eventually produce N_2 from the NH_3 and higher hydrocarbons — including those which condense as aerosols — from the CH_4, so we could end up with the atmosphere forming much as we see it today. Titan should then have the same isotopic composition as Saturn, for the ratio of deuterium to hydrogen in particular, possibly somewhat modified by the impact of volatile-rich comets, or of debris resulting from the apparent damage to Hyperion. Deuterium is found on Titan primarily in the form of mono-deuterated methane, CH_3D, the abundance of which suggests a D/H ratio of about 1.3×10^{-4}, which means that Titan is enriched in deuterium compared to Saturn and the other giant planets.

The argon abundance on Titan provides a further test for the different hypotheses since laboratory experiments by show that N_2 and Ar are similarly trapped in ice forming at temperatures near 75 K. Then, if nitrogen on Titan was originally in the form of N_2, it would have been trapped to about the same extent as argon (or perhaps slightly less). On the other hand, if the nitrogen originally came in the form of ammonia, the atmospheric value of Ar/N_2 is expected to be relatively low, about 1%. Thus a strong clue as to whether the nitrogen arrived as N_2 or NH_3 can be found in measurements of the Ar/N_2 ratio in the present-day atmosphere. Because molecular nitrogen and argon have similar physical properties, freezing points and so forth, a large argon abundance suggests N_2, while a small one favours NH_3, as the source of the present-day nitrogen. The very low value of primordial Ar with respect to N_2 measured by the Huygens GCMS indicates that the latter theory is correct.

10.3.2.2 Methane

Huygens *in situ* measurements finally allowed firm determinations for the methane mole fraction, which is 1.41% in the stratosphere, increasing below the tropopause and reaching 4.9% near the surface. These GCMS measurements are in good agreement with the stratospheric CH_4 value inferred from infrared spectroscopy by CIRS on the Cassini orbiter and the surface estimate given by the Huygens DISR spectra (~5%). This adds up to a huge amounts of methane, with most of the carbon in the atmosphere of Titan in this form, which is surprising when we consider that other carbon compounds, in particular CO, are more abundant in comets, the interstellar medium, and also, it is believed, in the protosolar nebula from which Saturn and Titan formed. Methane ice is very volatile and difficult to trap during planetary formation, except at very low temperatures, and its lifetime in the atmosphere is relatively short compared to Titan's age. Despite all these difficulties, the ratio CH_4/CO on Titan is $> 1,000$.

Scenarios have been suggested in which CO was converted into CH_4 in a dense subnebula of Saturn before it was trapped as clathrate hydrates by the condensing water vapour in the planetesimals that formed Titan and its primitive atmosphere.

Much of the methane ends up in Titan's interior, and some is available to replenish the atmospheric supply through a porous regolith. This is an essential part of the model, since solar photolysis irreversibly converts all of the methane in hydrocarbons in a relatively short time compared to Titan's age.

We saw above that the difference between the D/H ratio in water in comets on the one hand and the D/H in methane in Titan on the other hand tends to rule out a formation of the atmosphere of the satellite by impacts from external icy planetesimals (of which present-day comets are surviving examples) alone. Exterior contribution of volatiles may have played a part in the formation of Titan's atmosphere, but it does not explain the large excess of CH_4 with respect to CO observed in Titan, since CH_4 is currently less abundant than CO in comets.

On the other hand, the deuterium enrichment observed on Titan in methane with respect to the protosolar value is consistent with the formation of the atmosphere by degassing from the interior of the satellite if the CO conversion to CH_4 that took place in Saturn's subdisc was accompanied by isotopic exchange between CH_4 and HD. If no substantial conversion of CO to CH_4 occurred in the subnebula, it is possible that methane isotopically exchanged deuterium with hydrogen in the nebula instead. In this scenario, the temperature eventually becomes low enough at Saturn's distance form the Sun to permit the formation of solid clathrate hydrates of CH_4 which were incorporated in the planetesimals which formed Titan. Again, CH_4 could escape to the atmosphere through cracks, replenishing the atmospheric methane from the interior of the satellite through cryovolcanism. Both of these scenarios reproduce the large atmospheric CH_4 mass, the low CO/CH_4 ratio and the mass estimates of N_2 converted from NH_3 in the primitive Titanian atmosphere.

A different scenario for the origin of CH_4 in Titan has been recently proposed by S. Atreya and colleagues, who speculated that CH_4 was not trapped at all in planetesimals which formed Titan, but only H_2O, NH_3 and CO_2, since Titan was formed prior to the cooling of the subnebula to temperatures where CH_4 could be trapped in ices. In this model, CH_4 was produced from H_2O and CO_2 in the interior of Titan through a process called serpentinization, which releases hydrogen from water while oxidizing minerals and produces methane through a Fischer Tropsch reaction of the H_2 with CO_2 in the presence of a catalyst. Once formed, CH_4 may subsequently escape in the atmosphere as before.

10.3.2.3 Water and Other Oxygen Compounds

Three compounds of oxygen have been detected on Titan to date: CO, CO_2 and H_2O. The vapour pressure of all three is very low at the surface temperature of Titan, so their presence probably requires an external source for oxygen. This is probably the flux of meteoritic and cometary material that impinges continuously on Titan and all Solar System bodies, most of it in very small pieces and the volatile content is mostly in the form of H_2O. There must be a balance between the oxygen input, the loss of O atoms to space and by condensation, and the abundances of the oxygen compounds present. The fact that CO apparently has a homogeneous profile throughout the atmosphere suggests that no external source is needed and that CO is being delivered into Titan's atmosphere by outgassing from the interior, where it may also contribute to the formation of methane via the release of hydrogen through the serpentinization process, followed by Fischer–Tropsch catalysis.

The rate of influx deduced from water observations at 700 km (the expected ablation level for most micrometeorites) is close to the water influx inferred on Saturn from similar observations, suggesting that interplanetary sources dominate over the local candidates, like the ring system, in supplying water to the atmospheres of both Saturn and Titan.

10.3.2.4 *Isotopic Ratios*

The $^{14}N/^{15}N$ isotopic ratio in Titan's atmosphere had been obtained in HCN from ground-based heterodyne measurements and found to be 4 times less than the terrestrial one, while similar estimates found the $^{12}C/^{13}C$ isotopic ratio on Titan to be essentially terrestrial. The carbon isotopic ratio was also found to be terrestrial by the Huygens GCMS (82.3 ± 1), while the nitrogen isotopic ratio measured in N_2 instead of HCN came out only about 1.5 times less than terrestrial. Despite the uncertainties, the nitrogen data points to a denser primitive atmosphere, with models suggesting that at least 1.5 times and as much as 10 times the present atmospheric mass has been lost over Titan's whole history. The carbon isotope ratio supports the idea, required on other grounds as we have seen, that the methane in the atmosphere is secondary, having been outgassed recently from the interior. Rather more dramatically, the acquisition by Huygens of a precise value for the $^{12}C/^{13}C$ isotopic measurement, which is smaller than the inorganic standard value of 89.9, was trumpeted recently as apparently precluding the presence of Earthlike active bio-organisms on Titan.

We have also seen that the D/H ratio is another important measurement in connection with Titan's atmosphere and its origin and evolution. The observed deuterium enrichment, relative to the primitive material from which Titan formed, could originate from different processes acting during or after the satellite's formation, including fractionation involving deuterium exchange with cloud particles, ocean or crust; isotopic exchange catalyzed by metal grains in the Saturnian nebula; outgassing of already D-enriched grains of interstellar origin, and a number of others. Confirmed recent values for the D/H ratio in methane are around 1.2×10^{-4} with a good precision (Section 6.4.1.4). Considerably lower than the value found in the comets, this tends to suggest that Titan's atmosphere is in great part due to outgassing from the interior of the satellite.

More recently, the D/H ratio has been measured in other media: GCMS measured D/H in HD, and found it to be $2.3 \pm 0.5 \times 10^{-4}$, while in the CIRS data, the detection of C_2HD, brought another piece to the puzzle by adding a new D/H estimate from yet another medium. The different values reflect differences in the mechanisms which enriched each species in deuterium. Possibly the D/H in CH_4 is a relict of the solar nebula value in methane. On the other hand, D/H in hydrogen and in acetylene in Titan results from the photodissociation of methane by solar radiation. This produces additional deuterium fractionation with respect to that in methane contained in the solar nebula.

10.4 Titan and Life

"Titan's environment provides significant insight into the conditions which prevailed on Earth during its early history and into the chemistry which then gave rise to life." Quotes like this one (unattributed, from NASA publicity literature) are seen so often in connection with Titan that we have to consider explicitly whether Titan resembles Earth enough to have any relevance to terrestrial biology. As an extreme possibility, we might ask whether life might exist on Titan in some primitive form.

To recap, the similarities between Titan and the Earth include the prevalence of nitrogen, some qualitative similarities in temperature structure, importance of organic chemistry, condensation and rain in the lower atmosphere, and possible liquid reservoirs on the surface. These similarities undoubtedly present us with additional exciting possibilities for study.

However, conditions on present-day Titan do not replicate with precision those on primordial Earth, most importantly the fact that Titan is too cold for liquid water, and for some important chemical reactions to occur, including those that produce free oxygen. This single factor — the low temperature — is so overwhelmingly important that it essentially dominates all others — the word 'life' should be used very sparingly in connection with Titan, and perhaps never.

Yet we do know that there is organic chemistry going on in the upper atmosphere. This apparent contradiction arises because cold Titan is linked to a hot body — the Sun — by the latter's radiation field. Particles of light (photons) with an energy corresponding to the Sun's emission temperature of nearly 6,000 K pour down into Titan's upper atmosphere. These energetic photons dissociate methane and other molecules and allow them to recombine into the more complex hydrocarbons and nitriles that the Voyager spectrometers detected. Then, they probably condense and drizzle or rain out onto the cold surface, where they remain, in deep freeze, until some future astronaut scoops them up and analyses their composition.

We do not really know what that composition will be. How far does the organic chemistry go along the chain that, under more congenial conditions, led all the way to life on Earth? This is the real question. The answer depends on a better understanding of atmospheric photochemistry in general (remember that we are still struggling to understand the production and loss of ozone on Earth, which involves roughly analogous processes, despite a mass of data and decades of effort) but also on questions like whether charged particles, lightning and other energy-rich phenomena which can also drive chemical reactions along, have an important role.

Whatever the answer to these questions, and the certainty that more complex compounds than any that have already been detected certainly spawn from organic chemistry and form in Titan's atmosphere to eventually condense on the surface, we do not expect it to produce the molecules capable of giving rise to life. The complex chemistry is almost certainly confined to the relatively warm reaches of the upper atmosphere and we cannot picture life arising other than in a liquid environment on the surface. So that, in spite of all the similarities between Titan and the Earth (temperature structure, organic chemistry, condensation and rain in the lower atmosphere, possible liquid reservoir on the surface), the cold temperatures (which slow the chemical reactions), the low sunlight input (about 1% of that reaching Earth), and the lack of sufficient amounts of oxygen should forbid the miracle of life to happen in this distant satellite. Nevertheless, Titan's environment does provide significant insight into the conditions which prevailed on Earth during its early history and into the chemistry which then gave rise to life, and the satellite remains the best model we have, so far, for conditions on the early Earth, and undeniably a valuable tool to the study of some chemical and physical processes.

One other possibility is worth mentioning, before we leave the life issue altogether. We saw in Chapter 9 that Titan's icy crust may conceal a deep liquid layer at some depth, where the combination of high pressure, insulation from radiative cooling to space, and the release of interior heat (primordial or radiogenic) is enough to prevent freezing. Is it possible that this ocean is the home of Titanian life forms?

Until recently, such a question would not have received serious consideration. This has changed somewhat with the discovery by the Galileo spacecraft, in orbit around Jupiter, of such a subsurface ocean on the large satellite Europa. Tidal heating in Jupiter's gravitational field, induced by the gravitational attraction of the other Galilean satellites, provides a significant source of interior heat on Europa and the geological evidence on the surface suggests that the ice crust which separates the ocean from the vacuum of space is relatively thin, perhaps only 10km thick. This, and the recent appreciation that life on Earth flourishes wherever there is liquid water, has raised the idea that there might be basic forms of life in the Europan oceans, and prompted plans for space missions to take a closer look. If this turns out to be so, then there is hope for something similar on Titan also, in particular with the water jets discovered on Enceladus, suggesting that liquid water pockets exist as far as the Saturnian system below the surfaces.

10.5 Open Questions

We conclude this chapter with a summary of how the sort of comparison we have made above is helping to answer some of the key questions related to the issue of Titan's origin and evolution which were brought up in earlier chapters. In some cases, all we can do is to re-pose the question in somewhat different terms and hope that new data, from new missions, will soon bring answers closer.

Does the nitrogen in Titan's, Triton's and Pluto's atmospheres have a common origin? The presence of large amounts of gaseous and frozen nitrogen on the two much colder worlds tends to suggest that it derived from a gas containing molecular nitrogen, rather than chemical processing of ammonia, since ammonia is a very volatile gas which is less likely to have been available on the small, distant bodies for the long periods of time required for dissociation. However, we do not know enough about their early histories, when they may have been warmer, to be sure about this. When we discussed the origin of the nitrogen on Titan earlier in this chapter, we concluded in favour of ammonia as the primary source.

Was methane brought into these bodies as frozen condensate, entrapped in clathrate or produced in the interior? The answer may well be a combination of mechanisms. The methane in comets is mostly present as clathrate, but if the material which formed Pluto, say, was always as cold as it is now then methane ice may have formed part of its make-up, and may still be present.

Similar arguments apply to carbon monoxide, which is also a very volatile species, and one which is very common throughout the universe. It is observed to be present in Titan's atmosphere and on Pluto's surface. If we could but see how much there is, where and in what form on all of the icy bodies of the outer Solar System, but especially Titan, Triton and Pluto, we would have a first-class clue to their individual and family histories. We would know, for example, whether most of Titan's atmosphere was added by impact of volatile-rich comets after accretion.

We should also like to have precise knowledge of the abundances of the noble gases, such as neon, plus the isotopic ratios such as deuterium to hydrogen, in the volatile components of Titan, Triton, and Pluto, as diagnostic tests of the competing hypotheses on the origin of nitrogen and methane on Titan. The observed deuterium enrichment on Titan, relative to Saturn, could have many origins, acting during or after the satellite's formation, including fractionation processes involving deuterium exchange with cloud particles, ocean or crust; isotopic exchange catalysed by metal grains in the Saturnian nebula; and outgassing of already D-enriched grains of interstellar origin. A precise measurement of the D/H value on all three bodies would help discriminate these and tell us more in general about the ways in which hydrogen-containing compounds were incorporated in planets and satellites. The answer to the origin of the various nitrogen atmospheres may require more studies and measurements, especially for Triton and Pluto.

What we already know about Triton, and to a lesser extent Pluto, is consistent with the idea that they would be more Titan-like if they existed nearer to the Sun. They would certainly have more substantial atmospheres, although they would not retain them as well as Titan does because of their smaller masses. In the other radial direction, it becomes a little clearer why Titan has an atmosphere, while the Galilean satellites do not. In spite of their impressive sizes, the large Jovian satellites must have been too warm to hang on to the atmospheres they produced. Whether this was mainly due to the extra heat from the Sun, or whether the giant proto-Jupiter was more responsible, remains to be established.

CHAPTER 11

Beyond Cassini/Huygens: The Future Exploration of Titan

And maybe the fever shall have a cure, the true journey an end.

W.H. Auden

11.1 Returning to Titan

Following the success of Cassini/Huygens, space mission planners are already at work on plans for a sequel. Titan is clearly shown to be a very interesting place, and we can already see that there are important questions that are better defined by the results from current exploration, but which will not be solved without new investigations of a different kind. In this chapter, we talk about some of those objectives, and speculate on what shape the new mission is likely to take. This can be a fairly relaxed exercise — nothing is likely to arrive at Titan before the decade 2020–2030 and perhaps not even then. Also, new technology, new discoveries and revised priorities are likely to greatly alter any plan we can foresee today. Still, plan we must, or nothing will happen, ever.

The NASA scientists who met to identify objectives for what they called the Solar System Roadmap produced 24 possible missions that can address key objectives in the outer Solar System. Not surprisingly, they chose to emphasise prebiotic chemistry as the scientific focus of many of these, particularly at Europa, Enceladus and Titan. One of the example studies proposed by APL and JPL teams for a Flagship mission is Titan Explorer, which would make *in situ* measurements of the distribution and composition of organic material on the surface and in the atmosphere, while searching for evidence of prebiological or protobiological activity. Other forward-looking study groups have reached similar conclusions. In Europe, there has been an effort that produced a proposal in response to ESA's Cosmic Vision 2015–2025 for a combined mission called "TandEM" — for Titan and Enceladus Mission (Coustenis *et al.*, 2008b) — that would explore both Titan and Enceladus. Titan Explorer and TandEM were merged in a combined effort by ESA and NASA to send a new mission into the Saturnian system for the 2025–2030 time frame. But how would such a mission be carried out?

The top priorities for any new spacecraft design is that it should operate *in situ* — that is, inside Titan's atmosphere and on the surface, and that it should have

Figure 11.1   Future exploration of Titan with balloon over Titan's landscape and orbiter in the sky.
Artists: T. Balint and S. Cnudde.

mobility — the ability to move around and seek out and explore multiple interesting locations. One way of doing this that might well come about in the foreseeable future is to use a Titan 'aerobot', or airborne robot. These have been flown in the Earth's atmosphere, and the Russian-French VEGA missions deployed a version at Venus in the 1980s. New missions of this type are being planned for Venus and Mars; a Titan version could actually be implemented quite soon and probably a lot more cheaply than Cassini.

As the name implies, the aerobot is intended to be an autonomous aerial vehicle — a Montgolfière or hot-air balloon — with the intelligence to steer itself along a pre-planned trajectory, riding the winds at different levels to visit different parts of Titan. Obvious uses of such a mission would be to survey the surface of Titan and photograph it in much more detail than Huygens can, and to sample the atmosphere and aerosols at different levels over an extended period. The designers believe that the aerial platform could explore the atmosphere from 3 to 30 km altitude to investigate *in situ* the photochemical reactions and their products that are important to understanding prebiotic synthesis on Titan. Exploration of the surface could include aerobot descents to multiple sites, and of course unlike other types of rovers, the aerobot can easily traverse liquid as well as solid surfaces. Additional Huygens-like mini-probes could be deployed at different locations of the surface.

For a surface-oriented mission, hydrogen would be used as the working fluid in the balloon. The vehicle could visit hundreds of sites well distributed over the surface of the satellite, staying for several hours or days before rising to a higher altitude and drifting to another location. Unlike Venus, temperature variations in the Titan atmosphere are not large and thermal control considerations do not limit the duration of surface stay time.

For a mission focussed on the atmosphere, a higher altitude capability is needed. This would be done with a super-pressure balloon, which would descend to the lower atmosphere of Titan for inflation and then float at an altitude of 115 to 125 km to implement its atmospheric mission. The vehicle would then vent gas, descend to near the surface and skim the terrain using a guide rope. In this way, it could conduct observations at a single surface site for a distance of several tens of kilometres, as well as its atmospheric mission.

Later in this chapter, we will look at a future mission concept in more detail. First, it is necessary to look at the likely scientific background of such a programme more carefully, in order to establish what the robotic craft would actually need to do to make sure of addressing, or at least progressing, the most important goals. Missions to Titan occur only once or twice per generation — they have to be carefully planned and specified to make sure they represent maximum value, scientifically as well as financially.

11.2 Titan as a Target of Astrobiological Interest

The thing that makes Titan special, among Saturn's 60 known moons, and indeed among all the moons in the Solar System, is partly its massive size (just less than half the volume of Mars) but mainly, of course, the unique atmosphere. The chemistry going on in its dense envelope of nitrogen, methane (a proto-organic molecule) and hydrogen, is thought to be similar to the chemistry that went on in Earth's early atmosphere before life transformed it into the air we breathe today.

The organic compounds that have been detected in Titan's atmosphere are produced by the recombination of the molecular fragments that themselves result from the dissociation of methane and nitrogen in Titan's upper atmosphere. In addition to photolysis by solar ultraviolet radiation, the precipitation of electrons and other charged particles from Saturn's magnetosphere, and galactic cosmic rays, all contribute to the dissociation process and drive the chemical production of more complex molecules. As a result, Titan's atmosphere has a huge array of organic compounds and derivatives, some of which have been observed and measured and others, particularly the rarer, more complex molecules, which still await discovery.

As a result of the production of larger molecules that condense at the low temperatures prevailing on Titan, even in the relatively warm upper atmosphere, the haze layers form and organics rain, or drizzle, out onto the surface where they form shallow lakes of organic-rich liquid. This is likely to be similar to the "organic soup" in which, it is generally believed, life formed on Earth. The lakes seem to dry out during Titan's seasonal cycle, raising the attractive prospect of mineral and carbonaceous materials in deposits on the shorelines or lake beds. It is also likely that ancient layers and interior material brought to the surface are exposed in places, perhaps in the rims of impact craters.

Further chemistry, relevant to — if not actually involved in — pre-biotic synthesis, must occur at the surface, acting on the precipitated material and powered by sources including but not limited to direct solar radiation, some of which diffuses down through the clouds and haze. The biggest contrast with the Earth, apart of course from the very low temperatures and modest sunfall, is the almost complete absence in the gaseous form of oxygen or oxygen-containing compounds like water. There is plenty of water ice on Titan, of course, but current, and presumably past biological activity on Earth requires liquid water as a solvent, and involves organic molecules that contain not only carbon but also oxygen atoms. In the laboratory, synthetic tholins and nitriles that might be similar to those found on Titan are readily hydrolyzed by liquid water into amino acid. An interesting experiment would be to see what happens when a sample of the accumulated organics on Titan's surface are exposed to liquid water in a controlled experiment. Even better, there may be material to be found there that has already been in contact with liquid water, for example if the latter was expelled from an underground source and survived temporarily on the surface, or if surface organics and water ice were briefly heated by a meteoritic impact. It may be particularly rewarding to explore features that show some evidence of flow patterns, possibly the record of past episodes of liquid water.

11.3 Science Drivers and Measurements Needed

From discussions like that above, numerous study groups and conferences have converged on the idea that a science-driven mission will focus on a robotic search for evidence of pre-biotic chemistry and 'biomarkers' ranging from chemical evidence of processes that were active under special conditions in the past, all the way to rather unlikely but particularly interesting artifacts like fossilized bacteria. Sample collection and geologic mapping of Titan's surface are also of massive interest, partly as an adjunct to the above (finding the best places to land and search) and partly, of course, because characterizing the surface and atmosphere, and understanding Titan's origin and evolution, are key planetary science objectives in their own right.

The European Space Agency collected some quotes from scientists working on this problem that capture the excitement they feel about the topic:

When you study Titan, it's a bit like going back in a time machine to Earth four billion years ago. The atmosphere is a natural laboratory for studying prebiotic chemistry on early Earth — the chemistry that led to life. There's a very low probability of finding real life there (it's far too cold and all the water is deep frozen on or below the surface), but the chemistry in the atmosphere may be very similar to the chemistry that preceded life on Earth.

Methane in Titan's atmosphere is continuously destroyed by ultraviolet light. To explain the amount of the gas present in the atmosphere, we think there must be a large source either on or under Titan's surface. It could be in the form of lakes or oceans, or subsurface reservoirs.

Of course, we may not find the precise transformation that turned complex organic compounds into living things. But the better we understand the chemistry, the better our chance of working out how it led to life.

The more prosaic science objectives of the NASA and ESA study teams include:

• Investigate Titan's upper atmosphere, its ionosphere and exosphere as well as its plasma and magnetic field environment. The altitude range from ~400 km to 950 km, that is below the reach of Cassini *in situ* measurements but above the region of intense remote sensing and most Huygens measurements, is a broad transition region that affects the chemistry and physics of the entire atmosphere. Understanding this "agnostosphere" is a key to understanding the entire atmosphere together with its relevance for astrobiology.

- Understand Titan's atmospheric chemistry, including pre-biotic and proto-biotic chemistry: determine if self-organizing chemistry occurs or has occurred on Titan.

- Obtain data on temperature, pressure, CH_4 and C_2H_6 humidity; wind fields; internal structure and evolution of clouds; haze characteristics; evaporation rates and temperature over lakes; surface composition and thermal properties; volumetric changes in lakes and river systems; solar partitioning.

- Determine if complex organics exist involving C, H, O, N: determine what organic chemical processes are occurring on Titan.

- Measure the various isotopic ratios such as $^{36}Ar/^{38}Ar$ (for escape processes), D/H in surface H_2O (for the origin of methane in Titan's atmosphere), ^{16}O, ^{17}O, ^{18}O in water ice, Xe isotopes, $^{15}N/^{14}N$ in NH_3 on the surface (for the NH_3 amount available during the satellite's formation); also noble gases ratios: $^{36}Ar/Kr/Xe$ (relevant to the origin of the atmosphere); and data on the densities and gravity fields.

- Obtain infrared stereo and radar surface mapping on Titan with resolution < 100 m; highest-resolution (< 1 m) compositional context and infrared imaging from a near-surface platform (also required for selecting sampling sites for surface chemistry); high-resolution *in situ* measurements of surface material from a variety of locations; global compositional mapping with resolutions < 1 km; ground penetrating radar to determine the depth and vertical structure of near sub-surface deposits.

- Understand Titan's surface organics (distribution, composition, organic and chemical processes and context, energy sources); examine the chemistry taking place on the surface and near-surface.

- Measure elemental, isotopic and molecular composition, and chirality of Titan's aerosols and of its various surface materials; chemical analysis of the ejecta produced by cryovolcanism; seasonal and long-term variations for Titan global climate models. Titan's peculiar stratospheric polar hood has many analogies with Earth's ozone hole and its methane hydrological cycle with probable sparse but violent downpours is a greenhouse weather pattern taken to extremes. The climate/weather/hazard links to our own planet are thus obvious.

- Understand Titan's formation and the implications for Saturn's formation.

Since Titan's surface is so heterogeneous, sampling at different locations, or having mobility is critical to achieving the science goals and so also counts as a key science driver. Multiple surface sites will need to be sampled, possibly quite a long way apart, and it will be necessary for any lander to have local mobility at its landing site in order to access the most interesting samples. The aerobot will need to make atmospheric as well as surface measurements, and the landers may have to work in a liquid environment as well as on solid surfaces in some very variable terrain. An autonomous roving vehicle or several landing probes are required that can collect and analyze samples of (a) the atmosphere near the surface; (b) wet and dry lakes, at every level including the bottom material; (c) below the surface by drilling, and (d) the putative organic-rich deposits in cliffs, craters and shorelines.

Meteorological investigations and geologic mapping of Titan's surface are relatively straightforward, if technically demanding, goals to address, but a robotic recovery and analysis of samples to explain the surface composition and to search for biomarkers and pre-biotic chemistry is altogether in a different league, as our experience on Mars has already shown. What measurements and tools are actually needed to address such an esoteric goal, bearing in mind that we have only a very imperfect knowledge of what may be there to find, or where it is?

Clearly, the analysis of the samples obtained, whether done on the surface or back in the floating platform where larger instruments can be accommodated, must be capable of identifying the atomic and molecular composition of the airborne organic material we loosely refer to, in our present ignorance, as 'tholins'. On the surface, the tholins may have evolved in melt sheets to contain water-tholin interaction products such as amino acids or higher-order compounds of exobiological interest. Surface chemical sampling of frozen material needs to be accompanied by subsurface profiling, looking for layering of species and searching for the presence of near-surface liquids. Any that are found require a complete gas/liquid/solid analysis of their constituents, which should also focus on chirality (the 'handedness' of the molecules) and other signatures of aqueous organic evolution.

It is asking rather a lot of any design for a lander that can sample the solid surface, drill to depths of at least 10 cm below, and collect and analyze samples that it should also be able to explore the organic-rich methane lakes that Cassini's radar imaged on the surface. Probably, a separate kind of probe, perhaps a 'harpoon' lowered or dropped on a cable from an airborne floating platform, will be required to measure the depth of the lakes, to obtain profiles of their physical properties (temperature, pressure, density, conductivity, turbidity etc) and composition, and to collect samples of the sedimentary bottom material. Or perhaps directly deposited very small and light probes can do the job.

High-Resolution imaging from the floating platform must direct the surface and sub-surface operations and place them in their local context. A search for interesting geomorphological features, especially fluvial ones but also volcanic and aeolian phenomena at scales much smaller than Cassini could achieve, will likely precede the deployment of surface assets.

11.4 Advanced Titan Mission Concepts

Having laid out what we would like to do, we now need to work out how to do it. In order to be fairly specific, we follow closely the concepts derived during the different studies of such missions on both sides of the Atlantic, with the common aims described above. The scientists and engineers involved set out to summarise the desired science measurements and determine the types of samples that would be needed, and the instruments to collect and analyze them. Designs are needed for aerial platforms and sondes to take the sample-acquisition systems to the surface, splitting the science payload between components that would travel to the surface with the sondes and larger devices that would remain on the platform and wait for samples to be delivered. How might a fleet of platforms and multiple sondes be deployed and controlled? And of course, the whole thing must be launched, delivered to Titan, and the data collected and relayed back to Earth. Finally, reliability issues for such a complex mission in a very hostile requirement — at temperatures just above the freezing point of nitrogen, remember — would call for a very robust scheme for autonomously managing anomalies and faults.

Once a start has been made on fleshing out such a concept, a number of other decisions have to be made, and constraints accepted or adopted. For instance, it is no use assuming that infinite amounts of money are available to be spent. That is only true if the launch date of the mission is an infinite time in the future. The Titan Organics Explorer Mission Concept designers, for instance, sought to develop an extended-lifetime (e.g., months) surface science package that would fit within NASA's Flagship Mission cost cap — about $3 billion, about the cost of Cassini–Huygens. Moderate cost implies a rapid timescale, which reduces options in terms of flight time as well as launch date. Some science questions, as we see below, require technology that has not yet been developed, meaning that they should be planned in to a longer-term mission, but the development started now. Launch capabilities also have to be considered: vehicles that are available now, under development, or requiring special development. The same applies to communications infrastructure: what equipment is available on Earth to track the spacecraft and communicate with it at Titan? This might involve one or more relay satellites in orbit while the *in situ* explorers are on the surface or floating in the atmosphere, or conceivably the data could be transmitted directly back to Earth if a sufficiently sensitive receiver was available. (Although not part of the official project, a radio telescope on earth was just able to receive the carrier signal from Huygens while it was alive on the surface of Titan. NASA's and ESA's data reception scheme involved, of course, a relay link via the Cassini orbiter). Such a receiver could exist in a not so far future. The Square Kilometre Array (SKA) is expected to be fully operational by 2019, in time to help return data from any mission at Titan planned after that date. This interferometric array of individual antenna stations, will synthesize an aperture of a diameter of up to several 1,000 km. An orbiter is, however, still the preferred option. Most observations require a 3-axis stabilized spacecraft but some would benefit from a spinning orbiter. Whichever way this is resolved, the orbiter would also provide telecom relay between the different elements and Earth.

11.5 Technology Requirements

A list of some of the key aspects that have been spun out of the various studies as still needing to be thought through and then designed in detail include:

- Orbit insertion, using rockets and/or aerobraking or aerocapture.
- Construction, delivery and deployment of one or more airborne floating platforms.
- Surface delivery and deployment of a mobile explorer ('sonde').
- Design and deployment of an atmosphere/lake profiler ('harpoon').
- Sonde and harpoon atmosphere/surface /subsurface sampling concepts.
- Sonde and aerial platform designs/interface.
- Low-mass, high efficiency radioisotope power systems.
- Thermal control systems for sondes and aerial platform.
- Strategy for multiple device deployment in an unknown environment.
- Component failure and accident strategies.
- Control and data relay systems including a direct-to-Earth backup.
- Structural/electronics packaging for extremely tight volumes.
- Extreme cold sensors/electronics/batteries.
- Trade studies on solar electric propulsion.

This is a formidable list, but of course not all of these elements are unique to a Titan mission, and many of them form part of the ongoing technology development programmes of the space agencies and relevant industrial organisations.

11.6 Mission Architecture and Design

The biggest constraint on the engineering design is the payload mass that can be delivered to the Titan surface, and the total time envelope allowed for the mission. These are both a strong function of cost and therefore not easily predictable for a mission to be flown at some point in the future. A reasonable assumption for a "short-term trip to Titan", i.e. one that uses the propulsion systems currently available, is that a maximum allowance of 1,000 kg useful mass is to be delivered into the near-surface environment of Titan. This rises to 2,600 kg when the parachutes and the aeroshell that protect the science payload on its descent through Titan's thick atmosphere are included. The minimum practical trip time with current propulsion technology for the voyage to Titan with a payload of that size is about seven years.

Given a mass allocation, do we plan on a single aerial platform or several? Obviously, more means better coverage of the globe, and also offers some redundancy in case of technical failure with one platform. On the other hand, with a fixed mass, a single platform would be more capable in terms of the observations it could make. A similar decision has to be made whether to include an orbiter. This makes communications between Earth and the aerial platform and the sondes on the ground very much easier and allows a higher data rate, but it represents a great deal of mass.

Direct-to-Earth communication is possible from the platform if it is deployed and designed carefully so that sufficient power, a reasonable antenna, and a line-of sight is available. It is obvious, however, that a large part of the planned science can only be made from an orbiter, thus a combination of remote and *in situ* exploration of Titan is the most desirable scenario to date. The *in situ* elements could be a balloon and several landing probes. Since mobility for sampling of multiple surface sites is essential, a dirigible platform is called for, and not a simple balloon. It would have to be capable of navigation using only the Sun, the Earth and Titan itself as references, since the cloudy atmosphere does not permit stellar sightings.

11.7 Getting to Titan: Launch and Propulsion

Alternative combinations of propulsion that could be used to get a substantial payload to Titan and into orbit around the satellite always require a substantial "braking" capability to slow the spacecraft down when it reaches its destination. Cassini, of course orbited Saturn and not Titan; the smaller body offers a significantly greater challenge in this regard. The options, which can be used in various combinations, are chemical rockets, solar-electric propulsion, and (in the long term only, as it is not yet sufficiently developed) nuclear electric propulsion. It is also possible to use the gravity-assist that results from flying around the Sun and encountering Venus, Earth, Mars or Jupiter on a carefully calculated trajectory. This gives a considerable boost at little cost in terms of fuel (just a small amount for navigation, but it adds substantially to the length of the journey and therefore the flight time). On arrival at Titan, the spacecraft can achieve orbit either by a long rocket burn, or by the technique known as aerocapture, in which the spacecraft steers close to titan in order to use the drag of the atmosphere to slow down and enter orbit. Aerocapture saves a lot of fuel, but is just as risky as it sounds, especially if the properties of the atmosphere are not well known. Too much drag produces a crash; too little, and the spacecraft slows down but carries on past Titan and on towards Uranus.

Depending entirely on chemical rockets to send several tonnes to Titan requires both a very large rocket and a long trip time. Chemical plus aerocapture saves on both, and if a solar-electric ion propulsion engine is also used to give continuous thrust throughout the flight, the combination offers the shortest trip time and the use of a smaller launch vehicle. The new technology of nuclear-electric propulsion offers a more powerful alternative to solar electric, but this technology is still not ready, and in particular requires a large mass that tends to negate the advantages of using it for all but the most massive payloads. Developing compact nuclear sources and upping the mission payload could lead to the best approach for a mission in the longer term. The plan must also include "survivability" mechanisms against unfamiliar factors such as low temperatures and exotic forms of corrosion, as well as more usual problems of hardware and software failure. Autonomous Fault Diagnosis, fault tolerance, recovery and reconfiguration are needed for long missions to remote targets like Titan, along with systems that can tolerate and intelligently accommodate component failures and precautions to eliminate the possibility of biological contamination and to ensure protection of the environment on Titan. The availability of suitable radioisotope power generators is also a key issue.

With some fundamental trade-offs out of the way, the Titan mission studies could get down to details, and a vision of the future slowly emerged. The hope is for a successful collaboration as in the case of Cassini–Huygens.

11.8 The Voyage of the *Titania*

It is June 7, 2028, and the spacecraft Titania has just reached the Saturn system after an eight-year journey from the Earth. Its trip began on August 14, 2020, on board a huge Ares-V rocket, newly developed by NASA to ferry cargo to the Moon for the new manned American base there.

The ten-tonne spacecraft was thrown into an orbit around the Sun by the big rocket, there to continue its journey to Titan under its own power. *Titania* uses a nuclear electric propulsion system derived from the *Prometheus* program, started 20 years earlier to fly a large payload called the Jupiter Icy Moons Orbiter. JIMO was cancelled in 2005, but Prometheus was resurrected in 2009 to provide the power for the flight of Titania to Saturn, the first launch to the ringed planet since Cassini in 1997.

As the giant spacecraft circled the Sun, 100 kW of electricity generated in the small nuclear reactor in its nose powers eight electric *Herakles* ion thrusters. These expel Xenon gas at high speed from the tail, and gradually accelerate the spacecraft further from the Sun and out towards the outer solar system and its final destination, Saturn and Titan. Waste heat is dissipated by a triangular array of radiators 10 metres long. On its way out to Saturn, Titania flies close to Jupiter, obtaining a boost from the gravitational field of the largest planet that shortens its journey by more than a year.

Even so, it is nearly eight years before Saturn's disk fills the sky and Titan is a small orange blob in the near distance. *Titania* is decelerating now, and preparing to finish the job with a burst of chemical rocket propulsion, placing the large ship in Saturn orbit. Before that, on command from its on-board computer, *Titania* deploys three separate spacecraft, one orbiter and two entry probes, each in their individual lightweight, ultrastrong aeroshells. Each is targeted differently: one lander heads for the equator, and the other for high latitudes in the hemisphere that is currently enjoying summer conditions. The orbiter enters Titan's atmosphere at a tangent to the surface, aiming to use atmospheric drag to shed its high velocity — relative to Titan, it is travelling at nearly 6,500 metres per second. It is aimed to pass above the terrain at an altitude of just 65 km, before emerging from the atmosphere and heading off into space again. The drag on the aeroshell causes it to become incandescent; the visitor from Earth passes across Titan's skies like a meteor, glowing brightly

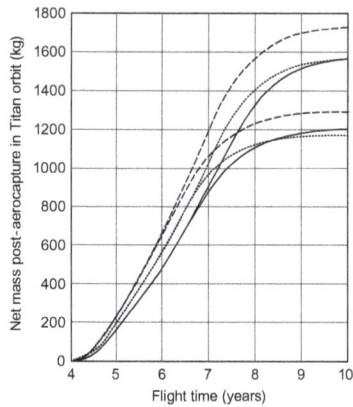

Figure 11.2   The trade-off between the mass available in Titan orbit and the flight time from the Earth, for various different launch vehicles and routes to Saturn. The graph applies to a net payload of about 1,000 kg, which is on the small side for the kind of mission the scientists would like in order to explore Titan properly, and address the complex exobiological objectives fully.

Figure 11.3   On Aug. 14, 2021, an Ares-V rocket takes off from Florida, carrying an international mission, en route for a landing on Titan more than 8 years later (NASA).

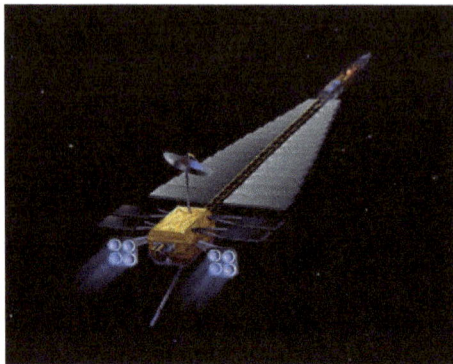

Figure 11.4   Titania cruises towards Saturn, its nuclear electric power system coupled to eight xenon ion drive motors at full stretch. The Titan orbiter and lander spacecraft, with a combined mass of 2.8 tonnes, are contained in the gold box near the tail, and are deployed on approach to Titan (NASA).

and leaving a trail of hot gas ablated from the specially-developed materials in the heat shield. When the orbiter sheds its blackened aeroshell and emerges from the atmosphere on the other side, it has lost most of its velocity, and is moving slowly enough to be captured by Titan's gravity in a long, elliptical orbit. Steered by short bursts of chemical propulsion motors at the distant end of this, the spacecraft dips into the upper atmosphere, again and again and less and less deeply, until the orbit is nearly circular, 1,700 km above the surface of Titan. Less than one tonne of mass reach the orbit; the rest has been dissipated as fuel on the long journey, or belongs to the now-separated probes and the Prometheus spacecraft, with its heavy nuclear propulsion system. The mother craft has vanished from the vicinity of Titan, for the time being, and is decelerating and manoeuvring to commence its 10-year tour of the rest of the Saturnian system.

The spacecraft now orbiting Titan is actually a carrier for five identical microsats, which now separate using low-thrust gas jets until they are equally spaced along their common orbit. They then form a dedicated communications and navigation network, resembling the GPS system that orbits Earth, but with mutual inter-communication. While these are taking up their stations, the Titan aerobots have steered deeper into the atmosphere. Once friction with the atmosphere has shed most of the momentum, the aeroshell is detached and a series of increasingly large parachutes is deployed. In the dense lower atmosphere, the descent is very slow and the inflation of the aerial platform can begin. It is fully inflated at 10 km altitude, the parachute is detached, and the platform — a giant blimp — drifts free, the hills of Utopia distantly visible below through the haze. 4,000 km away, its identical twin has completed a similar deployment in the north polar realm of Titan, where the Sun is high in the sky and dense methane cumulus clouds lurk on the horizon.

11.9 Explorers on Titan

A system of buoyancy control chambers and pumps inflates the blimp with hydrogen gas and stabilises its flight about 5 km above the surface of Titan. Miniature high-efficiency RTGs and thin film batteries power the control avionics, communications, and the science instruments in the gondola. Both aerobots have propeller-driven mobility systems with full control of their flight, and systems of fins, heaters and valves that allow them to vary their float altitude. Except for their use of advanced materials, to handle the extreme conditions on Titan, and the use of high-reliability, miniaturized electronic control systems and instruments, they superficially resemble the dirigible 'blimps' still flown on Earth, most famously by the Goodyear tyre company. Their design lifetime, as they navigate in Titan's atmosphere, allows for a mission duration of at least 40 days. Sometime not too long after that period expires, the special low-temperature balloon materials, or the cryogenic actuators, valves and seals, all of which have a limited lifetime, will give out and end the mission of the aerobots.

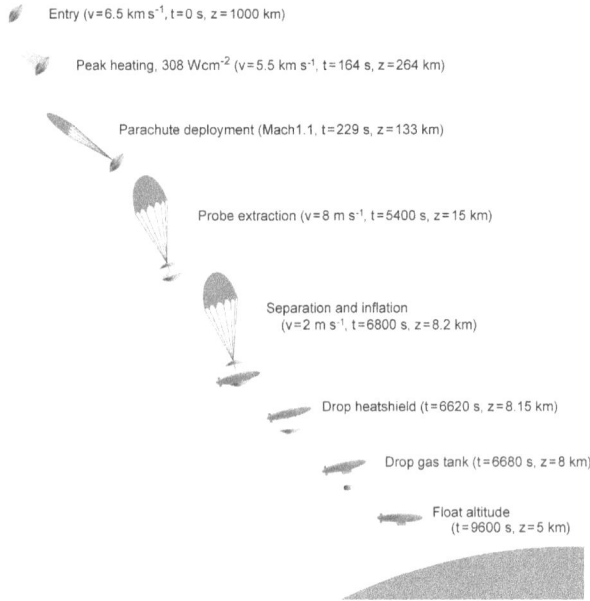

Figure 11.5   The deployment sequence for each Titan aerobot (NASA).

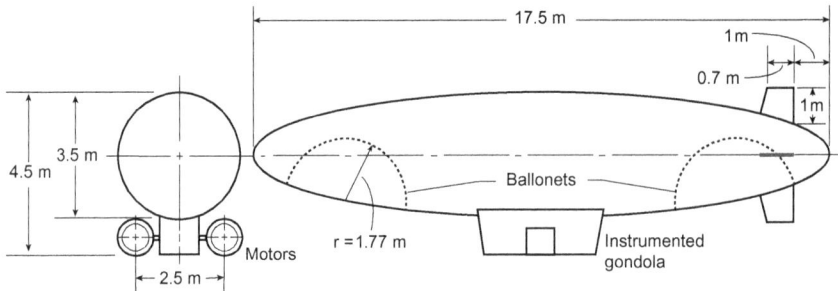

Figure 11.6   A technical drawing of the Titan aerobot, showing its dimensions (NASA).

While they live, the platforms use radio links with the orbiters for global positioning and undertake continuous surface feature mapping, not only for reconnaissance but also for additional information on the location, speed and orientation of the aerobot. Its downward-viewing radar also maps the surface below, and gives precise information on the cruising altitude. The imaging system that maps the surface does so over a range of wavelengths in the infrared, providing night vision and mineralogical information as well as navigational support. Noting its movement relative to the features below, the computer that pilots the blimp is constantly compensating for the wind that blows at a brisk 10 m per second at 5 km altitude, which is the preferred flight level for high-altitude surveillance of the landscape below.

In addition to optical studies, other science experiments on the platforms obtain and analyse atmosphere samples, and get ready to support the surface probes that soon will be released. First, the communication system has to be checked out. Each aerobot is equipped with an antenna in the form of an array of six 35 cm-square patches with mass 0.75 kg each on the upper surface of the blimp. The on-board pilot computer chooses pairs of these to form a dipole for transmission and reception, the choice depending on the orientation of the craft and the direction to the receiver. Normally, science data is sent first to one of the relay satellites in orbit, but direct communication to Earth is possible when a line-of sight is available, which is usually only for a few hours each day (with the Square Kilometer Array, for example). Under optimum conditions, the Direct-to-Earth link is possible with only about 20 W of transmitted power, but the data volume, at about 3.6 Mbytes/day, is much too small for good science and is kept as a back-up should the satellite link fail. However, since twelve satellites have steered themselves into different orbits, a satellite link is nearly always available, and even if one or more microsats fails the aerobots can store their data until another one comes over the horizon. The antenna of multiple, non-steerable patch arrays on the outer surface of the aerobot, using electronic switching between the patches, can transmit several Mbits per second to the nearest orbiter and so back to Earth, sufficient for high-resolution imaging.

Aerial platforms have the obvious advantage, over stationary or even mobile landers, of a much greater survey distance and surface coverage from each location, but there is a need for near-surface measurements and sampling of the atmosphere surface and sub-surface over multiple regions. Once both aerobots' systems have been checked out, and a sufficiently wide area surveyed from the air, they can begin to descend on selected targets, chosen for their scientific interest by the science team, watching back on Earth through the multispectral cameras and the radar mapper. The places of greatest interest initially will include dark areas on ancient shorelines, where rich organic deposits might occur, dried riverbeds, existing lakes and craters, and fresh-looking volcanic features. Also, of course there will be a continuous search for features unseen by Cassini or Huygens: they showed that Titan's surface is clearly diverse enough (with cryovolcanoes, lakes in the northern areas, dunes, etc,...) that there must be many surprises still waiting for the more advanced mission to discover.

The aerial platforms never land, but instead hover over the targets and provide guidance, communications, and control to deploy smaller probes, each typically with a mass of about 35 kg. Each blimp carries six each of two different kinds of these. The first types are autonomous rovers, with mobility on the surface to search a local area. These are lowered to the target area and released, communicating with the aerobot by radio. The rovers are amphibious and are equipped with drills and integrated science experiments for subsurface and surface exploration, so are very versatile provided they are used under relatively benign surface conditions of smooth terrain, gentle slopes, shallow lakes and so on.

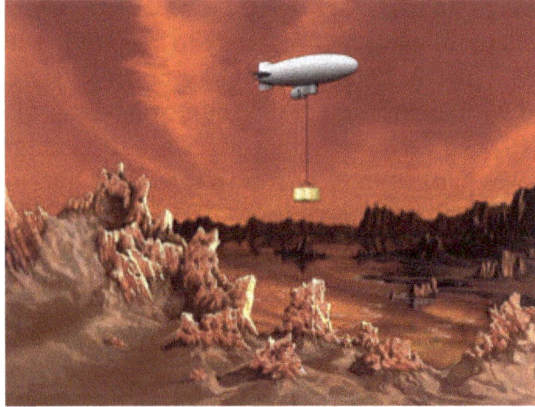

Figure 11.7   Having identified a target area for study, the Titan airship hovers while it lowers a sonde towards the surface (NASA).

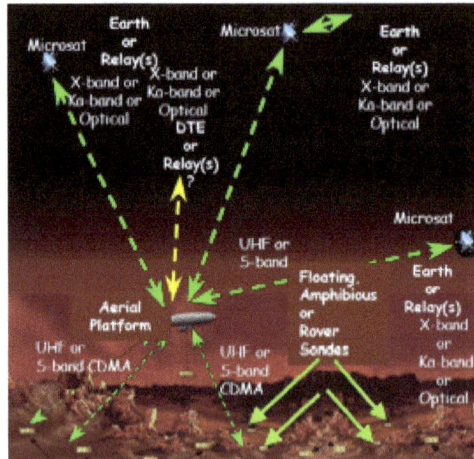

Figure 11.8   Diagram showing the deployment of rovers, sondes and harpoons and their communications links (NASA).

The other six probes are 'sondes', equipped with retractable harpoons for sample collection and a limited amount of *in situ* analysis, on tethers that serve as communications and power conduits. The sondes themselves are on retractable tethers, so they can be carried to various locations and used and reused at multiple geographical sites. The harpoons are equipped with a cryogenic microextraction sampling system for the passive ingestion of liquid and/or solid-phase samples that are hauled back up for analysis by the blimp's instruments. The tethered sondes are used for difficult terrain, some of them lowered on tethers into lakes and volcanic calderas, others firing harpoons into the sub-surface using pressurized helium gas guns, achieving a depth of several metres where the ground is soft. For taking samples from solid crater rims, the gas harpoon is again the only practical solid-core sampler.

For less dense, lower-viscosity materials, like the liquid in Titan's lakes, and at the bottoms of those lakes, a tethered probe is lowered into the lake and winched out again by the blimp, hovering overhead. However, trial and error is involved — key properties like the depth and viscosity of the hydrocarbon lakes are unknown at the outset. The autonomous, fully-amphibious sondes are the most useful in low-viscosity environments, where they can crawl or swim, while harpoons and non-mobile sondes are best in higher-viscosity situations. Under the cold conditions on Titan, the heat dissipated by the miniature isotopic power source on the mobile sonde tends to vaporise the dense cryo-organic liquids it encounters and, in addition to conducting their own analysis of the samples *in situ*, communicating wirelessly with the aerial platform, the boil-off is a target for the cameras and spectrometers on the gondola.

While the sensors and instruments that make the scientific measurements are spread between the platforms, the sondes, the rovers, and the harpoons, the main scientific instrument payload is in the gondola, since its greater payload mass and space allows more flexibility in instrument packaging. Tethered sample chambers are reeled in, and the solid, liquid and gas samples are transferred to the instrument suite in the gondola for analysis. The procedure for solid samples uses advanced solid phase micro-extraction techniques, involving controlled adsorption and desorption of volatile substances. These were originally developed for robot explorers on Mars but refined by the space agencies for the more difficult problem on Titan, where a range of unknown materials, including various ices and tarry organics, have to be dealt with. The sample acquisition methods and the sophisticated control devices and purging technologies needed to remove residues before the next sample arrives, have many applications on the Earth, ranging from surgery to sub-surface prospecting.

Where space, mass and power are at a premium, particularly in the tips of the harpoons, specially developed miniature *in situ* extremely high-resolution micro-wet chemical analysis instruments are used. For organic chemical analysis of airborne, rained-out and deposited tholins, an essential tool is the gas chromatograph-mass spectrometer. This disperses the constituents in a sample so they can be individually identified by their composition and organic signatures, including the chirality of the molecules. The integrated micromachined GCMS on the sondes and rovers has a 4 m effective chromatograph column length but measures just 10 cm on a side overall, weighs less than 1 kg and uses only a few watts of power. A larger, more powerful machine on the gondola is capable of analyzing a greater range of heavier, more complex molecules. It weighs several kg (on Earth) and uses a few hundred watts of electrical power. Other instruments include UV fluorescence and fibre-optic Raman spectrometers, piezo-electric chemical detectors and Nuclear Magnetic Resonance spectrometers to detect specific important organic molecules, such as amino acids, with high resolution between different types. Extremely small but very capable wide field and microscopic near-field imagers for examining the detailed structure of a sample, and engineering sensors measuring pressure, temperature etc. in the local environment, are fitted to all of the deployed devices including the harpoons.

The helium propulsion system for a harpoon needs just 0.4 g of He to achieve an impact velocity in excess of $100 \, \text{m s}^{-1}$. This achieves a penetration depth of around 5 cm, even in hard material like very cold ice, depending of course on the composition, producing a recoil of 1 to 5 m in the sonde launching the projectile. A sample of 5–10 mg is obtained by moving a length of polymer-coated silica fibre in and out of a needle with a plunger. The sample material adsorbs onto the polymer coating of the fibre, which is then introduced into the heated inlet of the gas chromatograph. This then analyses the material desorbed from the coating, the whole process taking about 15 minutes. Different thicknesses and types of coatings are used to optimise the take up of different substances.

The rovers must operate in liquid and solid surface environments and acquire gas, liquid and solid samples. They need planar thruster and lifting surfaces for descent/rise control, and crawler tracks for traction. Again a miniature RTG provides heat and power to the small vehicle — just 60 cm long and with a total mass of 33 kg, including a few times 100 g for the samples of solid and liquid material which it ingests for analysis. On shore lines, lake bottoms and in river beds, a scoop collects wet, loose surface solid particulates and small ($\sim$0.2 g) samples of surface or sub-surface liquid, with similar-sized samples from solid ground or ice (a core measuring $4 \, \text{cm} \times 0.3 \, \text{cm}$ diameter of water ice has a mass of 250 mg). The tethered vehicles have enough capacity for around 25 such samples before they return them to the larger instruments in the gondola for analysis.

The use of so many platforms and probes requires an advanced software system for coordinating and controlling them, a task that resembles those encountered in military or search-and-rescue operations. The pilot computer on the gondola oversees the whole operation, running algorithms for cooperative, distributed sensing and simultaneous control of multiple probes and the deployment and retrieval of the harpoons. The two gondolas also communicate with each other: since they are half a world apart, this link is also through the microsatellites. Each gondola can control not only its own probes, but also those deployed by the other gondola in an emergency, a hedge against computer failure or any other catastrophe. The polar airship has an extra hazard to face: seeking to probe Titan's summer monsoon meteorology, it must strive not to get too close to the massive updrafts and torrential downpours associated with the summer polar climate.

Sooner or later, after a month or so of frenetic activity, all the studies will be over. The probes will be expended and lifeless, the airships crashed on the surface. Only the still-glowing radioisotope generators, which will continue to dissipate measurable amounts of heat into the frigid landscape for centuries, remain alive, a beacon for future Titan explorers. If there are any; Titan is such an inhospitable place for astronauts that, at the moment, it is hard to imagine anyone seeking to live, or even walk, upon the surface. There will be more airborne platforms and surface probes, certainly, but these will probably be manned by robots and will be increasingly more sophisticated versions of the Titania mission we have described, until Titan is fully explored and understood. Then, we can just about imagine astronauts descending in an airborne vehicle, a futuristic helicopter perhaps, to look around and plant a flag, and perhaps to deploy some sort of long-lived, quasi-permanent robotic base.

But humans living on Titan? Perhaps one day, but for now, that remains in the realm of science fiction.

Actinic flux or radiation   This is the solar flux used in calculating photodissociation rates, corresponding to the mean intensity at a given point in the atmosphere.

Adaptive optics   A new observational technique where the phase perturbations induced by the Earth's atmospheric turbulence, responsible for the blur in the images recovered, is corrected in real time on the incident wavefront reaching the telescope. These perturbations are measured by a wavefront sensor. Opposite phase corrections are then applied using a thin deformable mirror in the pupil plane.

Adiabatic Lapse Rate   The lapse rate is the drop of temperature with altitude. Frequently this quantity is close or equal to that for a parcel of air moving without transfer of energy ('adiabatic'). As a parcel moves upwards, it expands as the atmospheric pressure falls. The expansion requires work to be done, this work coming from the internal energy of the air parcel. Thus the parcel cools as it rises.

Albedo   Albedo is a ratio of scattered to incident electromagnetic radiation power, most commonly light. It is a unitless measure of a surface or body's reflectivity. The geometric albedo of an astronomical body is the ratio of its total brightness at zero phase angle to that of an idealised fully reflecting, diffusively scattering (Lambertian) disk with the same cross-section. Zero phase angle corresponds to looking along the direction of illumination. For Earth-bound observers, this occurs when the body in question is at opposition. The visual geometric albedo refers to this quantity when calculated taking into account only electromagnetic radiation in the visual range. The Bond albedo is the fraction of power in the total electromagnetic radiation incident on an astronomical body that is scattered back out into space. This takes into account all wavelengths. It is an important quantity for characterising a planetary body's energy balance. For objects in the solar system, the main contribution comes from visible light because the majority of solar output is in this range. The Bond albedo (A) is related to the geometric albedo (p) by the expression $A = pq$, where (q) is termed the phase integral.

AMU   Atomic mass unit, defined so that the mass of the commonest carbon isotope is 12 amu, making the mass of the hydrogen atom approximately equal to 1 amu.

Apoapsis   The point on an orbit when the spacecraft is furthest from the planet it is orbiting.

350

Arcmin   An arc minute, or minute of arc. An angle equal to one-sixtieth of a degree, or sixty arc-seconds. The field of view of the Hubble Space Telescope Planetary Camera is 34×34 arc seconds. Titan as seen from Earth subtends an angle of about 1 arc-second.

Astronomical Unit (AU)   The mean distance of the Earth from the Sun, or about 150 million kilometres.

Atmospheric layers   By analogy with the Earth, Titan's atmosphere is subdivided into layers defined by the temperature variations with height (or pressure, through the hydrostatic law). On both Titan and the Earth, the mean temperature profile is characterised by two temperature inversions (locations above which the temperature increases with altitude, while the opposite takes place beneath). The boundaries between regions are identified by terms ending with -pause (from the Greek $\pi\alpha\upsilon\sigma\eta\varsigma$ meaning end).

Bar   A unit of pressure, defined to be equal to 10^5 pascals (Pa). The mean surface pressure of the Earth (often called one atmosphere) is approximately 1 bar (actually 1.01325×10^5 pascals). Atmospheric pressures are often quoted in millibars (mbar). 1 bar is also equal to 1,000,000 dynes per square centimetre; 1 mbar $= 0.001$ bar $= 100\,\mathrm{Pa} = 1,000\,\mathrm{dyn/cm}^2$.

Baroclinic, barotropic   Barotropic is a region of uniform temperature distribution; a lack of fronts. Everyday being similar (hot and humid with no cold fronts to cool things off) would be a barotropic type atmosphere, as we find at tropical latitudes. In a baroclinic region, on the other hand, distinct air mass regions exist. Fronts separate warmer from colder air. In a synoptic scale baroclinic environment, you will find the polar jet in the vicinity, troughs of low pressure (mid-latitude cyclones) and frontal boundaries. There are clear density gradients in a baroclinic environment caused by the fronts. Any time you are near a mid-latitude cyclone, you are in a baroclinic environment.

Central Meridian   The longitude of the horizontal center of a coordinate system. This longitude value is often the longitude origin of the coordinate system, or the chosen meridian plane from which longitude is measured. On Earth, the central, or zero meridian, is commonly accepted since 1884 as the Greenwich Meridian, which runs through Greenwich, England. On Titan, the Eastern Elongation corresponds to 90° Longitude of the Central Meridian — LCM — as opposed to geographical longitude, which is about 210°, since Titan rotates synchronously with Saturn.

Clathrate   A clathrate or clathrate compound or cage compound is a chemical substance consisting of a lattice of one type of molecule trapping and containing a second type of molecule. A clathrate therefore is a material which is a weak composite, with molecules of suitable size captured in spaces which are left by the other compounds.

Cryovolcanism   Cryovolcanism is the existence of cryovolcanoes and their effects. A cryovolcano is, literally, an icy volcano. Cryovolcanoes form on icy moons, and possibly on other low temperature astronomical objects (e.g. Kuiper belt objects). Rather than molten rock, these volcanoes erupt volatiles such as water, ammonia or methane. Collectively referred to as cryomagma or ice-volcanic melt, these substances are usually liquids and form flumes, but can also be in vapour form. After eruption, cryomagma condenses to a solid form when exposed to the very low surrounding temperature. The Cassini–Huygens mission has found cryovolcanism on Enceladus and a methane-spewing cryovolcano on Titan, and such volcanism is now believed to be a significant source of the methane found in Titan's atmosphere.

Collision-induced absorption   Absorption bands of molecules which occur because collisions between molecules induce dipole moments which are not otherwise present. On Titan, collisions between CH_4 and N_2 (in every combination) make an important contribution to the opacity of the atmosphere at middle and long infrared wavelengths and moderate to low altitudes.

Cyclostrophic balance   On Earth, the surface rotation typically surpasses the zonal winds, and the pressure gradient force generated by the unequal solar fluxes at low and high latitudes is balanced by the Coriolis force, in what is called a "geostrophic balance". On Titan (and Venus) the opposite is true, and pressure gradients are balanced by strong centrifugal forces arising from the rapid rotation of the atmosphere. This balance, typical of cyclones, is called cyclostrophic.

Dayglow   A fluorescent emission of radiation from the upper atmosphere, due to excitation by high-energy sunlight photons.

Deuterium   The heavy isotope of hydrogen, with a proton and neutron in its atomic nucleus. Deuterium (and hydrogen compounds like methane and water) participates in chemical reactions and physical processes such as diffusion and evaporation at different rates from normal hydrogen. The deuterium to hydrogen ratio (D/H) in an atmosphere is therefore an indicator of the separation processes that have occurred in it.

Dynamical inertia   The increase in radiative time constant due to mixing of more massive deeper layers of an atmosphere. The thin atmosphere at high altitudes would be expected to respond rapidly to changes in sunlight, but if there is substantial vertical circulation, the changes will occur more slowly as the mixing increases the effective mass of the layer under consideration.

Eccentricity The eccentricity of this conic section, the orbit's eccentricity, is an important parameter of the orbit that defines its absolute shape. Eccentricity may be interpreted as a measure of how much this shape deviates from a circle. Under standard assumptions, eccentricity (e) is strictly defined for all circular, elliptic, parabolic and hyperbolic orbits (for a circular orbit, the eccentricity is 0).

Exobiology The study of life beyond the Earth. More broadly, the term encompasses study of environments favourable for life, the origin and evolution of life in the universe, and the evolution of intelligence and technology.

Exosphere The outermost part of the atmosphere, extending to outer space from about 1,500 km altitude on Titan, where light elements, especially hydrogen, can escape the planet's grip and are lost to space.

Geometric Albedo The ratio of the total reflected flux in all directions by a planet to the incident solar flux received by a sphere of unit radius, divided by the phase integral. For a highly reflecting atmosphere or surface, the geometric albedo is in the range 0.65-0.80, depending on the phase function.

HST Hubble Space Telescope.

Hadley circulation A meridional (North-South) flow pattern in the atmosphere, transporting heat away from warmer to colder regions, which on the Earth rises at low latitudes and descends at high latitudes.

Homopause A fairly small distance up into the thermosphere, diffusion takes over as the dominant process and the atmosphere starts to separate into its lighter and heavier components. For many practical purposes, this level (the *homopause*) may be considered to be the effective top of the atmosphere.

Ionosphere The region of a planetary atmosphere where charged particles, ions and electrons, are present in significant numbers. The ionosphere is usually a low-density region; on Titan, its peak occurs roughly 1,000 km above the surface.

Infrared spectroscopy The dispersal and measurement of infrared light intensity. Near-infrared (0.7 to 5 microns) spectroscopy is useful for mineral and ice identification; mid-infrared (5 to 30 microns) is more useful for measuring surface temperatures, and measuring gas abundances.

IRIS The Infrared Interferometer Spectrometer, an instrument on the Voyager missions. See Chapter 2.

ISO The Infrared Space Observatory. See Chapter 3.

Jeans escape   The process by which fast (energetic, or hot) molecules of an atmo-sphere escape into space. The energy distribution of a gas at a given temperature has a hot tail — a few atoms moving faster than the rest. Therefore if, at an altitude where collisions between molecules are rare, the molecules in the hot tail move faster than the local escape velocity, they can escape to space. This process is fastest for hot atmospheres of light gases (hydrogen, helium) on bodies with low gravity.

Limb-darkening   The darkening of the edges of a planetary disk. This may be due to the scattering properties of the surface (if, for example, it is a strongly backscat-tering surface, like an icy one), or more usually to the presence of an optically thick atmosphere. It is often characterised by an exponent k, the Minnaert exponent, for a scattering law of the form $I = Io\mu k \mu ok - 1$, where μ and μo are the cosines of the angle between the normal at a given point and the observer and Sun respectively and Io is the brightness of the centre of the disk. $k = 0.5$ corresponds to a flat disk (rather like the moon), while $k = 1$ is a Lambertian disk with strong limb-darkening. $k < 0.5$ corresponds to limb-brightening, typical of a scattering but optically thin region above an absorbing (dark) region in the atmosphere.

Mesosphere   Above the stratopause, the temperature declines again, reaching a minimum at the *mesopause*, where the second temperature inversion occurs, signify-ing the end of the *mesosphere*. The pressure at the mesopause is about one microbar on Titan, and occurs approximately 600 km above the surface.

Micron (or micrometre)   A micrometre (symbol μm) is one millionth of a metre.

Mixing ratio   The fraction of the total volume due to a given compound. Also expressed as a column abundance (m-amagat).

Mixing ratio scale height   If the mixing ratio of a given compound in an atmo-sphere changes with altitude, due to its production or removal at given altitudes, its variation may be conveniently described by a function of the form $\exp(-h/H)$, where h is the change in altitude and H is the mixing ratio scale height, the vertical distance over which the mixing ratio changes by a factor e.

Oblateness   The flattening of a planet or satellite, usually due at least in part to its spin. The oblateness is the difference between the polar and equatorial radii, divided by the equatorial radius. The Earth's oblateness is 1/298; Titan's atmosphere has an oblateness of about 1/250 at an altitude of 250 km, from which the rapid rotation of the upper atmosphere has been inferred.

Obliquity   The tilt of a planet's axis of rotation, relative to the direction perpen-dicular to the ecliptic plane (the plane containing the Sun's equator and the orbits of most of the planets). A planet with near-zero obliquity, like Venus, has no seasons. The obliquity of Saturn is 26.4 degrees, so Titan has significant seasons.

Opacity The ability of an atmosphere to absorb (or sometimes scatter) radiation. Also called optical depth. A beam of monochromatic radiation passing through an atmosphere with an optical depth of one will have its intensity reduced by a factor $e(=2.718\cdots)$, while an optical depth of 4 absorbs 99% of the radiation. Opacity is a function of wavelength as well as the pressure, temperature and composition of the region of the atmosphere under consideration.

Optical depth A measure of how opaque a layer of atmospheric gas or cloud is, defined as the logarithm of the transmission. It can take values from 0 (perfectly transparent) to infinity (perfectly opaque), and is of course a function of wavelength.

Optically thin Absorbing or scattering relatively little radiation, i.e. 'clear'. The usual interpretation of this term is that the optical depth is less than one.

Periapsis The point on an orbit when the spacecraft is closest to the planet it is orbiting.

Phase angle The phase angle is the angle between the incoming and outgoing beams. High phase angle observations are those in which the incident and scattered rays are almost in a straight line, i.e. forward scattering or grazing incidence. Because Titan is so far from the Sun, the illumination always comes from behind an observer on or near the Earth, so she sees only backscattered rays. High phase angle observations are only obtainable from a spacecraft passing around the other side of Titan.

Phase function The phase function is the mathematical expression which describes the distribution of scattered light from an object, as a function of phase angle. It can be analysed in terms of the optical properties (refractive index) and microphysics (particle size and shape, or surface roughness) of the medium doing the scattering, which on Titan is the high-level haze in the visible, plus a contribution from the surface at longer wavelengths.

Photolysis The breaking of a chemical bond by a light photon (usually at ultraviolet wavelengths). The molecular fragments may recombine in a number of ways. Among easily-photolysed gases are methane, ammonia and oxygen, but even 'unreactive' gases such as nitrogen may react photochemically to produce other compounds (such as HCN on Titan, or nitrogen oxides on Earth).

Prebiotic Molecules or conditions which are thought to have been necessary or likely precursors for the processes which led to the origins of life on Earth.

Radio-occultation The passing of a radio beam through a planet's atmosphere. Attenuation and refraction (bending) of the beam — generally by phase delay — can be used to measure the density of electrons in the planets ionosphere, and the density of the gas in its atmosphere. The abrupt cut-off of the signal can also be used to make a precise measurement of the planet's surface radius.

Remote sensing   The determination of atmospheric and surface properties by measurement and analysis of emitted or reflected radiation, usually at infrared or longer (microwave, radio) wavelengths.

Scale height   The vertical distance over which atmospheric pressure changes by a factor e. The property is described by a function of the form $\exp(-h/H)$, where h is the change in altitude and H is the scale height. Four scale heights above the surface, the pressure is about 1/100 of its surface value. For an isothermal atmosphere, the pressure scale height is equal to (RT/g), where g is the local gravity, R the gas constant appropriate to the atmospheric composition and T the local temperature in K. As all of these quantities vary with altitude, the scale height also varies with altitude, but it is often a convenient approximation to assume it to be constant. For the Earth, the scale height is around 10 km, while for Titan's upper atmosphere, the value is nearer 40 km.

Stratosphere   The *stratosphere* is the region where each layer is heated by radiation from the optically thick atmosphere below, and cooled by radiating to space; here density decreases monotonically with height, and therefore the layers do not try to move up or down through each other as in the troposphere. The temperature is also maximum at the upper boundary, which is known as the *stratopause*, due to the absorption and conversion of UV solar radiation by different gases and aerosols. Titan's stratosphere lies between about 50 and 250 km altitude (roughly 100 to 0.1 mbar pressure).

Super-rotation   The rapid (prograde) rotation of the upper layers of the atmosphere, notable in both Venus and Titan. The atmosphere of Titan at a few hundred kilometres altitude may make around twenty revolutions for every revolution Titan's solid body makes.

Scale height   The vertical distance over which an atmospheric property (usually pressure, or density) changes by a factor e. The property is described by a function of the form $\exp(-h/H)$, where h is the change in altitude and H is the scale height. Four scale heights above the surface, the pressure is about 1/100 of its surface value. For an isothermal atmosphere, the pressure scale height is equal to $(c_p T/gM)$, where g is the local gravity, M the relative molecular mass, c_p the specific heat at constant pressure and T the local temperature in K. As all of these quantities vary with altitude, the scale height also varies with altitude, but it is often a convenient approximation to assume it to be constant. For the Earth, the scale height is around 10 km, while for Titan's upper atmosphere, the value is nearer 40 km.

Thermosphere   With such low densities of gas above, very energetic solar photons in the extreme ultraviolet, and particles too, penetrate into the region causing ionisation and dissociation and releasing kinetic energy. The heating thus produced causes the temperature to increase rapidly with height, leading to the name *thermosphere* ($\theta\varepsilon\rho\mu o$=warm), the most extensive part of the atmosphere, in which the energy is transported by thermal conduction.

Terminator   The boundary between day and night on a planet.

Troposphere   Convective instability exists in the lowest 40 km or so of the atmosphere (that is to say that since the temperature decreases with altitude, the warmer air lying under colder air is unstable and so it rises while the colder air sinks), which is known as the *troposphere* ('turning-region'). The upper boundary here is the level where the overlying atmosphere is of such a low density that a substantial amount of radiative cooling to space can occur in the thermal infrared region of the spectrum. At this level, called the *tropopause,* radiation cools rising air so efficiently that the temperature tends to become constant with height and convection ceases.

Wavenumber, wavelength   Often expressed also as a wavenumber (e.g. a wavenumber of $10,000 \, \text{cm}^{-1}$ indicates 10,000 wavelengths fit into one centimetre, so the wavelength is 1 micron. A wavelength of 10 microns therefore has a wavenumber of $1,000 \, \text{cm}^{-1}$.) Wavenumber is the spatial analogue of angular frequency. In spectroscopy, the wavenumber n of electromagnetic radiation is defined as $\nu = 1/l$, where λ as a length in the SI-system is measured in meters (m) and commonly quantified in centimetres ($\text{cm} = 10^{-2}$ m) refers to the wavelength in vacuum. The unit of this quantity is cm^{-1}, pronounced as reciprocal centimetre, or "inverse centimetre".

Window   A spectral region which is relatively transparent, between two regions which have higher opacity. A window region can be important for remote sensing of a planetary surface, and for limiting the extent of a greenhouse effect.

References and Bibliography

Several articles in Science and Nature issues of 2004 and 2005 describe in detail the Cassini–Huygens mission findings. We recommend in particular: Science **307**, No. 5713, 25 February 2005, "Cassini arrives at Saturn"; Nature 10 March 2005 "Imaging of Titan from the Cassini spacecraft"; Nature December 8, 2005, all the papers from the first Huygens results. Also: Geophysical Research Letters Vols. **32** and **33**, Icarus Vols. **173** and **182–183** and **186**, Nature Vols. **438** and **442**, Journal of Geophysical Research Vol. **111**, Science Vols. **308** and **310**; Earth Moon and Planets Vol. **96**; Planetary and Space Sciences Vols. **47, 49, 53** and **54**; Space Science Reviews Vols. **114** and **115**; Advances in Space Research Vols. **26, 28, 33** and **36**.

Volume **104** of Space Science Reviews includes articles with descriptions of the Huygens instruments and the mission in general, plus articles by the Huygens Interdisciplinary Scientists.

Books

Recent books published in relation to Titan include:

Coustenis, A., Taylor, F., 1999. *Titan: the Earth-like Moon*, World Scientific Publishers, Singapore.

Lorenz, R. D., Mitton, J., 2002. *Lifting Titan's Veil: Exploring the Giant Moon of Saturn*, Cambridge University Press.

Lunine, J. I., 2004. *Astrobiology: Multi Disciplinary Approach*. B. Cummings, Publ., ISBN: 0805380426.

Encrenaz, T., Kallenbach, R., Owen, T. C. and Sotin, C., eds., 2005, *The Outer Planets and Their Moons*. Space Science Series of ISSI, Vol. 19, Springer.

Web sites

http://www.nasa.gov/mission_pages/cassini/main/index.html
http://www.spaceflightnow.com/cassini/
http://saturn.jpl.nasa.gov/multimedia/images/index.cfm
http://www.esa.int/SPECIALS/Cassini-Huygens/
http://solarviews.com/cap/index/spacecraft-cassini1.html
http://cisas.unipd.it/hasi/welcome.html
http://filer.case.edu/~sjr16/advanced/saturn_moons.html
http://spaceplace.nasa.gov/en/kids/sse_flipflop2.shtml
http://www.aero.jussieu.fr/experience/ACP/
http://www.astro.uni-bonn.de/~dwe/
http://www.esa.int/export/esaSC/120378_index_0_m.html
http://www.planetary.org/explore/topics/our_solar_system/saturn/
http://ssd.jpl.nasa.gov/?sat_phys_par

http://www.lesia.obspm.fr/cosmicvision/tandem/index.php
http://www.lpi.usra.edu/opag/announcements.html

Other References

Hereafter some significant references on Titan with emphasis on recently published papers (mainly since 2000).

Abbott, A., 2005. Titan team claims just deserts as probe hits moon of crème brûlée. Nature **433**, 181.

Achterberg, R. K., Conrath, B. J., Gierasch, P. J., Flasar, F. M., Nixon, C. A., 2008. Titan's middle-atmospheric temperatures and dynamics observed by the Cassini Composite Infrared Spectrometer. Icarus **194**, 263–277.

Ádámkovics, M., de Pater, I., Roe, H. G., Gibbard, S. G., Griffith, C. A., 2004. Spatially-resolved spectroscopy at 1.6 μm of Titan's atmosphere and surface. Geophys. Res. Let. **31**, CiteID L17S05.

Ádámkovics, M., de Pater, I., Hartung, M., Eisenhauer, F., Genzel, R., Griffith, C. A., 2006. Titan's bright spots: Multiband spectroscopic measurement of surface diversity and hazes. J. Geophys. Res. **111**, CiteID E07S06.

Adriani, A., Moriconi, M. L., Liberti, G. L., Gardini, A., Orosei, R., D'Aversa, E., Filacchione, G., Coradini, A., 2005. Titan's ground reflectance retrieval from Cassini-IMS data taken during the July 2nd, 2004 Fly-By at 2 AM UT. Earth, Moon, and Planets **96**, 109–117.

Alibert, Y., Mousis, O., 2007. Formation of Titan in Saturn's subnebula: Constraints from Huygens probe measurements. Astron. Astrophys. **465**, 1051–1060.

Allen, M., Pinto, I. P., Yung, Y. L., 1980. Titan: Aerosol photochemistry and variations related to the sunspot cycle, Astrophys. J. **242**, L125–L128.

Allison, M. A., 1992. Preliminary Assessment of the Titan planetary boundary layer. In: Proceedings of the Symposium on Titan, ESA Special Publication **338**, 113–118.

Anders, E., Grevesse, N., 1989. Abundances of the elements: Meteoritic and solar. Geochim. Cosmochim. Acta **53**, 197–214.

Anderson, C. M., Chanover, N. J., McKay, C. P., Rannou, P., Glenar, D. A., Hillman, J. J., 2004. Titan's haze structure in 1999 from spatially-resolved narrowband imaging surrounding the 0.94 μm methane window. Geophys. Res. Let. **31**, CiteID L17S06.

Andrews, D. G., Holton, J. R., Leovy, C. B., 1987. Middle Atmosphere Dynamics. New York: Academic Press.

Anicich, Vincent G., Milligan, Daniel B., Fairley, David A., McEwan, Murray J., 2000. Termolecular ion-molecule reactions in Titan's atmosphere, I Principal Ions with Principal Neutrals. Icarus **146**, 118–124.

Anicich, Vincent G., McEwan, Murray J., 2002. Termolecular ion-molecule reactions in Titan's atmosphere. III. Clustering of Ions. Icarus **154**, 522–530.

Artemieva, N., Lunine, J., 2003. Cratering on Titan: Impact melt, ejecta, and the fate of surface organics. Icarus **164**, 471–480.

Artemieva, N., Lunine, J. I., 2005. Impact cratering on Titan II. Global melt, escaping ejecta, and aqueous alteration of surface organics. Icarus **175**, 522–533.

Atkins, P. W., 1983. Molecular Quantum Mechanics. Oxford University Press.

Atreya, S. K., Donahue, T. M., Kuhn, W. R., 1978. Evolution of a nitrogen atmosphere on Titan, Science **201**, 611–613.

Atreya, S. K., 1986 Atmospheres and Ionospheres of the Outer Planets and their Satellites. Springer-Verlag. New York.

Atreya, S. K., Adams, E. Y., Niemann, H. B., Demick-Montelara, J. E., Owen, T. C., Fulchignoni, M., Ferri, F., Wilson, E. H., 2006. Titan's methane cycle. Plan. Space Sci. **54**, 1177–1187.

Awal, M., Lunine, J. I., 1994. Moist convective clouds in Titan's atmosphere. Geophys. Res. Lett **21**, 2491–2494.

Backes, H., Neubauer, F. M., Dougherty, M. K., Achilleos, N., André, N., Arridge, C. S., Bertucci, C., Jones, G. H., Khurana, K. K., Russell, C. T., Wennmacher, A., 2005. Titan's magnetic field signature during the first Cassini encounter. Science **308**, 992–995.

Bakes, E. L. O., Lebonnois, S., Bauschlicher, Charles W., McKay, Christopher P., 2003. The role of submicrometer aerosols and macromolecules in H_2 formation in the Titan haze. Icarus **161**, 468–473.

Balsinger, H., Altwegg, K., Geiss, J., 1995. D/H and $^{18}O/^{16}O$ ratio in the hydronium ions and neutral water from *in situ* measurements in comet Halley. J. Geophys. Res. **100**, 5827–5834.

Banaszkiewicz, M., Lara, L. M., Rodrigo, R., López-Moreno, J. J., Molina-Cuberos, G. J., 2000. A coupled model of Titan's atmosphere and ionosphere. Icarus **147**, 386–404.

Banwell, C. N., 1986. Fundamentals of Molecular Spectroscopy. McGraw-Hill.

Baines, K. H., Brown, R. H., Matson, D. L., Nelson, R. M., Buratti, B. J., Bibring, J. P., Langevin, Y., Sotin, C., Carusi, A., Coradini, A., Clark, R. N., Combes, M., Drossart, P., Sicardy, B., Cruikshank, D. P., Formisano, V., Jaumann, R., 1992. VIMS/Cassini at Titan: Scientific objectives and observational scenarios. In: Proceedings of the Symposium on Titan, ESA Special Publication **338**, 137–148.

Baines, K. H., Drossart, P., Momary, T. W., Formisano, V., Griffith, C., Bellucci, G., Bibring, J. P., Brown, R. H., Buratti, B. J., Capaccioni, F., Cerroni, P., Clark, R. N., Coradini, A., Combes, M., Cruikshank, D. P., Jaumann, R., Langevin, Y., Matson, D. L., McCord, T. B., Mennella, V., Nelson, R. M., Nicholson, P. D., Sicardy, B., Sotin, C., 2005. The atmospheres of Saturn and Titan in the near-infrared first results of Cassini/VIMS. Earth, Moon, and Planets **96**, 119–147.

Baines, K. H., Drossart, P., Lopez-Valverde, M. A., Atreya, S. K., Sotin, C., Momary, Th. W., Brown, R. H., Buratti, B. J., Clark, R. N., Nicholson, Ph. D., 2006. On the discovery of CO nighttime emissions on Titan by Cassini/VIMS: Derived stratospheric abundances and geological implications. Plan. Space Sci. **54**, 1552–1562.

Bakes, E. L. O., McKay, Christopher, P., Bauschlicher, Charles, W., 2002. Photoelectric charging of submicron aerosols and macromolecules in the Titan Haze. Icarus **157**, 464–475.

Balucani, N., Asvany, O., Osamura, Y., Huang, L. C. L., Lee, Y. T., Kaiser, R. I., 2000. Laboratory investigation on the formation of unsaturated nitriles in Titan's atmosphere. Plan. Space Sci. **48**, 447–462.

Barnes, J. W., Brown, R. H., Turtle, E. P., McEwen, A. S., Lorenz, R. D., Janssen, M., Schaller, E. L., Brown, M. E., Buratti, B. J., Sotin, C., Griffith, C., Clark, R., Perry, J., Fussner, S., Barbara, J., West, R., Elachi, C., Bouchez, A. H., Roe, H. G., Baines, K. H., Bellucci, G., Bibring, J.-P., Capaccioni, F., Cerroni, P., Combes, M., Coradini, A., Cruikshank, D. P., Drossart, P., Formisano, V., Jaumann, R., Langevin, Y., Matson, D. L., McCord, T. B., Nicholson, Ph. D., Sicardy, B., 2005. A 5-micron-bright spot on Titan: Evidence for surface diversity. Science **310**, 92–95.

Barnes, Jason W., Brown, Robert H., Radebaugh, Jani, Buratti, Bonnie J., Sotin, Christophe, Le Mouelic, Stephane, Rodriguez, Sebastien, Turtle, Elizabeth P., Perry, Jason, Clark, Roger, Baines, Kevin H., Nicholson, Phillip D., 2006. Cassini observations of flow-like features in western Tui Regio, Titan. Geophys. Res. Let. **33**, CiteID L16204.

Barnes, Jason W., Brown, Robert H., Soderblom, Laurence, Buratti, Bonnie J., Sotin, Christophe, Rodriguez, Sebastien, Le Mouèlic, Stephane, Baines, Kevin H., Clark, Roger, Nicholson, Phil, 2007. Global-scale surface spectral variations on Titan seen from Cassini/VIMS. Icarus **186**, 242–258.

Bar-Nun, A., Kleinfeld, I., Ganor, E., 1988. Shape and optical properties of aerosols formed by photolysis of acetylene, ethylene and hydrogen cyanide J. Geophys. Res. **93**, 8383–8387.

Barth, Erika L., Toon, Owen B., 2003. Microphysical modeling of ethane ice clouds in Titan's atmosphere. Icarus **162**, 94–113.

Barth, Erika L., Toon, Owen B., 2004. Properties of methane clouds on Titan: Results from microphysical modelling. Geophys. Res. Let. **31**, CiteID L17S07.

Barth, Erika L., Toon, Owen B., 2006. Methane, ethane, and mixed clouds in Titan's atmosphere: Properties derived from microphysical modeling. Icarus **182**, 230–250.

Barth, E. L., Rafkin, S. C. R., 2007. TRAMS: A new dynamic cloud model for Titan's methane clouds. Geophys. Res. Let. **34**, CiteID L03203.

Beisker, W., Bittner, C., Bode, H. J., Buechner, R., Denzau, H., Dunham, D., Nezel, M., Reidel, E., 1989. The occultation of 28-Sgr by Titan — First results. Occultation News **4**(13), 324–326.

Bénilan, Y., Jolly, A., Raulin, F., Guillemin, J.-C., 2006. IR band intensities of DC_3N and $HC_3^{15}N$: Implication for observations of Titan's atmosphere. Plan. Space Sci. **54**, 635–640.

Bernard, J.-M., Coll, P., Coustenis, A., Raulin, F., 2003. Experimental simulation of Titan's atmosphere: Detection of ammonia and ethylene oxide. Plan. Space Sci. **51**, 1003–1011.

Bernard, J.-M., Quirico, E., Brissaud, O., Montagnac, G., Reynard, B., McMillan, P., Coll, P., Nguyen, M.-J., Raulin, F., Schmitt, B., 2006. Reflectance spectra and chemical structure of Titan's tholins: Application to the analysis of Cassini Huygens observations. Icarus **185**, 301–307.

Bézard, B., Coustenis, A., McKay, C. P., 1995. Titan's stratospheric temperature asymmetry: A radiative origin? Icarus **113**, 267–276.

Bézard, B., Marten, A., Paubert, G., 1993. Detection of acetonitrile on Titan. Bull. Am. Astron. Soc. **25**, 3, 1100.

Bird, M. K., Allison, M., Asmar, S. W., Atkinson, D. H., Avruch, I. M., Dutta-Roy, R., Dzierma, Y., Edenhofer, P., Folkner, W. M., Gurvits, L. I., Johnston, D. V., Plettemeier, D., Pogrebenko, S. V., Preston, R. A., Tyler, G. L., 2005. The vertical profile of winds on Titan. Nature **438**, 800–802.

Borucki, W. J., Giver, L. P., Mckay, C. P., Scattergood, T., Parris, J. E., 1988. Lightning production of hydrocarbons and HCN on Titan: Laboratory measurements. Icarus **76**, 125–134.

Borucki, W. J., McKay, C. P., Whitten, R. C., 1984. Possible production by lightning of aerosols and trace gases in Titan's atmosphere. Icarus **60**, 260–273.

Borucki, W. J., Whitten, R. C., Bakes, E. L. O., Barth, E., Tripathi, S., 2006. Predictions of the electrical conductivity and charging of the aerosols in Titan's atmosphere. Icarus **181**, 527–544.

Borysow, A., Frommhold, L., 1986. Theoretical collision-induced roto-translational absorption spectra for modelling Titan's atmosphere: H_2-N_2 pairs. Astron. Astrophys. J. **303**, 495–510.

Bouchez, A. H., Brown, M. E., 2005. Statistics of Titan's south polar tropospheric clouds. Astrophys. J. **618**, L53–L56.

Brasseur, G., Solomon, S., 1986. Aeronomy of the Middle Atmosphere. Dortrecht: D. Reidel.

Brecht, Stephen H., Luhmann, Janet G., Larson, David J., 2000. Simulation of the Saturnian magnetospheric interaction with Titan. J. Geophys. Res. **105**, 13119–13130.

Broadfoot, A. L., Belton, M. J. S., Takacs, P. Z., Sandel, B. R., Shemansky, D. E., Holberg, J. B., Ajello, J. M., Atreya, S. K., Donahue, T. M., Moos, H. W., Bertaux, J. L., Blamont, J. E., Strobel, D. F., McConnell, J. C., Dalgarno, A., Goody, R., McElroy, M. B., 1981. Extreme ultraviolet observations from Voyager 1 encounter with Saturn. Science **212**, 206–211.

Brown, Michael E., Bouchez, Antonin H., Griffith, Caitlin A., 2002. Direct detection of variable tropospheric clouds near Titan's south pole. Nature **420**, 795–797.

Bruston, P., Poncet, H., Raulin, F., Cossart-Marcos, C., Courtin, R., 1989. UV spectroscopy of Titan's atmosphere, planetary organic chemistry, and prebiological synthesis, I, Absorption spectra of gaseous propynenitrile ad^2-butynenitrile in the 185- to 250-nm region. Icarus **78**, 38–54.

Burr, D. M., Emery, J. P., Lorenz, R. D., Collins, G. C., Carling, P. A., 2006. Sediment transport by liquid surficial flow: Application to Titan. Icarus **181**, 235–242.

Cabane, M., Chassefière, E., 1993. Growth of aerosols in Titan's atmosphere and related time scales: A stochastic approach. Geophys. Res. Let. **20**, 967–970.

Cabane, M., Chassefière, E., 1995. Laboratory simulations of Titan's atmosphere: Organic gases and aerosols. Plan. Space Sci. **43**, 47–65.

Cabane, M., Chassefière, E., Israel, G., 1992. Formation and growth of photochemical aerosols in Titan's atmosphere. Icarus **96**, 176–189.

Cabane, M., Chassefière, E., Botet, R., McKay, C. P., Courtin, R., 1995. Titan's geometric albedo: Role of the fractal structure of the aerosols. Icarus **118**, 355–372.

Cabane, M., Rannou, P., Chassefière, E., Israel, G., 1993. Fractal aggregates in Titan's atmosphere. Plan. Space Sci. **41**, 257–267.

Calcutt, S. B., Taylor, F. W., Ade, P., Kunde, V. G., Jennings, D., 1992. The composite infrared spectrometer. J. Brit. Interplanetary Soc. **45**, 811–816.

Caldwell, J., Cunningham, C. C., Anthony, D., White, H. P., Groth, E. J., Hasan, H., Noll, K., Smith, P. H., Tomasko, M. G., Weaver, H. A., 1992. Titan: Evidence for seasonal change — A comparison of Hubble Space Telescope and Voyager images. Icarus **97**, 1–9.

Caldwell, J., Owen, T., Rivolo, A. R., Moore, V., Hunt, G. E., Butterworth, P. S., 1981. Observations of Uranus, Neptune and Titan by the International Ultraviolet Explorer. Astron. Astrophys. J. **86**, 298–305.

Campbell, Donald B., Black, Gregory J., Carter, Lynn M., Ostro, Steven J., 2003. Radar evidence for liquid surfaces on Titan. Science **302**, 431–434.

Campbell, B. A., 2007. A rough-surface scattering function for Titan radar studies. Geophys. Res. Let. **34**, CiteID L14203.

Canosa, A., Páramo, A., Le Picard, S. D., Sims, I. R., 2007. An experimental study of the reaction kinetics of $C_2(X^1\Sigma_g^+)$ with hydrocarbons (CH_4, C_2H_2, C_2H_4, C_2H_6 and C_3H_8) over the temperature range 24–300 K: Implications for the atmospheres of Titan and the Giant Planets. Icarus **187**, 558–568.

Capone, L. A., Dubach, J., Prasad, S. S., Whitten, R. C., 1983. Galactic cosmic rays and N_2 dissociation on Titan. Icarus **55**, 73–82.

Carrasco, N., Dutuit, O., Thissen, R., Banaszkiewicz, M., Pernot, P., 2007. Uncertainty analysis of bimolecular reactions in Titan ionosphere chemistry model. Plan. Space Sci. **55**, 141–157.

Carlson, R., Smythe, W., Bained, K., Barbinis, E., Becker, K., Burns, R., Calcutt, S., Calvin, W., Clark, R., Danielson, G., Davies, A., Drossart, P., Encrenaz, Th., Fanale, F., Granahan, J., Hansen, G., Herrera, P., Hibbitts, C., Hui, J., Irwin, P., Johnson, T., Kamp, L., Kieffer, H., Leader, F., Lellouch, E., Lopes-Gautier, R., Matson, D., McCord, T., Mehlman, R., Ocampo, A., Orton, G., Roos-Serote, M., Segura, M., Shirley, J., Soderblom, L., Stevenson, A., Torson, J., Taylor, F., Weir, A., Weissman, P., 1996. Near-infrared spectroscopy and spectral mapping of Jupiter and the Galilean Satellites: First results from Galileo's initial orbit. Science **274**, 385–388.

Cassini, J. D., 1671. Observations concerning Saturn. Phil. Trans. **6**, 3024–3027.

Cassini, J. D., 1673a. Découverte de deux Nouvelles Planètes autour de Saturne, Paris.

Cassini, J. D., 1673b. Discovery of two new planets about Saturn, made in the Royal Parisian Observatory by Signor Cassini. Phil. Trans. **8**, 5178–5185.

Cassini, J. D., 1676. An extract of Signor Cassini's letter concerning a spot lately seen in the Sun: Together with a remarkable observation of Saturn by the same. Phil. Trans. **11**, 689–690.

Cassini, J. D., 1686. An extract of the Journal des Scavans. of April 22nd 1686. Giving an account of two new satellites of Saturn, discovered lately by Mr. Cassini at the Royal Observatory at Paris. Phil. Trans. **16**, 79–85.

Celnikier, L. M., 1993. Basics of Space Flight. Editions Frontières.

Cerceau, F., Raulin, F., Courtin, R., Gautier, D., 1985. Infrared spectra of gaseous mononitriles: Application to the atmosphere of Titan. Icarus **62**, 207–220, 1985.

Cess, R. D., Owen, T., 1973. Titan: The effect of noble gases on an atmospheric greenhouse. Nature **244**, 272–273.

Chamberlain, J. W., Hunten, D. M., 1987. Theory of Planetary Atmospheres. Academic Press, San Diego, Calif.

Chanover, N. J., Anderson, C. M., McKay, C. P., Rannou, P., Glenar, D. A., Hillman, J. J., Blass, W. E., 2003. Probing Titan's lower atmosphere with acousto-optic tuning. Icarus **163**, 150–163.

Chassefière, E, Cabane, M., 1995. Two formation regions for Titan's Hazes: Indirect clues and hypothesised chemical synthesis processes. Plan. Space Sci. **43**, 91–103.

Chiu, Wang-Ting, Hsu, Hui-Chun, Kopp, Andreas, Ip, Wing-Huen, 2001. On ion outflows from Titan's exosphere. Geophys. Res. Let. **28**, 3405–3408.

Clark, R. N., Fanale, F. P., Gaffey, M. J., 1986. Surface composition of natural satellites. In: Satellites (Burns, J. A. and Matthews, M. S., eds.), The Univ. of Arizona Press, Tucson, 437–491.

Clark, R. N., Brown, R. H., Owensby, P. D., Steele, A., 1984. Saturn's satellites: Near-infrared spectrophotometry (0.65–2.5 μm) of the leading and trailing sides and compositional implications. Icarus **58**, 265–281, 1984.

Clarke, D. W., Ferris, J. P., 1995. Photodissociation of cyanoacetylene: Application to the atmospheric chemistry of Titan. Icarus **115**, 119–126.

Clarke, David W., Joseph, Jeffrey C., Ferris, James P., 2000. The design and use of a photochemical flow reactor: A laboratory study of the atmospheric chemistry of cyanoacetylene on Titan. Icarus **147**, 282–291.

Clausen, K., Sainct, H., 1994. The Huygens probe and mission design. Adv. Space Res. **14**(12), 189–195.

Coll, P., Coscia, D., Gazeau, M. C., de Vanssay, E., Guillemin, J. C., Raulin, F., 1995. Organic chemistry in Titan's atmosphere: New data from laboratory simulations at low temperature. Adv. Space Res. **16**(2), 93–103.

Coll, P., Coscia, D., Gazeau, M. C., Guez, L., Raulin, F., 1998. Review and latest results of laboratory investigation of Titan's aerosol. Origins of Life **28**, 195–213.

Coll, P., Coscia, D., Smith, N., Gazeau, M.-C., Ramírez, S. I., Cernogora, G., Israël, G., Raulin, F., 1999. Experimental laboratory simulation of Titan's atmosphere: Aerosols and gas phase. Plan. Space Sci. **47**, 1331–1340.

Coll, P., Guillemin, J.-C., Gazeau, M.-C., Raulin, F., 1999. Report and implications of the first observation of C4N2 in laboratory simulations of Titan's atmosphere. Plan. Space Sci. **47**, 1433–1440.

Coll, P., Bernard, J.-M., Navarro-González, R., Raulin, F., 2003. Oxirane: An exotic oxygenated organic compound on Titan? The Astrophys. J. **598**, 700–703.

Collins, G. C., 2005. Relative rates of fluvial bedrock incision on Titan and Earth. Geophys. Res. Let. **32**, CiteID L22202.

Comas Solá, J., 1908. Observationes des Satellites Principaux de Jupiter et de Titan. Astron. Nachr. **179**, 289–290.

Combes, M., Vapillon, L., Gendron, E., Coustenis, A., Lai, O., Wittemberg, R., Sirdey, R., 1997. Spatially resolved images of Titan by means of adaptive optics. Icarus **129**, 482–497.

Consolmagno, G. J., Lewis, J. S., 1978. The evolution of icy satellites interiors and surfaces. Icarus **34**, 280–293.

Courtin, R., 1988. Pressure-induced absorption coefficients for radiative transfer calculations in Titan's atmosphere. Icarus **75**, 245–254.

Courtin, R., Gautier, D., McKay, C. P., 1995. Titan's thermal emission spectrum: Re-analysis of the Voyager infrared measurements. Icarus **114**, 144–163.

Courtin, R., Wagener, R., McKay, C. P., Caldwell, J., Fricke, K.-H., Raulin, F., Bruston, P., 1991. UV spectroscopy of Titan's atmosphere, planetary organic chemistry and prebiological synthesis. II. Interpretation of new IUE observations in the 220–335 nm range. Icarus **90**, 43–56.

Courtin, R., Gautier, D., McKay, C. P., 1996. Titan's thermal emission spectrum: Reanalysis of the Voyager infrared measurements. Icarus **114**, 144–162.

Courtin, R., Kim, S. J., 2002. Mapping of Titan's tropopause and surface temperatures from Voyager IRIS spectra. Plan. Space Sci. **50**, 309–321.

Courtin, R., 2005. Aerosols on the giant planets and Titan. Space Science Reviews **116**, 185–199.

Coustenis, A., 1990. Spatial variations of temperature and composition in Titan's atmosphere: Recent results. Ann. Geophysicae **8**, 645–652.

Coustenis, A., 1991 Titan's atmosphere: Recent developments. Vistas in Astronomy **34**, 11–50.

Coustenis, A., 1995. Titan's atmosphere and surface: Parallels and differences with the primitive Earth. In: "Comparative planetology with an Earth prospective", Earth, Moon, and Planets **67**, Chahine, M. T., A'Hearn, M. F. and Rahe, J. eds., 95–100.

Coustenis, A., 1998. Titan in the Solar System. Planet. Space Sci. **46**, 1085–1097.

Coustenis, A., 2005. Formation and evolution of Titan's atmosphere. Space Sci. Rev. **116**, 171–184.

Coustenis, A., 2007. What Cassini-Huygens has revealed about Titan. Astronomy and Geophysics **48**, 2.14–2.20.

Coustenis, A., Bézard, B., Gautier, D., 1989a. Titan's atmosphere from Voyager infrared observations: I. The gas composition of Titan's equatorial region. Icarus **80**, 54–76.

Coustenis, A., Bézard, B., Gautier, D., 1989b. Titan's atmosphere from Voyager infrared observations: II. The CH_3D abundance and D/H ratio from the 900–1200 cm^{-1} spectral region. Icarus **82**, 67–80.

Coustenis, A., Bézard, B., Gautier, D., Marten, A., Samuelson, R. E., 1991. Titan's atmosphere from Voyager infrared observations. III. Vertical distributions of hydrocarbons and nitriles near Titan's north pole. Icarus **89**, 152–167.

Coustenis, A., Encrenaz, Th., Bézard, B., Bjoraker, G., Graner, G., Dang-Nhu, M., Arie, E., 1993. Modelling Titan's thermal infrared spectrum for high-resolution space observations. Icarus **102**, 240–260.

Coustenis, A., Bézard, B., 1995a. Titan's atmosphere from Voyager infrared observations. IV. Latitudinal variations of temperature and composition. Icarus **115**, 126–140.

Coustenis, A., Lellouch, E., Maillard, J.-P., McKay, C. P., 1995b. Titan's surface: Composition and variability from the near-infrared albedo. Icarus **118**, 87–104.

Coustenis, A., Schneider, J., Bockelée-Morvan, D., Rauer, H., Wittenberg, R., Chassefière, E., Greene, T., Penny, A., Guillot, T., 1997. Spectroscopy of 51 Peg B: Search for atmospheric signatures. In: Planets beyond the Solar System and the Next Generation of Space Missions. ASP Conference Series, **119**, 101–105.

Coustenis, A., Salama, A., Lellouch, L., Encrenaz, Th., Bjoraker, G., Samuelson, R., de Graauw, Th., Feuchtgruber, H., Kessler, M. F., 1998. Evidence for water vapor in Titan's atmosphere from ISO/SWS data. Astron. Astrophys. **336**, L85–L89.

Coustenis, A., Schmitt, B., Khanna, R., Trotta, F., 1999. Plausible condensates in Titan's stratosphere from Voyager IR spectra. Plan. Space Sci. **47**, 1305–1329.

Coustenis, A., Gendron, E., Lai, O., Véran., J.-P. Woillez, J., Combes, M., Vapillon, L., Fusco, Th., Mugnier, L., Rannou, P., 2001. Images of Titan at 1.3 and 1.6 microns with adaptive optics at the CFHT. Icarus **154**, 501–515.

Coustenis, A., Salama, A., Schulz, B., Ott, S., Lellouch, E., Encrenaz, Th., Gautier, D., Feuchtgruber, H. 2003. Titan's atmosphere from ISO mid-infrared spectroscopy. Icarus **161**, 383–403.

Coustenis, A., 2005. Formation and evolution of Titan's atmosphere. Space Sci. Rev., **116**, 171–184.

Coustenis, A., Hirtzig, M., Gendron, E., Drossart, P., Lai, O., Combes, M., Negrao, A., 2005. Maps of Titan's surface from 1 to 2.5 micron. Icarus **177**, 89–105.

Coustenis, A., Negrao, A., Salama, A., Schulz, B., Lellouch, E., Rannou, P., Drossart, P., Encrenaz, Th., Schmitt, B., Boudon, V., Nikitin, A., 2006. Titan's 3-micron spectral region from ISO high-resolution spectroscopy. Icarus **180**, 176–185.

Coustenis, A., Achterberg, R., Conrath, B., Jennings, D., Marten, A., Gautier, D., Bjoraker, G., Nixon, C., Romani, P., Carlson, R., Flasar, M., Samuelson, R. E., Teanby, N., Irwin, P., Bézard, B., Orton, G., Kunde, V., Abbas, M., Courtin, R., Fouchet, Th., Hubert, A., Lellouch, E., Mondellini, J., Taylor, F. W., Vinatier, S., 2007. The composition of Titan's stratosphere from Cassini/CIRS mid-infrared spectra. Icarus **189**, 35–62.

Coustenis, A., Jennings, D., Jolly, A., Bénilan, Y., Nixon, C., Gautier, D., Vinatier, S., Bjoraker, G., Romani, P., 2008a. Detection of C_2HD and the D/H ratio on Titan. Icarus, in press.

Coustenis, A., Atreya, S., Balint, T., Brown, R. H., Dougherty, M., Ferri, F., Fulchignoni, M., Gautier, D., Gowen, R., Griffith, C., Gurvits, L., Jaumann, R., Langevin, Y., Leese, M., Lunine, J., McKay, C. P., Moussas, X., Müller-Wodarg, I., Neubauer, F., Owen, T., Raulin, F., Sittler, E., Sohl, F., Sotin, C., Tobie, G., Tokano, T., Turtle, E, Wahlund, J.-E., Waite, H., Baines, K., Balmont. J., Dandouras, I., Krimigis, T., Lellouch, E., Lorenz, R., Morse, A., Porco, C., Hirtizig, M., Saur, J., Coates, A., Spilker, T., Zarnecki, J. and 113 co-authors, 2008b. TandEM: Titan and Enceladus mission. Astrophysical Instruments and Methods, in press.

Cravens, T. E., Robertson, I. P., Clark, J., Wahlund, J.-E., Waite, J. H., Ledvina, S. A., Niemann, H. B., Yelle, R. V., Kasprzak, W. T., Luhmann, J. G., McNutt, R. L., Ip, W.-H., De La Haye, V., Müller-Wodarg, I., Young, D. T., Coates, A. J., 2005. Titan's ionosphere: Model comparisons with Cassini Ta data. Geophys. Res. Let. **32**, CiteID L12108.

Cravens, T. E., Robertson, I. P., Waite, J. H., Yelle, R. V., Kasprzak, W. T., Keller, C. N., Ledvina, S. A., Niemann, H. B., Luhmann, J. G., McNutt, R. L., Ip, W.-H., De La Haye, V., Mueller-Wodarg, I., Wahlund, J.-E., Anicich, V. G., Vuitton, V., 2006. Composition of Titan's ionosphere. Geophys. Res. Let. **33**, CiteID L07105.

Crespin, A., Lebonnois, S., Vinatier, S., Bézard, B., Coustenis, A., Teanby, N. A., Achterberg, R. K., Rannou, P., 2008. Diagnostics of Titan's stratospheric dynamics using CIRS/Cassini data and the IPSL General Circulation Model. Icarus, in press.

Cruikshank, D. P., Morgan, J. S., 1980. Titan: Suspected near infrared variability. Astrophys. J. **235**, L53–L54.

Dandouras, J., Amsif, A., 1999. Production and imaging of energetic neutral atoms from Titan's exosphere: A 3-D model. Plan. Space Sci. **47**, 1355–1369.

Danehy, R. G., Owen, T., Lutz, B. L., Woodman, J. H., 1978. Detection of the Kuiper Bands in the Spectrum of Titan. Icarus **35**, 247–251.

Danielson, R. E., Caldwell, J. J., Larach, D. R., 1973. An inversion in the atmosphere of Titan. Icarus **20**, 437–443, 1973.

De Bergh, C., Lutz, B. L., Owen, T., Chauville, J., 1988. Monodeuterated methane in the outer solar system. III. Its abundance on Titan. Astron. Astrophys. J. **329**, 951–955.

De Kok, R., Irwin, P. G. J., Teanby, N. A., Lellouch, E., Bézard, B., Vinatier, S., Nixon, C. A., Fletcher, L., Howett, C., Calcutt, S. B., Bowles, N. E., Flasar, F. M., Taylor, F. W., 2007. Oxygen compounds in Titan's stratosphere as observed by Cassini CIRS. Icarus **186**, 354–363.

De La Haye, V., Waite, J. H., Johnson, R. E., Yelle, R. V., Cravens, T. E., Luhmann, J. G., Kasprzak, W. T., Gell, D. A., Magee, B., Leblanc, F., Michael, M., Jurac, S., Robertson, I. P., 2007. Cassini ion and neutral mass spectrometer data in Titan's upper atmosphere and exosphere: Observation of a suprathermal corona. J. Geophys. Res. **112**, CiteID A07309.

De Pater, I., Ádámkovics, M., Bouchez, A. H., Brown, M. E., Gibbard, S. G., Marchis, F., Roe, H. G., Schaller, E. L., Young, E., 2006. Titan imagery with Keck adaptive optics during and after probe entry. J. Geophys. Res. **111**, CiteID E07S05.

Delitsky, M. L., Thompson, W. R., 1987. Chemical processes in Triton's atmosphere and surface. Icarus **70**, 354–365.

Desch, M. D., Kaiser, M. L., 1990. Upper limit set for level of lightning activity on Titan. Nature **343**, 442–444.

Dimitrov, Vasili I., Bar-Nun, Akiva, 1999. Admissible height of local roughness of Titan's landscape. J. Geophys. Res. **104**, 5933–5938.

Dimitrov, Vasili, Bar-Nun, Akiva, 2002. Aging of Titan's aerosols. Icarus **156**, 530–538.

Dimitrov, Vasili, Bar-Nun, Akiva, 2003. Hardening of Titan's aerosols by their charging. Icarus **166**, 440–443.

Dire, J. R., 2000. Seasonal photochemical and meridional transport model for the stratosphere of Titan. Icarus. **145**, 428–444.

Dóbé, Zoltán, Szegõ, Károly, 2005. Wave activity above the ionosphere of Titan: Predictions for the Cassini mission. J. Geophys. Res. **110**, CiteID A03224.

Dóbé, Z., Szego, K., Quest, K. B., Shapiro, V. D., Hartle, R. E., Sittler, E. C., 2007. Nonlinear evolution of modified two-stream instability above ionosphere of Titan: Comparison with the data of the Cassini Plasma Spectrometer. J. Geophys. Res. **112**, CiteID A03203.

Dollfus, A., 1961. Visual and Photographic Studies of Planets at the Pic Du Midi. In: The Solar System and Planets III. Planets and Satellites. (Kuiper, Middlehurst, eds.), 534–571. University of Chicago Press.

Dubouloz, N., Raulin, F., Lellouch, E., Gautier, D., 1989. Titan's hypothesized ocean properties: The influence of surface temperature and atmospheric composition uncertainties, Icarus **82**, 81–94.

Eberhardt, P., Reber, M., Krankowsky, D., Hidges, R., 1995. The D/H and $^{18}O/^{16}O$ ratios in water from comet P/Halley. Astron. Astrophys. **302**, 301–316.

Ehrenfreund, P., Boon, J. J., Commandeur, J., Sagan, C., Thompson, W. R., Khare, B. N., 1995. Analytical pyrolysis experiments of Titan aerosol analogues in preparation for the Cassini-Huygens Mission. Adv. Space Res. **15**(3), 335–342.

Elachi, C., Wall, S., Allison, M., Anderson, Y., Boehmer, R., Callahan, P., Encrenaz, P., Flamini, E., Franceschetti, G., Gim, Y., Hamilton, G., Hensley, S., Janssen, M., Johnson, W., Kelleher, K., Kirk, R., Lopes, R., Lorenz, R., Lunine, J., Muhleman, D., Ostro, S., Paganelli, F., Picardi, G., Posa, F., Roth, L., Seu, R., Shaffer, S., Soderblom, L., Stiles, B., Stofan, E., Vetrella, S., West, R., Wood, C., Wye, L., Zebker, H., 2005. Cassini radar views the surface of Titan. Science **308**, 970–974.

Elachi, C., Wall, S., Janssen, M., Stofan, E., Lopes, R., Kirk, R., Lorenz, R., Lunine, J., Paganelli, F., Soderblom, L., Wood, C., Wye, L., Zebker, H., Anderson, Y., Ostro, S., Allison, M., Boehmer, R., Callahan, P., Encrenaz, P., Flamini, E., Franceschetti, G., Gim, Y., Hamilton, G., Hensley, S., Johnson, W., Kelleher, K., Muhleman, D., Picardi, G., Posa, F., Roth, L., Seu, R., Shaffer, S., Stiles, B., Vetrella, S., West, R., 2006. Titan Radar Mapper observations from Cassini's T3 fly-by. Nature **441**, 709–713.

Elachi, C., Wall, S., Janssen, M., Stofan, E., Lopes, R., Kirk, R., Lorenz, R., Lunine, J., Paganelli, F., Soderblom, L., Wood, C., Wye, L., Zebker, H., Anderson, Y., Ostro, S., Allison, M., Boehmer, R., Callahan, P., Encrenaz, P., Flamini, E., Franceschetti, G., Gim, Y., Hamilton, G., Hensley, S., Johnson, W., Kelleher, K., Muhleman, D., Picardi, G., Posa, F., Roth, L., Seu, R., Shaffer, S., Stiles, B., Vetrella, S., West, R., 2006. Corrigendum: Titan Radar Mapper observations from Cassini's T3 fly-by. Nature **442**, 322.

Encrenaz, Th., Gulkis, S., Lellouch., E., 1991. Heterodyne spectroscopy of planetary and satellite atmospheres in the millimeter and submillimeter range. In: Coherent detection techniques at millimeter wavelength and their applications. (Encrenaz, P., Kollberg, E., Gulkis, S., Winnewisser, G., eds.), Nova Science.

Encrenaz, Thérèse, 2003. ISO observations of the giant planets and Titan: What have we learnt? Plan. Space Sci. **51**, 89–103.

Engel, S., Lunine, J. I., Hartmann, W. K., 1995. Cratering on Titan and implications for atmospheric history. Planet. Space Sci., **43**(9), 1059–1066.

Engel, S., Lunine, J. I., Norton, D. L., 1994. Silicate interactions with ammonia-water fluids on early Titan. J. Geophys. Res. **99**, E2, 3745–3752.

English, M. A., Lara, L. M., Lorenz, R. D., Ratcliff, P. R., Rodrigo, R., 1996. Ablation and chemistry of meteoric materials in the atmosphere of Titan. Adv. Space Res. **12**, 157–160.

Eshleman, V. R., Lindal, G. F., Tyler, G. L., 1983. Is Titan Wet or Dry? Science **221**, 53–55.

Farinella, P., Paolicchi, P., Strom, R. G., Kargel, J. S., Zappalà, V., 1990. The fate of Hyperion's fragments. Icarus **83**, 186–204.

Farinella, P., Marzari, F., Matteoli, S., 1997. The disruption of Hyperion and the origin of Titan's atmosphere. Astron. J. **113**, 2312–2316.

Fegley, Jnr. B., Prinn, R. G., 1989. Solar Nebula chemistry: Implications for volatiles in the solar system. In: The Formation and Evolution of Planetary Systems. (Weaver, Danley, eds.), 171–211. Cambridge University Press.

Feuchtgruber, H., Lellouch, E., de Graauw, Th., Bézard, B., Encrenaz, T., Griffin, M., 1997. External supply of oxygen to the atmospheres of the giant planets. Nature **389**, 159–162.

Fink, U., Larson, H. P., 1979. The infrared spectra of Uranus, Neptune, Titan from 0.8 to 2.5 microns. Astron. Astrophys. **233**, 1021–1040.

Fink, U., Larson, H. P., 1978. Deuterated methane observed on Titan. Science **201**, 343–345.

Fink, U, Larson, H. P., 1979. The infra-red spectra of Uranus, Neptune and Titan from 0.8 to 2.5 Microns. Astrophys. J. **233**, 1021–1040.

Fischer, G., Tokano, T., Macher, W., Lammer, H., Rucker, H. O., 2004. Energy dissipation of possible Titan lightning strokes. Plan. Space Sci. **52**, 447–458.

Flasar, F. M., Conrath, B. J., 1990. Titan's stratospheric temperatures: A case for dynamical inertia? Icarus **85**, 346–354.

Flasar, F. M., Samuelson, R. E., Conrath, B. J., 1981. Titan's atmosphere: Temperature and dynamics. Nature **292**, 693–698.

Flasar, F. M., 1983. Oceans on Titan? Science **221**, 55–57.

Flasar, F. M., 1998. The composition of Titan's atmosphere: A meteorological perspective. Planet. Space Sci. **46**, 1109–1124.

Flasar, F. M., 1998. The dynamic meteorology of Titan. Planet. Space Sci. **46**, 1125–1148.

Flasar, F. M., Kunde, V. G., Achterberg, R. K., Conrath, B. J., Simon-Miller, A. A., Nixon, C. A., Gierasch, P. J., Romani, P. N., Bézard, B., Irwin, P., Bjoraker, G. L., Brasunas, J. C., Jennings, D. E., Pearl, J. C., Smith, M. D., Orton, G. S., Spilker, L. J., Carlson, R., Calcutt, S. B., Read, P. L., Taylor, F. W., Parrish, P., Barucci, A., Courtin, R., Coustenis, A., Gautier, D., Lellouch, E., Marten, A., Prangé, R., Biraud, Y., Fouchet, T., Ferrari, C., Owen, T. C., Abbas, M. M., Samuelson, R. E., Raulin, F., Ade, P., Césarsky, C. J., Grossman, K. U., Coradini, A., 2004. An intense stratospheric jet on Jupiter. Nature **427**, 132–135.

Flasar, F. M., Achterberg, R. K., Conrath, B. J., Bjoraker, G. L., Jennings, D. E., Pearl, J. C., Romani, P. N., Simon-Miller, A. A., Kunde, V. G., Nixon, C. A., Bézard, B., Orton, G. S., Spilker, L. J., Irwin, P., Teanby, N. A., Spencer, J. A., Owen, T. C., Brasunas, J. C., Segura, M. E., Carlson, R., Matmoukine, A., Giearasch, P. J., Schinlder, P. J., Ferrari, C., Showalter, M. R., Barucci A., Courtin R., Coustenis A., Fouchet T., Gautier D., Lellouch E., Marten A., Prangé, R., Strobel, D. F., Calcutt S. B., Read P. L., Taylor F. W., Bowles, N., Samuelson R. E., Abbas M. M., Raulin F., Ade P., Edgington, S., Pilorz, S., Wallis, B., Wishnow, E. 2005. Temperatures, winds, and composition in the Saturnian system. Science **307**, 1247–1251.

Flasar, F. M., Achterberg, R. K., Conrath, B. J., Gierasch, P. J., Kunde, V. G., Nixon, C. A., Bjoraker, G. L., Jennings, D. E., Romani, P. N., Simon-Miller, A. A., Bézard, B., Coustenis, A., Irwin, P. G. J., Teanby, N. A., Brasunas, J., Pearl, J. C., Segura, M. E., Carlson, R., Matmoukine, A., Schinder, P. J., Barucci, A., Courtin, R., Fouchet, T., Gautier, D., Lellouch, E., Marten, A., Prangé, R., Vinatier, S., Strobel, D. F., Calcutt S. B., Read, P. L., Taylor, F. W., Bowles, N., Samuelson, R. E., Orton, G. S., Spilker, L. J., Owen, T. C., Spencer, J. A., Showalter, M. R., Ferrari, C., Abbas, M. M., Raulin, F., Edgington, S., Ade P., Wishnow, E. H., 2005. Titan's atmospheric temperatures, winds, and composition. Science **308**, 975–978.

Folkner, W. M., Asmar, S. W., Border, J. S., Franklin, G. W., Finley, S. G., Gorelik, J., Johnston, D. V., Kerzhanovich, V. V., Lowe, S. T., Preston, R. A., Bird, M. K., Dutta-Roy, R., Allison, M., Atkinson, D. H., Edenhofer, P., Plettemeier, D., Tyler, G. L., 2006. Winds on Titan from ground-based tracking of the Huygens probe. J. Geophys. Res. **111**, CiteID E07S02.

Fortes, A. D., 2000. Exobiological implications of a possible Ammonia-Water ocean inside Titan. Icarus **146**, 444–452.

Fortes, A. Dominic, Grindrod, Peter M., 2006. Modelling of possible mud volcanism on Titan. Icarus **182**, 550–558.

Fortes, A. D., Grindrod, P. M., Trickett, S. K., Voèadlo, L., 2007. Ammonium sulfate on Titan: Possible origin and role in cryovolcanism. Icarus **188**, 139–153.

Fouchet, Th., Bézard, B., Encrenaz, Th., 2005. The planets and Titan observed by ISO. Space Science Reviews **119**, 123–139.

Frère, C., Raulin, R., Israel, G., Cabane, M., 1990. Microphysical modelling of Titan's aerosols: Application to the *in situ* analysis. Adv. Space Res. **10**, 159–163.

Fujii, Toshihiro, Arai, Norihisa, 1999. Analysis of N-containing hydrocarbon species produced by a CH_4/N_2 microwave discharge: Simulation of Titan's atmosphere. Astrophys. J. **519**, 858–863.

Friedson, A. J. Yung., Y. L. 1984. The thermosphere of Titan. J. Quant. Spectrosc. Radiat. Transfer **89**, 85–90.

Fulchignoni, M., Ferri, F., Angrilli, F., Ball, A. J., Bar-Nun, A., Barucci, M. A., Bettanini, C., Bianchini, G., Borucki, W., Colombatti, G., Coradini, M., Coustenis, A., Debei, S., Falkner, P., Fanti, G., Flamini, E., Gaborit,V., Grard, R., Hamelin, M., Harri, A. M., Hathi, B., Jernej, I., Leese, M. R., Lehto, A., Lion Stoppato, P. F., Lopez-Moreno, J. J., Mäkinen, T., McDonnell, J. A. M., McKay, C. P., Molina-Cuberos, G., Neubauer, F. M., Pirronello, V., Rodrigo, R., Saggin, B., Schwingenschuh, K., Seiff, A.,Simoes, F., Svedhem, H., Tokano, T., Towner, M. C., Trautner, R., Withers, P., Zarnecki, J. C., 2005. Titan's physical characteristics measured by the Huygens Atmospheric Instrument (HASI). Nature **438**, 785–791.

Galand, Marina, Lilensten, Jean, Toublanc, Dominique, Maurice, Sylvestre, 1999. The ionosphere of Titan: Ideal diurnal and nocturnal cases. Icarus **140**, 92–105.

Galand, M., Yelle, R. V., Coates, A. J., Backes, H., Wahlund, J.-E., 2006. Electron temperature of Titan's sunlit ionosphere. Geophys. Res. Let. **33**, CiteID L21101.

Garnier, P., Dandouras, I., Toublanc, D., Brandt, P. C., Roelof, E. C., Mitchell, D. G., Krimigis, S. M., Krupp, N., Hamilton, D. C., Waite, H., 2007. The exosphere of Titan and its interaction with the kronian magnetosphere: MIMI observations and modelling. Plan. Space Sci. **55**, 165–173.

Gautier, D., 1995. Titan's atmosphere composition: certainties and speculations. Adv. Space Res. **15**(3), 295–301.

Gautier, D., 1997. The aeronomy of Titan. In: Huygens: Science payload and mission (Lebreton, J.-P., ed.), ESA Special report SP-**1177**, 359–364.

Gautier, D., Owen, T., 1985. Observational constraints on models for giant planet formation. In: Protostars and Planets II (Black, D. C., Matthews, M. S., eds.), 832–846. Univ. of Arizona Press, Tucson.

Gautier, D., Owen, T., 1988. The composition of outer planet atmospheres. In: Origin and evolution of planetary and satellite atmospheres (Atreya, S. Pollack, J. Matthews, M. S., eds.), 487–512 Univ. of Arizona Press, Tucson.

Gautier, D., Owen, T., 1989. Titan: Some new results. Adv. Space Res. **9**, 73–78.

Gautier, D., Raulin, F. 1997, 'Chemical composition of Titan's atmosphere', in Wilson, A. (ed.), Huygens: Science, Payload and Mission, ESA Special report SP-1177, 359–364.

Gautier, D., Hersant, F., 2005. Formation and composition of planetesimals: Trapping volatiles by clathration. Space Sci. Rev. **116**, 25–52.

Gazeau, M.-C., Cottin, H., Vuitton, V., Smith, N., Raulin, F., 2000. Experimental and theoretical photochemistry: Application to the cometary environment and Titan's atmosphere. Plan. Space Sci. **48**, 437–445.

Ge, Su-Hong, Cheng, Xin-Lu, Yang, Xiang-Dong, Liu, Zi-Jiang, Wang, Wei, 2006. Calculations of the thermochemistry of six reactions leading to ammonia formation in Titan's atmosphere. Icarus **183**, 153–158.

Geballe, T. R., Kim, S. J., Noll, K. S., Griffith, C. A., 2003. High-resolution 3 micron spectroscopy of molecules in the mesosphere and troposphere of Titan. Astrophys. J. **583**, L39–L42.

Geiss, J., Gloecker, G. 1998, 'Abundances of deuterium and helium-3 in the proto-solar cloud. Space Sci. Rev. **82**, 239–250.

Gendron, E., Coustenis, A., Drossart, P., Combes, M., Hirtzig, M., Lacombe, F., Rouan, D., Collin, C., Pau, S., Lagrange,A.-M., Mouillet, D., Rabou, P., Fusco,Th., Zins, S., 2004. VLT/NACO adaptive optics imaging of Titan. Astron. Astroph. **417**, L21–L24.

Ghafoor, Nadeem A.-L., Zarnecki, John C., Challenor, Peter, Srokosz, Meric A., 2000. Wind-driven surface waves on Titan. J. Geophys. Res. **105**, 12077–12092.

Gibbard, S. G., Macintosh, B., Gavel, D., Max, C. E., de Pater, I., Ghez, A. M., Young, E. F., McKay, C. P., 1999. Titan: High-resolution speckle images from the Keck telescope. Icarus **139**, 189–201.

Gibbard, S. G., Macintosh, B., Gavel, D., Max, C. E., de Pater, I., Roe, H. G., Ghez, A. M., Young, E. F., McKay, C. P., 2004a. Speckle imaging of Titan at 2 microns: Surface albedo, haze optical depth, and tropospheric clouds 1996–1998. Icarus **169**, 429–439.

Gibbard, S. G., de Pater, I., Macintosh, B. A., Roe, H. G., Max, C. E., Young, E. F., McKay, C. P., 2004b. Titan's 2 μm surface albedo and haze optical depth in 1996–2004. Geophys. Res. Let. **31**, CiteID L17S02.

Gill, A. E., 1982. Atmosphere-Ocean Dynamics. New York: Academic Press, 1982.

Gillet, F. C, Forrest W. J, Merril K. M., 1973. 8–13 micron observations of Titan. Astrophys. J. **184**, L93–L95.

Gillet, F. C., 1975. Further observations of the 8–13 micron spectrum of Titan. Astrophys. J. **201**, L41–L43.

Goldstein, R. M., Jurgens, R. F., 1992. DSN observations of Titan. TDAPR 42–109, Jet Propulsion Laboratory.

Grard, R., Hamelin, M., López-Moreno, J. J., Schwingenschuh, K., Jernej, I., Molina-Cuberos, G. J., Simões, F., Trautner, R., Falkner, P., Ferri, F., Fulchignoni, M., Rodrigo, R., Svedhem, H., Béghin, C., Berthelier, J.-J., Brown, V. J. G., Chabassière, M., Jeronimo, J. M., Lara, L. M., Tokano, T., 2006. Electric properties and related physical characteristics of the atmosphere and surface of Titan. Plan. Space Sci. **54**, 1124–1136.

Grasset, O., Sotin, C., 1996. The cooling rate of a liquid shell in Titan's interior. Icarus **123**, 101–112.

Grasset, O., Sotin, C., Deschamps, F., 2000. On the internal structure and dynamics of Titan. Plan. Space Sci. **48**, 617–636.

Grasset, O., Pargamin, J., 2005. The ammonia water system at high pressures: Implications for the methane of Titan. Plan. Space Sci. **53**, 371–384.

Greeley, R., Iverson, J. P., 1985. Wind as a geological process on Earth, Mars, Venus, and Titan. Cambridge University Press.

Grieger, B., Lemmon, M. T., Markiewicz, W. J., Keller, H. U., 2003. Inverse radiation modeling of Titan's atmosphere to assimilate solar aureole imager data of the Huygens probe. Plan. Space Sci. **51**, 147–158.

Grieger, B., Rodin, A. V., Salinas, S. V., Keller, H. U., 2003. Simultaneous retrieval of optical depths and scattering phase functions in Titan's atmosphere from Huygens/DISR data. Plan. Space Sci. **51**, 991–1001.

Grieger, B., 2005. Shading under Titan's sky. Plan. Space Sci. **53**, 577–585.

Griffith, C. A., Owen, T., Wagener, R., 1991. Titan's surface and troposphere, investigated with ground-based, near-infrared observations. Icarus **93**, 362–378.

Griffith, C. A., 1993. Evidence for surface heterogeneity on Titan. Nature **364**, 511–514.

Griffith, C. A., Zahnle, K., 1995. Influx of cometary volatiles to planetary moons: The atmospheres of 1000 possible Titans. J. Geophys. Res. **100**, 16, 907–16, 922.

Griffith, C. A., Owen, T., Miller, G. A., Geballe, T., 1998. Transient clouds in Titan's lower atmosphere. Nature **395**, 575–578.

Griffith, Caitlin A., Hall, Joseph L., Geballe, Thomas R., 2000. Detection of daily clouds on Titan. Science **290**, 509–513.

Griffith, Caitlin A., Owen, Tobias, Geballe, Thomas R., Rayner, John, Rannou, Pascal, 2003. Evidence for the exposure of water ice on Titan's surface. Science **300**, 628–630.

Griffith, C. A., Penteado, P., Greathouse, T. K., Roe, H. G., Yelle, R. V., 2005. Observations of Titan's mesosphere. Astrophys. J. **629**, L57–L60.

Griffith, C. A., Penteado, P., Baines, K., Drossart, P., Barnes, J., Bellucci, G., Bibring, J., Brown, R., Buratti, B., Capaccioni, F., Cerroni, P., Clark, R., Combes, M., Coradini, A., Cruikshank, D., Formisano, V., Jaumann, R., Langevin, Y., Matson, D., McCord, T., Mennella, V., Nelson, R., Nicholson, P., Sicardy, B., Sotin, C., Soderblom, L. A., Kursinski, R., 2005. The Evolution of Titan's mid-latitude clouds. Science **310**, 474–477.

Griffith, C. A., Penteado, P., Rannou, P., Brown, R., Boudon, V., Baines, K., Clark, R., Drossart, P., Buratti, B., Nicholson, P., Jaumann, R., McKay, C. P., Coustenis, A., Negrão, A., 2006. Evidence for ethane clouds on Titan from Cassini VIMS observations. Science **313**, 1620–1622.

Griffith, G., 2006. Planetary science: Titan's exotic weather. Nature **442**, 362–363.

Grossman, A. W., Muhleman, D. O., 1991. Observations of Titan's radio light-curve at 3.5 cm (abstract) Bull. Am. Astron. Soc. 27, 1104.

Gupta, S., Ochiai, E., Ponnaperuma, C., 1981. Organic synthesis in the atmosphere of Titan. Nature **293**, 725–727.

Gurwell, M. A., 2004. Submillimeter observations of Titan: Global measures of stratospheric temperature, CO, HCN, HC3N, and the isotopic ratios 12C/13C and 14N/15N. Astrophys. J. **616**, L7–L10.

Gurwell, M. A., Muhleman, D. O., 1995. CO on Titan: Evidence for a well-mixed vertical profile. Icarus **117**, 375–382.

Gurwell, M. A., Muhlemann, D. O., 2000. CO on Titan: More evidence for a well-mixed vertical profile. Icarus **145**, 653–656.

Hagermann, A., Ball, A. J., Hathi, B., Leese, M. R., Lorenz, R. D., Rosenberg, P. D., Towner, M. C., Zarnecki, J. C., 2006. Inferring the composition of the liquid surface on Titan at the Huygens probe landing site from Surface Science Package measurements. Adv. Space Res. **38**, 794–798.

Hagermann, A., Zarnecki, J. C., 2006. Virial treatment of the speed of sound in cold, dense atmospheres and application to Titan. Monthly Notices of the Royal Astronomical Society **368**, 321–324.

Hagermann, A., Rosenberg, P. D., Towner, M. C., Garry, J. R. C., Svedhem, H., Leese, M. R., Hathi, B., Lorenz, R. D., Zarnecki, J. C., 2007. Speed of sound measurements and the methane abundance in Titan's atmosphere. Icarus **189**, 538–543.

Hall, J. L., Kerzhanovich, V. V., Yavrouian, A. H., Jones, J. A., White, C. V., Dudik, B. A., Plett, G. A., Mennella, J., Elfes, A., 2006. An aerobot for global *in situ* exploration of Titan. Adv. Space Res. **37**, 2108–2119.

Hanel, R., Conrath, B., Flasar, F. M., Kunde, V., Maguire, W., Pearl, J. C., Pirraglia, J., Samuelson, R., Herath, L., Allison, M., Cruikshank, D. P., Gautier, D., Gierasch, P. J., Horn, L., Koppany, R., Ponnamperuma, C., 1981. Infrared observations of the Saturnian system from Voyager 1. Science **212**, 192–200.

Hanel, R., Pearl, J. C., Mayo, L. A., 1982. In: Voyager Infrared Interferometer Spectrometer and Radiometer (IRIS). Documentation for Reduced Data Records (RDR) for the Saturnian System, NASA SP X-693–82–30. GSFC, Greenbelt, Maryland 20771.

Hapke, B., 1990. Coherent backscatter and the radar characteristics of the outer planet satellites. Icarus **88**, 407–419.

Hapke, B., 1986. On the sputter alteration of regoliths of outer solar system bodies. Icarus **66**, 270–279.

Harri, A.-M., Mäkinen, T., Lehto, A., Kahanpää, H., Siili, T., 2006. Vertical pressure profile of Titan — Observations of the PPI/HASI instrument. Plan. Space Sci. **54**, 1117–1123.

Hartle, R. E., Sittler, E. C., Neubauer, F. M., Johnson, R. E., Smith, H. T., Crary, F., McComas, D. J., Young, D. T., Coates, A. J., Simpson, D., Bolton, S., Reisenfeld, D., Szego, K., Berthelier, J. J., Rymer, A., Vilppola, J., Steinberg, J. T., Andre, N., 2006. Preliminary interpretation of Titan plasma interaction as observed by the Cassini plasma spectrometer: Comparisons with Voyager 1. Geophys. Res. Let. **33**, CiteID L08201.

Hartle, R. E., Sittler, E. C., Neubauer, F. M., Johnson, R. E., Smith, H. T., Crary, F., McComas, D. J., Young, D. T., Coates, A. J., Simpson, D., Bolton, S., Reisenfeld, D., Szego, K., Berthelier, J. J.,

Rymer, A., Vilppola, J., Steinberg, J. T., Andre, N., 2006. Initial interpretation of Titan plasma interaction as observed by the Cassini plasma spectrometer: Comparisons with Voyager 1. Plan. Space Sci. **54**, 1211–1224.

Hartung, M., Herbst, T. M., Close, L. M., Lenzen, R., Brandner, W., Marco, O., Lidman, C., 2004a. A new VLT surface map of Titan at 1.575 microns. Astron. Astrophys. **421**, L17–L20.

Hartung, M., Herbst, T. M., Dumas, C., Coustenis, A., 2006. Limits to the abundance of CO_2 ice on Titan. J. Geophys. Res. Planets **111**, E07S09.

Hayashi, C., Nakazawa, K., Nakagawa, Y., 1985 Formation of the Solar System. In: Protostars and Planets II (Black, Matthews, eds.), 1100–1153. University of Arizona Press.

Hébrard, E., Dobrijevic, M., Bénilan, Y., Raulin, F., 2007. Photochemical kinetics uncertainties in modeling Titan's atmosphere: First consequences. Plan. Space Sci. **55**, 1470–1489.

Hersant, F., Gautier, D., Lunine, J. I., 2004. Enrichments in volatiles in the giant planets of the solar system. Plan. Space Sci. **52**, 623–624.

Herschel, J., 1847. Results of astronomical observations made during the year 1834–8 at the Cape of Good Hope. London.

Hidayat, T., Marten, A., Bézard, B., Gautier, D., Owen, T., Matthews, H. E., Paubert, G., 1997. Millimetre and submillimetre heterodyne observations of Titan: Retrieval of the vertical profile of HCN and the $^{12}C/^{13}$C Ratio. Icarus **126**, 170–182.

Hidayat, T., Marten, A., Bézard, B., Gautier, D., Owen, T., Matthews, H. E., Paubert, G., 1998. Millimeter and submillimeter heterodyne observations of Titan: The vertical profile of carbon monoxide in its stratosphere. Icarus **133**, 1, 109–133.

Hestroffer, D., 2003. Photometry with a periodic grid. II. Results for J2 Europa and S6 Titan. Astron. Astrophys. **403**, 749–756.

Hinson, D. P., Tyler, G. L., 1983. Internal gravity waves in Titan's atmosphere observed by Voyager radio occultation. Icarus **54**, 337–352.

Hirtzig, M., Coustenis, A., Lai, O., Emsellem, E., Pecontal-Rousset, A., Rannou, P., Negrao, A., Schmitt, B., 2005. Near-infrared study of Titan's resolved disk in spectro-imaging with CFHT/OASIS. Plan. Space Sci. **53**, 535–556.

Hirtzig, M., Coustenis, A., Gendron, E., Drossart, P., Negrao, A., Combes, M., Lai, O., Rannou, P., Lebonnois, S., Luz, D., 2006. Monitoring atmospheric phenomena on Titan. Astron. Astrophys. **456**, 761–774.

Hirtzig, M., Coustenis, A., Gendron, E., Drossart, P., Hartung, M., Negrao, A., Rannou, Combes, M., 2007. Titan: Atmospheric and surface features as observed with NAOS/CONICA at the time of the Huygens' landing. J. Geophys. Res. Planets **112**, E02S91.

Hodyss, R., McDonald, G., Sarker, N., Smith, M. A., Beauchamp, P. M., Beauchamp, J. L., 2004. Fluorescence spectra of Titan tholins: *In-situ* detection of astrobiologically interesting areas on Titan's surface. Icarus **171**, 525–530.

Houghton, J. T., 1986. The Physics of Atmospheres. Cambridge University Press.

Houghton, J. T., Taylor, F. W., Rodgers, C. D., 1986. Remote Sounding of Atmospheres. Cambridge University Press.

Hourdin, F., Talagrand, O., Sadourny, R., Courtin, R., Gautier, D., McKay, C. P., 1995. Numerical simulation of the general circulation of the atmosphere of Titan. Icarus **117**, 358–374.

Hourdin, F., Lebonnois, S., Luz, D., Rannou, P., 2004. Titan's stratospheric composition driven by condensation and dynamics. J. Geophys. Res. **109**, CiteID E12005.

Hubbard, W. B., Hunten, D. M., Reitsema, H. J. Brosch, N., Nevo, Y., Carreira, E., Rossi, F., Wasserman, L. H., 1990. Results of Titan's atmosphere from its occultation of 28 Sagittari. Nature **343**, 353–355.

Hubbard, W. B., Porco, C. C., Hunten, D. M., Rieke, G. H., Rieke, M. J., McCarthy, D. W., Haemmerle, V., Clark, R., Turtle, E. P., Haller, J., McLeod, B., Lebofsky, L. A., Marcialis, R., Holberg, J. B., Landau, R., Carrasco, L., Elias, J., Buie, M. W., Persson, S. E., Boroson, T., West, S., Mink, D. J., 1993. The occultation of 28 Sgr by Titan. Astron. Astrophys. **269**, 541–563.

Hudson, R. L., Moore, M. H., 2004. Reactions of nitriles in ices relevant to Titan, comets, and the interstellar medium: Formation of cyanate ion, ketenimines, and isonitriles. Icarus **172**, 466–478.

Hueso, R., Sánchez-Lavega, A., 2006. Methane storms on Saturn's moon Titan. Nature **442**, 428–431.

Hunten, D. M., 1977. Titan's atmosphere and surface. In: Planetary Satellites (Burns, J. A., ed.) Tucson: Univ. of Arizona Press, pp. 420–437.

Hunten, D. M., Tomasko, M. G, Flasar, F. M, Samuelson, R. E., Strobel, D. F., Stevenson, D. J., 1984. Titan. In: Saturn (Gehrels, T., Matthews, M., eds.), 671–759. University of Arizona Press.

Hunten, D. M., 2006. The sequestration of ethane on Titan in smog particles. Nature **443**, 669–670.

Hutzell, W. T., McKay, C. P., Toon, O. B., 1993. Effects of time-varying haze production on Titan's geometric albedo. Icarus **105**, 162–174.

Hutzell, W. T., McKay, C. P., Toon, O. B., Hourdin, F., 1996. Simulations of Titan's brightness by a two-dimensional haze model. Icarus **119**, 112–129.

Huygens, C., 1659. Systema Saturnium.

Huygens, C., 1698. The Celestial Worlds Discover'd.

Huygens, C., 1888–1950. Oeuvres completes de Christiaan Huygens, 22 Vols. Martinus Nijhoff, The Hague.

Imanaka, H., Khare, B. N., Elsila, J. E., Bakes, E. L. O., McKay, C. P., Cruikshank D. P., Sugita, S., Matsui, T., Zare, R. N., 2004. Laboratory experiments of Titan tholin formed in cold plasma at various pressures: Implications for nitrogen-containing polycyclic aromatic compounds in Titan haze. Icarus **168**, 344–366.

Imanaka, H., Smith, M. A., 2007. Role of photoionization in the formation of complex organic molecules in Titan's upper atmosphere. Geophys. Res. Let. **34**, CiteID L02204.

Israel, G., Cabane, M., Raulin, F., Chassefière, E., Boon, J. J., 1991. Aerosols in Titan's atmosphere: Models, sampling techniques and chemical analysis Ann. Geophys. **9**, 1–13.

Israël, G., Szopa, C., Raulin, F., Cabane, M., Niemann, H. B., Atreya, S. K., Bauer, S. J., Brun, J.-F., Chassefière, E., Coll, P., Condé, E., Coscia, D., Hauchecorne, A., Millian, P., Nguyen, M.-J., Owen, T., Riedler, W., Samuelson, R. E., Siguier, J.-M., Steller, M., Sternberg, R., Vidal-Madjar, C., 2005. Complex organic matter in Titan's atmospheric aerosols from *in situ* pyrolysis and analysis. Nature **438**, 796–799.

Jaffe, W., Caldwell, J., Owen, T., 1979. The brightness of Titan at 6 centimetres from the Very Large Array. Astrophys. J. **232**, L75–L76.

Jacquinet-Husson, N., Scott, N. A., Chédin, A., Crépeau, L., Armante, R., Capelle, V., Orphal, J., Coustenis, A., Barbe, A., Birk, M., Brown, L. R., and 40 co-authors, 2008. The GEISA spectroscopic database: Current and future archive for Earth's planetary atmosphere studies. JQSRT, in press.

Jaumann, R., Stephan, K., Brown, R. H., Buratti, B. J., Clark, R. N., McCord, T. B., Coradini, A., Capaccioni, F., Filacchione, G., Cerroni, P., Baines, K. H., Bellucci, G., Bibring, J.-P., Combes, M., Cruikshank, D. P., Drossart, P., Formisano, V., Langevin, Y., Matson, D. L., Nelson, R. M., Nicholson, P. D., Sicardy, B., Sotin, C., Soderbloom, L. A., Griffith, C., Matz, K.-D., Roatsch, Th., Scholten, F., Porco, C. C., 2006. High-resolution CASSINI-VIMS mosaics of Titan and the icy Saturnian satellites. Plan. Space Sci. **54**, 1146–1155.

Jones, T. D., Lewis, J. S., 1987. Estimated impact shock production of N_2 and organic compounds on early Titan. Icarus **72**, 381–393.

Justus, C. G., Duvall, A., Keller, V. W., Spilker, T. R., Kae Lockwood, M., 2005. Connecting atmospheric science and atmospheric models for aerocapture at Titan and the outer planets. Plan. Space Sci. **53**, 601–605.

Kabin, K., Gombosi, T. I., De Zeeuw, D. L., Powell, K. G., Israelevich, P. L., 1999. Interaction of the Saturnian magnetosphere with Titan: Results of a three-dimensional MHD simulation. J. Geophys. Res. **104**, 2451–2458.

Kabin, K., Israelevich, P. L., Ershkovich, A. I., Neubauer, F. M., Gombosi, T. I., De Zeeuw, D. L., Powell, K. G., 2000. Titan's magnetic wake: Atmospheric or magnetospheric interaction. J. Geophys. Res. **105**, 10761–10770.

Kallio, E., Sillanpää, I., Janhunen, P., 2004. Titan in subsonic and supersonic flow. Geophys. Res. Let. **31**, CiteID L15703.

Karatekin, Ö., Van Hoolst, T., 2006. The effect of a dense atmosphere on the tidally induced potential of Titan. Icarus **183**, 230–232.

Karkoschka, E., Lorenz, R. D., 1997. Latitudinal variation of aerosol sizes inferred from Titan's shadow. Icarus **125**, 369–379.

Kazeminejad, B., Lammer, H., Coustenis, A., Fischer, G., Schwingenschuh, K., Rucker, H. O., 2005. Temperature variations in Titan's upper atmosphere: Impact on Cassini/Huygens. Ann. Geophys. **23**, 1183–1189.

Keller, C. N., Anicich, V. G., Cravens T. E., 1998. Model of Titan's ionosphere with detailed hydro-carbon chemistry. Planet. Space Sci. **46**, 1157–1174.

Khanna, R. K., Perera-Jarmer, M. A., Ospina, M. J., 1987. Vibrational infared and Raman spectra of dicyanoacetylene. Spectroch. Acta **43A**, 421–425.

Khanna, R. K., 2005. Condensed species in Titan's stratosphere: Identification of crystalline propionitrile (C_2H_5CN, CH_3CH_2CN) based on laboratory infrared data. Icarus **177**, 116–121.

Khanna, R. K., 2005. Condensed species in Titan's stratosphere: Confirmation of crystalline cyanoacetylene (HC_3N) and evidence for crystalline acetylene (C_2H_2) on Titan. Icarus **178**, 165–170.

Khanna, R. K., 2007. Corrigendum to "Condensed species in Titan's stratosphere: Confirmation of crystalline cyanoacetylene (HC_3N) and evidence for crystalline acetylene (C_2H_2) on Titan" [Icarus **178** (2005) 165–170]. Icarus **186**, 589–589.

Khare, B. N., Sagan, C., Zumberge, J. F., Sklarew, D. S., Nagy, A., 1982. Organic solids produced by electrical discharge in reducing atmospheres: Tholin molecular analysis. Icarus **48**, 290–297.

Khare, B. N., Sagan, C., Thompson, W. R., 1987. Solid hydrocarbon aerosols produced in simulated Uranian and Neptunian stratosphere. J. Geophys. Res. **92**, 15, 067–15, 082.

Khare, B. N., Sagan, C. Arakawa, E. T. Suits, F., Callcott,T. A., Williams, M. W., 1984. Optical constants of organic tholins produced in a simulated Titanian atmosphere: From soft X-ray to microwave frequencies. Icarus **60**, 127–137.

Khare, B. N., Sagan, C., Ogino, H., Nagy, B., Er. C., Schram, K. H., Arakawa, E. T., 1986. Amino acids derived from Titan tholins. Icarus **68**, 176–184.

Khare, B. N., Sagan, C., Thompson, W. R., Arakawa, E. T., Suits, F, Callcott, T. A., Williams, M. W., Shrader, S., Ogino, H., Willingham, T. O., Nagy, B., 1984. The organic aerosols of Titan. Adv. Space Res. **4**, 59–68.

Khare, Bishun N., Bakes, E. L. O., Imanaka, Hiroshi, McKay, Christopher P., Cruikshank, Dale P., Arakawa, Edward T., 2002. Analysis of the time-dependent chemical evolution of Titan haze tholin. Icarus **160**, Issue 1, pp. 172–182.

Kim, S. J., King, W. T. 1984. Integrated infrared intensities in cyanogen. J. Chem. Phys. **80**, 974–977.

Kim, S. J, Caldwell, J., 1982 The Abundance of CH_3D in the Atmosphere of Titan derived from 8 to 14 mm thermal emission. Icarus **52**, 473–482.

Kim, S. J., Geballe, T. R., Noll, Keith S., 2000. NOTE: Three-micrometer CH_4 line emission from Titan's high-altitude atmosphere. Icarus **147**, 588–591.

Kim, S. J., Lee, Y. S., Kim, Y. H., 2001. Spectroscopic studies of the atmospheres of giant planets, Titan, and comets. Plan. Space Sci. **49**, 117–141.

Kopp, A., Ip, W.-H., 2001. Asymmetric mass loading effect at Titan's ionosphere. J. Geophys. Res. **106**, 8323–8332.

Korycansky, D. G., Zahnle, K. J., 2005. Modeling crater populations on Venus and Titan. Plan. Space Sci. **53**, 695–710.

Kostiuk, T., Fast, K. E., Livengood, T. A., Hewagama, T., Goldstein, J. J., Espenak, F., Buhl, D., 2001. Direct measurement of winds of Titan. Geophys. Res. Let. **28**, 2361–2364.

Kostiuk, T., Livengood, T. A., Hewagama, T., Sonnabend, G., Fast, K. E., Murakawa, K., Tokunaga, A. T., Annen, J., Buhl, D., Schmülling, F., 2005. Titan's stratospheric zonal wind, temperature, and ethane abundance a year prior to Huygens insertion. Geophys. Res. Let. **32**, CiteID L22205.

Kostiuk, T., Livengood, T. A., Sonnabend, G., Fast, K. E., Hewagama, T., Murakawa, K., Tokunaga, A. T., Annen, J., Buhl, D., Schmülling, F., Luz, D., Witasse, O., 2006. Stratospheric global winds on Titan at the time of Huygens descent. J. Geophys. Res. **111**, CiteID E07S03.

Kress, M. E., McKay, C. P., 2004. Formation of methane in comet impacts: Implications for Earth, Mars, and Titan. Icarus **168**, 475–483.

Krimigis, S. M., Armstrong, T. P., Axford, W. I., Bostrom, C. O., Gloeleckler, G., Keath, E. P., Lanzerotti, L. J., Carbarry, J. F., Hamilton, D. C., Roelof, E. C., 1982. Low energy hot plasma and particles in Saturn's magnetosphere. Science **215**, 571–577.

Kuiper, G. P., 1944. Titan: A satellite with an atmosphere. Astrophys J., 100, 378–383.

Kunde, V. G., Aikin, A. C., Hanel, R. A., Jennings, D. E., Maguire, W. C., Samuelson, R. E., 1981. C_4H_2, HC_3N and C_2N_2 in Titan's atmosphere, Nature **292**, 686–688, 1981.

Lammer, H., Stumptner, W., 1999. High altitude haze: Influence of monomer particles on Titan's temperature profile. Plan. Space Sci. **47**, 1341–1346.

Lammer, H., Stumptner, W., Molina-Cuberos, G. J., Bauer, S. J., Owen, T., 2000. Nitrogen isotope fractionation and its consequence for Titan's atmospheric evolution. Plan. Space Sci. **48**, 529–543.

Lammer, H., Tokano, T., Fischer, G., Stumptner, W., Molina-Cuberos, G. J., Schwingenschuh, K., Rucker, H. O., 2001. Plan. Space Sci. **49**, 561–574.

Lara, L. M., Lorenz, R. D., Rodrigo, R., 1994. Liquids and solids on the surface of Titan. Planet. Space Sci. **42**, 5–14.

Lara, L. M., Lellouch, E., López-Moreno, J. J, Rodrigo, R., 1996. Vertical distribution of Titan's atmospheric neutral constituents. J. Geophys. Res. **101**, 23262–23283.

Lara, L.-M., Lellouch, E., Shematovich, V., 1999. Titan's atmospheric haze: The case for HCN incorporation. Astron. Astrophys. **341**, 312–317.

Lara, L. M., Banaszkiewicz, M., Rodrigo, R., Lopez-Moreno, J. J., 2002. The CH4 density in the upper atmosphere of Titan. Icarus **158**, 191–198.

Lavvas, P. P., Coustenis, A., Vardavas, I. M., 2007a. Coupling photochemistry with haze formation in Titan's atmosphere. Part I: Model description. Plan. Space Sci., in press.

Lavvas, P. P., Coustenis, A., Vardavas, I. M., 2007b. Coupling photochemistry with haze formation in Titan's atmosphere. Part II: Results and validation with Cassini/Huygens data. Plan. Space Sci., in press.

Lebonnois, S., Toublanc, D., 1999. Actinic fluxes in Titan's atmosphere, from one to three dimensions: Application to high-latitude composition. J. Geophys. Res. **104**, 22025–22034.

Lebonnois, S., Toublanc, D., Hourdin, F., Rannou, P., 2001. Seasonal variations of Titan's atmospheric composition. Icarus **152**, 384–406.

Lebonnois, S., Bakes, E. L. O., McKay, C. P., 2002. Transition from gaseous compounds to aerosols in Titan's atmosphere. Icarus **159**, 505–517.

Lebonnois, S., Bakes, E. L. O., McKay, C. P., 2003a. Atomic and molecular hydrogen budget in Titan's atmosphere. Icarus **161**, 474–485.

Lebonnois, S., Hourdin, F., Rannou, P., Luz, D., Toublanc, D., 2003b. Impact of the seasonal variations of composition on the temperature field of Titan's stratosphere. Icarus **163**, 164–174.

Lebonnois, S., 2005. Benzene and aerosol production in Titan and Jupiter's atmospheres: A sensitivity study. Plan. Space Sci. **53**, 486–497.

Lebreton, J.-P., Witasse, O., Sollazzo, C., Blancquaert, T., Couzin, P., Schipper, A.-M., Jones, J. B., Matson, D. L., Gurvits, L. I., Atkinson, D. H., Kazeminejad, B., Pérez-Ayúcar, M., 2005. An overview of the descent and landing of the Huygens probe on Titan. Nature **438**, 758–764.

Lécluse, C., Robert, F., 1994. Hydrogen isotope exchange rates: Origin of water in the inner Solar System. Geochim. Cosmochim. Acta **58**, 2297–2939.

Lécluse, C., Robert, F., Gautier, D., Guiraud, M., 1996. Deuterium enrichment in giant planets. Plan. Space Sci. **44**, 1579–1592.

Ledvina, Stephen A., Brecht, Stephen H., Luhmann, Janet G., 2004. Ion distributions of 14 amu pickup ions associated with Titan's plasma interaction. Geophys. Res. Let. **31**, CiteID L17S10.

Ledvina, S. A., Cravens, T. E., Kecskeméty, K., 2005. Ion distributions in Saturn's magnetosphere near Titan. J. Geophys. Res. **110**, CiteID A06211.

Lellouch, E., Coustenis, A., Gautier, D., Raulin, F., Dubouloz, N., Frère, C., 1989. Titan's atmosphere and hypothesized ocean: A reanalysis of the Voyager 1 radio-occultation and IRIS 7.7 μm data. Icarus **79**, 328–349.

Lellouch, E., Hunten, D., Kockarts, G., Coustenis, A., 1990. Titan's thermosphere profile. Icarus **83**, 308–224.

Lellouch, E., Coustenis, A., Sebag, B., Cuby, J.-G., Lopez-Valverde, M., Fouchet, T., Crovisier, J., Schmitt, B., 2003. Titan's 5-micron window: Observations with the very large telescope. Icarus **162**, 125–142.

Lellouch, E., Schmitt, B., Coustenis, A., Cuby, J.-G. 2004. Titan's 5-micron lightcurve. Icarus **168**, 209–214.

Lemmon, M. T., Karkoschka E., Tomasko, M., 1993. Titan's rotation: Surface feature observed. Icarus **103**, 329–332.

Lemmon, M. T., Karkoshka, E., Tomasko, M., 1995. Titan's rotational lightcurve. Icarus **113**, 27–38.

Lemmon, M. T., Smith, P. H., Lorenz, R. D., 2002. Methane abundance on Titan, measured by the space telescope imaging spectrograph. Icarus **160**, 375–385.

Letourneur, B., Coustenis, A., 1993. Titan's atmospheric structure from Voyager 2 infrared spectra. Planet. Space Sci. **41**, 593–602.

Levine, J. S. (ed.), 1985. Photochemistry of Atmospheres. New York: Academic Press.

Lewis. J. S., 1998. Physics and Chemistry of the Solar System. Academic Press, Second edition.

Liang, M.-C., Heays, A. N., Lewis, B. R., Gibson, S. T., Yung, Y. L., 2007. Source of nitrogen isotope anomaly in HCN in the atmosphere of Titan. Astrophys. J. **664**, L115–L118.

Liang, M.-C., Yung, Y. L., Shemansky, D. E., 2007. Photolytically generated aerosols in the mesosphere and thermosphere of Titan. Astrophys. J. **661**, L199–L202.

Lilensten, J., Simon, C., Witasse, O., Dutuit, O., Thissen, R., Alcaraz, C., 2005. A fast computation of the diurnal secondary ion production in the ionosphere of Titan. Icarus **174**, 285–288.

Lilensten, J., Witasse, O., Simon, C., Soldi-Lose, H., Dutuit, O., Thissen, R., Alcaraz, C., 2005. Prediction of a N_2^{++} layer in the upper atmosphere of Titan. Geophys. Res. Let. **32**, CiteID L03203.

Lindal, G. F., Wood, G. E., Hotz, H. B., Sweetnam, D. N., Eshelman, V. R., Tyler, G. L., 1983. The atmosphere of Titan: An analysis of the Voyager 1 radio-occultation measurements. Icarus **53**, 348–363.

Livengood, T. A., Hewagama, T., Kostiuk, T., Fast, K. E., Goldstein, J. J., 2002. Improved determination of ethane (C_2H_6) abundance in Titan's stratosphere. Icarus **157**, 249–253.

Livengood, T. A., Kostiuk, T., Sonnabend, G., Annen, J. N., Fast, K. E., Tokunaga, A., Murakawa, K., Hewagama, T., Schmülling, F., Schieder, R., 2006. High-resolution infrared spectroscopy of ethane in Titan's stratosphere in the Huygens epoch. J. Geophys. Res. **111**, CiteID E11S90.

Lopes, R. M. C., Mitchell, K. L., Stofan, E. R., Lunine, J. I., Lorenz, R., Paganelli, F., Kirk, R. L., Wood, C. A., Wall, S. D., Robshaw, L. E., Fortes, A. D., Neish, C. D., Radebaugh, J., Reffet, E., Ostro, S. J., Elachi, C., Allison, M. D., Anderson, Y., Boehmer, R., Boubin, G., Callahan, P., Encrenaz, P., Flamini, E., Francescetti, G., Gim, Y., Hamilton, G., Hensley, S., Janssen, M. A., Johnson, W. T. K., Kelleher, K., Muhleman, D. O., Ori, G., Orosei, R., Picardi, G., Posa, F., Roth, L. E., Seu, R., Shaffer, S., Soderblom, L. A., Stiles, B., Vetrella, S., West, R. D., Wye, L., Zebker, H. A., 2007. Cryovolcanic features on Titan's surface as revealed by the Cassini Titan Radar Mapper. Icarus **186**, 395–412.

Lopez-Valverde, M. A., Lellouch, E., Coustenis, A., 2005. Carbon monoxide fluorescence from Titan's atmosphere. Icarus **175**, 503–521.

Lorenz, R. D., 1996. Pillow Lava on Titan: Expectations and constraints on cryovolcanic processes. Planet. Space Sci. **44**(9), 1021–1028.

Lorenz, R. D., 2002. Thermodynamics of geysers: Application to Titan. Icarus **156**, 176–183.

Lorenz, R. D., 2001. Erratum: "Titan, Mars and Earth: Entropy production by latitudinal heat transport". Geophys. Res. Let. **28**, 3169–3170.

Lorenz, R. D., Zarnecki, J. C., 1992. Precipitation on Titan and the methane icing hazard to the Huygens descent probe. Ann. Geophys. **10**(3), C487.

Lorenz, R. D., 1993a. The life, death, afterlife of a raindrop on Titan. Planet. Space Sci. **41**(9), 647–655.

Lorenz, R. D., 1993b. Raindrops on Titan. Adv. Space Res. **15**(3), 317–320.

Lorenz, R. D., 1994. Crater lakes on Titan: Rings, horseshoes and bullseyes. Planet. Space Sci. **42**(1), 1–4.

Lorenz, R. D., Lunine, J. I., Grier, J. A., Fisher, M. A., 1995. Predicting Aeolian activity on planets: Application to Titan palaeoclimatology. J. Geophys. Res. **100**, 26377–26386.

Lorenz, R. D., Lunine, J. I., 1996. Erosion on Titan: Past and Present. Icarus **122**, 79–91.

Lorenz, R. D., Smith, P. H., Lemmon, M. T., Karkoschka, E, Lockwood, G. W., Caldwell, J. D., 1997a. Titan's north south asymmetry from Hubble space telescope and voyager imaging: Comparison with models and ground based photometry. Icarus **127**, 173–189.

Lorenz, R. D., McKay, C. P., Lunine, J. I., 1997b. Photochemically-induced collapse of Titan's atmosphere. Science **275**, 642–644.

Lorenz, R. D., Lemmon, M. T., Smith, P. H., Lockwood, G. W., 1999a. Seasonal change on Titan observed with the Hubble space telescope WFPC-2. Icarus **142**, 391–401.

Lorenz, R. D., McKay, C. P., Lunine, J. I., 1999b. Analytic investigation of climate stability on Titan: Sensitivity to volatile inventory. Plan. Space Sci. **47**, 1503–1515.

Lorenz, R. D., Shandera, S. E., 2001a. Physical properties of ammonia-rich ice: Application to Titan. Geophys. Res. Let. **28**, 215–218.

Lorenz, R. D., Lunine, J. I., Withers, P. G., McKay, C. P., 2001b. Titan, Mars and Earth: Entropy production by latitudinal heat transport. Geophys. Res. Let. **28**, 415–418.

Lorenz, R. D., Young, E. F., Lemmon, M. T., 2001. Titan's smile and collar: HST observations of Seasonal Change 1994–2000. Geophys. Res. Let. **28**, 4453–4456.

Lorenz, R. D., Lunine, J. I., 2002. Titan's snowline. Icarus **158**, 557–559.

Lorenz, R. D., Dooley, J. M., West, J. D., Fujii, M., 2003. Backyard spectroscopy and photometry of Titan, Uranus and Neptune. Plan. Space Sci. **51**, 113–125.

Lorenz, R. D., Biolluz, G., Encrenaz, P., Janssen, M. A., West, R. D., Muhleman, Duane O., 2003. Cassini RADAR: Prospects for Titan surface investigations using the microwave radiometer. Plan. Space Sci. **51**, 353–364.

Lorenz, R. D., Smith, P. H., Lemmon, M. T., 2004. Seasonal change in Titan's haze 1992–2002 from Hubble Space Telescope observations. Geophys. Res. Let. **31**, Issue 10, CiteID L10702.

Lorenz, R. D., Griffith, C. A., Lunine, J. I., McKay, C. P., Rennò, N. O., 2005. Convective plumes and the scarcity of Titan's clouds. Geophys. Res. Let. **32**, Issue 1, CiteID L01201.

Lorenz, R. D., Lunine, J. I., 2005. Titan's surface before Cassini. Plan. Space Sci. **53**, 557–576.

Lorenz, R. D., Kraal, E. R., Eddlemon, E. E., Cheney, J., Greeley, R., 2005. Sea-surface wave growth under extraterrestrial atmospheres: Preliminary wind tunnel experiments with application to Mars and Titan. Icarus **175**, 556–560.

Lorenz, R. D., Wall, S., Radebaugh, J., Boubin, G., Reffet, E., Janssen, M., Stofan, E., Lopes, R., Kirk, R., Elachi, C., Lunine, J., Mitchell, K., Paganelli, F., Soderblom, L., Wood, C., Wye, L., Zebker, H., Anderson, Y., Ostro, S., Allison, M., Boehmer, R., Callahan, P., Encrenaz, P., Ori, G. G., Francescetti, G., Gim, Y., Hamilton, G., Hensley, S., Johnson, W., Kelleher, K., Muhleman, D., Picardi, G., Posa, F., Roth, L., Seu, R., Shaffer, S., Stiles, B., Vetrella, S., Flamini, E., West, R., 2006. The sand seas of Titan: Cassini RADAR observations of longitudinal dunes. Science **312**, 724–727.

Lorenz, R. D., 2006. Thermal interactions of the Huygens probe with the Titan environment: Constraint on near-surface wind. Icarus **182**, 559–566.

Lorenz, Ralph D., Lemmon, Mark T., Smith, Peter H., 2006. Seasonal evolution of Titan's dark polar hood: Midsummer disappearance observed by the Hubble Space Telescope. Monthly Notices of the Royal Astronomical Society **369**, 1683–1687.

Lorenz, R. D., Wood, C. A., Lunine, J. I., Wall, S. D., Lopes, R. M., Mitchell, K. L., Paganelli, F., Anderson, Y. Z., Wye, L., Tsai, C., Zebker, H., Stofan, E. R., 2007. Titan's young surface: Initial impact crater survey by Cassini RADAR and model comparison. Geophys. Res. Let. **34**, CiteID L07204.

Loveday, J. S., Nelmes, R. J., Guthrie, M., Belmonte, S. A., Allan, D. R., Klug, D. D., Tse, J. S., Handa, Y. P., 2001. Stable methane hydrate above 2GPa and the source of Titan's atmospheric methane. Nature **410**, 661–663.

Luna, H., Michael, M., Shah, M. B., Johnson, R. E., Latimer, C. J., McConkey, J. W., 2003. Dissociation of N2 in capture and ionization collisions with fast H^+ and N^+ ions and modeling of positive ion formation in the Titan atmosphere. J. Geophys. Res. (Planets) **108**, 14–1.

Lunine, J. I., 1989. The Urey prize lecture: Volatile processes in the outer solar system. Icarus **81**, 1–13.

Lunine, J. I., 1993. Does Titan have an ocean? A review of current understanding of Titan's surface. Rev. of Geophys. **31**, 133–149.

Lunine, J. I., Stevenson, D. J., Yung, Y. L. 1983. Ethane ocean on Titan, Science **222**, 1229–1230.

Lunine, J., Tittemore, W. C., 1983. Origins of outer-planet satellites. In: Protostars and Planets III (Levy and Lunine, eds.), University of Arizona Press, 1177–1252.

Lunine, J. I., Stevenson, D. J., 1985. Evolution of Titan's coupled ocean-atmosphere system and interaction of ocean with bedrock. In: Ices in the Solar System (Klinger, J., Benest, D., Dollfus, A., Smoluchowski, R., eds.), Dordrecht: D. Reidel 741–757.

Lunine, J. I., Stevenson, D. J., 1987. Clathrate and ammonia hydrates at high pressure: Application to the origin of methane on Titan. Icarus **70**, 61–77.

Lunine, J. I., Rizk, B., 1989. Thermal evolution of Titan's atmosphere. Icarus **80**, 370–389.

Lunine, J. I., Atreya, S. K., Pollack., J. B., 1989. Present state and chemical evolution of the atmospheres of Titan, Triton, and Pluto. In: Origin and evolution of planetary and satellite atmospheres (Atreya, S. K., Pollack, J. B., Matthews, M. S., eds.), Tucson: University of Arizona Press, 605.

Lunine, J., Tittemore, W. C., 1993. Origins of outer-planet satellites. In: Protostars and Planets III (Levy, Lunine, eds.), 1177–1252. University of Arizona Press.

Lunine, J. I., McKay, C. P., 1995 Surface-atmosphere interactions on Titan compared with those on the pre-biotic Earth. Adv. Space Res. **15**(3), 303–311.

Lunine, J. I., Lorenz, R. D., Hartmann, W. K., 1998. Some speculation about Titan's past, present, and future. Planet. Space Sci. **46**, 1099–1108.

Lunine, J. I., Yung, Y. L., Lorenz, R. D., 1999. On the volatile inventory of Titan from isotopic abundances in nitrogen and methane. Plan. Space Sci. **47**, 1291–1303.

Lutz, B. L., de Bergh, C., Owen, T., 1983.Titan: Discovery of carbon monoxide in its atmosphere. Science **220**, 1374–1375.

Lutz, B. L., De Bergh, C., Maillard, J. P., Owen, T., Brault, J., 1981. On the possible detection of CH_3D on Titan and Uranus. Astrophys. J. **248**, L141–L145.

Lutz, B. L., Owen, T., Cess, R. D., 1976. Laboratory band strengths of methane and their application to the atmospheres of Jupiter, Saturn, Uranus, Neptune and Titan. Astron. Astrophys. J. **203**, 541–551.

Lutz, B. L., Owen, T., Cess, R. D., 1982. Laboratory band strengths of methane and their application to the atmospheres of Jupiter, Saturn, Uranus, Neptune and Titan. Astron. Astrophys. J. **258**, 886–898.

Luz, D., Hourdin, F., Rannou, P., Lebonnois, S., 2003a. Latitudinal transport by barotropic waves in Titan's stratosphere. I. General properties from a horizontal shallow-water model. Icarus **166**, 328–342.

Luz, D., Hourdin, F., Rannou, P., Lebonnois, S., 2003b. Latitudinal transport by barotropic waves in Titan's stratosphere. II. Results from a coupled dynamics-microphysics-photochemistry GCM. Icarus **166**, 343–358.

Luz, D., Civeit, T., Courtin, R., Lebreton, J.-P., Gautier, D., Rannou, P., Kaufer, A., Witasse, O., Lara, L., Ferri, F., 2005. Characterization of zonal winds in the stratosphere of Titan with UVES. Icarus **179**, 497–510.

Luz, D., Civeit, T., Courtin, R., Lebreton, J.-P., Gautier, D., Witasse, O., Kaufer, A., Ferri, F., Lara, L., Livengood, T., Kostiuk, T., 2006. Characterization of zonal winds in the stratosphere of Titan with UVES: 2. Observations coordinated with the Huygens Probe entry. J. Geophys. Res. **111**, CiteID E08S90.

Ma, Y.-J., Nagy, A. F., Cravens, T. E., Sokolov, I. V., Clark, J., Hansen, K. C., 2004. 3-D global MHD model prediction for the first close flyby of Titan by Cassini. Geophys. Res. Let. **31**, CiteID L22803.

Ma, Y., Nagy, A. F., Cravens, Th. E., Sokolov, I. V., Hansen, K. C., Wahlund, J.-E., Crary, F. J., Coates, A. J., Dougherty, M. K., 2006. Comparisons between MHD model calculations and observations of Cassini flybys of Titan. J. Geophys. Res. **111**, CiteID A05207.

Maguire, W. C., Hanel, R. A., Jennings, D. E., Kunde, V. G., Samuelson, R. E., 1981. C_3H_8 and C_3H_4 in Titan's atmosphere. Nature **292**, 683–686.

Mahaffy, P. R., Donahue, T. M., Atreya, S. K., Owen, T. C., Niemann, H. B., 1998. Galileo probe measurements of D/H and $^3He/^4He$ in Jupiter's atmosphere. Space Sci. Rev. **84**, 251–263.

Mahaffy, P. R., 2005. Intensive Titan exploration begins. Science **308**, 969–970.

Makalkin, A. B., Dorofeeva, V. A., 2006. Models of the protosatellite disk of Saturn: Conditions for Titan's formation. Solar System Research **40**, 441–455.

Mäkinen, J. Teemu T., Harri, A.-M., Tokano, T., Savijärvi, H., Siili, T., Ferri, F., 2006. Vertical atmospheric flow on Titan as measured by the HASI instrument on board the Huygens probe. Geophys. Res. Let. **33**, CiteID L21803.

Marten, A., Gautier, D., Tanguy, L., Lecacheux, A., Rosolen, C., Paubert, G., 1988. Abundance of carbon monoxide in the stratosphere of Titan from millimeter heterodyne observations, Icarus **76**, 558–562, 1988.

Marten, A., Hidayat, T., Moreno, R., Paybert, G., Bézard, B., Gautier, D., Owen, T., 1997. Saturn VI (Titan). IAU Circular **6702**, 19 July.

Marten, A., Hidayat, T., Biraud, Y., Moreno, R., 2002. New millimeter heterodyne observations of Titan: Vertical distributions of Nitriles HCN, HC_3N, CH_3CN, and the isotopic ratio $^{15}N/^{14}N$ in its atmosphere. Icarus **158**, 532–544.

Mayo, L. A., Samuelson, R. E., 2005. Condensate clouds in Titan's north polar stratosphere. Icarus **176**, 316–330.

Mayor, M., Queloz, D., 1995. A Jupiter-mass companion to a solar-type star. Nature **378**, 355–259.

McCord, T. B., Hansen, G. B., Buratti, B. J., Clark, R. N., Cruikshank, D. P., D'Aversa, E., Griffith, C. A., Baines, E. K. H., Brown, R. H., Dalle Ore, C. M., Filacchione, G., Formisano, V., Hibbitts, C. A., Jaumann, R., Lunine, J. I., Nelson, R. M., Sotin, C., the Cassini VIMS Team, 2006. Composition of Titan's surface from Cassini VIMS. Plan. Space Sci. **54**, 1524–1539.

McDonald, G. D., Thompson, W. R., Heinrich, M., Khare, B. N., Sagan, C., 1994. Chemical investigation of Titan and Triton tholins. Icarus **108**, 137–145.

McDonough, T., Brice, N., 1973. A Saturnian gas ring and the recycling of Titan's atmosphere. Icarus **20**, 136–145.

McGrath, M. A., Courtin, R., Smith, T. E., Feldman, P. D., Strobel, D. F., 1998. The ultraviolet albedo of Titan. Icarus **131**, 382–392.

McKay, C. P., Pollack, J. B., Courtin, R., 1989. The thermal structure of Titan's atmosphere, Icarus **80**, 23–53.

McKay, C. P., Pollack, J. B., Courtin, R., 1991. The greenhouse and anti-greenhouse effects on Titan. Science **253**, 1118–1121.

McKay, C. P, Pollack, J. B., Lunine, J. I., Courtin, R., 1993. Coupled atmosphere-ocean models of Titan's past. Icarus **102**, 88–98.

McKay, C. P., 1996. Elemental composition, solubility, and optical properties of Titan's organic haze, Planet. Space Sci. **44**, 8, 741–747.

McKay, C. P., Martin, S. C., Griffin, C. A., Keller, R. M., 1997. Temperature lapse rate and methane in Titan's troposphere. Icarus **129**, 498–505.

McKay, C. P., Lorenz, R. D., Lunine, J. I., 1999. Analytic solutions for the antigreenhouse effect: Titan and the early earth. Icarus **137**, 56–61.

McKay, C. P., Coustenis, A., Samuelson, R. E., Lemmon, M. T., Lorenz, R. D., Cabane, M., Rannou, P., Drossart, P., 2001. Physical properties of the organic aerosols and clouds on Titan. Plan. Space Sci. **49**, 79–99.

McKay, C. P., Smith, H. D., 2005. Possibilities for methanogenic life in liquid methane on the surface of Titan. Icarus **178**, 274–276.

Meier, R., *et al.*, 1998a. A determination of the HDO/H_2O ratio in comet C/1995 01 (Hale-Bopp). Science **279**, 842–844.

Meier, R., Smith, B. A., Owen, T. C., Terrile, R. J., 2000. The surface of Titan from NICMOS observations with the Hubble Space Telescope. Icarus **145**, 462–473.

Michael, M., Johnson, R. E., Leblanc, F., Liu, M., Luhmann, J. G., Shematovich, V. I., 2005. Ejection of nitrogen from Titan's atmosphere by magnetospheric ions and pick-up ions. Icarus **175**, 263–267.

Michael, M., Johnson, R. E., 2006. Energy deposition of pickup ions and heating of Titan's atmosphere. Plan. Space Sci. **53**, 1510–1514.

Mitchell, D. G., Brandt, P. C., Roelof, E. C., Dandouras, J., Krimigis, S. M., Mauk, B. H., 2005. Energetic neutral atom emissions from Titan interaction with Saturn's magnetosphere. Science **308**, 989–992.

Mitri, Giuseppe, Showman, Adam P., Lunine, Jonathan I., Lorenz, Ralph D., 2007. Hydrocarbon lakes on Titan. Icarus **186**, 385–394.

Molina-Cuberos, G. J., López-Moreno, J. J., Rodrigo, R., Lara, L. M., 1999a. Chemistry of the galactic cosmic ray induced ionosphere of Titan. J. Geophys. Res. **104**, 21997–22024.

Molina-Cuberos, G. J., López-Moreno, J. J., Rodrigo, R., Lara, L. M., O'Brien, K., 1999b. Ionization by cosmic rays of the atmosphere of Titan. Plan. Space Sci. **47**, 1347–1354.

Molina-Cuberos, G. J., López-Moreno, J. J., Rodrigo, R., 2000. Influence of electrophilic species on the lower ionosphere of Titan. Geophys. Res. Let. **27**, 1351.

Molina-Cuberos, G. J., Lammer, H., Stumptner, W., Schwingenschuh, K., Rucker, H. O., López-Moreno, J. J., Rodrigo, R., Tokano, T., 2001. Ionospheric layer induced by meteoric ionization in Titan's atmosphere. Plan. Space Sci. **49**, 143–153.

Molina-Cuberos, G. J., Schwingenschuh, K., López-Moreno, J. J., Rodrigo, R., Lara, L. M., Anicich, V., 2002. Nitriles produced by ion chemistry in the lower ionosphere of Titan. J. Geophys. Res. (Planets) **107**, 9–1.

Monks, P. S., Romani, P. N., Neshitt, F. L., Scanlon, M., Stief, L. J., 1993. The kinetics formation of nitrile compounds in the atmospheres of Titan and Neptune, Geophys. Res. **98**, 17,115–17, 123.

Moore, P., Hardy, D. A., 1972. In: Challenge of the Stars, pp. 32–33. Mitchell Beazley Ltd.

Moreno, R., Marten, A., Hidayat, T., 2005. Interferometric measurements of zonal winds on Titan. Astron. Astrophys. **437**, 319–328.

Morente, Juan A., Molina-Cuberos, Gregorio J., Portí, Jorge A., Schwingenschuh, Korand, Besser, Bruno P., 2003. A study of the propagation of electromagnetic waves in Titan's atmosphere with the TLM numerical method. Icarus **162**, 374–384.

Mori, K., Tsunemi, H., Katayama, H., Burrows, D. N., Garmire, G. P., Metzger, A. E., 2004. An X-ray measurement of Titan's atmospheric extent from its Transit of the Crab Nebula. Astrophys. J. **607**, 1065–1069.

Morrison, D., Owen, T., Soderblom, L. A., 1986. The satellites of Saturn. In: Satellites (Burns, J., Matthews, M. S., eds.), Univ. of Arizona Press.

Mousis, O., Gautier, D., Bockelée-Morvan, D., 2002a. An evolutionary turbulent model of Saturn's subnebula: Implications for the origin of the atmosphere of Titan. Icarus **156**, 162–175.

Mousis, O., Gautier, D., Coustenis, A., 2002b. The D/H ratio in methane in Titan. Origin and history. Icarus **159**, 156–169.

Mousis, O., 2004. An estimate of the D/H ratio in Jupiter and Saturn's regular icy satellites — Implications for the Titan Huygens mission. Astron. Astrophys. **414**, 1165–1168.

Muhleman, D. O., Berge, G. L., Clancy, R. T., 1984. Microwave measurements of carbon monoxide on Titan. Science **223**, 393–396.

Muhleman, D. O., Grossman, A. W., Butler, B. J., Slade, M. A., 1990. Radar reflectivity of Titan. Science **248**, 975–980.

Muhleman, D. O, Grossman, A. W., Butler, B. J., 1995. Radar investigation of Mars, Mercury, and Titan. Ann. Rev. Earth Planet. Sci. **23**, 337–374.

Müller-Wodarg, I. C. F., Yelle, R. V., 2002. The effect of dynamics on the composition of Titan's upper atmosphere. Geophys. Res. Let. **29**, 54–1.

Müller-Wodarg, I. C. F., Yelle, R. V., Mendillo, M. J., Aylward, A. D., 2003. On the global distribution of neutral gases in Titan's upper atmosphere and its effect on the thermal structure. J. Geophys. Res. **108**, SIA 18–1.

Müller-Wodarg, I. C. F., Yelle, R. V., Borggren, N., Waite, J. H., 2006. Waves and horizontal structures in Titan's thermosphere. J. Geophys. Res. **111**, CiteID A12315.

Mumma, M. J., Weissman, P. R., Stern, S. A., 1993. Comets and the Origin of the Solar System: Reading the Rosetta Stone. In: Protostars and Planets III. (Levy, Lunine, eds.), 1177–1252. University of Arizona Press.

Nagy, A. F., Cravens, T. E., 1998. Titan's ionosphere: A review. Plan. Space Sci. **46**, 1149–1156.

Nagy, A. F., Liu, Y., Hansen, K. C., Kabin, K., Gombosi, T. I., Combi, M. R., DeZeeuw, D. L., Powell, K. G., Kliore, A. J., 2001. The interaction between the magnetosphere of Saturn and Titan's ionosphere. J. Geophys. Res. **106**, 6151–6160.

Neff, J. S., Humm, D. C., Bergstralh, J. T., Cochran, A. L., Cochran, W. D., Barker, E. S., Tull, R. G., 1984. Absolute spectrophotometry of Titan, Uranus, and Neptune: 3,500–10,500 Angstroms. Icarus **60**, 221–235.

Neff, J. S., Ellis, T. A., Apt, J., Bergstralh, J. T., 1985. Bolometric albedos of Titan, Uranus, and Neptune. Icarus **62**, 425–432.

Negrão, A., Roos-Serote, M., Rannou, P., Rages, K., Lourenço, B., 2005. On the latitudinal distribution of Titan's haze at the Voyager epoch. Plan. Space Sci. **53**, 526–534.

Negrão, A., Coustenis, A., Lellouch, E., Maillard, J.-P., Rannou, Combes, M., Schmitt, B., McKay, C. P., Boudon, V., 2006. Titan's surface albedo from near-infrared CFHT/FTS spectra: Modeling dependence on the methane absorption. Plan. Space Sci. **54**, 1225–1246.

Negrão, A., Hirtzig, M., Coustenis, A., Gendron, E., Drossart, P., Rannou, Combes, M., Boudon, V., 2007. 2-micron spectroscopy of Huygens' landing site on Titan with VLT/NACO. J. Geophys. Res. Planets **112**, E02S92.

Neish, C. D., Lorenz, R. D., O'Brien, D. P., and the Casini RADAR Team, 2006. The potential for prebiotic chemistry in the possible cryovolcanic dome Ganesa Macula on Titan. International J. Astrobiology **5**, 57–65.

Nelson, R. M., Brown, R. H., Hapke, B. W., Smythe, W. D., Kamp, L., Boryta, M. D., Leader, F., Baines, K. H., Bellucci, G., Bibring, J.-P., Buratti, B. J., Capaccioni, F., Cerroni, P., Clark, R. N., Combes, M., Coradini, A., Cruikshank, D. P., Drossart, P., Formisano, V., Jaumann, R., Langevin, Y., Matson, D. L., McCord, T. B., Mennella, V., Nicholson, P. D., Sicardy, B., Sotin, C.,

2006. Photometric properties of Titan's surface from Cassini VIMS: Relevance to Titan's hemispherical albedo dichotomy and surface stability. Plan. Space Sci. **54**, 1540–1551.

Neubauer, F. M., Backes, H., Dougherty, M. K., Wennmacher, A., Russell, C. T., Coates, A., Young, D., Achilleos, N., André, N., Arridge, C. S., Bertucci, C., Jones, G. H., Khurana, K. K., Knetter, T., Law, A., Lewis, G. R., Saur, J., 2006. Titan's near magnetotail from magnetic field and electron plasma observations and modeling: Cassini flybys TA, TB, and T3. J. Geophys. Res. **111**, CiteID A10220.

Nickolaenko, A. P., Besser, B. P., Schwingenschuh, K., 2003. Model computations of Schumann resonance on Titan. Plan. Space Sci. **51**, 853–862.

Niemann, H. B., Atreya, S. K., Bauer, S. J., Carignan, G. R., Demick, J. E., Frost, R. L., Gautier, D., Haberman, J. A., Harpold, D. N., Hunten, D. M., Israel, G., Lunine, J. I., Kasprzak, W. T., Owen, T. C., Paulkovich, M., Raulin, F., Raaen, E., Way, S. H., 2005. The abundances of constituents of Titan's atmosphere from the GCMS instrument on the Huygens probe. Nature **438**, 779–784.

Nixon, C. A., Achterberg, R. K., Vinatier, S., Bézard, B., Coustenis, A., Teanby, N. A., de Kok, R., Romani, P. N., Jennings, D. E., Bjoraker, G. L., Flasar, F. M. 2007. The ^{12}C/^{13}C ratio in Titan hydrocarbons from Cassini/CIRS Infrared Spectra. Icarus, in press.

Noll, K. S., Geballe, T. R., Knacke, R. F., Pendleton, Y. J., Titan's 5 μm spectral window: Carbon monoxide and the albedo of the surface. Icarus **126**, 625–631, 1996.

Noll, K. S., Knacke, R. F., Titan: 1–5 mm photometry and spectrophotometry and a search for variability. Icarus **101**, 272–281, 1993.

O'Brien, D. P., Lorenz, R. D., Lunine, J. I., 2005. Numerical calculations of the longevity of impact oases on Titan. Icarus **173**, 243–253.

Ori, G. G., Marinangeli, L., Baliva, A., Bressan, M., Strom, R. G., 1998. Fluid dynamics of liquids on Titans surface. Plan. Space Sci. **46**, 1417–1421.

Orton, G., 1992. Ground-based observations of Titan's thermal spectrum, In: Symposium on Titan, ESA-SP **338** (Kaldeich, B., ed.), 81–85.

Osegovic, J. P., Max, M. D., 2005. Compound clathrate hydrate on Titan's surface. J. Geophys. Res. **110**, CiteID E08004.

Ostro, S. J., Campbell, D. B., Simpson, R. A., Hudson, R. S., Chandler, J. F., Rosema, K. D., Shapiro, I. I., Standish, E. M., Winkler, R., Yeomans, D. K., Velez, R., Goldstein, R. M., 1992. Europa, Ganymede and Callisto: New radar results from Arecibo and Goldstone. J. Geophys. Res. **97**, 18,227–18,244.

Owen, T., 1982a. Titan. Scientific American, 76–85.

Owen, T., 1982b. The composition and origin of Titan's atmosphere. Planet. Space Sci. **30**, 833–838.

Owen, T., Cess, R. D., 1975. Methane absorption in the visible spectra of the outer planets and Titan. Astrophys. J. **197**, L37–L40.

Owen, T., Gautier, D., 1989. Titan: Some new results. Adv. Space Res. **9**, 73–78.

Owen, T., Lutz, B. L., de Bergh, C., 1986. Deuterium in the outer Solar system: Evidence for two distinct reservoirs. Nature **320**, 244–246.

Owen, T., Bar-Nun, A., 1995, Comets, impacts, and atmospheres. Icarus **116**, 215–226.

Owen T. C., 2000a. On the origin of Titan's atmosphere. Plan. Space Sci. **48**, 747–752.

Owen T., 2005. Planetary science: Huygens rediscovers Titan. Nature **438**, 756–757.

Paillou, Ph., Crapeau, M., Elachi, Ch., Wall, S., Encrenaz, P., 2006. Models of synthetic aperture radar backscattering for bright flows and dark spots on Titan. J. Geophys. Res. **111**, CiteID E11011.

Paubert, G., Gautier, D., Courtin, R., 1984. The millimeter spectrum of Titan: Detectability of HCN, HC$_3$N and CH$_3$CN and the CO abundance. Icarus **60**, 599–612.

Peale, S. J., Lassen, P., Reynolds, R. T., 1980. Tidal dissipation, orbital evolution and the nature of Saturn's inner satellites. Icarus **43**, 65–72.

Penteado, P. F., Griffith, C. A., Greathouse, T. K., de Bergh, C., 2005. Measurements of CH3D and CH4 in Titan from infrared spectroscopy. Astrophys. J. **629**, L53–L56.

Penz, T., Lammer, H., Biernat, H. K., 2005. The influence of the solar particle and radiation environment on Titan's atmosphere evolution. Adv. Space Res. **36**, 241–250.

Pérez-Ayúcar, M., Lorenz, R. D., Floury, N., Prieto-Cerdeira, R., Lebreton, J.-P., 2006. Bistatic observations of Titan's surface with the Huygens probe radio signal. J. Geophys. Res. **111**, CiteID E07001.

Perron, J. T., de Pater, I., 2004. Dynamics of an ice continent on Titan. Geophys. Res. Let. **31**, CiteID L17S04.

Perron, J. T., Lamb, M. P., Koven, C. D., Fung, I. Y., Yager, E., Ádámkovics, M., 2006. J. Geophys. Res. **111**, CiteID E11001.

Pettengill, G., 1965. Lunar radar reflections, In: Solar System Radio Astronomy (Aarons, J., ed.), p. 355. Plenum Press, New York.

Petculescu, A., Lueptow, R. M., 2007. Atmospheric acoustics of Titan, Mars, Venus, and Earth. Icarus **186**, 413–419.

Petrie, S., 2001. Hydrogen isocyanide, HNC: A key species in the chemistry of Titan's ionosphere? Icarus **151**, 196–203.

Pétrie, S., 2004. Products of meteoric metal ion chemistry within planetary atmospheres. 1. Mg^+ at Titan. Icarus **171**, 199–209.

Pinto, J. P., Lunine, J. I., Kim, S. J., Yung, Y. L., 1986. D to H ratio and the origin and evolution of Titan's atmosphere. Nature **319**, 388–390.

Plankensteiner, K., Reiner, H., Rode, B. M., Mikoviny, T., Wisthaler, A., H., Armin, M., Tilmann, D., Fischer, G., Lammer, H., Rucker, H. O., 2007. Discharge experiments simulating chemical evolution on the surface of Titan. Icarus **187**, 616–619.

Podolak, M., Bar-Nun, A., Tvoy, N., Giver, L. P., 1984. Inhomogeneous models of Titan's aerosol distribution, Icarus **57**, 72–82.

Pollack, J. B., Rages, K., Toon, O. B., Yung, Y. L., 1980. On the relationship between secular brightness changes of Titan and solar variability. Geophys. Res. Lett. **7**, 829–832.

Pollack, J. B., Bodenheimer, P., 1989. Theories of the origin and evolution of the Giant Planets. In: Origin and Evolution of Planetary and Satellite Atmospheres (Atreya, Pollack, Matthews, eds.), 564–602. University of Arizona Press.

Porco, C. C., Baker, E., Barbara, J., Beurle, K., Brahic, A., Burns, J. A., Charnoz, S., Cooper, N., Dawson, D. D., Del Genio, A. D., Denk, T., Dones, L., Dyudina, U., Evans, M. W., Fussner, S., Giese, B., Grazier, K., Helfenstein, P., Ingersoll, A. P., Jacobson, R. A., Johnson, T. V., McEwen, A., Murray, C. D., Neukum, G., Owen, W. M., Perry, J., Roatsch, T., Spitale, J., Squyres, S., Thomas, P., Tiscareno, M., Turtle, E. P., Vasavada, A. R., Veverka, J., Wagner, R., West, R., 2005. Imaging of Titan from the Cassini spacecraft. Nature **434**, 159–168.

Prinn, R. G., Fegley, B. Jr., 1981. Kinetic inhibition of CO and N_2 reduction in circumplanetary nebulae — Implications for satellite composition. Astrophys. J. **249**, 308–317.

Prockter, L., 2005. Planetary science: Shades of Titan. Nature **435**, 749–750.

Rages, K., Pollack, J. B., Smith, P. H., 1983. Size estimates of Titan's aerosols based on Voyager 1 high-phase-angle images. J. Geophys. Res. **88**, 8721–8728.

Rages, K. A., Pollack, J. B., 1983. Vertical distribution of scattering hazes in Titan's upper atmosphere. Icarus **55**, 50–62.

Rages, K., Pollack, J. B. 1980. Titan aerosols: Optical properties and vertical distribution, Icarus **41**, 119–130.

Rannou, P., Cabane, M., Chassefière, E., 1993. Growth of aerosols in Titan's atmosphere and related time scales: A stochastic approach. Geophys. Res. Lett. **20**, 967–970.

Rannou, P., Cabane, M., Chassefière, E., Botet, R., McKay, C. P., Courtin, R., 1995. Titan's geometric albedo: Role of the fractal structure of the aerosols. Icarus **118**, 355–372.

Rannou, P., Cabane, M., Botet, R., Chassefière, E., 1997. A new interpretation of the scattered light at Titan's limb. J. Geophys. Res. **102**, 10997–11013.

Rannou, P., Ferrari, C., Rages, K., Roos-Serote, M., Cabane, M., 2000. Characterization of aerosols in the detached haze layer of Titan. Icarus **147**, 267–281.

Rannou, P., Hourdin, F., McKay, C. P., 2002. A wind origin for Titan's haze structure. Nature **418**, 853–856.

Rannou, P., McKay, C. P., Lorenz, R. D., 2003. A model of Titan's haze of fractal aerosols constrained by multiple observations. Plan. Space Sci. **51**, 963–976.

Rannou, P., Hourdin, F., McKay, C. P., Luz, D., 2004. A coupled dynamics-microphysics model of Titan's atmosphere. Icarus **170**, 443–462.

Rannou, P., Montmessin, F., Hourdin, F., Lebonnois, S., 2006. The latitudinal distribution of clouds on Titan. Science **311**, 201–205.

Rappaport, N., Bertotti, B., Giampieri, G., Anderson, J. D., 1997. Doppler measurements of the quadrupole moments of Titan. Icarus **126**, 313–323.

Raulin, F., 1987. Organic chemistry in the oceans of Titan. Adv. Space Res. **7**, 571–581.

Raulin, F., 2007. Astrobiology and habitability of Titan. Space Sci. Rev., DOI: 10.1007/s11214–006–9133–7.

Raulin, F., 2007. Question 2: Why an astrobiological study of Titan will help us understand the origin of life. In: Origins of Life and Evolution of Biospheres, DOI : 10.1007/s11084–007–9077–2.

Raulin, F., Mourey, D., Toupance, G., 1982. Organic synthesis from CH4 — N2 atmospheres: Implications for Titan. Orig. Life **12**, 267–279.

Raulin, F., Accaoui, B. Razaghi, A., Dang-Nhu, M., Coustenis, A., Gautier, D., 1990. Infrared spectra of gaseous organics: Application to the atmosphere of Titan. II C4 alkanenitriles and benzene. Spectrochimica Acta **46**, 671–683.

Raulin, F., Bruston, P., Paillous, P., Sternberg, R., 1995. The low temperature organic chemistry of Titan's geofluid. Adv. Space Res. **15**(3), 321–333.

Raulin, F., 2005. Exo-astrobiological aspects of Europa and Titan: From observations to speculations. Space Sci. Rev. **116**, 471–487.

Redondo, P., Pauzat, F., Ellinger, Y., 2006. Theoretical survey of the NH+CH3 potential energy surface in relation to Titan atmospheric chemistry. Plan. Space Sci. **54**, 181–187.

Rees, M. H., 1989. Physics and Chemistry of the Upper Atmosphere, Cambridge: Cambridge University Press.

Ricca, A., Bauschlicher, C. W., Bakes, E. L. O., 2002. A computational study of the mechanisms for the incorporation of a Nitrogen atom into polycyclic aromatic hydrocarbons in the Titan haze. Icarus **154**, 516–521.

Richardson, J., Lorenz, R. D., McEwen, A., 2004. Titan's surface and rotation: New results from Voyager 1 images. Icarus **170**, 113–124.

Rodriguez, S., Le Mouélic, S., Sotin, C., Clénet, H., Clark, R. N., Buratti, B., Brown, R. H., McCord, T. B., Nicholson, P. D., Baines, K. H., the VIMS Science Team, 2006. Cassini/VIMS hyperspectral observations of the HUYGENS landing site on Titan. Plan. Space Sci. **54**, 1510–1523.

Rodriguez, S., Paillou, P., Dobrijevic, M., Ruffié, G., Coll, P., Bernard, J. M., Encrenaz, P., 2003. Impact of aerosols present in Titan's atmosphere on the CASSINI radar experiment. Icarus **164**, 213–227.

Roe, H. G., de Pater, I., Macintosh, B. A., Gibbard, S. G., Max, C. E., McKay, C. P., 2002a. Titan's atmosphere in late southern spring observed with adaptive optics on the W. M. Keck II 10-Meter Telescope. Icarus **157**, 254–258.

Roe, H. G., de Pater, I., Macintosh, B. A., McKay, C. P., 2002b. Titan's clouds from Gemini and Keck adaptive optics imaging. Astrophys. J. **581**, 1399–1406.

Roe, H. G., Greathouse, T. K., Richter, M. J., Lacy, J. H., 2003. Propane on Titan. Astrophys. J. **597**, L65–L68.

Roe, H. G., de Pater, I., Gibbard, S. G., Macintosh, B. A., Max, C. E., Young, E. F., Brown, M. E., Bouchez, A. H., 2004a. A new 1.6-micron map of Titan's surface. Geophys. Res. Let. **31**, CiteID L17S03.

Roe, H. G., de Pater, I., McKay, C. P., 2004b. Seasonal variation of Titan's stratospheric ethylene observed. Icarus **169**, 440–461.

Roe, H. G., Bouchez, A. H., Trujillo, C. A., Schaller, E. L., Brown, M. E., 2005a. Discovery of temperate latitude clouds on Titan. Astrophys. J. **618**, L49–L52.

Roe, H. G., Brown, M. E., Schaller, E. L., Bouchez, A. H., Trujillo, C. A., 2005b. Geographic control of Titan's mid-latitude clouds. Science **310**, 477–479.

Roos-Serote, M., 2005. The changing face of Titan's haze: Is it all dynamics? Space science reviews **116**, 201–210.

Roush, T. L., Dalton, J. B., 2004. Reflectance spectra of hydrated Titan tholins at cryogenic temperatures and implications for compositional interpretation of red objects in the outer Solar System. Icarus **168**, 158–162.

Saint-Pé, O., Combes, M., Rigaut, F., Tomasko, M., Fulchignoni, M., 1993. Demonstration of adaptive optics for resolved imagery of solar system objects: Preliminary results on Pallas and Titan. Icarus **105**, 263–270.

Sagan, C., 1973. The greenhouse of Titan. Icarus **18**, 649–656.

Sagan, C., Dermott, S. F., 1982. The tide in the seas of Titan. Nature **300**, 731–733.

Sagan, C., Thompson, W. R., 1984. Production and condensation of organic gases in the atmosphere of Titan. Icarus **59**, 133–161.

Sagan, C., Khare, B. N., Lewis, J. S., 1984. Organic matter in the Saturn system. In: Saturn (Gehrels, T. and M. S. Matthews, eds.), 788–807, Univ. Arizona Press, Tucson.

Sagan, C., Thompson, W. R., Khare, B. N., 1992. Titan: A laboratory for prebiological organic chemistry. Accounts Chem. Res. **25**, 286–292.

Salinas, S. V., Grieger, B., Markiewicz, W. J., Keller, H. U., 2003. A spherical model for computing polarized radiation in Titan's atmosphere. Plan. Space Sci. **51**, 977–989.

Samuelson, R. E., Hanel, R. A., Kunde, V. G., Maguire, W. C., 1981. Mean molecular weight and hydrogen abundance of Titan's atmosphere. Nature **292**, 688–693.

Samuelson, R. F., Maguire, W. C., Hand, R. A., Kunde, V. G., Jennings, D. F., Yung, Y. L., Aikin, A. C., 1983. CO_2 on Titan, J. Geophys. Res. **88**, 8709–8715.

Samuelson, R. F., Mayo, L. A., 1991. Thermal infrared properties of Titan's stratospheric aerosol, Icarus **91**, 207–219.

Samuelson, R. E., Mayo, L. A., Knuckles, M. A., Khanna, R. J., 1997a. C_4N_2 ice in Titan's north polar stratosphere. Planet. Space Sci. **45**, 941–948.

Samuelson, R. E., Mayo, L. A., 1997b. Steady-state model for methane condensation in Titan's troposphere. Planet. Space Sci. **45**, 949–958.

Samuelson, R. E., Nath, N. R., Borysow, A., 1997c. Gaseous abundances and methane supersaturation in Titan's troposphere. Planet. Space Sci. **45**, 959–980.

Samuelson, R. E., 2003. Titan's atmospheric engine: An overview. Planet. Space Sci. **51**, 127–145.

Samuelson, R. E., Smith, M. D., Achterberg, R. K., Pearl, J. C., 2007. Cassini CIRS update on stratospheric ices at Titan's winter pole. Icarus 189, 63–71.

Scattergood, T. W, Lau, E. Y., Stone, B. M., 1992. Titan's aerosols. I. Laboratory investigations of shapes, size distributions, and aggregation of particles produced by UV photolysis of model Titan atmospheres. Icarus **99**, 98–105.

Schaller, E. L., Brown, M. E., Roe, H. G., Bouchez, A. H., 2006. A large cloud outburst at Titan's south pole. Icarus **182**, 224–229.

Schaller, E. L., Brown, M. E., Roe, H. G., Bouchez, A. H., Trujillo, C. A., 2006. Dissipation of Titan's south polar clouds. Icarus **184**, 517–523.

Shemansky, D. E., Stewart, A. I. F., West, R. A., Esposito, L. W., Hallett, J. T., Liu, X., 2005. The Cassini UVIS Stellar Probe of the Titan atmosphere. Science **308**, 978–982.

Shematovich, V. I., Johnson, R. E., Michael, M., Luhmann, J. G., 2003. Nitrogen loss from Titan. J. Geophys. Res. **108**, 6–1.

Schulze-Makuch, D., Grinspoon, D. H., 2005. Biologically enhanced energy and carbon cycling on Titan? Astrobiology **5**, 560–567.

Sears, W. D., 1995. Tidal dissipation in oceans on Titan. Icarus **113**, 39–56.

Seiff, A., Stoker, C. R., Young, R. E., Mihalov, J. D., McKay, C. P., Lorenz, R. D., 2005. Determination of physical properties of a planetary surface by measuring the deceleration of a probe upon impact: Application to Titan. Plan. Space Sci. **53**, 594–600.

Sekine, Y., Sugita, S., Shido, T., Yamamoto, T., Iwasawa, Y., Kadono, T., Matsui, T., 2005. The role of Fischer Tropsch catalysis in the origin of methane-rich Titan. Icarus **178**, 154–164.

Shindo, F., Benilan, Y., Guillemin, J.-C., Chaquin, P., Jolly, A., Raulin, F., 2003. Ultraviolet and infrared spectrum of C6H2 revisited and vapor pressure curve in Titan's atmosphere. Plan. Space Sci. **51**, 9–17.

Sicardy, B., Brahic, A., Ferrari, C., Gautier, D., Lecacheux, J., Lellouch, E., Roques, F., Arlot, J. E., Thuillot, W., Colas, F., Sevres, F., Vidal, J.-L., Blanco, C., Cristaldi, S., Buile, C., Klotz, A., Thouvenot, E., 1989. The July 3, 1989, occultation of 28 Sagitarii: Probing Titan's atmosphere. Nature **343**, 350–353.

Sicardy, B., Colas, F., Widemann, T., Bellucci, A., Beisker, W., Kretlow, M., Ferri, F., Lacour, S., Lecacheux, J., Lellouch, E., Pau, S., Renner, S., Roques, F., Fienga, A., Etienne, C., Martinez, C., Glass, I. S., Baba, D., Nagayama, T., Nagata, T., Itting-Enke, S., Bath, K.-L., Bode, H.-J., Bode, F., Lüdemann, H., Lüdemann, J., Neubauer, D., Tegtmeier, A., Tegtmeier, C., Thomé, B., Hund, F., deWitt, C., Fraser, B., Jansen, A., Jones, T., Schoenau, P., Turk, C., Meintjies, P., Hernandez, M., Fiel, D., Frappa, E., Peyrot, A., Teng, J. P., Vignand, M., Hesler, G., Payet, T., Howell, R. R., Kidger, M., Ortiz, J. L., Naranjo, O., Rosenzweig, P., Rapaport, M., 2006. The two Titan stellar occultations of 14 November 2003. J. Geophys. Res. **111**, CiteID E11S91.

Siegert, M. J., Hodgkins, R., 2000. A stratigraphic link across 1100 km of the Antarctic Ice Sheet between the Vostok ice-core site and Titan Dome (near South Pole). Geophys. Res. Let. **27**, 2133–2136.

Sillanpää, I., Kallio, E., Janhunen, P., Schmidt, W., Mursula, K., Vilppola, J., Tanskanen, P., 2006. Hybrid simulation study of ion escape at Titan for different orbital positions. Adv. Space Res. **38**, 799–805.

Sittler, E. C., Hartle, R. E., Viñas, A. F., Johnson, R. E., Smith, H. T., Mueller-Wodarg, I., 2005. Titan interaction with Saturn's magnetosphere: Voyager 1 results revisited. J. Geophys. Res. **110**, CiteID A09302.

Smith, B. A., Soderblom, L., Batson, R., Bridges, P., Inge, J., Masursky, H., Shoemaker, E., Beebe, R., Boyce, J., Briggs, G., Buncer, A., Collins, S. A., Hansen, C. J., Johnson, T. V., Mitchell, J. L., Terrile, R. J., Cook Ii, A. F., Cuzzi, J., Pollack, J. P., Hunt, G. E., Danielson, G., Morrison, D., Owen, T., Sagan, C., Veverka, J., Strom, R., Suomi, V. E., 1982. A new look at the Saturn system: The Voyager 2 images. Science **215**, 504–537.

Smith, B. A., Soderblom, L., Beebe, R., Boyce, J., Briggs, G., Buncer, A., Collins, S. A., Hansen, C. J., Johnson, T. V., Mitchell, J. L., Terrile, R. J., Carr, M., Cook Ii, A. F., Cuzzi, J., Pollack, J. P., Danielson, G. E., Ingersoll, A., Davies, M. E., Hunt, G. E., Masursky, H., Shoemaker, E., Morrison, D., Owen, T., Sagan, C., Veverka, J., Strom, R., Suomi, V. E., 1981. Encounter with Saturn: Voyager 1 imaging science results. Science **212**, 163–191.

Smith, G. R., Strobel, D. F., Broadfoot, A. L., Sandel, B. R. Shemansky, D. F., Holberg, J. B., 1982. Titan's upper atmosphere: Composition and temperature from the EUV solar occultation results, J. Geophys. Res. **87**, 1351–1359.

Smith, H. T., Johnson, R. E., Shematovich, V. I., 2004. Titan's atomic and molecular nitrogen tori. Geophys. Res. Let. **31**, CiteID L16804.

Smith, P. H., Lemmon, M. T., Lorenz, R. D., Sromovsky, L. A., Caldwell, J. J., Allison, M. D., 1996. Titan's surface revealed by HST imagery. Icarus **119**, 336–349.

Smyth, W. H., 1981. Titan's hydrogen torus, Astrophys. J. **246**, 344–353.

Sohl, F., Sears, W. D., Lorenz, R. D., 1995. Tidal dissipation on Titan. Icarus **115**, 278–294.

Sohl, F., Hussmann, H., Schwentker, B., Spohn, T., Lorenz, R. D., 2003. Interior structure models and tidal love numbers of Titan. J. Geophys. Res. **108**, 4–1.

Sotin, C., Jaumann, R., Buratti, B. J., Brown, R. H., Clark, R. N., Soderblom, L. A., Baines, K. H., Bellucci, G., Bibring, J.-P., Capaccioni, F., Cerroni, P., Combes, M., Coradini, A., Cruikshank, D. P., Drossart, P., Formisano, V., Langevin, Y., Matson, D. L., McCord, T. B., Nelson, R. M., Nicholson, P. D., Sicardy, B., Lemouelic, S., Rodriguez, S., Stephan, K., Scholz, C. K., 2005. Release of volatiles from a possible cryovolcano from near-infrared imaging of Titan. Nature **435**, 786–789.

Sotin, C., 2007. Titan's lost seas found. Nature **445**, 29–30.

Spilker, T. R., Significant science at Titan and Neptune from aerocaptured missions. Plan. Space Sci. **53**, 606–616.

Sromovsky, L. A., Suomi, V. E., Pollack, J. B., Krauss, R. J., Limaye. S.S, Owen, T., Revercomb, H. E., Sagan, C., 1981. Implications of Titan's north south brightness asymmetry. Nature **292**, 698–702.

Stahl, F., Schleyer, P. V. R., Schaefer, H. F., III, Kaiser, R. I., 2002. Reactions of ethynyl radicals as a source of C4 and C5 hydrocarbons in Titan's atmosphere. Plan. Space Sci. **50**, 685–692.

Steinfeld, J. I. 1985. Molecules and Radiation. MIT Press.

Stevens, Michael, H., 2001. The EUV airglow of Titan: Production and loss of N_2 c'$_4$(0)-X. J. Geophys. Res. **106**, 3685–3690.

Stevenson, D. J., 1992. Interior of Titan. In: Symposium on Titan, ESA-SP **338** (Kaldeich, B., ed.), 29–33.

Stevenson, D. J., Harris, A. W., Lunine, J. I., 1986. Origins of satellites. In: Satellites (Burns, J. A. and Matthews, M. S. Eds.), 39–88, University of Arizona Press, Tucson.

Stevenson, D. J., Lunine, J. I., 1986a. Mobilisation of cryogenic ices in Outer Solar System satellites. Nature **323**, 46–48.

Stevenson, D. J., Potter, B. E., 1986b. Titan's latitudinal temperature distribution and seasonal cycle. Geophys. Res. Lett. **17**, 93–96.

Stofan, E. R., Lunine, J. I., Lopes, R., Paganelli, F., Lorenz, R. D., Wood, C. A., Kirk, R., Wall, S., Elachi, C., Soderblom, L. A., Ostro, S., Janssen, M., Radebaugh, J., Wye, L., Zebker, H., Anderson, Y., Allison, M., Boehmer, R., Callahan, P., Encrenaz, P., Flamini, E., Francescetti, G., Gim, Y., Hamilton, G., Hensley, S., Johnson, W. T. K., Kelleher, K., Muhleman, D., Picardi, G., Posa, F., Roth, L., Seu, R., Shaffer, S., Stiles, B., Vetrella, S., West, R., 2006. Mapping of Titan: Results from the first Titan radar passes. Icarus **185**, 443–456.

Stofan, E. R., Elachi, C., Lunine, J. I., Lorenz, R. D., Stiles, B., Mitchell, K. L., Ostro, S., Soderblom, L., Wood, C., Zebker, H., Wall, S., Janssen, M., Kirk, R., Lopes, R., Paganelli, F., Radebaugh, J., Wye, L., Anderson, Y., Allison, M., Boehmer, R., Callahan, P., Encrenaz, P., Flamini, E., Francescetti, G., Gim, Y., Hamilton, G., Hensley, S., Johnson, W. T. K., Kelleher, K., Muhleman, D., Paillou, P., Picardi, G., Posa, F., Roth, L., Seu, R., Shaffer, S., Vetrella, S., West, R., 2007. The lakes of Titan. Nature **445**, 61–64.

Strobel, D. F., 1974. The photochemistry of hydrocarbons in the atmosphere of Titan, Icarus **21**, 466–470.

Strobel, D. F., 1982. Chemistry and evolution of Titan's atmosphere. Planet. Space Sci. **30**, 839–848.

Strobel, D. F., 1985. The photochemistry of hydrocarbons in the atmosphere of Titan. In: The atmospheres of Saturn and Titan, Proc. Int. Workshop, Alpbach, Austria, 16–19 September 1985, ESA SP-241.

Strobel, D. F., 2005. Photochemistry in outer solar system atmospheres. Space Sci. Rev. **116**, 155–170.

Strobel, D. F., Summers, M. F., Zhu, X., 1992. Titan's upper atmosphere: Structure and ultraviolet emissions, Icarus **100**, 512–526.

Strobel, D. F., Hall, D. T., Zhu, X., Summers, M. F., 1993. Upper limit on Titan's atmospheric argon abundance. Icarus **103**, 333–336.

Strobel, D. F., 2006. Gravitational tidal waves in Titan's upper atmosphere. Icarus **182**, 251–258.

Sultan-Salem, A. K., Tyler, G. L., 2007. Revisiting Titan's Earth-based scattering data at 13 cm-λ. Geophys. Res. Let. **34**, doi: 10.1029/2007GL029928.

Sultan-Salem, A. K., Tyler, G. L., 2007. Modeling quasi-specular scattering from the surface of Titan. J. Geophys. Res. **112**, CiteID E05012.

Szego, K., Bebesi, Z., Erdos, G., Foldy, L., Crary, F., McComas, D. J., Young, D. T., Bolton, S., Coates, A. J., Rymer, A. M., Hartle, R. E., Sittler, E. C., Reisenfeld, D., Bethelier, J. J., Johnson, R. E., Smith, H. T., Hill, T. W., Vilppola, J., Steinberg, J., Andre, N., 2005. The global plasma environment of Titan as observed by Cassini Plasma Spectrometer during the first two close encounters with Titan. Geophys. Res. Let. **32**, CiteID L20S05.

Szopa, C., Cernogora, G., Boufendi, L., Correia, J. J., Coll, P., 2006. PAMPRE: A dusty plasma experiment for Titan's tholins production and study. Plan. Space Sci. **54**, 394–404.

Tanguy, L., Bézard, B., Marten, A., Gautier, D., Gérard, E., Paubert, G., Lecacheux, A., 1990. Stratospheric profile of HCN on Titan from millimeter observations. Icarus **85**, 43–57.

Taylor, F. W., 1972. Temperature sounding experiments for the Jovian planets. J. Atmos. Sci. **29**, 950–958.

Taylor, F. W., S. B. Calcutt, P. G. J., Irwin, C. A., Nixon, P. L., Read, P. J. C., Smith, T. J., 1998. Vellacott. Investigation of Saturn's atmosphere by CASSINI. Plan. Space Sci. **46**, 1315–1324.

Taylor, F. W., 2006. Climate variability on Venus and Titan. Space Sci. Rev. **125**, 445–455.

Taylor, F. W., Coustenis, A., 1998. Titan in the Solar system. Planet. Space Sci. **46**, 1085–1098.

Teanby, N. A., Irwin, P. G. J., de Kok, R., Nixon, C. A., Coustenis, A., Bézard, B., Calcutt, S. B., Bowles, N. E., Flasar, F. M., Fletcher, L., Howett, C., Taylor F. W., 2006. Latitudinal variations of HCN, HC_3N and C_2N_2 in Titan's stratosphere derived from Cassini CIRS data. Icarus **181**, 243–255.

Teanby, N. A., Irwin, P. G. J., de Kok, R., Vinatier, S., Bézard, B., Nixon, C. A., Flasar, F. M., Calcutt, S. B., Bowles, N. E., Fletcher, L., Howett, C., Taylor, F. W., 2007. Vertical profiles of HCN, HC_3N, and C_2H_2 in Titan's atmosphere derived from Cassini/CIRS data. Icarus **186**, 364–384.

Teanby, N. A., Irwin, P. G. J., de Kok, R., Nixon, C. A., Coustenis, A., Calcutt, S. B., Bowles, N. E., Fletcher, L., Howett, C., Taylor, F. W., 2007. Global variations of C_2H_2, C_3H_4, C_4H_2, HCN and HC_3N in Titan's stratosphere for early northern winter. Submitted to Icarus.

The Cassini Vims Team, Hansen, G. B., Buratti, B. J., Clark, R. N., Cruikshank, D. P., D'Aversa, E., Griffith, C. A., Baines, E. K. H., Brown, R. H., Dalle Ore, C. M., Filacchione, G., Formisano, V., Hibbitts, C. A., Jaumann, R., Lunine, J. I., Nelson, R. M., Sotin, C., 2006. Composition of Titan's surface from Cassini VIMS. Plan. Space Sci. **54**, 1524–1539.

The Vims Science Team, Le Mouélic, S., Sotin, C., Clénet, H., Clark, R. N., Buratti, B., Brown, R. H., McCord, T. B., Nicholson, P. D., Baines, K. H., 2006. Cassini/VIMS hyperspectral observations of the HUYGENS landing site on Titan. Plan. Space Sci. **54**, 1510–1523.

Thomas-Osip, J. E., Gustafson, B. Å. S., Kolokolova, L., Xu, Y.-L., 2005. An investigation of Titan's aerosols using microwave analog measurements and radiative transfer modelling. Icarus **179**, 511–522.

Thompson, W. R., Sagan, C., 1984. Titan: Far-infrared and microwave remote sensing of methane clouds and organic haze. Icarus **60**, 236–259.

Thompson, W. R., Sagan, C., 1991. Plasma discharge in N_2 + CH_4 at low pressures: Experimental results and applications to Titan. Icarus **90**, 57–73.

Tobie, G., Grasset, O., Lunine, J. I., Mocquet, A., Sotin, C., 2005a. Titan's internal structure inferred from a coupled thermal-orbital model. Icarus **175**, 496–502.

Tobie, G., Mocquet, A., Sotin, C., 2005b. Tidal dissipation within large icy satellites: Applications to Europa and Titan. Icarus **177**, 534–549.

Tobie, G., Lunine, J. I., Sotin, C., 2006. Episodic outgassing as the origin of atmospheric methane on Titan. Nature **440**, 61–64.

Tokano, T., 2005. Meteorological assessment of the surface temperatures on Titan: Constraints on the surface type. Icarus **173**, 222–242.

Tokano, T., Neubauer, F. M., Laube, M., McKay, C. P., 1999. Seasonal variation of Titan's atmospheric structure simulated by a general circulation model. Planet. Space Sci. **47**, 493–520.

Tokano, T., Molina-Cuberos, G. J., Lammer, H., Stumptner, W., 2001a. Modelling of thunderclouds and lightning generation on Titan. Plan. Space Sci. **49**, 539–560.

Tokano, T., Neubauer, F. M., Laube, M., McKay, C. P., 2001b. Three-dimensional modeling of the tropospheric methane cycle on Titan. Icarus **153**, 130–147.

Tokano, T., Neubauer, F. M., 2002. Tidal winds on Titan caused by Saturn. Icarus **158**, 499–515.

Tokano, T., Neubauer, F. M., 2005. Wind-induced seasonal angular momentum exchange at Titan's surface and its influence on Titan's length-of-day. Geophys. Res. Lett. **32**, CiteID L24203.

Tokano, T., Ferri, F., Colombatti, G., Mäkinen, T., Fulchignoni, M., 2006. Titan's planetary boundary layer structure at the Huygens landing site. J. Geophys. Res. **111**, CiteID E08007.

Tokano, T., Lorenz, R. D., 2006. GCM simulation of balloon trajectories on Titan. Plan. Space Sci. **54**, 685–694.

Tokano, T., McKay, C. P., Neubauer, F. M., Atreya, S. K., Ferri, F., Fulchignoni, M., Niemann, H. B., 2006. Methane drizzle on Titan. Nature **442**, 432–435.

Tomasko, M., Smith, P. H., Photometry and polarimetry of Titan: Pioneer 11 observations and their implications for aerosol properties. Icarus **51**, 65–95, 1982.

Tomasko, M. G., West, R. A., Orton, G. S., Tejfel, V. G., 1984. Clouds and aerosols in Saturn's atmosphere. In: Saturn (Gehrels, Matthews, eds.), Univ. of Arizona Press.

Toon, O. B., Turco, R. P., Pollack, J. B., 1980. A physical model of Titan's clouds. Icarus **43**, 260–282.

Toon, O. B., McKay, C. P., Courtin, R., Ackerman, T., 1988. Methane rain on Titan, Icarus **75**, 255–284.

Toon, O. B., McKay, C. P., Griffith, C. A., Turco, R. P., 1992. A physical model of Titan's aerosols. Icarus **95**, 24–53.

Toublanc, D., Parisot, P., Brillet, J., Gautier, D., Raulin, F., McKay, C. P., 1995. Photochemical modelling of Titan's atmosphere. Icarus **113**, 2–16.

Towner, M. C., Garry, J. R. C., Lorenz, R. D., Hagermann, A., Hathi, B., Svedhem, H., Clark, B. C., Leese, M. R., Zarnecki, J. C., 2006. Physical properties of Titan's surface at the Huygens landing site from the Surface Science Package Acoustic Properties sensor (API-S). Icarus **185**, 457–465.

Tracadas, P. W., Hammel, H. B., Thomas-Osip, J. E., Elliot, J. L., Olkin, C. B., 2001. Probing Titan's atmosphere with the 1995 August Stellar Occultation. Icarus **153**, 285–294.

Trafton, L., 1972. The bulk composition of Titan's atmosphere. Astrophys. J. **175**, 295–306.

Trainer, M. G., Pavlov, A. A., Jimenez, J. L., McKay, C. P., Worsnop, D. R., Toon, O. B., Tolbert, M. A., 2004. Chemical composition of Titan's haze: Are PAHs present? Geophys. Res. Let. **31**, CiteID L17S08.

Tran, Buu N., Ferris, James P., Chera, John J., 2003a. The photochemical formation of a titan haze analog. Structural analysis by x-ray photoelectron and infrared spectroscopy. Icarus **162**, 114–124.

Tran, B. N., Joseph, J. C., Ferris, J. P., Persans, P. D., Chera, J. J., 2003b. Simulation of Titan haze formation using a photochemical flow reactor: The optical constants of the polymer. Icarus **165**, 379–390.

Tran, B. N., Joseph, J. C., Force, M., Briggs, R. G., Vuitton, V., Ferris, J. P., 2005. Photochemical processes on Titan: Irradiation of mixtures of gases that simulate Titan's atmosphere. Icarus **177**, 106–115.

Tyler, G. L., Eshleman, V. R., Anderson, J. D., Levy, G. S., Lindal, G. F., Wood, G. E., Croft, T. A., 1981. Radio science investigations of the Saturn System with Voyager 1. Preliminary results. Science **212**, 201–205.

Vacher, J. R., Le Duc, E., Fitaire, M., 2000. Clustering reactions of $HCNH^+$, $HCNH^+(N_2)$ and $HCNH^+(CH_4)$ with ethane: Application to Titan atmosphere. Plan. Space Sci. **48**, 237–247.

Vervack, R. J., Sandel, B. R., Strobel, D. F., 2004. New perspectives on Titan's upper atmosphere from a reanalysis of the Voyager 1 UVS solar occultations. Icarus **170**, 91–112.

Veverka, J. Titan: Polarimetric evidence for an optically thick atmosphere. Icarus **18**, 657–660, 1973.

Vinatier, S., Bézard, B., Fouchet, T., Teanby, N. A., de Kok, R., Irwin, P. G. J., Conrath, B. J., Nixon, C. A., Romani, P. N., Flasar, F. M., Coustenis, A., 2006. Vertical abundance profiles of hydrocarbons in Titan's atmosphere at 15°S and 80°N retrieved from Cassini/CIRS spectra. Icarus **188**, 120–138.

Voss, L. F., Henson, B. F., Robinson, J. M., 2007. Methane thermodynamics in nanoporous ice: A new methane reservoir on Titan. J. Geophys. Res. **112**, CiteID E05002.

Vuitton, V., Gée, C., Raulin, F., Bénilan, Y., Crépin, C., Gazeau, M.-C., 2003. Intrinsic lifetime of metastable excited C_4H_2: Implications for the photochemistry of C_4H_2 in Titan's atmosphere. Plan. Space Sci. **51**, 847–852.

Vuitton, V., Doussin, J.-F., Bénilan, Y., Raulin, F., Gazeau, M.-C., 2006a. Experimental and theoretical study of hydrocarbon photochemistry applied to Titan stratosphere. Icarus **185**, 287–300.

Vuitton, V., Yelle, R. V., Anicich, V. G., 2006b. The nitrogen chemistry of Titan's upper atmosphere revealed. Astrophys. J. **647**, L175–L178.

Wahlund, J.-E., Boström, R., Gustafsson, G., Gurnett, D. A., Kurth, W. S., Pedersen, A., Averkamp, T. F., Hospodarsky, G. B., Persoon, A. M., Canu, P., Neubauer, F. M., Dougherty, M. K., Eriksson, A. I., Morooka, M. W., Gill, R., André, M., Eliasson, L., Müller-Wodarg, I., 2005. Cassini measurements of cold plasma in the ionosphere of Titan. Science **308**, 986–989.

Waite, J. H., Niemann, H., Yelle, R. V., Kasprzak, W. T., Cravens, Th. E., Luhmann, J. G., McNutt, R. L., Ip, W.-H., Gell, D., De La Haye, V., Müller-Wordag, I., Magee, B., Borggren, N., Ledvina, S., Fletcher, G., Walter, E., Miller, R., Scherer, S., Thorpe, R., Xu, J., Block, B., Arnett, K., 2005. Ion neutral mass spectrometer results from the first flyby of Titan. Science **308**, 982–986.

Waite, J. H., Young, D. T., Cravens, T. E., Coates, A. J., Crary, F. J., Magee, B., Westlake, J., 2007. The process of tholin formation in Titan's upper atmosphere. Science **316**, 870–875.

Walterscheid, R. L., Schubert, G., 2006. A tidal explanation for the Titan haze layers. Icarus **183**, 471–478.

Wayne, R. P., 1985. Chemistry of Atmospheres, Oxford: Clarendon Press.

West, R. A., 1991. Optical properties of aggregate particles whose outer diameter is comparable to the wavelength. Appl. Opt. **30**, 5316–5324.

West, R. A., Lane, A. L., Hart, H., Simmons, K. E., Hord, C. W., Coffeen, D. L., Esposito, L. W., Sato, M., Pomphrey, R. B., 1983. Voyager 2 photopolarimeter observations of Titan. J. Quant. Spectrosc. Radiat. Transfer **88**, 8699–8708.

West, R. A., Smith, P. H., 1991. Evidence for aggregate particles in the atmospheres of Titan and Jupiter. Icarus **90**, 330–333.

West, R. A., Brown, M. E., Salinas, S. V., Bouchez, A. H., Roe, H. G., 2005. No oceans on Titan from the absence of a near-infrared specular reflection. Nature **436**, 670–672.

Whitten, R. C., Borucki, W. J., Tripathi, S., 2007. Predictions of the electrical conductivity and charging of the aerosols in Titan's nighttime atmosphere. J. Geophys. Res. **112**, CiteID E04001.

Wilson, E. H., Atreya, S. K., 2000. Sensitivity studies of methane photolysis and its impact on hydrocarbon chemistry in the atmosphere of Titan? J. Geophys. Res. **105**, 20,263–20,274.

Wilson, E. H., Atreya, S. K., Coustenis, A., 2003a. Mechanisms for the formation of benzene in the atmosphere of Titan. J. Geophys. Res. — Planets **108**(E2), 5014–5024.

Wilson, E. H., Atreya, S. K., 2003b. Chemical sources of haze formation in Titan's atmosphere. Plan. Space Sci. **51**, 1017–1033.

Wilson, E. H., Atreya, S. K., 2004. Current state of modeling the photochemistry of Titan's mutually dependent atmosphere and ionosphere. J. Geophys. Res. **109**, CiteID E06002.

Witasse, O., Lebreton, J.-P., Bird, M. K., Dutta-Roy, R., Folkner, W. M., Preston, R. A., Asmar, S. W., Gurvits, L. I., Pogrebenko, S. V., Avruch, I. M., Campbell, R. M., Bignall, H. E., Garrett, M. A., van Langevelde, H. Jan, Parsley, S. M., Reynolds, C., Szomoru, A., Reynolds, J. E., Phillips, C. J., Sault, R. J., Tzioumis, A. K., Ghigo, F., Langston, G., Brisken, W., Romney, J. D., Mujunen, A., Ritakari, J., Tingay, S. J., Dodson, R. G., van 't Klooster, C. G. M., Blancquaert, T., Coustenis, A., Gendron, E., Sicardy, B., Hirtzig, M., Luz, D., Negrao, A., Kostiuk, T., Livengood, T. A., Hartung, M., de Pater, I., Ádámkovics, M., Lorenz, R. D., Roe, H., Schaller, E., Brown, M., Bouchez, A. H., Trujillo, C. A., Buratti, B. J., Caillault, L., Magin, T., Bourdon, A., Laux, C., 2006. Overview of the coordinated ground-based observations of Titan during the Huygens mission. J. Geophys. Res. Planets **111**, E07S01.

Wong, A., Morgan, C. G, Yung, Y. L., Owen, T. O., 2000. Evolution of CO on Titan. Icarus **155**, 382–392.

Wye, L. C., Zebker, H. A., Ostro, S. J., West, R. D., Gim, Y., Lorenz, R. D., The Cassini Radar Team, 2007. Electrical properties of Titan's surface from Cassini RADAR scatterometer measurements. Icarus **188**, 367–385.

Yelle, R. V., 1991. Non-LTE models of Titan's upper atmosphere, Astrophys. J. **383**, 380–400.

Yelle, R. V., Griffith, C. A., 2003. HCN fluorescence on Titan. Icarus **166**, 107–115.

Yelle, R. V., Borggren, N., de La Haye, V., Kasprzak, W. T., Niemann, H. B., Müller-Wodarg, I., Waite, J. H., 2006. The vertical structure of Titan's upper atmosphere from Cassini Ion Neutral Mass Spectrometer measurements. Icarus **182**, 567–576.

Young, E. F., Rannou, P., McKay, C. P., Griffith, C. A., Noll, K., 2002. A three-dimensional map of Titan's tropospheric haze distribution based on Hubble Space Telescope Imaging. Astron. J. **123**, 3473–3486.

Young, E. F., Puetter, R., Yahil, A., 2005. Direct imaging of Titan's extended haze layer from HST observations. Geophys. Res. Let. **31**, L17S09.

Yung, Y. L., 1987. An update of nitrile photochemistry on Titan, Icarus **72**, 468–472.

Yung, Y. L., Allen, M., Pinto, J. P., 1984. Photochemistry of the atmosphere of Titan: Comparison between model and observation. Astrophys. J. **55**, 465–506.

Zahnle, K., Pollack, J. B., Grinspoon, D., Done, L., 1992. Impact generated atmospheres over Titan, Ganymede and Callisto. Icarus **95**, 1–23.

Zarnecki, J. C., Leese, M. R., Hathi, B., Ball, A. J., Hagermann, A., Towner, M. C., Lorenz, R. D., McDonnell, J. A. M., Green, S. F., Patel, M. R., Ringrose, T. J., Rosenberg, P. D., Atkinson, K. R., Paton, M. D., Banaszkiewicz, M., Clark, B. C., Ferri, F., Fulchignoni, M., Ghafoor, N. A. L., Kargl, G., Svedhem, H., Delderfield, J., Grande, M., Parker, D. J., Challenor, P. G., Geake, J. E., 2005. A soft solid surface on Titan as revealed by the Huygens Surface Science Package. Nature **438**, 792–795.

Zhu, X., Strobel, D. F., 2005. On the maintenance of thermal wind balance and equatorial superrotation in Titan's stratosphere. Icarus **176**, 331–350.

Zhu, X., 2006. Maintenance of equatorial superrotation in the atmospheres of Venus and Titan. Plan. Space Sci. **54**, 761–773.

Acetylene (C$_2$H$_2$), 14, 27–30, 32, 55, 98, 152, 162–164, 175–178, 185–188, 300

Aerosol composition, 14, 100–101, 104–105, 193, 213–221

Ammonia (NH$_3$), 5, 15, 286, 300

Argon, 5, 23, 155, 262, 322, 326

Ariel (satellite of Uranus), 1

Benzene (C$_6$H$_6$), 43, 98, 152, 158–159, 163, 165

Callisto (satellite of Jupiter), 1, 63, 286

Carbon dioxide (CO$_2$), 22, 29, 158, 169

Carbon monoxide (CO), 29, 159, 171–173, 189–192

Cassini, Giovanni Domenico, 1–2, 6–7

Cassini (spacecraft), 22, 53, 71–116, 130, 135–136, 141, 145, 149–150, 152, 155–160, 163–166, 172, 176–177, 179–183, 197–212, 230, 238, 241, 246, 251–252, 254

Climate, 129, 147–149

Cloud and haze, 20, 32–35, 52, 153, 192–230

Composite Infrared Spectrometer (CIRS), 91–94, 141–144, 160, 168, 175–181

Condensate clouds, 206–210

Deuterated methane (CH$_3$D), 14, 29, 159, 161, 173–175, 328

Dione (satellite of Saturn), 1, 2, 8, 311

Doppler wind experiment, 88, 103–104, 245, 249

Dunes, 236, 255, 272–279

Dynamics, 210–213, 231–250

Enceladus (satellite of Saturn), 1, 5, 8, 316–317, 332

Energy balance, 137, 139, 211

Ethane (C$_2$H$_6$), 14, 24, 27–30, 32, 68, 140, 162, 185, 195, 202, 207–210, 216, 220–221, 239, 295–296, 300

Ethylene (C$_2$H$_4$), 14, 28, 68, 98, 145, 152, 158, 160, 163–164, 178, 185–186, 300

Europa (satellite of Jupiter), 1, 315, 318

Fractal models for haze, 195–197, 223–228

Galileo Galilei, 1, 318

Galileo (spacecraft), 109, 174, 255, 285, 299, 302–303

Ganymede (satellite of Jupiter), 1–2, 36, 56, 58, 258, 285–286, 305

Ground-based observatories, 44–51, 71

Herschel, William, 1, 313–314

Herschel, John, 3, 5

Hubble Space Telescope (HST), 21, 39–40, 64–66

Hunten, Don, 15, 22

Huygens, Christiaan, 1–4, 6–7, 18

Huygens (spacecraft), 22, 65, 71–116, 256–257, 261–266

Hydrocarbons, 158–159, 161–166, 176–181, 185–188, 192, 246

Hydrogen (H$_2$), 14–15, 25, 29, 98–99, 140, 147, 152, 155, 163, 172–174, 325–329, 333

Hydrogen cyanide (HCN), 29–31, 144, 166

Hyperion (satellite of Saturn), 1, 8, 63, 111, 316

Iapetus (satellite of Saturn), 1, 63, 308–310

Infrared Interferometer Spectrometer (IRIS), 22, 25, 30–31, 158

Infrared Space Observatory (ISO), 38, 41–44, 158

Io (satellite of Jupiter), 1, 235, 236, 288, 307, 318, 322

Isotopic ratios, 78, 94, 105, 130, 164, 290, 299–300, 328–329, 331, 336

James Webb Space Telescope, 40–41

Jet Propulsion Laboratory (JPL), 17–18, 332

Jupiter, 1, 3, 5, 12, 14, 16–18, 36, 73, 79–80, 82, 84, 94, 109, 174, 235, 255, 258, 285, 286, 288, 290, 297–304, 308, 317–318, 323, 330–331, 340

Jupiter Icy Moons Orbiter, 341

Kuiper, Gerard, 13–14, 44

Lakes, 33, 37, 54, 68, 147, 175, 254, 258, 262,
    265–266, 271–272, 275–277, 285, 292, 294,
    296, 316, 334–337, 345–347
Life, 285, 329–330, 334–335
Lightning, 37, 78, 91, 95, 99, 105, 237,
    255–257, 339

Magnetospheric measurements, 97–99,
    112–113, 144–147
Mars, 9, 12, 288, 296–297
Mercury, 12, 287, 309, 310
Meridional circulation, 65, 178, 243–244
Methane (CH_4), 5, 14–15, 21–25, 27–30,
    32–33, 37, 40, 44, 54–57, 59–62, 64–65, 67,
    70–71, 96, 98, 103, 130, 133–135, 137,
    139–142, 144–145, 147–156, 161, 172,
    174–175, 190, 195–196, 199, 204–205,
    207–211, 218, 220–223, 230, 235, 243, 248,
    251–254, 256, 259–262, 264–265, 273–275,
    277–278, 280, 282–285, 291–295, 299–300,
    310, 318, 320–328, 331, 334
Meteorology, 66, 231–232, 250–257
Mimas (satellite of Saturn), 1, 5, 311–315
Moon (satellite of Earth), 9–12, 23, 41, 56, 58,
    89, 281, 287, 290, 307, 309, 310, 313, 316,
    318, 341

Neon, 5, 105, 324, 331
Neptune, 1, 12, 18, 297–304
Nitriles, 22, 29, 158, 166–168, 176–181,
    188–189
Nitrogen (N_2), 5, 14–15, 22–23, 150–152, 286,
    325–326
North-south asymmetry, 20–21, 64–65, 200,
    206, 212, 244–245

Oberon (satellite of Uranus), 1
Oceans or Seas, 24, 54–57, 153–154, 259
Oxygen, 43, 99, 105, 153, 164, 169, 183,
    189–192, 293, 325, 327–328

Photochemistry, 31–33, 54, 165, 181–185
Phoebe (satellite of Saturn), 317
Pioneer (spacecraft), 16–17
Pluto, 13, 286, 315, 318–321
Propane (C_3H_8), 27–29, 161

Questions, outstanding, 36–37, 78, 229–230,
    335–336

Radar observations, 54–57, 96, 261, 270, 272,
    275–277, 279, 336
Rain, 24, 33–34, 37, 54, 71, 153, 190, 212,
    252–254, 261, 269, 271, 273–274, 278, 293,
    316, 329–330
Rhea (satellite of Saturn), 1, 8, 111, 310–311

Sagan, Carl, 16, 32–33, 193
Saturn, 1, 12, 107–109, 297–304
Snow, 33, 70, 106, 153, 263, 273
Solar and thermal radiation, 137
Solar system, 289–290
Sun, 8–9, 12, 25, 192, 198, 217, 235, 243, 275,
    286, 289–292, 298–299, 323, 324, 329,
    340–341
Surface and landscape, 35–37, 66–71, 258–280
Surface composition, 277–280

Tethys (satellite of Saturn), 1, 2, 8
Tholins, 33, 193, 217–223
Titan (satellite of Saturn)
    albedo, 57–64
    atmospheric composition, 28–31, 154, 159,
        163–164
    discovery, 2, 13
    energy balance, 26
    interior, 37, 281–285
    ionosphere, 155
    magnetic field, 145
    mythology, 5
    origin and evolution, 321–329
    name, 4
    physical data, 23
    pressure profile, 129–131
    radius, 22–23
    surface temperature, 22
    temperature profile, 23–28, 53, 131–144
Titania (satellite of Uranus), 1
Titania (spacecraft), 341–343
Triton (satellite of Neptune), 1, 286, 315,
    318–321, 331

Ultraviolet (UV) observations, 19–21, 25–28,
    31–32, 94–95
Umbriel (satellite of Uranus), 1
Uranus, 1, 12, 18, 297–304

Venus, 12, 287, 290–296

Visible and Infrared Mapping Spectrometer (VIMS), 53, 95, 172, 208, 261, 272, 274, 278–279, 313

Volcanism, 33, 57, 66, 148, 174, 208, 236, 262, 266, 271–274, 277, 282–283, 285, 288, 294, 296, 307, 317, 322–323, 327, 336, 345–346

Vortex, polar, 245–247, 304

Voyager (spacecraft), 2, 13, 16–38, 42–43, 64–65, 73, 81, 93, 140, 150–152, 155, 158–162, 167–169, 176, 180–183, 206, 213, 224, 232, 239, 248, 255, 309, 310, 318

Water vapour, 43, 158, 171, 327–328

Water ice, 57, 60–64

Winds, 52–53, 78, 103–104, 231, 238–242

Xanadu, 68, 86, 232–234, 260–261, 266–267, 270

Xenon, 341–342

Yung, Yuk, 27, 146, 182–184, 224, 246

Zarnecki, John, 106, 263, 282

Zonal winds, 52–53, 233, 237–242

www.ingramcontent.com/pod-product-compliance
Lightning Source LLC
Chambersburg PA
CBHW070150240326

41458CB00126B/2665